CURRENT TOPICS IN MEMBRANES, VOLUME 81

Series Editor

IRENA LEVITAN
Professor of Medicine, Pharmacology and Bioengineering
Division of Pulmonary and Critical Care Medicine
Scholarly Activities Council
University of Illinois at Chicago
Chicago, USA

VOLUME EIGHTY ONE

Current Topics in MEMBRANES

Cell Volume Regulation

VOLUME EIGHTY ONE

CURRENT TOPICS IN MEMBRANES

Cell Volume Regulation

Edited by

IRENA LEVITAN
Professor of Medicine, Pharmacology and Bioengineering
Division of Pulmonary and Critical Care Medicine
Scholarly Activities Council
University of Illinois at Chicago
Chicago, USA

ERIC DELPIRE
Professor of Anesthesiology, Molecular Physiology & Biophysics
Vanderbilt University School of Medicine
Nashville, Tennessee, USA

HECTOR RASGADO-FLORES
Professor of Physiology and Biophysics
Rosalind Franklin University of Medicine and Science
North Chicago, USA

ELSEVIER

ACADEMIC PRESS
An imprint of Elsevier

Academic Press is an imprint of Elsevier
125 London Wall, London EC2Y 5AS, United Kingdom
525 B Street, Suite 1650, San Diego, CA 92101, United States
50 Hampshire Street, 5th Floor, Cambridge, MA 02139, United States
The Boulevard, Langford Lane, Kidlington, Oxford OX5 1GB, United Kingdom

First edition 2018

Notices

Knowledge and best practice in this field are constantly changing. As new research and experience broaden our understanding, changes in research methods, professional practices, or medical treatment may become necessary.

Practitioners and researchers must always rely on their own experience and knowledge in evaluating and using any information, methods, compounds, or experiments described herein. In using such information or methods they should be mindful of their own safety and the safety of others, including parties for whom they have a professional responsibility.

To the fullest extent of the law, neither the Publisher nor the authors, contributors, or editors, assume any liability for any injury and/or damage to persons or property as a matter of products liability, negligence or otherwise, or from any use or operation of any methods, products, instructions, or ideas contained in the material herein.

ISBN: 978-0-12-815456-4
ISSN: 1063-5823

Publisher: Zoe Kruze
Acquisition Editor: Ashlie Jackman
Editorial Project Manager: Alina Cleju
Production Project Manager: Abdulla Sait
Cover Designer: Alan Studholme

Typeset by TNQ Technologies

CONTENTS

Section 2
Regulation of Intracellular Chloride and Water Homeostasis

Section 3
Cell Volume Regulation in the Airways

CONTRIBUTORS

Manuela A.A. Ayee
University of Illinois at Chicago, Chicago, IL, United States

Francis Boudreault
Centre de recherche, Centre hospitalier de l'Université de Montréal (CRCHUM), Montreal, QC, Canada

Neil A. Bradbury
Department of Physiology and Biophysics, Chicago Medical School, North Chicago, IL, United States

Pei-Chuan Chao
Department of Civil, Structural and Environmental Engineering, University at Buffalo, The State University of New York, Buffalo, NY, United States

Paola de los Heros
División de Investigación, Facultad de Medicina, Universidad Nacional Autónoma de México, Coyoacán, Mexico City, Mexico

Eric Delpire
Department of Anesthesiology, Vanderbilt University School of Medicine, Nashville, TN, United States

Isha Dey
Department of Physiology and Biophysics, Chicago Medical School, North Chicago, IL, United States

Kenneth B. Gagnon
Department of Medicine, University of Louisville School of Medicine, Louisville, KY, United States

Gerardo Gamba
Molecular Physiology Unit, Instituto de Investigaciones Biomédicas, Universidad Nacional Autónoma de México, Tlalpan, Mexico City, Mexico; Department of Nephrology and Mineral Metabolism, Instituto Nacional de Ciencias Médicas y Nutrición Salvador Zubirán, Tlalpan, Mexico City, Mexico

Ryszard Grygorczyk
Centre de recherche, Centre hospitalier de l'Université de Montréal (CRCHUM), Montreal, QC, Canada; Department of Medicine, Faculty of Medicine, University of Montreal, Montreal, QC, Canada

Md. Rafiqul Islam
Division of Cell Signaling, National Institute for Physiological Sciences (NIPS), Okazaki, Japan

Irena Levitan
University of Illinois at Chicago, Chicago, IL, United States

Michael A. Model
Kent State University, Kent, OH, United States

Alexander A. Mongin
Department of Neuroscience and Experimental Therapeutics, Albany Medical College, Albany, NY, United States; Department of Biophysics and Functional Diagnostics, Siberian State Medical University, Tomsk, Russian Federation

Catherine E. Morris
Senior Scientist Emeritus, Neuroscience, Ottawa Hospital Research Institute, Ottawa, ON, Canada

Toshiaki Okada
Division of Cell Signaling, National Institute for Physiological Sciences (NIPS), Okazaki, Japan

Yasunobu Okada
Department of Physiology and Systems Bioscience, Kyoto Prefectural University of Medicine, Kyoto, Japan; Division of Cell Signaling, National Institute for Physiological Sciences (NIPS), Okazaki, Japan

Sergei N. Orlov
Faculty of Biology, M.V. Lomonosov Moscow State University, Moscow, Russia; Siberian State Medical University, Tomsk, Russia; National Research Tomsk State University, Tomsk, Russia

James Osei-Owusu
Department of Physiology, Johns Hopkins University School of Medicine, Baltimore, MD, United States

Diana Pacheco-Alvarez
Escuela de Medicina, Universidad Panamericana, Mexico City, Mexico

Jonathan C. Petruccelli
University at Albany, Albany, NY, United States

Olga Ponomarchuk
Faculty of Biology, M.V. Lomonosov Moscow State University, Moscow, Russia; Centre de recherche, Centre hospitalier de l'Université de Montréal (CRCHUM), Montreal, QC, Canada

Zhaozhu Qiu
Department of Physiology, Johns Hopkins University School of Medicine, Baltimore, MD, United States; Solomon H. Snyder Department of Neuroscience, Johns Hopkins University School of Medicine, Baltimore, MD, United States

Ravshan Z. Sabirov
Division of Cell Signaling, National Institute for Physiological Sciences (NIPS), Okazaki, Japan; Laboratory of Molecular Physiology, Institute of Bioorganic Chemistry, Academy of Sciences of Uzbekistan, Tashkent, Uzbekistan

Frederick Sachs
Department of Physiology and Biophysics, University at Buffalo, The State University of New York, Buffalo, NY, United States

Aleksandra Shiyan
Faculty of Biology, M.V. Lomonosov Moscow State University, Moscow, Russia

Mettupalayam Sivaselvan
Department of Civil, Structural and Environmental Engineering, University at Buffalo, The State University of New York, Buffalo, NY, United States

Robert Tarran
Marsico Lung Institute, University of North Carolina at Chapel Hill, Chapel Hill, NC, United States; Department of Cell Biology & Physiology, University of North Carolina at Chapel Hill, Chapel Hill, NC, United States

Maria del Carmen Vitery
Department of Physiology, Johns Hopkins University School of Medicine, Baltimore, MD, United States

Megan J. Webster
Marsico Lung Institute, University of North Carolina at Chapel Hill, Chapel Hill, NC, United States

Corinne S. Wilson
Department of Neuroscience and Experimental Therapeutics, Albany Medical College, Albany, NY, United States

Junhua Yang
Department of Physiology, Johns Hopkins University School of Medicine, Baltimore, MD, United States

PREFACE

The journey from discovering cell volume regulation of euryhaline fish to the identification of key mechanisms of cell volume homeostasis

Following the observation in the 1960s that cells from euryhaline fish species recover their volume during adaptation from sea water to fresh water, and that this basic property could be found in all cells throughout evolution, an international community of scientists formed around the theme of *Cell Volume Regulation* (CVR). Meetings dedicated to this research area started in the early 1980s and have been held regularly for the past 30 years. The latest meeting was organized by us in August 2016 in Chicago. This volume of *Current Topics in Membranes* gathers a group of experts that contributed to the success of the Chicago meeting. Note that a volume (volume 30) of *Current Topics in Membranes,* titled "Cell Volume Control: Fundamental and Comparative Aspects in Animal Cells," was published in 1987, some 30 years ago. It was edited by Raymond Gilles, Arnost Kleinzeller, and Liana Bolis. While this first volume details the basic phenomenology of cell volume regulation, many of the transport processes and regulatory molecules had yet to be identified at the molecular level. Obviously, in 30 years, significant advances have been made in the field, even if some fundamental questions still remain, including the identity of the volume sensing mechanism or process that links cell volume to the activation of ion channels and transporters.

The current volume summarizes the phenomenology, highlights landmark papers, and provides ample historical data (e.g., in Chapter by Delpire & Gagnon). It then discusses in detail our current understanding the mechanisms at play, including the properties of biological membranes and the sponge-like nature of the cell's interior. As thoroughly discussed in Sergei Orlov's chapter, the central question that remains a "holy grail" in our field is whether changes in cell volume are detected at the plasma membrane, at membrane–cytoskeleton interaction sites, or at the level of the cytoplasm and macromolecules. In addition, the role of intracellular ions and interacting proteins as "sensors" cannot be excluded. To this latest possibility, we know that phosphorylation and dephosphorylation are the mechanisms by which volume-sensitive transporters are activated or inhibited, and different kinases have now been identified that seem to respond differentially to changes in intracellular ions (e.g., Cl^-) and changes in cell volume (see Chapter by de los Heros, Pacheco-Alvarez, & Gamba). The chapter of Model & Petrucelli is dedicated to methods to quantify intracellular macromolecules.

One of the major advances in the field in the last decade highlighted in the two chapters of this volume, by the groups of Yasunobu Okada and Zhaozhu Qiu, was establishing the molecular identity of the volume-sensitive anion channel, a key mechanism in regulatory volume decrease, the ability of cells to recover their volumes under hypotonic stress. The question of the molecular identity of these channels has been a source of major controversy for decades with multiple molecular candidates being identified and then discovered to be false positives. The pioneering work of Qiu and coworkers leading to the discovery of the new channel, *Swell1*, is described in detail in the chapter by Osei-Owusu et al.

The volume also addresses a current controversy about the role of mechanical cues in cell volume regulation. The original premise that cell volume regulation is initiated by mechanical signals was based on the expectation that cell swelling should result in membrane stretch and increase in membrane tension, which in turn would activate mechanosensitive ion channels and other mechanosensitive mechanisms. However, a new detailed analysis of membrane tension and elastic modulus during cell swelling described here by Ayee & Levitan shows that this is not the case and that cell swelling is not accompanied with an increase in membrane tension. Instead, the authors discuss work from their lab and from other investigators showing that cell swelling increases cell stiffness/elastic moduli raising the question of whether intracellular hydrostatic pressure rather than membrane tension may be critical for cell volume regulation. A novel idea of membrane tension homeostasis and its role in cytotoxic swelling is introduced by Catherine Morris in her chapter describing this relationship in different types of excitable cells. Two chapters, by Fred Sachs and Sergei Orlov, also described growing evidence that cytoskeleton has a gel-like properties and may absorb water during cell swelling thus serving as a protective reservoir. Further studies are needed to establish whether incorporation of water into the cytoskeleton gel may also be a signaling event and initiate enzymatic cascades.

In addition, one chapter is dedicated to cell volume regulation in nervous system. Wilson & Mongin describe unique aspects of cell volume maintenance in neurons and astrocytes and cover the consequences of water and ion dysregulation in conditions of hyponatremia, epilepsy, hepatic encephalopathy, and stroke. Two chapters also address the role of extracellular water around epithelial cells: Dey & Bradbury focus on intestinal cells, whereas Webster & Taran highlight the importance of cell surface and mucus hydration in the airway epithelium.

Together, these assays provide a comprehensive analysis of what is known today about Cell Volume Regulation.

Irena Levitan
Eric Delpire

SECTION 1

General Principles of Cell Volume Regulation

CHAPTER ONE

Water Homeostasis and Cell Volume Maintenance and Regulation

Eric Delpire*^,[1] and Kenneth B. Gagnon[§]
*Department of Anesthesiology, Vanderbilt University School of Medicine, Nashville, TN, United States
[§]Department of Medicine, University of Louisville School of Medicine, Louisville, KY, United States
[1]Corresponding author: E-mail: eric.delpire@vanderbilt.edu

Contents

Abstract

From early unicellular organisms that formed in salty water environments to complex organisms that live on land away from water, cells have had to protect a homeostatic internal environment favorable to the biochemical reactions necessary for life. In this chapter, we will outline what steps were necessary to conserve the water within our

Current Topics in Membranes, Volume 81
ISSN 1063-5823
https://doi.org/10.1016/bs.ctm.2018.08.001

cells and how mechanisms have evolved to maintain and regulate our cellular and organismal volume. We will first examine whole body water homeostasis and the relationship between kidney function, regulation of blood pressure, and blood filtration in the process of producing urine. We will then discuss how the composition of the lipid-rich bilayer affects its permeability to water and salts, and how the cell uses this differential to drive physiological and biochemical cellular functions. The capacity to maintain cell volume is vital to epithelial transport, neurotransmission, cell cycle, apoptosis, and cell migration. Finally, we will wrap up the chapter by discussing in some detail specific channels, cotransporters, and exchangers that have evolved to facilitate the movement of cations and anions otherwise unable to cross the lipid-rich bilayer and that are involved in maintaining or regulating cell volume.

In memory of Raymond Gilles (1940–2018), my first mentor and scientific hero

Eric Delpire

1. INTRODUCTION

The conventional theory of how life began some 3.8 billion years ago involves a "primordial soup" of inanimate matter, present in the oceans of early Earth, that upon addition of external energy in the form of lightning or ultraviolet light caused simple molecules to combine into more complex compounds (Haldane, 1929). Haldane further postulated that these organic compounds accumulated to concentrations sufficient to produce macromolecules and ultimately fully fledged cells. However, the ionizing UV radiation required to stimulate these early chemical reactions would inherently destroy as much as it creates, indicating that a "surface" primordial soup would not be the best environment for the emergence of life.

Volcanic deep submarine hydrothermal vents discovered in the late 1970s provided another potential source of primordial life-promoting energy (Corliss, et al., 1979). However, their extreme temperatures, low pH, and short lifespans made them problematic as sites for the origin of life (Bada, 2004; Orgel, 2008). Russell and colleagues proposed in 1993 that emergence of life in the depths of the oceans would require warm alkaline hydrothermal fluids rich in dissolved hydrogen and light hydrocarbons such as methane and acetate (Russell, Daniel, & Hall, 1993; Russell, Daniel, Hall, & Sherringham, 1994). The discovery of a second type of hydrothermal vent in 2000 (Kelley, et al., 2001) produced by the geochemical process of olivine hydroxylated to serpentine provided just such an alkaline hydrothermal environment (Bach, et al., 2006; Sleep, Meibom, Fridriksson, Coleman, & Bird, 2004). These "first cells" were most likely autotrophs that derived their energy needs and carbon from the chemical reduction of CO_2 with H_2 electrons (Lane & Martin, 2012; Martin & Russell, 2007; Russell & Martin, 2004). Indeed, the hydrogenation of carbon dioxide to form methane or acetate via the reductive acetyl CoA pathway is present in the most ancient prokaryotes (Shock & Schulte, 1998).

Seawater contains in mequiv/L: 540 Cl^-, 460 Na^+, 50 Mg^{2+}, 30 SO_4^{2-}, 10 Ca^{2+}, and 10 K^+. Thus, the first biochemical reactions occurred in a salt-rich environment. Primordial cells isolated themselves from this "salty" outside environment by creating a lamellar phospholipid bilayer matrix to protect their biological processes. Depending on the molecular structure, water content, pH, ionic strength, temperature, and pressure, this lipid-rich barrier allowed these cells to keep an inside environment that is conducive to protein synthesis and enzymatically driven biochemical reactions (Daniel, Oger, & Winter, 2006). This basic cellular architecture has remained intact all throughout evolution, from simple unicellular to complex multicellular organisms. As unicellular cells combined and evolved into specific organ systems within multicellular organisms, this lipid-rich membrane acquired additional components and properties, such as a membrane potential that cells have espoused to communicate between themselves, and/or an apical-basolateral polarity that cells utilize for *trans*-membrane solute transport.

By mass, water is the most abundant molecule inside a cell and consequently, the volume of water constitutes the major parameter in the volume of a cell. The water content of animal tissues obtained by drying ranges from 70% to 80% (Cook, Cramer, & Kenyon, 1952), which corresponds to some 3 kg of water per kg of dry weight ($3/(3+1) = 75\%$). The water content of the human body is a bit lower (60%), due to the lower water content of bones (30%). As animals have transitioned from aquatic ponds, rivers, lakes, and oceans to terrestrial landmasses of earth, they had to find ways to conserve this water (Fig. 1A). In the next section, we will examine whole body water homeostasis, a critical function of all creatures, especially terrestrial species subjected to continuous water loss. In Section 3, we will discuss cell volume homeostasis. Despite a relatively stable internal environment in the bodies of terrestrial species, individual cells have retained throughout evolution mechanisms that allow them to maintain and regulate their volume.

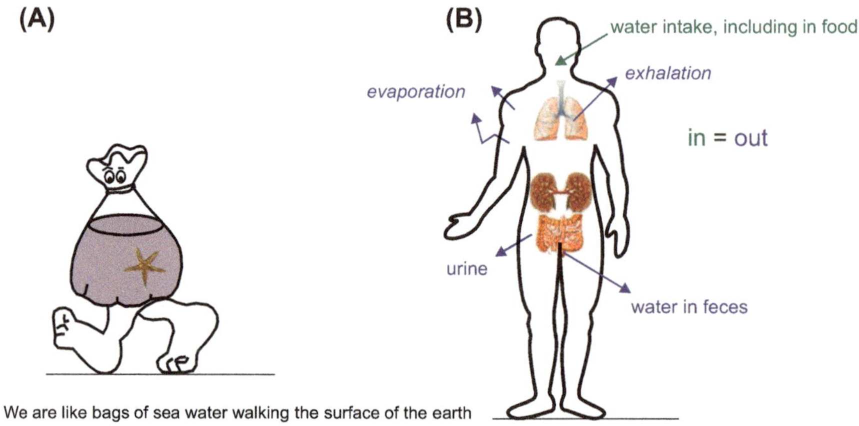

Figure 1 ***Terrestrial species evolved from creatures that originated in the ocean.*** (A) As life invaded land, it carried with it some of the ocean water. The drawing was taken from the Delpire laboratory website at Vanderbilt University Medical center: https://www.vumc.org/delpire-lab/. (B) In humans, and most terrestrial species, water is constantly lost through perspiration/evaporation from the skin, exhalation from the lungs, urination from the bladder, and feces from the colon. To maintain water homeostasis (in = out), an equivalent amount of water needs to be consumed through drinking and food.

2. WHOLE BODY WATER HOMEOSTASIS

In 1885, the Belgian physiologist Léon Fredericq noted that as you climb the evolutionary ladder, the number of regulatory mechanisms became greater and more complex and tended to free organisms from harmful influences and changes of their environment (Fredericq, 1885). He noted that for invertebrates, this independence to the external milieu was only relative. Experiments showing a lower salt content in the blood of fresh water versus seawater crustaceans supported this observation, whereas "the blood from sea water fish was not saltier to the taste than the blood of fresh water fish." Our reason to mention salt in the first paragraph of this section is that salt and water are tightly interconnected.

The vast majority (70%—75%) of water in the human body is intracellular or located within cells. The remaining 25%—30% is extracellular water, with 1/5 found in plasma and 4/5 found in the interstitial space. As a terrestrial creature, humans lose water through urine (1.5 L), feces (0.1 L), respiration and perspiration (0.5—0.8 L). To maintain water balance, this loss is countered by water intake through eating and drinking (Fig. 1B).

How exactly does the body sense the loss of water? Specialized "sensors" in the vasculature and other organs detect a decrease in extracellular volume as well as an increase in the plasma osmolarity.

2.1 Blood Pressure

Blood pressure refers to the pressure of blood in the large arteries of the circulatory system. Cardiac output, arterial stiffness, and peripheral resistance all influence the body's blood pressure. One can simplify the circulatory system by considering it as a "closed" circuit consisting of a pump (heart) and pipes (arteries and veins). The pressure that exists in the system depends upon the cardiac output, vascular resistance, and the amount of liquid in the system. The reason that "closed" is written with quotations is because unlike the non-permeable plumbing in our house, blood vessels are able to exchange fluid, gases, and solutes with the interstitial milieu. Moreover, the circulatory system is in contact with the outside through lung, intestinal tract, and kidney endothelia and epithelia, and as such, has the capacity for environmental exchange of fluids, gases, and solutes. Also, note that in contrast to the non-permeable solid pipes in our house, the vessels in our body can also contract or dilate and therefore changes in vascular wall tension will modify the volume of the "closed" circuit.

2.2 The Kidney and Blood Pressure

The human kidney is central to fluid and salt homeostasis. Each kidney is composed of around one million nephrons, each independently operating as a filtration, reabsorption, and secretion unit. Individual nephrons adjust blood composition and form urine beginning with the glomerulus (filtration unit), followed by proximal convoluted tubule, loop of Henle, distal convoluted tubule (reabsorption units), and the common collecting duct (secretion unit). Numerous collecting ducts merge into the renal pelvis that ultimately becomes the ureter connecting the kidneys to the bladder. Unfiltered blood enters the glomerulus via afferent arterioles where small molecular weight substances pass through fenestrations in glomerular capillaries and filtration slits formed by the foot processes of podocyte cells wrapped around these capillaries. Larger molecular weight proteins remain in the plasma and exit the glomerulus via efferent arterioles. The principal role of the renal tubule is to reabsorb water and key solutes (e.g., sugars, amino acids, and ions) while eliminating waste products and toxins in the urine. In the opening paragraph, we indicated that salt and water were interconnected and they directly affected whole body water homeostasis. While the bulk of Na^+ reabsorption occurs in early segments (e.g., proximal tubule and thick ascending limb), finely tuned reabsorption occurs more distally by the sodium-chloride cotransporter (NCC) in the distal convoluted tubule and the epithelial Na^+ channel (ENaC) in the connecting segment and the cortical collecting duct. Because NCC and ENaC are the last mechanisms participating in Na^+ reabsorption, they each play critical roles in fine-tuning plasma Na^+ levels and systemic blood pressure.

Multiple observations support the role of NCC and ENaC functions in regulating blood pressure. First, thiazides and amiloride are compounds that inhibit with great specificity NCC and ENaC, respectively (Garty & Benos, 1988; Stokes, Lee, & D'Amico, 1984). These drugs are widely used in clinical medicine, often in combination, to treat volume expansion (e.g., congestive heart failure) and high blood pressure (Padilla, Armas-Hernández, Hernández, Israili, & Valasco, 2007). Second, loss-of-function mutations in NCC results in Gitelman's syndrome, a renal salt wasting disorder with many patients suffering from arterial hypotension (Simon, et al., 1996), whereas loss-of function-mutations in ENaC results in pseudohypoaldosteronism type I (PHA-I), a disease that also involves the loss of salts leading to hyponatremia (Chang, et al., 1996). The disease is called pseudohypoaldosteronism, because it mimics the absence of aldosterone production (or hypo-aldosteronism). Aldosterone is a key hormone produced by the adrenal gland that enhances Na^+ reabsorption through NCC and ENaC in the distal tubule. Absence of aldosterone causes urinary Na^+ wasting, leading to volume depletion and hypotension (Merriam & Baer, 1980). In contrast, there are genetic conditions that lead to *increased* salt reabsorption, volume retention, and high blood pressure. Liddle syndrome is due to a mutation in ENaC that prevents its recycling from the plasma membrane and degradation (Schild, et al., 1996; Tamura, et al., 1996). The consequence of this lack of recycling is increased channel density at the plasma membrane and increased ENaC function. Type II pseudohypoaldosteronism (PHA-II) is caused by mutations in two "with no lysine kinases" (WNK1 or WNK4) leading to increased phosphorylation and activity of NCC (Wilson, et al., 2001). Mutations in Kelch-3 or CUL3, genes products that are part of the ubiquitination and degradation of NCC, also manifest as PHA-II (Boyden, et al., 2012).

In summary, decreased NCC or ENaC function results in volume depletion and hypotension, whereas increased NCC or ENaC function lead to volume retention and hypertension. In the preceding paragraph, we focused on inhibitors and genetic disorders that affect salt transport and blood pressure. In a normal situation, however, how exactly does the body regulate "physiological" changes in blood volume?

2.3 Hypovolemia and Tubuloglomerular Feedback Mechanism

Afferent and efferent arterioles carry blood into and out of the glomerulus, respectively. Hypovolemia, or low blood volume, leads to a decrease in glomerular filtration, which in turn, leads to decreased flow of Cl^- at the transition between the end portion of the loop of Henle and the beginning of the distal convoluted tubule. Specialized epithelial cells (macula densa) in this transition zone sense the reduced flow of Cl^- and send signals to the adjacent afferent arteriole to vasodilate and increase glomerular filtration. In addition, the decreased flow of Cl^- stimulates the release of renin (Schnermann & Briggs, 2013) from other specialized cells that are located in the wall of the afferent arteriole (Fig. 2). Renin, in turn, stimulates the conversion of plasma angiotensinogen to angiotensin II, which is a potent vasoconstrictor. Thus, angiotensin II leads to increased vascular resistance and increased blood pressure. In addition to its vasoconstrictive effect, angiotensin II activates the release of the mineralocorticoid hormone aldosterone from the adrenal gland. Aldosterone as we discussed before stimulates salt and water reabsorption in the distal portion of the nephron.

2.4 Urinary Concentrating Mechanism

The human body, like the bodies of all mammals and birds, concentrates its urine. This important function stems from the fact that the body needs to eliminate the excess salts it consumes, but not at the expense of losing too much water. The combined activity of the apical Na−K−2Cl cotransporter, NKCC2, and basolateral Na^+/K^+ ATPase in the thick ascending limb of Henle (TALH) moves Na^+ from the tubular fluid to the renal interstitium (Castrop & Schießl, 2014). Because the TALH is impermeable to water, this transport of Na^+ contributes to a hyperosmotic interstitium. The juxtaposition of descending limb (permeable to water but not ions) and ascending limb (pumping ions and impermeable to water) creates a countercurrent multiplier mechanism that maintains high osmolarity in the medulla, while keeping normal the osmolarity in the cortex (Gottschalk & Mylle, 1958, 1959; Hargitay & Kuhn, 1951). Comparative physiologists have observed

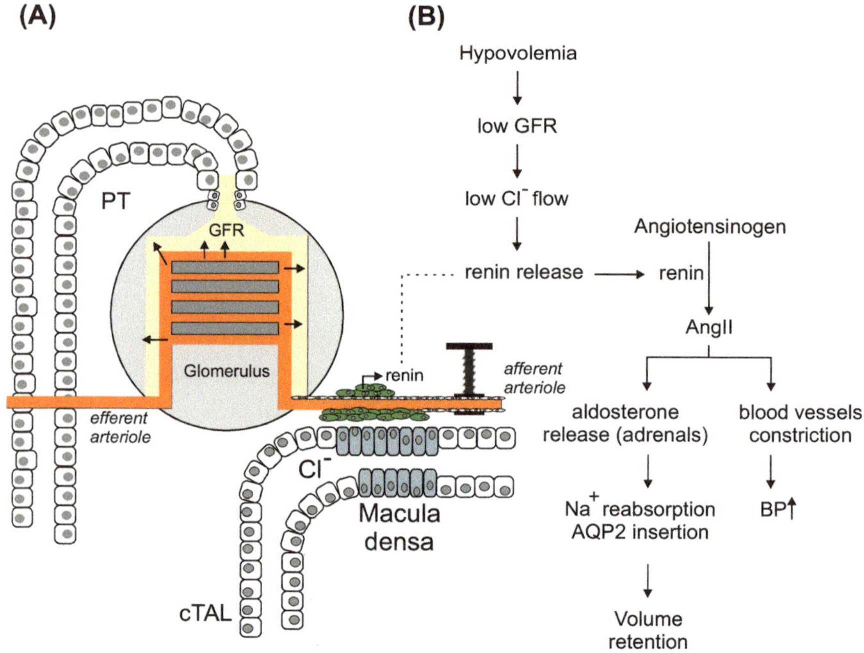

Figure 2 ***The kidney nephron is a rheostat for plasma volume.*** (A) The ultrafiltration of blood into urine that begins in the glomerulus depends upon the wall pressure of afferent and efferent blood vessels. The rate of filtration at the glomerulus is one component that defines the amount of Cl^- that flows at the macula densa. Macula densa cells serve as "sensors" of the amount of Cl^--containing filtrate that passes along the cortical thick ascending limb (cTAL). These cells relay information to specialized renin cells located in the wall of the afferent arteriole to release renin when necessary. (B) Low blood volume (hypovolemia) ultimately results in renin release that acts on the renin-angiotensin-aldosterone system. Angiotensin II and aldosterone restore blood volume (and blood pressure) by affecting Na^+ and water reabsorption (increased AQP2 insertion) and constriction of blood vessels.

that in species that typically survive with minimal water and produce a very concentrated urine, the majority of their nephrons possess very long loops of Henle (Jamison, 1987). Along with Na^+, urea is another key osmolyte that contributes to a hyperosmotic medullary interstitium (Sands, 2004). The urea cycle converts ammonia (NH_3), a very toxic compound, into urea, a water-soluble non-toxic compound easily excreted. Some of the urea present in urine is re-absorbed in the inner medullary collecting ducts thereby raising the osmolarity of the medullary interstitium. Anti-diuretic hormone, also called vasopressin or arginine vasopressin (AVP), controls this re-absorptive mechanism (Nielsen & Knepper, 1993, 3586).

2.5 Circumventricular Organ, Hypothalamus and Vasopressin

The vascular organ of lamina terminalis (VOLT) and the subfornical organ are highly vascularized circumventricular organs that sense changes as small as 1% (3 mOsmol/L) in plasma osmolarity. They project to hypothalamic neurons that produce AVP when increased osmolarity is detected (Prager-Khoutorsky, 2017). They also send signals to the cortex where conscious water craving is processed. AVP produced in the somata of magnocellular neurosecretory cells (located in the hypothalamic supraoptic (SON) and paraventricular nuclei) then travels down the axons of these neurons towards the posterior pituitary and is released into the circulation. AVP has multiple sites of action. The primary role of AVP is to increase the number of water channels (aquaporin 2) on the apical membrane of collecting duct principal cells, leading to increased water reabsorption and conservation. Second, AVP also increases the number of urea transporters located in the inner medullary collecting duct, increasing interstitial osmolarity and the capacity to concentrate urine, thereby also contributing to water conservation. Third, AVP stimulates Na^+ reabsorption in the TALH, adding to the countercurrent multiplication phenomenon further helping in water conservation. Fourth, AVP constricts blood vessels, leading to increase in arterial blood pressure. Thus, in conditions of dehydration when plasma osmolarity increases, release of AVP into the circulation results in several water conservation mechanisms to maintain whole body water homeostasis.

2.6 Back to the Relationship Between Na^+ and Water

Arthur Guyton proposed that under normal (i.e., steady-state) conditions, dietary Na^+ increase is matched by urinary Na^+ excretion with no net consequences on volume and blood pressure (Guyton, 1987). However, if the individual is unable to secrete the excess dietary Na^+, extracellular fluid volume will expand leading to an increase in blood pressure (Guyton, 1987). Natriuresis lowers the sodium concentration in the blood by excreting sodium in the urine. Osmotic forces draw water out of the body's blood circulation decreasing blood volume and restoring blood pressure to normal values. As stated by Alan Pao "high blood pressure is adaptive from the standpoint of maintaining Na^+ homeostasis, but this adaptation comes at the price of exposing end organs to the damaging effects of high blood pressure (Pao, 2014). Recent studies indicates that Na^+ and water are not always linked, as Na^+ can be stored in organs without increasing body fluid content (Kopp, et al., 2013). Furthermore, evidence has shown that the body can vary its Na^+ content while maintaining constant its water content (Rakova, et al., 2017). It is indeed clear than in healthy individuals, the kidney concentrating mechanism accommodates the disposal of any extra dietary Na^+, without the loss of any additional water.

3. CELL VOLUME HOMEOSTASIS

The activities that exist within a cell are intimately linked to the spatial organization of internal macromolecular structures. In this respect, any change to the volume of a cell potentially risks major disorganization of these structures and consequently perturbation of the many key activities of the cell.

The composition of the lipid bilayer greatly influences the movement of water and ions across the plasma membrane. Proteins embedded in the membrane provide pathways for the movement of water (e.g., aquaporins or water channels) and ions (e.g., ion channels and transporters). Even in the absence of these integral membrane proteins, the plasma membrane still possesses some intrinsic water permeability. Asymmetry between the nature and composition of the lipids between the inner and outer leaflets of the bilayer has a significant effect on this water permeability (Hill & Zeidel, 2000; Krylov, Pohl, Zeidel, & Hill, 2001). Indeed, membranes containing sphingolipids, glycosphingolipids, phospholipids, and cholesterol have much lower water permeabilities (3.4×10^{-4} cm/s) than membranes containing only phospholipids and cholesterol (34.4×10^{-4} cm/s).

Irrespective of its lipid composition, the permeability of a membrane to water will always be greater than the membrane permeability to ions and other solutes. Under these conditions, the cell is subjected to osmotic pressure differences that arise between the outside environment and the interior of the cell. If we exclude bacteria and plants that have a rigid membrane, all other cells have flexible membranes and thus undergo significant changes in cell volume proportional to the intensity of the osmotic gradient. Before we consider the behavior of cells under anisosmotic conditions, we will examine the challenge of cells in maintaining their volume under isosmotic conditions.

3.1 Isosmotic Cell Volume Maintenance

Cells can grow and change their size while still maintaining their water content constant. In this section, we will not be considering cell growth, but instead, the volume of a cell occupied by water. Cells contain in their cytoplasm macromolecules that are negatively charged. The presence of these macromolecules creates a colloido-osmotic disequilibrium first described by British chemist Frederick G. Donnan in 1911 (Donnan, 1911). While some ionic species can cross cellular membranes, larger anionic proteins are generally not permeable. These larger anions attract smaller cations while simultaneously repelling smaller anions creating an uneven electric potential called the Donnan potential.

Consider the following two simple scenarios: in the first scenario, consider a system composed of two compartments separated by a membrane that is permeable to small cations (Na^+, and K^+) but impermeable to large anions (A^-) (Fig. 3A). If we add NaA and KA salts to these compartments, respectively, and the salt concentration is different between the two compartments, the dissociative ions will re-distribute to create a new equilibrium (Fig. 3B and C). At equilibrium, the equations in Box 1 describe the work represented by the movement of Na^+ and K^+.

One can play with the numbers and dissolve 10 mM NaA in the first compartment and 100 mM KA in the second compartment and calculate

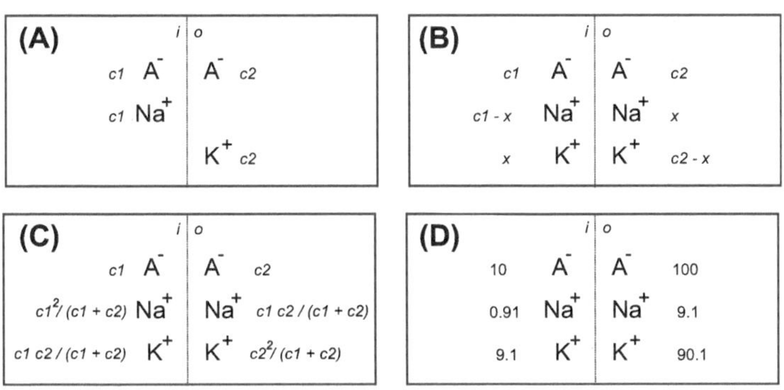

Figure 3 ***Physicochemical system of two compartments separated by a membrane permeable only to small cations.*** (A) Starting conditions with c1 and c2 concentrations of NaA and KA salts in their respective compartments. (B) Conditions after salt dissociation and cationic redistribution (impermeant anions remained in their respective compartments). (C) Concentrations obtained after x has been solved. (D) Example of ionic redistribution when starting with a 10-fold gradient difference between NaA and KA.

Box 1 Donnan Equilibrium for scenario 1 of Fig. 3

$$(\delta F)_{p,t} = 0 \quad \text{or} \quad F = RT[\delta n \log(Na^+{}_1/Na^+{}_2) + \delta n \log(K^+{}_1/K^+{}_2)] = 0 \qquad (1)$$

This translates to the Donnan equilibrium relationship of

$$(Na^+{}_1/Na^+{}_2) = (K^+{}_1/K^+{}_2) \qquad (2)$$

If we apply this equation to the newly formed equilibrium, we obtain

$$(c1 - x)/x = x/(c2 - x) \qquad (3)$$

Or

$$x = c1\ \ c2/(c1 + c2) \qquad (4)$$

Which leads at equilibrium to the concentrations written in Fig. 3C.

x to be 9.1 mM. At equilibrium, most of the Na^+ (91%) will have traveled to the second compartment, while only 9.1% of the K^+ will have moved to the first compartment (Fig. 3D). Thus, very different ion distributions across the membrane can exist due to the presence of the large impermeable anions. However, equal concentrations would be reached if either the system starts with equal concentrations of the large anion, or the anion can freely cross the membrane.

In a second scenario, consider a similar system composed of two compartments separated by a semi-permeable membrane where the large macromolecules are present only in the first compartment (called "i" for internal) with K^+ at a concentration of c1 and NaCl is only present in the second compartment (called "e" for external) at a concentration of c2 (Fig. 4A). At equilibrium, the amount of Cl^-, K^+ and Na^+ that moved are x, z, and y, respectively (Fig. 4B). At equilibrium, the values will be as shown in Fig. 4C. If we examine the case of c1 = c2, one can calculate that $[Cl^-]_e = 2\ [Cl^-]_i$, $[K^+]_i = 2\ [K^+]_o$, and $[Na^+]_i = 2\ [Na^+]_o$. Note that the majority of the Na^+, which started on the outside is now on the inside, while Cl^- is mostly prevented to move to the inside compartment. The calculated osmolalities will be 250 mOsM inside versus 200 mOsM outside, if the non-diffusible anions are osmotically inactive (i.e. large negatively charged macromolecules). If the non-diffusible charges are osmotically active then the calculated osmolalities will be 400 mOsM

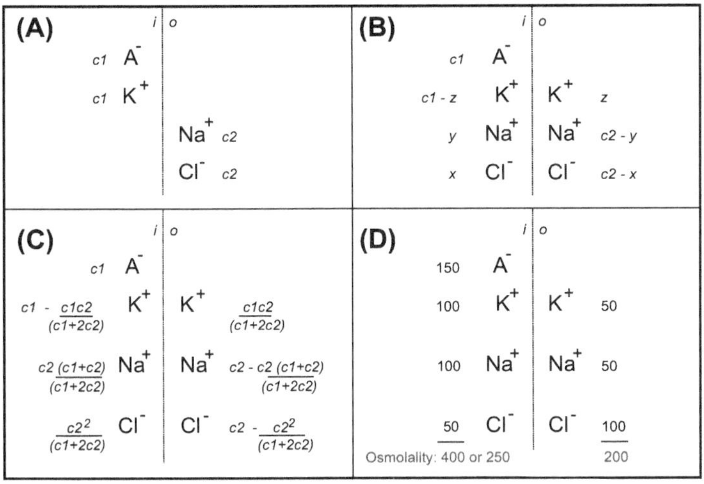

Figure 4 ***Physicochemical system closer to the situation of a cell.*** (A) Impermeant anions (A^-) were placed in one compartment with K^+, NaCl salt was placed in the other compartment. (B) Conditions after salt dissociation and ionic redistribution. (C) Concentrations obtained after x, y, and z have been solved. (D) Example of resulting osmolality differential between the two compartments after ionic redistribution when 150 mM NaCl and KA salts were placed in their respective compartments.

versus 200 mOsM. Thus, this scenario creates an osmotic gradient or "Gibbs-Donnan disequilibrium" that creates a driving force for the movement of water into the internal compartment.

Why is this physicochemical system relevant to a cell? As mentioned earlier, the membrane of a cell is permeable to water and far less permeable to ions. The cells also contain in their cytoplasm many macromolecules with anionic charges. Indeed, the majority of cytosolic proteins have amino and carboxyl moieties that are negatively charged at physiological pH (Gianazza & Righetti, 1980). Similarly, nucleic acids have an acidic isoelectric point making them also carry negative charges at physiological pH (Sherbet, Lakshmi, & Cajone, 1983). Thus, cells observe all the requirements for Gibbs-Donnan disequilibrium and they are under pressure to swell. The reason they maintain their volume is because they use cellular energy to fuel the Na^+/K^+-pump to constantly fight the leak of ions and the influx of water (Kay, 2017; Tosteson & Hoffman, 1960). The pump basically renders the membrane "impermeable" to Na^+ by actively extruding the Na^+ that enters the cell through leak pathways. The presence of non-diffusible anionic charges in cells is likely a key factor in the measured low intracellular

Cl^- concentration in most cells, including central neurons which have the lowest intracellular Cl^- concentrations (Delpire & Staley, 2014; Glykys, et al., 2014).

3.1.1 Epithelial Transport

Epithelia are specialized barriers that keep compartments separate within the body. The compartments can be both internal (e.g., choroid plexus separating the blood from CNS) or external (e.g., airway, intestinal, bladder epithelia separating the blood from the outside). In addition to forming a barrier, epithelial cells actively participate to the movement of solutes from one side to the other. This requires coordinated movement at both poles (apical and basolateral) of the membrane. Since transport is a dynamic process, any disruption at one membrane will create a disruption that will last until the second membrane responds proportionally. As water obligatorily follows the movement of solutes, any disruption in one membrane only will tend to disrupt cell volume.

As an example, we will consider the choroid plexus, a cuboidal epithelium that participates to CSF fluid secretion and K^+ reabsorption (Johanson, et al., 2008). As shown in Fig. 5A, cells isolated with mild collagenase treatment dissociate and keep their polarity with an apical pole of light density due to brush borders (white arrows), and a rough basolateral pole (black arrows) (Wu, Delpire, Hebert, & Strange, 1998). Staining with anti-NKCC1 antibody reveals staining on the apical membrane only (Fig. 5B) — consistent with staining in intact tissue (Plotkin, et al., 1997). When isolated cells were exposed to bumetanide as a NKCC1-specific inhibitor (see Fig. 5E), they demonstrated cell volume reduction, consistent with ion fluxes on the apical membrane not being balanced by ion movement on the basolateral side. If, however, the entry of salts and water on the apical pole is not balanced by a reduction of efflux on the basolateral pole, the net loss of ions and water will result in cell shrinkage. As the effect of bumetanide is reversible, when the inhibitor was removed, the cells started to regain their volume (Fig. 5E). The addition of bumetanide to the choroid plexus epithelial cells was a manipulation that disrupted the balance of transport between the two membranes. Under normal conditions, if an epithelium uptakes solute on the basolateral side, and extrudes them on the apical side, it just facilitates their *trans*-epithelial movement while keeping in check (homeostasis) the internal milieu. This is the case for Cl^- secreting epithelia (e.g., salivary gland, sweat gland, stomach), K^+ secreting epithelia (inner ear), and Na^+ reabsorbing epithelia (intestine, distal convoluted tubule, collecting duct).

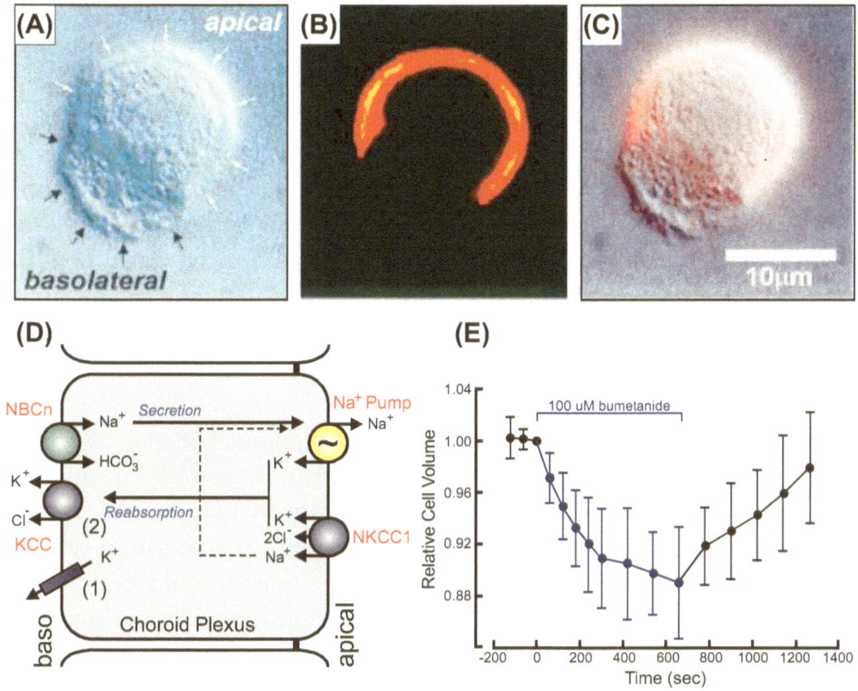

Figure 5 ***Ion and water transport in choroid plexus.*** (A) Mouse choroid plexus epithelial cell observed by Differential interference contrast (DIC) microscopy with *white arrows* identifying the apical and *black arrows* identifying the basolateral membrane, respectively. (B) Fluorescent microscopy image of NKCC1 staining of the same cell using a secondary Cy3-conjugated antibody targeting a primary anti-NKCC1 antibody. (C) Overlay of A & B showing presence of NKCC1 in apical membrane. (D) Model of choroid plexus epithelial cell with apical Na^+/K^+-ATPase (pump) and NKCC1 and basolateral Na-bicarbonate cotransporter (NBCn) and K^+ efflux mechanisms (KCC and K^+ channels). (E) Effect of bumetanide on relative cell volume, as measured by microscopy. *Materials were taken with permission from Wu, Q., Delpire, E., Hebert, S. C., & Strange, K. (1998). Functional demonstration of Na-K-2Cl cotransporter activity in isolated, polarized choroid plexus cells.* American Journal of Physiology, 275, *C1565-C1572.*

3.1.2 Neurotransmission

Charges move in and out of a neuron during synaptic transmission. Glutamate-induced excitation involves the movement of Na^+ and Ca^{2+} into a neuron. An action potential involves the entrance of Na^+, followed by the exit of K^+. GABA (γ-aminobutyric acid)-mediated inhibition involves the inward movement of Cl^- in central neurons (Rivera, et al., 1999; Zhu, Lovinger, & Delpire, 2005) or outward movement of Cl^- in the terminals of peripheral sensory fibers (Sung, Kirby, McDonald, Lovinger, & Delpire, 2000). Compared to the amount of charges (ions) existing in and out of cells, the amount of ions moving during these processes is relatively small and thus, the ion concentrations are not likely to be affected that much. If for instance during an action potential the membrane of a cell moves from −70 mV to +40 mV due to the inward movement of Na^+, this translates to a 110 mV difference. The charge associated with that change can be calculated to be 6875 ions per μm^2. If we hypothetically consider a round cell with a diameter of 10 μm, the calculated surface area is approximately 314 μm^2 and the amount of ions would be $6875 \times 314 = 2.16 \times 10^6$ ions. If the intracellular Na^+ concentration of a neuron is 10 mM, the cytoplasm of the cell contains 3.2×10^9 Na^+. Adding 2.6×10^6 Na^+ ions is equivalent to adding 0.08% to the total, which is negligible. This calculation is based on the overall volume of a cell, and not micro-domains like dendritic spines which can be as small as 0.01 μm^3 or 2×10^5 smaller than a 10 μm cell. In addition, a cell or a dendrite can experience repeated stimulation which could conceivably alter the ion concentrations. Using gramicidin-perforated patch, we demonstrated that long (6 s) application of GABA or repeated applications of GABA to a neuron in culture could lead to measurable changes in the Cl^- concentration (Zhu et al., 2005). These changes if not regulated, are likely to lead to changes in cell volume.

3.1.3 Cell Cycle

Cell volume is a major determinant of the timing of cell division in unicellular algae like *Chlamydomonas* (Craigie & Cavalier-Smith, 1982). The algae grow to a critical volume of 140 μm^3, which seems to be the factor that triggers cell division. In mammalian cells, cell volume changes during the different phases of the cell cycle. When Ehrlich Ascites tumor cells are synchronized, their average mean corpuscular cell volume increases during the S phase and plateau in the G2 phase (DuPre & Hempling, 1978). When mitosis is initiated the volume decreases significantly and this cell volume reduction continues for several hours after the mitosis phase has been completed (DuPre & Hempling, 1978). Interestingly the volume change seems to be related to an increase in osmotically inactive cell volume (Fig. 6) that we define as "b" or the intercept on the Y axis of the relationship between volume ratios over osmotic pressure ratios (see below).

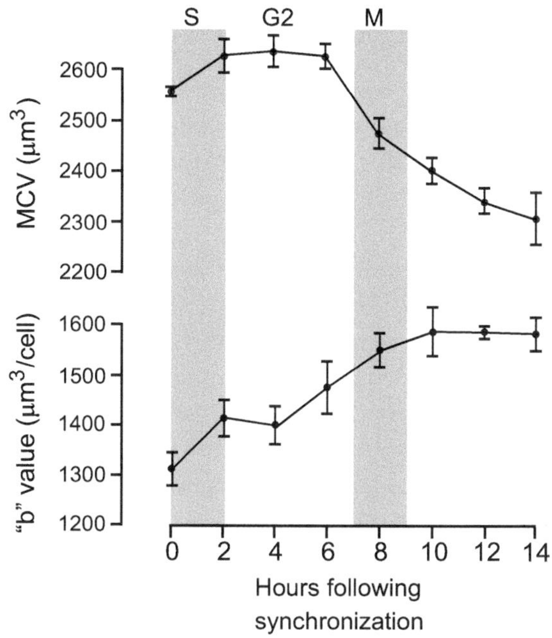

Figure 6 ***Cell volume changes with cell cycle phases.*** Measurement of mean corpuscular volume (MCV) and osmotically-inactive volume "b" in synchronized Ehrlich ascites tumor cells maintained in suspension culture during S, G2, and M phases of the cell cycle. Figure re-drawn (DuPre & Hempling, 1978).

3.1.4 Apoptosis

Cells die through the distinct processes of necrosis or apoptosis. In necrotic cell death, cells originally undergo ATP depletion, cell swelling, increased membrane permeability and lysis (Nieminen, 2003). In contrast, apoptotic cell death is associated with isosmotic cell shrinkage (Kerr, Wyllie, & Currie, 1972). As K^+ is the major osmolytes in the cell, loss of K^+ is a major component of apoptotic cell volume decrease (Bortner & Cidlowski, 2007; Bortner, Hughes, & Cidlowski, 1997). When mouse and human lymphoma cells are exposed to dexamethasone, A23187, thapsigargin, or staurosporine, they demonstrate a significant reduction in cell volume that is accompanied by a reduction in K^+ content. While the Na^+ content of a cell is much lower than that of K^+, there is also a significant reduction in Na^+ content (Bortner et al., 1997). Note that increase in Na^+ (in contrast to decrease) has also been observed in apoptotic cells (Yurinskaya, Rubashkin, & Vereninov, 2011). It is important to stress that loss of K^+ by itself is not a factor in the decision of a cell to undergo apoptotic cell death. Indeed, cells exposed to hypotonic solution, lose a sizable fraction of their K^+ content, but do not undergo apoptotic cell death. That being said, the K^+ loss is an important component of apoptosis, as cells induced to undergo apoptosis failed to follow through when they were exposed to high external K^+ environments (Bortner et al., 1997).

Interestingly, apoptosis can also be triggered by exposing cells to hypertonic cell shrinkage. Their ability to proceed through regulatory volume decrease (RVI) then determines their resistance to apoptosis (Bortner & Cidlowski, 2007; Numata, Sato, Okada, & Wehner, 2008). In 1985, we published two papers showing that rat pheochromocytoma (PC12) cells exposed to hypertonicity demonstrated condensation of chromatin (a hallmark of apoptotic cells (Kerr et al., 1972)), ruffling of the nuclear envelope with loosening of condensed chromatin from the lamina, and apparent loss of nucleolar fibrillar component which disappears in a background of diffuse granular material (Delpire, Duchêne, Gilles & Goessens, 1985; Delpire, Duchêne, Goessens & Gilles, 1985). These cells were unable to undergo RVI (Delpire, 1989). We did not observe the formation of small cytoplasmic fragments with or without pyknotic remnants of nuclei, indicating that if the cells were on the path to apoptosis, they were captured in the very early stages (Delpire, Duchêne, Gilles & Goessens, 1985; Delpire, Duchêne, Goessens & Gilles, 1985). Several studies have attempted to model apoptotic cell volume decrease (Model, 2014; Yurinskaya et al., 2011). Whether the loss of soluble inorganic (K^+ and Cl^-) and organic (taurine, etc.) osmolytes is enough to account for the extent of shrinkage during apoptosis is still a matter of debate. The participation of non-soluble negative (Donnan) charges and osmotically-inactive water molecules might also need to be considered (Model, 2014).

3.1.5 Cell Migration

Cell migration is a key component in the formation and maintenance of multicellular organisms. While this phenomenon occurs predominantly during development, it does continue to occur throughout life, as it is associated with processes such as inflammation and tissue repair. Unfortunately, cell migration is also largely facilitated by malignant cancer cells invading neighboring tissues. Cell migration involves the formation of membrane protrusions such as lamellipodia and membrane ruffles at the leading edge of the cell. These protrusions or localized cell swelling require the rapid movement of ions and water into the protrusion (Condeelis, 1993; Jakab & Ritter, 2006; Lauffenburger & Horwitz, 1996). Verkman (2005) hypothesized that "actin cleavage and ion uptake at the tip of the lamellipodium could create local osmotic gradients that drive the influx of water" across the membrane at the leading edge. Water channels have also been shown to be essential for T cell function and macrophage migration (Rump & Adamzik, 2018). In addition, aortic endothelial cells isolated from aquaporin-1 knockout mice demonstrated a deficit in cell migration (Saadoun, Papadopoulos, Hara-Chikuma, & Verkman, 2005). Interestingly, this deficit could be rescued by expressing non-endothelial cell aquaporin 1 or unrelated aquaporin-4 into these aortic endothelial cells (Saadoun et al., 2005). Aquaporin inhibition is currently being investigated in animal models to reduce metastatic cancer cell invasion (Simone, et al., 2018) or in culture to prevent cell migration (Chen, et al., 2017). The ion transport mechanisms at play in the swelling of the leading edge are multiple. They include the Na^+/H^+ and Cl^-/HCO_3 exchangers (Schwab, 2001), and the Na–K–2Cl cotransporter (Schwab, Wojnowski, Gabriel, & Oberleithner, 1994). Interestingly, these same mechanisms are involved in RVI (see below). Cell migration also involves cell retraction at the trailing edge of the cell (Jakab & Ritter, 2006). Transport mechanisms involved in regulatory volume decrease or RVD are thought to be at play for this process. This is the case of the K–Cl cotransporter (KCC) (Gagnon, 2012; Shen, Chou, & Ellory, 2000; Shen, et al., 2004), the volume activated Cl^- channel (Ransom, O'Neal, & Sontheimer, 2001; Soroceanu, Manning, & Sontheimer, 1999), as well as K^+ channels (Soroceanu et al., 1999).

3.2 Anisosmotic Cell Volume Regulation

In the previous section we considered the movement of water and changes to cell volume under conditions of constant external osmotic pressure. In this section, we will consider the behavior of cells following osmotic challenges.

3.2.1 The Osmotic Equilibration Phase

When two solutions of different osmotic pressures are separated by a semipermeable membrane, water will diffuse from the least concentrated solution to the more concentrated solution in order to equilibrate the pressures. Such a physical system was first described in 1748 by Jean-Antoine Nollet, a catholic priest, who used a pig bladder to close a vial containing spirit of wine that he placed in water. After several hours, he observed a bulged membrane that he pierced with a pin to witness a column of water shooting out (Nollet, 1748). The term osmosis (or endosmosis) was later introduced by French physiologist Dutrochet (1826). In his book, Dutrochet relates experiments done with chicken intestines filled with different liquid contents (milk, egg white, etc.) placed in buckets filled with rain water. In each case, the weight of the intestinal bag increased over time (see Box 2). He called this process endosmosis as the water moved from the bucket into the intestinal bag. He was also clever enough to repeat the experiment in reverse by placing the intestinal bag in buckets containing higher solute concentrations and observe exosmosis. Similar experiments were later performed by the German botanist (Pfeffer, 1877). He created a device (the Pfeffer Cell) that measured the osmotic pressure of a solution. Many studies in the latter part of the 19th century and early part of the 20th century described swelling and shrinking of cells, both vegetal and animal, when exposed to hypotonic or hypertonic solutions, respectively (Gough, 1924; Koeppe, 1895; Nägeli, 1855; Overton, 1907; Pringsheim, 1854; de Vries, 1884, 1888).

The Dutch chemist Jacobus H. van't Hoff, the recipient of the first (1901) Nobel Prize in chemistry, postulated that solutes in dilute solutions

Box 2 Experiment performed by Dutrochet and described on pages 123–124 of his book.

Je pris in coecum de poulet qui, dans l'état de distenstion, avait 12 millimètres de largeur, et qui, courbe en arc, avait 10 centimètres de longeur. Je le remplis à peu près à moitié avec de l'albumen. Dans cet état, il pesait 58 grains. Le coecum fut plongé dand de l'eau de pluie ... Huit heures et demi après le commencement de l'expérience, le coecum pesait 130 grains. Il avait gagné 72 grains, et était devenu tres-turgide ... Il me fut pleinement demontré que l'introduction de l'eau dand la cavité organique dépendait entièrement de la nature du fluide, plus dense que l'eau que contenait cette cavité. Tant que ce fluide conservait son integrité de composition, l'endosmose avait lieu.

Translation: "I took a chicken caecum which when distended measured 12 mm in diameter and 10 cm in length. I filled it half with egg white. In this state, it weighed 3.08 g (1 grain = 53.114 mg). The caecum was placed in rain water ... Eight and a half hour after the beginning of the experiment, the caecum weighed 6.9 g. It had gained 3.8 g and became very turgid ... It was fully demonstrated to me that the introduction of water in the cavity depended on the nature of the fluid, more dense than water existing in this cavity. As long as this fluid kept its integrity [these experiments had tendency to lead to putrefaction], endosmosis occurred."

obey the ideal gas laws, which led to the osmotic pressure formula $\pi = nRT/V$ where R is the gas constant and T the absolute temperature (Box 3).

The definition of the osmotically inactive volume (called "b") in Fig. 7 depends upon the method of cell volume measurement. If the cell volume measurement does not involve the measurement of cell water, b is the combination of the volume occupied by "dry" (or non-water) material added to the volume of water that does not participate in intracellular osmolality. If, however, the amount of water represents cell volume, b then represents the osmotically inactive water content. The ratio of osmotically active water to total water is sometimes referred to as Ponder's R factor (Ponder, 1948, pp. 83–114). In the muscle fiber of the giant barnacle, *Balanus nubilis*, solvent water was estimated to be 64%–72% of total water (Hinke, 1970). Osmotically inactive water is believed to be associated or bound to macromolecules. Note that if experiments with cells are performed with an infinite extracellular volume, in vivo, the extracellular space is often limited and tends to buffer volume changes. Let us consider a rigid system with two compartments designated *in* and *out*. At equilibrium, each compartment has an osmotic pressure π_0. If we modify the osmotic pressure of the

Box 3 Van'Hoff's cell volume versus osmotic pressure relationship

The Boyle/van't Hoff law of conservation stipulates that

$$\pi \cdot v = \pi_0 \cdot v_0 \tag{5}$$

If we consider that the osmotically active component of a cell does not occupy the entire volume of a cell, the equation becomes:

$$\pi \cdot (v - b) = \pi_0 \cdot (v_0 - b) \tag{6}$$

where b represents osmotically inactive volume.

If we then divide both terms by the initial volume (v_0), one obtains the linear relationship:

$$v/v_0 = \pi_0/\pi \cdot (v_0 - b)/v_0 + b/v_0 \tag{7}$$

The intercept on the Y axis does not cross 0 but results in a positive value that is equivalent to b, the osmotically inactive volume. As the osmolarity of the solution decreases, the volume of the cell increases proportionally (Fig. 7A).

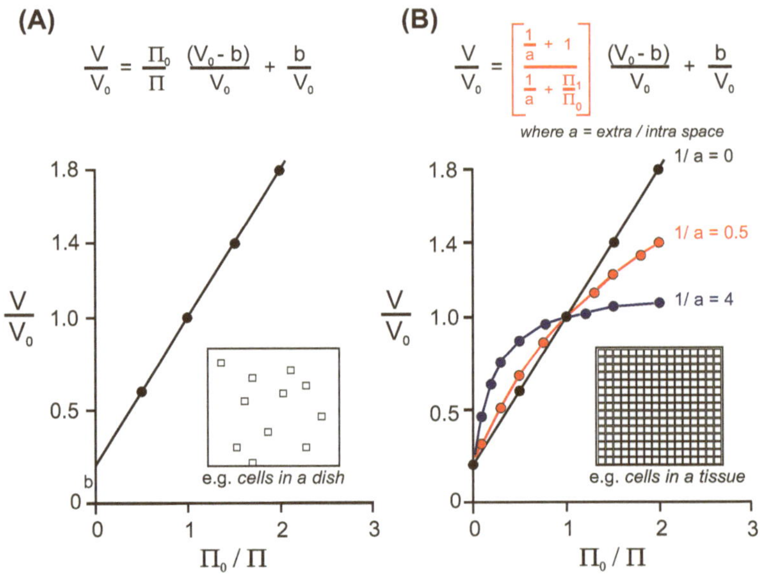

Figure 7 ***Relationship between cell volume and osmolarity.*** (A), Linear relationship of volume ratio divided by osmotic ratio is based on modified ideal gas law for dilute solutions. Note that the slope is always less than 1 and the intercept on the Y axis has a positive value "b". (B) Same relationship but considering the ratio of extracellular versus intracellular space (called "a"). Note that as "a" decreases (or 1/a increases), the relationship deviates from a linear relationship and volume changes are minimized or buffered. The insets visually represents cells in an infinite volume (in A) compared to cells packed in a tissue (in B).

"out" compartment (π_0 becomes π_1), a new equilibrium will develop leading to new volumes *in′* and *out′* with a new osmotic pressure π'.

If the extracellular volume (out) is infinite, 1/a is equal to 0 and Eq. (11) becomes Eq. (7) (straight line in Fig. 7B). If, however the ratio of extracellular to intracellular volume is small, the relationship is no longer linear and extracellular space plays the role of a buffer. Fig. 7B shows the relationship for intracellular to extracellular ratios of 0.5 and 4. Thus, in vivo, a confined extracellular volume buffers the effect of osmotic pressure changes. Note that in the mathematical model described in Box 4 below, we postulated that the sum of the "in" and "out" compartments was constant, which is not entirely accurate as tissues and organs are not fully rigid.

The behavior of isolated cells varies from cell type to cell type. While some cells at the peak response closely follow the van't Hoff linear relationship (Dellasega & Grantham, 1973; Olson & Holtzman, 1982),

Box 4 Role of the extracellular space in the relationship between cell volume and osmolality.

First, we can stipulate

$$\text{in} + \text{out} = \text{in}' + \text{out}'$$
$$\text{in}' = \text{in} \cdot \pi_0/\pi'$$
$$\text{out} = \text{out} \cdot \pi_1/\pi'$$

From these equations, we can extract π as

$$\pi = (\pi_0 \cdot \text{in} + \pi_1 \cdot \text{out})/(\text{in} + \text{out}) \quad (8)$$

Dividing both terms of the ratio by out:

$$\pi = (\pi_0 \cdot \text{in}/\text{out} + \pi_1)/(\text{in}/\text{out} + 1) \quad (9)$$

Defining a = out/in

$$\pi = (\pi_0/a + \pi_1)/(1/a + 1) \quad (10)$$

Inserting Eq. (10) into Eq. (8), we obtain:

$$v/v_0 = \pi_0/\pi[(1/a + 1)/(1/a + \pi'/\pi_0)] \cdot (v_0 - b)/v_0 + b/v_0 \quad (11)$$

where a = extracellular space/intracellular space.

others (e.g., chicken lymphocytes, Hela cells, and murine hepatocytes) demonstrate resistance to swelling during large hypotonic shocks (similar to channel rectification) (Doljanski, Ben-Sasson, Reich, & Grover, 1974; Howard & Wondergem, 1987; Tivey, Simmons, & Aiton, 1985). This likely indicates some participation of the cytoskeleton in creating resistance to cell swelling.

3.2.2 Cell Response to Hypotonicity

When cells are exposed to hypotonic conditions, their swelling induces the activation of separate K^+ and Cl^- channels, leading to a loss of K^+, Cl^-, and obligatory water, leading the restoration of cell volume (Fig. 8) (Hoffmann, Lambert, & Simonsen, 1986; Hoffmann, Simonsen, & Lambert, 1984). Because the conductive Cl^- permeability of cells is typically low, an increase in in conductive K^+ permeability alone cannot account for an observed net loss of KCI during RVD, an increase in conductive Cl^- permeability must also occur. We will now discuss the participation of K^+ channels and Cl^- channels in regulatory volume decrease.

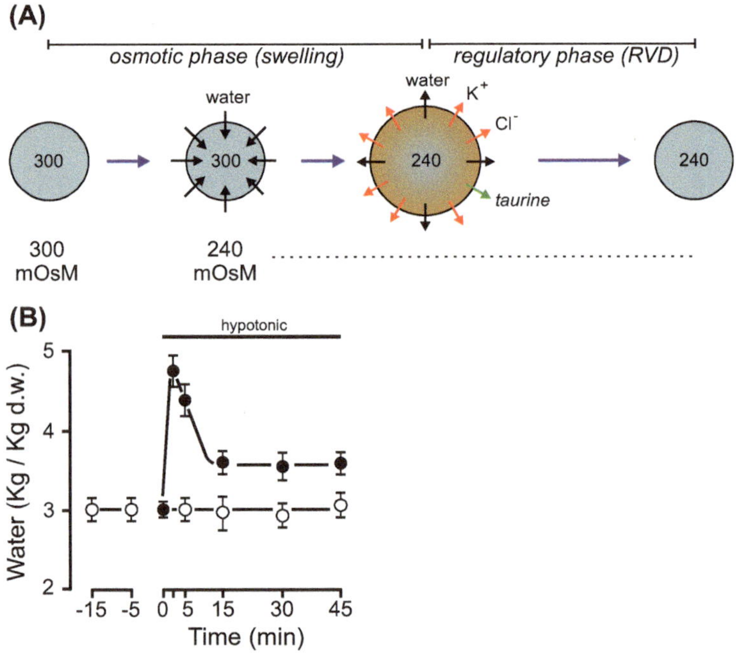

Figure 8 ***Behavior of mammalian cells exposed to a change in osmolalities from 300 mOsM to 240 mOsM.*** (A) separation between the osmotic (swelling) phase and the regulatory phase. As water enters the cell to equilibrate osmotic pressure, the cell swells. The loss of osmolytes such as K^+, Cl^-, and taurine lead to water loss and cell volume recovery. Note that the inside of the cell after RVD is at 240 mOsM. (B) Example of a cell (PC12 cells) undergoing cell swelling and regulatory volume decrease upon exposure to hypotonic conditions. Close symbols are cells exposed to hypotonic conditions, open symbols are cells remaining under isosmotic conditions. *Redrawn from Delpire, E. (1989).* Controle du volume cellulaire: Etudes au niveau de cellules PC12 en culture. *Université de Liège: Liège; Delpire, E., Cornet, M., & Gilles, R. (1991). Volume regulation in rat pheochromocytoma cultured cells submitted to hypoosmotic conditions.* Archiv International Physiology Biochemistry, *99, 71–76.*

3.2.2.1 Volume-Activated K^+ Channels

Early work has shown that in Ehrlich ascites tumor cells, the K^+ loss that is activated by cell swelling (Hendil & Hoffmann, 1974) is mediated by a Ca^{2+}-dependent K^+ channel sensitive to quinine (Hoffmann et al., 1984). In these cells, the loss of K^+ exceeds the loss of Cl^- loss by a factor of about 1.6 (Hendil & Hoffmann, 1974). Interestingly, while the permeability of the cells to Cl^- increases some 60-fold, the permeability of a cell for K^+ increases only 2-fold upon cell swelling (Hoffmann & Simonsen, 1989). This indicates that the Cl^- permeability is likely to be a greater limiting factor to RVD than the K^+ permeability (Hendil & Hoffmann, 1974). The conductance of the Ca^{2+}-activated K^+ channel in Ehrlich cells was measured under normal external K^+ concentration at 7 pS (Christensen & Hoffmann, 1992). Arachidonic acid metabolites such as leukotriene D_4 (LTD_4) have been shown to enhance KCl loss and RVD following hypotonic shock (Jorgensen, Lambert, & Hoffmann, 1996). While there is a concomitant increase in intracellular Ca^{2+}, the LTD_4 effect seemed to be independent of the transient rise in the divalent cation. The K^+ current activated by cell swelling was shown to be insensitive to charybdotoxin as well as to apamin, inhibitors of calcium-activated voltage-gated shaker K^+ channels and small conductance SK channels, respectively (Riquelme, Sepúlveda, Jørgensen, Pedersen, & Hoffmann, 1998). In contrast, clofilium, a class III antiarrhythmic drug blocked the volume-sensitive K^+ current in a voltage independent manner and acted as a strong inhibitor of the regulatory volume decrease response of Ehrlich cells (Niemeyer, et al., 2000). Because the volume-sensitive K^+ channel is pH-sensitive and has a small ion conductance, TASK-1 and TASK-2 channels expressed in Ehrlich cells (Niemeyer, et al., 2000) were proposed as candidates (Hougaard, Jørgensen, & Hoffmann, 2001). The TASK-2 channel was later shown to be a major component of the K^+ conductance activated by cell swelling (Kirkegaard, Lambert, Gammeltoft, & Hoffmann, 2010; Kirkegaard, Strøm, Gammeltoft, Hansen, & Hoffmann, 2016). KCNQ4, but not KCNQ2/3, were also shown to be involved in regulatory volume decrease in HEK293 cells overexpressing these channels (Hougaard, Klaerke, Hoffmann, Olesen, & Jorgensen, 2004).

3.2.2.2 Volume-Activated Cl^- Channels

Work from Else Hoffmann's group in Ehrlich Ascites tumor cells and of Sergio Grinstein and colleagues in human lymphocytes showed activation of a Cl^- conductance during RVD (Grinstein, Clarke, Dupre, & Rothstein, 1982; Grinstein, Clarke, Rothstein & Gelfand, 1983; Hoffmann, Simonsen, & Sjøholm, 1979). Because ionophores, which produce increases in K^+ permeability far greater than a hypotonic shock, failed to produce any changes in cell volume, Grinstein argued that the anion conductance was the rate-limiting factor in salt and water transport (Grinstein et al., 1982). However, because the gramicidin ionophore greatly accelerated RVD in Ehrlich cells, Hoffmann argued that the K^+ conductance was also limiting (Hoffmann et al., 1986). Thus, it can be concluded that both conductance are low under isotonic conditions, and the movement of one ion through one conductance is limited unless there is movement of the oppositely charged ion in the other conductance. In Ehrlich cells the Cl^- conductance was described as unselective carrying CI^-, Br^-, NO_3^-, and SCN^- (Hoffmann et al., 1986), while in lymphocytes the channel had the following anion series: $SCN^- = I^- > NO_3^- > Br^- >= Cl^- > acetate^- > SO_4^{2-} = gluconate^-$.

In 1983, Else Hoffman and Ian Lambert demonstrated that Ehrlich Ascites tumor cells exposed to a hypotonic solution lose significant amount of amino acids (Hoffmann & Lambert, 1983). In these cells, taurine (53 mM) is definitely the most abundant amino acid, followed by glycine (9 mM), and other amino acids at 1–2 mM each. During a decrease in osmotic pressure from 300 mOsM to 225, the vast majority (90%) of taurine is lost, whereas only 30% of glycine is lost, making taurine the most significant organic osmolytes in cells. The permeability (loss) of taurine is proportional to the amount of cell swelling (Hoffmann & Lambert, 1983). Ten years later, the authors demonstrated that the swelling-induced activation of the taurine leak involved arachidonic acid and leukotrienes, as did the activation of K^+ and Cl^- conductances (Lambert & Hoffmann, 1993). The authors indicated for the first time the possibility that the taurine efflux was directly linked to the activation of the Cl^- channels. While the Cl^- channel blocker MK196 inhibited both Cl^- efflux and taurine efflux pathways, the fact that 4,4′-Diisothiocyano-2,2′-stilbenedisulfonic acid (DIDS) minimally affected Cl^- efflux in their cells, prompted them to later conclude that the swelling-activated Cl^- channel and the taurine channel in the Ehrlich cells represented two distinct types of channels (Hoffmann & Lambert, 1994; Lambert & Hoffmann, 1994).

Demonstration that amino acid efflux and Cl^- efflux were mediated by one single channel was made by Kevin Strange (Jackson & Strange, 1993; Strange, Morrison, Shrode, & Putnam, 1992). They demonstrated that $P_{taurine}/P_{Cl}$ was ~ 0.2 meaning that the channel was preferentially permeable to Cl^-, but the permeability to taurine was not negligible. They called the channel Volume Sensitive Organic Anion Channel or VSOAC. In C6 cells, the channel has a unitary conductance of 35–40 pS (Jackson, Morrison, & Strange, 1994). Many attempts have been made to identify the molecular makup of this functional channel unit. Proteins like P-glycoprotein (Gill, et al., 1992), I_{Cln} (Paulmichl, et al., 1992), CLC2 (Gründer, Thiemann, Pusch, & Jentsch, 1992), and ClC3 (Duan, Winter, Cowley, Hume, & Horowitz, 1997) were all initially considered as candidates. Following these first attempts at identifying the Cl^- channels involved in RVD, came the demonstration that TMEM16, a Ca^{2+}-activated Cl^- channel is active during apoptosis and cell swelling (Martins, et al., 2011). Jurkat cells activated by Fas ligand express an outwardly rectifying Cl^- current, sensitive to NPPB (5-Nitro-2-(3-phenylpropylamino) benzoic acid) and to the anoctamin-blocker $CaCC_{inh}$-AO1 that was completely abrogated when TMEM16F was knocked down. Outwardly rectifying Cl^- currents, which were NPPB−, DIDS-, and $CaCC_{inh}$-AO1-sensitive, were also eliminated by TMEM16F siRNA in airway epithelial cells (Martins, et al., 2011). The authors demonstrated that regulatory volume decrease after hypotonic cell swelling was augmented in HEK293 cells overexpressing TMEM16F. Another Ca^{2+}-activated Cl^- channel, drosophila Bestrophin-1 (or dBest1) was proposed to mediate volume-stimulated Cl^- currents. Knockdown of dBest 1 in Drosophila S2 cells abolished volume-regulated Cl^- currents and significantly reduced regulatory volume decrease (Chien & Hartzell, 2007).

It took another few years for the molecular identification of the volume-activated Cl^- channel associated with amino acid efflux to be revealed in two landmark papers (Qiu, et al., 2014; Voss, et al., 2014). In the two papers, the authors used an RNA interference screen and swelling-induced iodine uptake and quenching of a yellow fluorescent protein, to identify LRRC8A (Leucine Rich Repeat Containing 8 VRAC Subunit A). The first paper showed that knockdown of LRRC8A protein strongly suppressed $I_{Cl\text{-}swell}$ in follow-up patch clamping experiments. When LRRC8A was overexpressed in HEK293 cells, the $I_{Cl\text{-}swell}$ current decreased, but when LRRC8A and LRRC8C were co-transfected, a significant increase in $I_{Cl\text{-}swell}$ current was observed. These data indicated that multiple subunits were needed to form a functional channel that traffics properly to the plasma membrane (Voss, et al., 2014). The authors also demonstrated that cas9-mediated inactivation of LRRC8A resulted in cells that could no longer efflux taurine upon hypotonic treatment. The specific participation of TMEMs, bestrophins, and LRRC8 channels to chloride loss during RVD will still need to be individually resolved from cell type to cell type. Whether these channels also regulate cell volume under isosmotic conditions is currently unknown.

3.2.2.3 K—Cl Cotransporters

A volume- and N-ethylmaleimide-stimulated Cl^--dependent K^+ flux was first described in human and sheep red blood cells (Dunham & Ellory, 1981; Dunham, Steward, & Ellory, 1980; Lauf & Theg, 1980). This Cl^--dependent K^+ transport was attributed to a K—Cl transport mechanism (now K−Cl cotransporter). Coincidentally, the function of a Na−K−2Cl cotransport mechanism was also described the same year (see below and (Geck, Pietrzyk, Burckhardt, Pfeiffer, & Heinz, 1980)). The red blood cell K−Cl cotransporter is functionally silent under control conditions (Lauf & Theg, 1980) and activated by sulfhydryl reagents as well as cell swelling. In his 1980 paper, Peter Lauf indicated that red blood cell precursors had 10 fold higher K^+ flux than mature red blood cells (Lauf & Theg, 1980). Similarly, reticulocytes which are larger young red blood cells, demonstrate greater Cl^--sensitive K^+ fluxes than mature erythrocytes (Lauf, 1983). The cells are thus able to reduce their volume by losing K^+, Cl^-, and water (Lauf and Bauer, 1987). K−Cl cotransport, through a still unknown signaling mechanism, is activated in red blood cell from sickle cell patients, contributing to the dehydration and sickling process (Berkowitz & Orringer, 1987; Brugnara, 1989; Quarmyne, Risinger, Linkugel, Frazier, & Joiner, 2011).

When healthy red blood cells are exposed to a hypotonic solution, they rapidly swell and then slowly recover their volume. This behavior was first demonstrated in erythrocytes isolated from *Pleuronectes flesus*, although this study investigated the role of amino acids loss and not the role of ions (Fugelli, 1967). It was later shown in duck red blood cells, that a loss of K^+ was associated with the "volume regulatory phase" (Kregenow, 1971), and this loss of K^+ was Cl^--dependent in marine teleost *Opsanus tau* erythrocytes (Lauf, 1982). KCC activation has also been shown in rat (Haas & Harrison, 1989), rabbit (Brugnara, Van Ha, & Tosteson, 1989; Jennings & Schultz, 1990), dog (Fujise, et al., 1991; Parker, Colclasure, & McManus, 1991), sheep (Bergh, Kelly, & Dunham, 1990), pig (Kim, Sergeant, Forte, Sohn, & Im, 1989) and human (Brugnara et al., 1989; Dunham & Ellory, 1981; Kaji, 1986; Sachs & Martin, 1993) erythrocytes.

Molecular cloning in the late 1990s led to the identification of four genes (*SLC12A4-7*) in mammals that encode four K−Cl cotransporters: KCC1-4 (Gillen, Brill, Payne, & Forbush, 1996; Hiki, et al., 1999; Mount, et al., 1999; Payne, Stevenson, & Donaldson, 1996; Race, et al., 1999). Interestingly, only KCC2 displays some basal transport activity under isotonic conditions (Mercado, Broumand, Zandi-Nejad, Enck, & Mount, 2006). Activation of all the other K−Cl cotransporters could only be demonstrated following cell swelling. After observing a lag phase of activation following cell swelling, Michael Jennings proposed in 1990 a two-state kinetic model, where KCC is inactivated by the action of a protein kinase and activated by the action of a phosphatase (Jennings & al-Rohil, 1990). The model and the data indicated that cell swelling inhibited a kinase with no effect on the conjugate phosphatase thereby resulting in an increase in the rate of dephosphorylation concurrent with a decrease in the rate of phosphorylation (Fig. 9). While we always assume that swelling is the factor that leads to KCC activation (because we utilize this manipulation to activate the transporter), any physiological factor that decreases transporter phosphorylation or increases transporter de-phosphorylation will lead to an active K–Cl cotransporter.

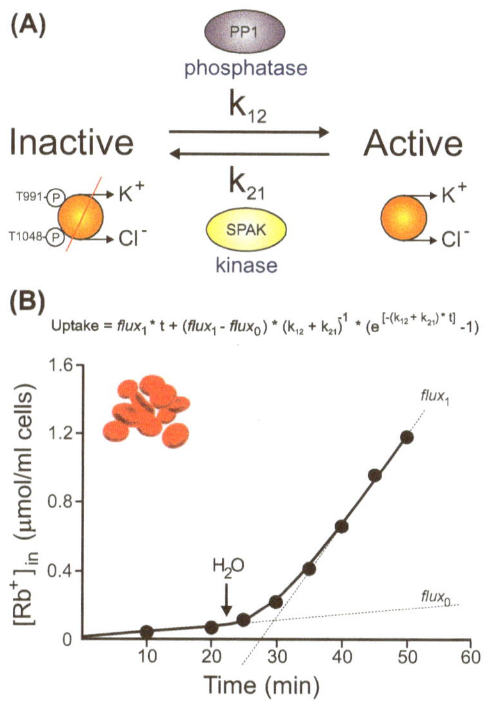

Figure 9 ***Diagram showing the two-state model of K–Cl cotransport activation/inactivation.*** (A) A protein phosphatase (PP1) mediates a de-phosphorylation (activation) reaction of specific threonine residues (T991 & T1048) with a rate constant of k_{12}, whereas a kinase (SPAK) mediates the opposite phosphorylation (inhibition) reaction of these same threonine residues with a different rate constant (k_{21}). (B) Measurement of ^{86}Rb uptake versus time in rabbit red blood cells illustrates a lag phase occurs in the stimulation of the flux from $flux_0$ to $flux_1$. Water is added at arrow to decrease the medium's osmolarity. *Figure adapted/redrawn from Jennings, M. L., & al-Rohil, N. (1990). Kinetics of activation and inactivation of swelling-stimulated K+/Cl– transport. The volume-sensitive parameter is the rate constant for inactivation.* The Journal of General Physiology, 95, *1021–1040.*

A role for KCC3 in cell volume maintenance of peripheral nerve fibers was evidenced by genetic disruptions of the human *SLC12A6* gene. Loss of function mutations in the gene are found in a few sporadic families throughout the world (Uyanik, et al., 2006) as well as in many Canadian families living in two regions around Quebec City (Howard, et al., 2002). Axonal swelling was found in KCC3-deficient patients (Auer, et al., 2016; Dupré, et al., 2003), consistent with this cotransporter isoform playing a role in regulatory volume decrease. Analysis of peripheral nerves in a knockout mouse model also found significant swelling of peripheral nerve fibers (Byun & Delpire, 2007). Interestingly, a single individual that carries a gain of function mutation in KCC3 was also identified (Kahle, et al., 2016). The mutated residue (Thr991) is one of the two key threonine (Thr991 and Thr1048), involved in shutting off KCC3 function (Fig. 9). Mutation of Thr991 into an alanine prevents phosphorylation of the residue, leading to a constitutively active transporter (Rinehart, et al., 2009). Our laboratory also created a KCC3-T991A knock-in mouse (Kahle, et al., 2016) and demonstrated the shrinkage of peripheral nerve fibers (Flores, Schornack, & Delpire, 2018). Thus, when KCC3 function is abolished the nerve fibers are swollen, and when KCC3 function is enhanced, the nerve fibers are shrunken. These data clearly establish a key role for KCC3 in cell volume maintenance.

There are two recently published cell volume experiments that provide puzzling data worth discussing in this section. A 2015 paper reported the behavior of HEK293 cells overexpressing a human KCC3a T991A-T1048A double mutant protein (Adragna, et al., 2015). Cell volume was measured using a well-established microscopic method based on fluorescent dye dilution (Blanco, Márquez, & Alvarez-Leefmans, 2013; Lenart, Kintner, Shull, & Sun, 2004). When wild-type HEK293 cells were exposed to a large 150 mOsM/Kg water osmotic shock, the cells increased their volume 3—4 times, then recovered their volume with a slope of 12% cell water volume recovery per min. When HEK293 cells overexpressing KCC3-T991A-T1048A were exposed to the same hypotonic shock, cell swelling was significantly attenuated (up to twofold) and the rate of recovery was also dampened (4% recovery/min). The authors did not attempt to decrease the extent of the osmotic shock to match the swelling. Such an experiment would have determined if the rate of regulatory volume decrease would have matched that of the mutant cells. The significant reduction in peak swelling observed in the KCC3-T991A-T1048A expressing cells was totally unexpected. Because the cells might have changed their phenotype (e.g., attachment to substrate), furosemide was also added to inhibit the overactive cotransporter. When treated acutely with furosemide, the cells expressing the constitutively active transporter showed significantly increased swelling (threefold) followed by minimal cell volume recovery. Note that native HEK293 cells treated with furosemide also showed a small increase in peak swelling. These data clearly indicated that the activity of the transporter was responsible in preventing cell swelling.

Note that the data presented in that paper did not comply with the van't Hoff relationship. If the osmotically inactive volume of the cells is negligible (or equal to zero), the van't Hoff relationship passes through the graph origin ($X = 0$ and $Y = 0$), and at best, a 300 mOsM to 150 mOsM (twofold) osmotic shock should lead to a two fold increase in cell volume. Why was a fourfold increase in cell volume measured with wild-type HEK293 cells? This error could be related to the methodology that overestimated the changes in cell volume. Irrespective of this discrepancy with the van' Hoff relationship, the fact remains that cells overexpressing the mutant KCC3 did not swell as much as wild-type cells. This could only be explained by cells dumping K^+, Cl^-, and water during the swelling phase. In contrast to a wild-type transporter which demonstrates a lag-phase of activation upon swelling (see above), the mutant transporter is already fully active when the swelling occurs. The data therefore suggest that with an active KCC3 transporter, the membrane can no longer be regarded as semi-permeable, i.e., far more permeable to water than solutes.

This unusual behavior was also observed with human fibroblasts isolated from the patient carrying the single T991A mutation in KCC3 (Kahle, et al., 2016). Fibroblasts from the patient as well as control human fibroblasts were exposed to the same hypotonic shock and cell volume was measured using the same methodology. In fact, the experiments were performed in the same laboratory. Wild-type fibroblasts increased their volume by 2.5 fold, whereas the patient fibroblasts swelled only marginally (Kahle, et al., 2016). No regulatory volume decrease was observed in the wild-type cells.

Along with red blood cells and peripheral nerve fibers, KCC has been implicated in cell volume regulation in mouse sperm (Klein, Cooper, & Yeung, 2006), and glial cells (Ringel & Plesnila, 2008). Interestingly, as with nerve fibers, specific inhibition of KCCs caused glial cell swelling under isotonic conditions indicating that either the cotransporter is active *in vivo* under isotonic conditions, or that the cotransporter responds to small increases in cell volume. Cell volume increase could be due to *Donnan* forces (see above) or to cell transport activities that leads to small water intake and swelling. In this regard, a study published in 1986 provides evidence that cells can maintain their volume constant under anisosmotic conditions provided that the changes in osmolarities are slow and gradual. Indeed, Lohr and Grantham exposed kidney tubules to a solution with an osmolarity that changed from 293 mOsM/Kg to 110 mOsM/Kg at a rate of 1.5 mOsM decrease per min. They observe no changes in cell volume over a large range of osmolarities (293–170 mOsM) (Lohr & Grantham, 1986). They named the portion of the study in which cell volume did not change as the osmolality decreased as the *isovolumetric regulation phase* (Lohr & Grantham, 1986). They determined that cells that did remain isovolumetric during a 100 mOsM change had lost significant amount of osmolytes, as these cells shrank abruptly when re-exposed to isosmotic conditions. A similar isovolumetric behavior was also observed in the direction of increased osmolarities.

3.2.3 Cell Response to Hypertonicity

3.2.3.1 Short Term – Regulatory Volume Increase

When cells are exposed to a hypertonic solution, they first lose water to equilibrate internal and external osmotic pressures (Fig. 10A). Following this initial cell volume reduction, some cells undergo a regulatory volume increase process (called RVI) to restore their volume, while other cells do not volume regulate and remain shrunken (Fig. 10B, Box 5). Several transport mechanisms, that include the Na^+/H^+ exchanger and the Na–K–2Cl cotransporter, have been implicated in RVI. If the cells remain in the hypertonic environment, they will adapt by accumulating organic osmolytes. This will be covered in more detail in the long-term accumulation of organic osmolytes section.

3.2.3.1.1 The Na^+/H^+ and Cl^-/HCO_3^- Exchangers When *Amphiuma* red blood cells are transferred from a 245 mOsM isosmotic saline to a 320 mOsM hypertonic saline, they first shrink, then undergo regulatory volume increase in response to net cellular uptake of Na^+ and Cl^- (Cala, 1980). The net influx of Na^+ associated with RVI is completely prevented by the addition of 1 mM amiloride (Cala, 1980; Siebens & Kregenow, 1978). Because the accumulation of Na^+ is pH-dependent

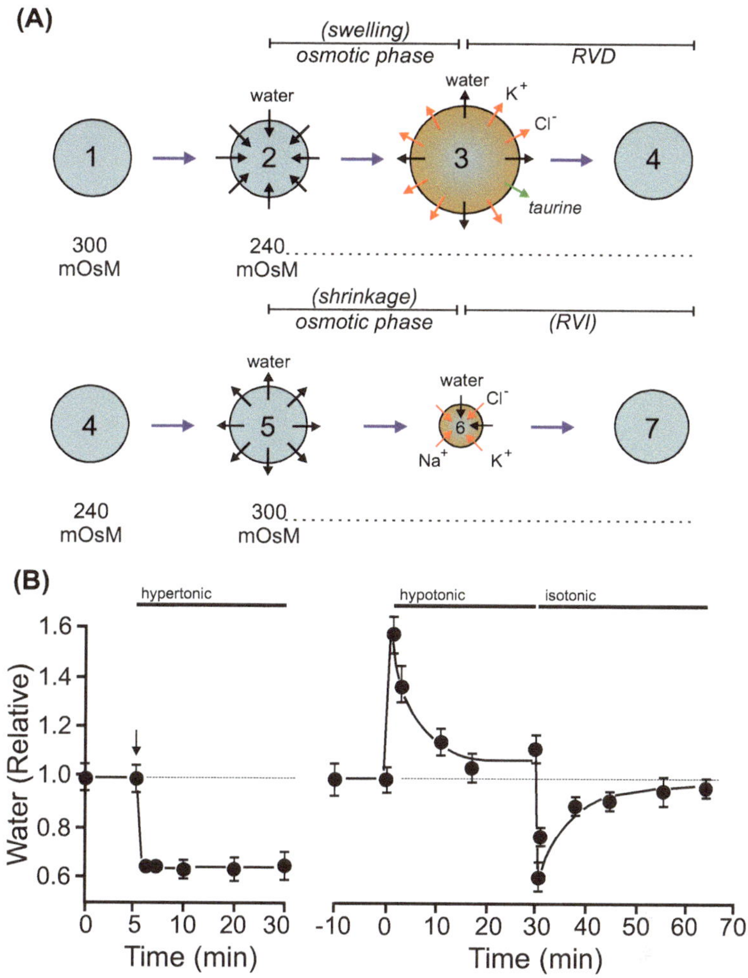

Figure 10 ***Behavior of mammalian cells exposed to changes in osmolality.*** (A) After the hypotonic (240 mOsM) swelling (2 to 3) and regulatory volume decrease (3 to 4) phases, the cells are re-exposed to isotonic (300 mOsM) conditions. They then lose water and shrink (5 to 6). This treatment triggers a secondary regulatory volume increase (6 to 7). (B) Example of a cell (PC12 cells) undergoing cell shrinkage with no volume recovery when exposed to hypertonic conditions. Upon return to isotonic conditions after regulatory volume decrease, the same cells now demonstrate post-RVD RVI. *Redrawn from Delpire, E. (1989).* Controle du volume cellulaire: Etudes au niveau de cellules PC12 en culture. *Université de Liège: Liège.*

Box 5 Behavior of different cell types when exposed to a hypertonic solution.

Cell Type	RVI – Yes	NKCC-mediated	RVI – No
Guinea pig jejunal villus cells	MacLeod and Hamilton (1990)	Yes	
Duck red blood cells	Schmidt and McManus (1977)	Yes	
Ehrlich Ascites cells	Geck and Pfeiffer (1985)	Yes	
Ehrlich Ascites cells	Hendil and Hoffmann (1974), Hoffmann and Lambert (1983)	Yes	
Amphiuma red blood cells	Cala (1980)	No	
Human lymphoid cells		No	Roti Roti and Rothstein (1973)
Human PBM		No	Grinstein, Clarke and Rothstein (1983)

and amiloride-sensitive, a Na^+/H^+ exchanger is involved. Due to the stimulation of Na^+ uptake by high HCO_3^- medium, it was postulated that Na^+/H^+ exchange is functionally coupled to Cl^-/HCO_3^- exchange. The net result is accumulation of NaCl and obligatory water (Cala, 1980). Similar transport systems are activated by hypertonicity if dog red blood cells (Parker, 1983), *Necturus* gallbladder (Ericson & Spring, 1982) and human lymphocytes (Grinstein, Clarke & Rothstein, 1983). In *Amphiuma* red blood cells, the accumulation of Na^+ triggers the exchange of some of that Na^+ with K^+ through activation of the Na^+/K^+-ATPase. This phenomenon is prevented by the application of ouabain, while the overall amount of Na^+ *plus* K^+ that enters during RVI is ouabain-independent (Siebens & Kregenow, 1985).

3.2.3.1.2 The Na–K–2Cl Cotransporter Na–K–2Cl cotransport was functionally defined by Peter Geck in a landmark 1980 paper (Geck et al., 1980). The molecular identification was done 14 years later with the cloning of a kidney-specific isoform, NKCC2 (Gamba, et al., 1994; Igarashi, Vanden Heuvel, Payne, & Forbush, 1995; Payne & Forbush, 1994), and a more widely-distributed isoform, NKCC1 (Delpire, Rauchman, Beier, Hebert, & Gullans, 1994; Payne, et al., 1995; Xu, et al., 1994). The Na−K−2Cl cotransporters are integral transmembrane proteins activated (in opposition to K−Cl cotransporters) by phosphorylation and inactivated by dephosphorylation (Lytle & Forbush, 1992; Palfrey & Pewitt, 1993). Several key regulatory phospho-residues are located in the cytosolic N-terminal of the cotransporter (Darman & Forbush, 2002; Flemmer, Gimenez, Dowd, Darman, & Forbush, 2002; Gimenez & Forbush, 2005). The two Ste20p-kinases, SPAK and OSR1, which constitute the germinal center kinase 6 subfamily, were identified as the kinases that bind and phosphorylate the cotransporter (Gagnon & Delpire, 2010b; Gagnon, England, & Delpire, 2007; Piechotta, Lu, & Delpire, 2002). These terminal kinases are themselves under the control of upstream kinases, such as WNKs (Gagnon, England, & Delpire, 2006; Moriguchi, et al., 2005; Piechotta, Garbarini, England, & Delpire, 2003; Vitari, Deak, Morrice, & Alessi, 2005).

Cells that fail to undergo RVI typically do express the Na−K−2Cl cotransporter, but the mechanism is functionally silent. This is evidenced by the ability of these cells to demonstrate RVI after they have been first exposed to hypotonicity and allowed to volume regulate, and then returned to the isosmotic solution (Grinstein, Clarke & Rothstein, 1983; Grinstein, Rothstein, Sarkadi, & Gelfand, 1984; Hoffmann, Sjoholm, & Simonsen, 1983). This secondary RVI is now possible because of the loss of Cl^- that occurred during the RVD, allowing the activation of NKCC1 by the shrinkage induced by the return to the isosmotic solution (Fig. 10). Alternatively, the rate of RVI greatly increased in simian eccrine clear cells exposed to a hypertonic solution when cells are co-treated with forskolin (Toyomoto, Knutsen, Soos, & Sato, 1997) (Fig. 11). This observation can be simply explained by forskolin inducing the opening of Cl^- channels (such as CFTR), the loss of intracellular Cl^-, and the activation of the WNK-SPAK signaling cascade, leading to stimulation of the Na−K−2Cl cotransport, and ultimately cell volume recovery (Fig. 11C). The reverse situation is also true. When eccrine cells were returned to isosmoticity after a hypotonic challenge, they readily underwent secondary RVI. This regulatory volume phase was inhibited with co-treatment with phorbol ester 12-O-Tetradecanoylphorbol-13-acetate (TPA) which activates protein kinase C and inhibits NKCC1 (Fig. 11B) (Toyomoto et al., 1997).

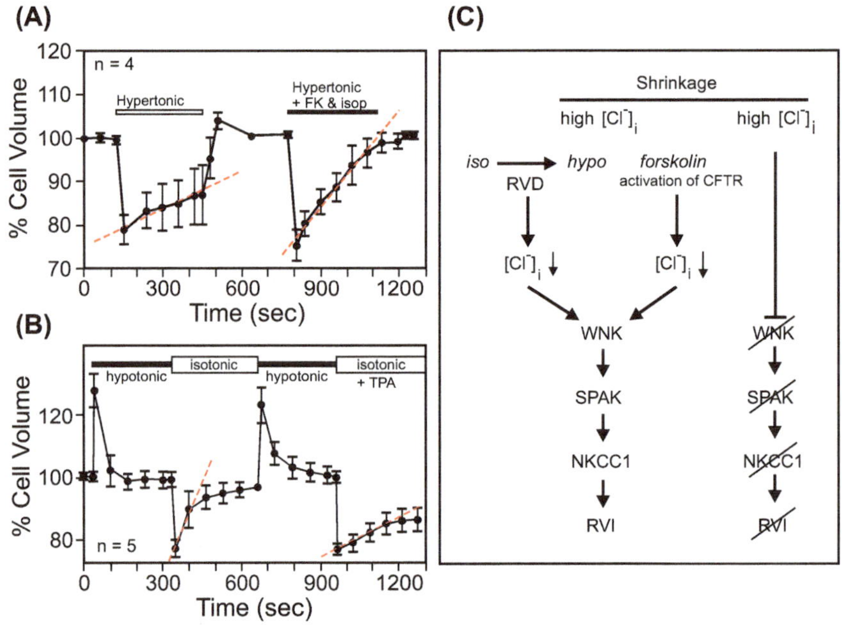

Figure 11 ***Intracellular Cl⁻ and NKCC1 account for RVI in cells.*** (A) Behavior of simian eccrine clear cells exposed to hypertonic solution. Minimal RVI is observed unless the cells were treated with forskolin (FK) and isoproterenol (isop). (B) Post-RVD RVI is strong in these cells but inhibited by phorbol ester (TPA). (C) Both RVD and forskolin through activation of Cl^- channels cause a loss of intracellular Cl^-. Through the activation of the protein kinases WNK and SPAK, this leads to an active NKCC1, which helps the cells recover their volume. Under non-stimulated conditions, the high intracellular Cl^- prevents activation of Na–K–2Cl cotransport and the cells remain shrunken. *(B) Data were redrawn and modified from Toyomoto, T., Knutsen, D., Soos, G., & Sato, K. (1997). Na-K-2Cl cotransporters are present and regulated in simian eccrine clear cells.* American Journal of Physiology, 273, *R270–R277*

The Na–K–2Cl cotransporter is often cited as a mechanism for regulatory volume increase (Hoffmann & Simonsen, 1989; Hoffmann, Lambert, & Pedersen, 2009) and indeed, its activity seems to be activated by hypertonicity. It is, however, important to note that the transporter in some conditions mediates K^+/K^+ exchange instead of Na–K–2Cl cotransport (Canessa, Brugnara, Cusi, & Tosteson, 1986; Gagnon & Delpire, 2010a; Haas, Starke, & McManus, 1984; Lauf, et al., 1987; Orlov, Tremblay, & Hamet, 1996). This is important because the preferred method to assay "cotransporter" function is unidirectional rubidium (Rb^+) uptake, where cold (^{85}Rb) or radiolabeled (^{86}Rb) rubidium is used as a tracer for K^+ movement. Many studies have demonstrated activation of Rb^+ influx under hypertonicity.

This is the case for vascular endothelial cells (Klein & O'Neill, 1995; O'Donnell, Martinez, & Sun, 1995), guinea pig jenual villus cells (MacLeod & Hamilton, 1990), HeLa cells (Ikehara, Yamaguschi, Hosokawa, & Miyamoto, 1993), turkey erythrocytes (Ueberschär & Bakker-Grunwald, 1985), duck red blood cells (Lytle, 1998), rat vascular smooth muscle cells (Orlov et al., 1996), and osteoblasts (Whisenant, Zhang, Khademazad, Loessberg, & Muallem, 1991). Whether increased ^{86}Rb uptake signifies in all cases activation of Na—K—2Cl cotransport is unclear. When NKCC1 mediates Na—K—2Cl cotransport, unidirectional K^+ uptake is proportional to the uptake of one Na^+ and 2 Cl^- ions. The entry of ions is followed by obligatory water movement. If, however, NKCC1 functions in K^+/K^+ exchange mode, the unidirectional movement of K^+ is not accompanied by the movement of Na^+, Cl^-, and obligatory water. In *Xenopus laevis* oocytes over-expressing mouse NKCC1 and subjected to hypertonicity, the protein functions preferentially in K^+/K^+ exchange mode with no net water movement (Delpire & Gagnon, 2011; Gagnon & Delpire, 2010a).

While shrinkage itself tends to increase the concentration of ions inside the cells, when cells undergo RVI, they also accumulate salt and water. Because salts are "toxic" to proteins (Arakawa & Timasheff, 1982a, 1982b; Gekko & Timasheff, 1981a, 1981b), long-term adaptation to high osmolality conditions involves accumulation of organic osmolytes and replacement of high intracellular salts by such neutral compounds.

3.2.3.2 Long Term — Accumulation of Organic Osmolytes

Early studies examining the inner medulla of the kidney demonstrated that mammalian cells accumulate osmolytes when exposed to hypertonic environments. These studies demonstrated that cells which live in a high urea environment accumulate organic osmolytes like trimethylamine (Balaban & Knepper, 1983), glycerylphosphorylcholine and inositol (Bagnasco, Balaban, Fales, Yang, & Burg, 1986), sorbitol (Bagnasco, Uchida, Balaban, Kador, & Burg, 1987), and amino acids (Law & Turner, 1987). High urea and salts environments perturb the structure of proteins (Arakawa & Timasheff, 1982a) and accumulation of organic osmolytes not only allow for respective decrease in intracellular ions, but also for the protection of proteins from denaturation (Arakawa & Timasheff, 1982b, 1985; Gekko & Timasheff, 1981; Gekko & Timasheff, 1981a, 1981b). Amino acids also stabilize chromatin from salt denaturation (Buche, et al., 1989). Accumulation of organic osmolytes is the result of uptake through membrane transport mechanisms (e.g., myo-inositol, betaine transporters) as well as synthesis (e.g., sorbitol). Transporters and enzymes are up-regulated at the transcriptional level by hypertonicity, due to the presence of a tonicity-responsive element (TonE) in the promoter of these genes (Ferraris & Burg, 2006; Takenaka, Preston, Kwon, & Handler, 1994).

The demonstration that marine organisms accumulate organic osmolytes was first made at the end of the 19th century by Leon Fredericq (1851–1935) at the University of Liège, Belgium, and expanded by work from distinguished physiologists/biochemists such as Marcel Florquin (1900–79), Ernest Schoffeniels (1927–92), and Raymond Gilles (1940–2018). The latter three individuals spent summers working in marine biology stations both in France (Roscoff and Concarneau in Brittany) and in the United States (Hopkins Marine Biology station in Pacific Grove, CA, and Duke marine biology station in Beaufort, NC). In several seminal papers, they demonstrated that in euryhaline species, or species that are able to adapt to a wide range of salinities, organic osmolytes represent a very sizable portion of the intracellular osmolarity (Gilles & Delpire, 1997; Gilles, 1979, 1987). For example, in the muscle of sea water acclimated *Callinectes sapidus*, amino acids account to 70% of the intracellular osmolality (Gilles, 1979). I (first author) was extremely fortunate to have been taught biochemistry and physiology by Ernest Schoffeniels and Raymond Gilles, respectively. I also completed my graduate work under the mentorship of Raymond Gilles. While I made my career in the United States, I do feel privileged to belong to this lineage of renowned Liège physiologists.

ACKNOWLEDGMENTS

E. Delpire is funded by NIH grants DK93501, DK110375, GM118944, a Transatlantic Network of Excellence grant from the Leducq Foundation (17CVD05), and by an endowed directorship from Vanderbilt University Medical Center. K.B. Gagnon is supported by the University of Louisville, School of Medicine—Division of Nephrology and Hypertension.

REFERENCES

Adragna, N. C., Ravilla, N. B., Lauf, P. K., Begum, G., Khanna, A. R., Sun, D., et al. (2015). Regulated phosphorylation of the K-Cl cotransporter KCC3 is a molecular switch of intracellular potassium content and cell volume homeostasis. *Frontiers in Cellular Neuroscience, 9*, 255.

Arakawa, T., & Timasheff, S. N. (1982a). Preferential interactions of protein with salts in concentrated solutions. *Biochemistry, 21*, 6545–6552.

Arakawa, T., & Timasheff, S. N. (1982b). Stabilization of protein structure by sugars. *Biochemistry, 21*, 6536–6544.

Arakawa, T., & Timasheff, S. N. (1985). The stabilization of proteins by osmolytes. *Biophysical Journal, 47*, 411–414.

Auer, R. N., Laganière, J. L., Robitaille, Y. O., Richardson, J., Dion, P. A., Rouleau, G. A., et al. (2016). KCC3 axonopathy: Neuropathological features in the central and peripheral nervous system. *Modern Pathology: an Official Journal of the United States and Canadian Academy of Pathology, Inc, 29*, 962–976.
Bach, W., Paulick, H., Garrido, C. J., Ildefonse, B., Meurer, W. P., & Humphris, S. E. (2006). Unraveling the sequence of serpentinization reactions: Petrography, mineral chemistry, and petrophysics of serpentinites from MAR 15N (ODP leg 209, site 1274). *Geophysical Research Letters, 33*, L13306.
Bada, J. L. (2004). How life began on earth: A status report. *Earth and Planetary Science Letters, 226*, 1–15.
Bagnasco, S. M., Balaban, R. S., Fales, H. M., Yang, Y. M., & Burg, M. D. (1986). Predominant osmotically active organic solutes in rat and rabbit renal medulla. *Journal of Biological Chemistry, 261*, 5872–5877.
Bagnasco, S. M., Uchida, S., Balaban, R. S., Kador, P. F., & Burg, M. D. (1987). Induction of aldose reductase and sorbitol in renal inner medullary cells by elevated extracellular NaCl. *Proceedings of the National Academy of Sciences of the United States of America, 84*, 1718–1720.
Balaban, R. S., & Knepper, M. A. (1983). Nitrogen-14 nuclear magnetic resonance spectroscopy of mammalian tissues. *American Journal of Physiology*, C439–C444.
Bergh, C., Kelly, S. J., & Dunham, P. B. (1990). K-Cl cotransport in LK sheep erythrocytes: Kinetics of stimulation by cell swelling. *The Journal of Membrane Biology, 117*, 177–188.
Berkowitz, L. R., & Orringer, E. P. (1987). Cell volume regulation in hemoglobin CC and AA erythrocytes. *American Journal of Physiology, 252*, C300–C306.
Blanco, V. M., Márquez, M. S., & Alvarez-Leefmans, F. J. (2013). Parallel changes in intracellular water volume and pH induced by NH(3)/NH(4)(+) exposure in single neuroblastoma cells. *Cellular Physiology and Biochemistry: International Journal of Experimental Cellular Physiology, Biochemistry, and Pharmacology, 32*, 57–76.
Bortner, C. D., & Cidlowski, J. A. (2007). Cell shrinkage and monovalent cation fluxes: Role in apoptosis. *Archives of Biochemistry and Biophysics, 462*, 176–188.
Bortner, C. D., Hughes, F. M. J., & Cidlowski, J. A. (1997). A primary role for K^+ and Na^+ efflux in the activation of apoptosis. *Journal of Biological Chemistry, 272*, 32436–32442.
Boyden, L. M., Choi, M., Choate, K. A., Nelson-Williams, C. J., Farhi, A., Toka, H. R., et al. (2012). Mutations in kelch-like 3 and cullin 3 cause hypertension and electrolyte abnormalities. *Nature, 482*, 98–102.
Brugnara, C. (1989). Characteristics of the volume- and chloride-dependent K transport in human erythrocytes homozygous for hemoglobin C. *The Journal of Membrane Biology, 111*, 69–81.
Brugnara, C., Van Ha, T., & Tosteson, D. C. (1989). Role of chloride in potassium transport through a K-Cl cotransport system in human red cells. *American Journal of Physiology, 256*, C994–C1103.
Buche, A., Ouassaidi, A., Hacha, R., Delpire, E., Gilles, R., & Houssier, C. (1989). Glycine and other amino compounds prevent chromatin precipitation at physiological ionic strength. *FEBS Letters, 247*, 367–370.
Byun, N., & Delpire, E. (2007). Axonal and periaxonal swelling precede peripheral neurodegeneration in KCC3 knockout mice. *Neurobiology of Disease, 28*, 39–51.
Cala, P. M. (1980). Volume regulation by *Amphiuma* red blood cells: The membrane potential and its implications regarding the nature of the ion-flux pathways. *The Journal of General Physiology, 76*, 683–708.
Canessa, M., Brugnara, C., Cusi, D., & Tosteson, D. C. (1986). Modes of operation and variable stoichiometry of the furosemide- sensitive Na and K fluxes in human red cells. *The Journal of General Physiology, 87*, 113–142.

Gagnon, K. B., England, R., & Delpire, E. (2007). A single binding motif is required for SPAK activation of the Na-K-2Cl cotransporter. *Cellular Physiology and Biochemistry: International Journal of Experimental Cellular Physiology, Biochemistry, and Pharmacology, 20*, 131–142.
Gamba, G., Miyanoshita, A., Lombardi, M., Lytton, J., Lee, W.-S., Hediger, M., et al. (1994). Molecular cloning, primary structure, and characterization of two members of the mammalian electroneutral sodium-(potassium)-chloride cotransporter family expressed in kidney. *Journal of Biological Chemistry, 269*, 17713–17722.
Garty, H., & Benos, D. J. (1988). Characteristics and regulatory mechanisms of the amiloride-blockable Na^+ channel. *Physiological Reviews, 68*, 309–373.
Geck, P., & Pfeiffer, B. (1985). $Na^+ + K^+ + 2Cl^-$cotransport in animal cells - its role in volume regulation. *Annals of the New York Academy of Annals of the New York Academy of Sciences, 456*, 166–182.
Geck, P., Pietrzyk, C., Burckhardt, B.-C., Pfeiffer, B., & Heinz, E. (1980). Electrically silent cotransport of Na^+, K^+ and Cl^- in Ehrlich cells. *Biochimica et Biophysica Acta, 600*, 432–447.
Gekko, K., & Timasheff, S. N. (1981a). Mechanism of protein stabilization by glycerol: Preferential hydration in glycerol-water mixtures. *Biochemistry, 20*, 4667–4676.
Gekko, K., & Timasheff, S. N. (1981b). Thermodynamic and kinetic examination of protein stabilization by glycerol. *Biochemistry, 20*, 4677–4686.
Gianazza, E., & Righetti, P. G. (1980). Size and charge distribution of macromolecules in living systems. *Journal of Chromatography, 193*, 1–8.
Gillen, C. M., Brill, S., Payne, J. A., & Forbush, B. I. (1996). Molecular cloning and functional expression of the K-Cl cotransporter from rabbit, rat, and human. A new member of the cation-chloride cotransporter family. *Journal of Biological Chemistry, 271*, 16237–16244.
Gilles, R. (1979). Intracellular organic osmotic effectors. In R. Gilles (Ed.), *Mechanisms of osmoregulation in animals* (pp. 111–153). London, New York: Wiley Interscience.
Gilles, R. (1987). Volume regulation in cells of euryhaline invertebrates. *Curr. Top. Membr. Transp., 30*, 205–247.
Gilles, R., & Delpire, E. (1997). Variations in salinity, osmolarity, and water availability: Vertebrates and invertebrates. In W. H. Danzler (Ed.), *Handbook of physiology: Section 13, comparative physiology* (vol. |Vol. 2, pp. 1523–1586). New York: American Physiological Society.
Gill, D. R., Hyde, S. C., Higgins, C. F., Valverde, M. A., Mintenig, G. M., & Sepúlveda, F. V. (1992). Separation of drug transport and chloride channel functions of the human multidrug resistance P-glycoprotein. *Cell, 71*, 23–32.
Gimenez, I., & Forbush, B. (2005). Regulatory phosphorylation sites in the NH_2 terminus of the renal Na-K-Cl cotransporter (NKCC2). *American Journal of Physiology - Renal Physiology, 289*, F1341–F1345.
Glykys, J., Dzhala, V., Egawa, K., Balena, T., Kuchibhotla, K. V., Bacskai, B. J., et al. (2014). Local impermeant anions set the GABAA reversal potential. *Science, 343*, 670–675.
Gottschalk, C. W., & Mylle, M. (1958). Evidence that the mammalian nephron functions as a countercurrent multiplier system. *Science, 128*, 594.
Gottschalk, C. W., & Mylle, M. (1959). Micropuncture study of the mammalian urinary concentrating mechanism: Evidence for the countercurrent hypothesis. *American Journal of Physiology, 196*, 927–936.
Gough, A. (1924). The nature of the red blood corpuscle. *The Biochemical Journal, 18*, 202–214.
Grinstein, S., Clarke, C. A., Dupre, A., & Rothstein, A. (1982). Volume-induced increase of anion permeability in human lymphocytes. *The Journal of General Physiology, 80*, 801–823.

Grinstein, S., Clarke, C. A., & Rothstein, A. (1983a). Activation of Na^+/H^+ exchange in lymphocytes by osmotically induced volume changes and by cytoplasmic acidification. *The Journal of General Physiology, 82*, 619–638.

Grinstein, S., Clarke, C. A., Rothstein, A., & Gelfand, E. W. (1983b). Volume-induced anion conductance in human B lymphocytes is cation independent. *American Journal of Physiology, 245*, C160–C163.

Grinstein, S., Rothstein, A., Sarkadi, B., & Gelfand, E. W. (1984). Responses of lymphocytes to anisotonic media: Volume-regulating behavior. *American Journal of Physiology, 246*, C204–C215.

Gründer, S., Thiemann, A., Pusch, M., & Jentsch, T. J. (1992). Regions involved in the opening of CIC-2 chloride channel by voltage and cell volume. *Nature, 360*, 759–762.

Guyton, A. C. (1987). Renal function curve–a key to understanding the pathogenesis of hypertension. *Hypertension, 10*, 1–6.

Haas, M., & Harrison, J. H., Jr. (1989). Stimulation of K-Cl cotransport in rat red cells by a hemolytic anemia-producing metabolite of dapsone. *American Journal of Physiology, 256*, C265–C272.

Haas, M., Starke, L. C., & McManus, T. J. (1984). Soium modulation of cathecholamine-stimulated K/K (K/Rb) exchange in duck red blood cells. *The Journal of General Physiology, 82*, 10a–11a.

Haldane, J. B. S. (1929). Origin of life. *Rationalist Annual, 148*, 3–10.

Hargitay, B., & Kuhn, W. (1951). Das multiplikationsprinzip als Grudnlage der harkonzentrier in der Niere. *Zeitshrift fur Electrochemie und angewandte physikalische Chemie, 55*, 539–558.

Hendil, K. B., & Hoffmann, E. K. (1974). Cell volume regulation in Ehrlich ascites tumor cells. *Journal of Cellular Physiology, 84*, 115–125.

Hiki, K., D'Andrea, R. J., Furze, J., Crawford, J., Woollatt, E., Sutherland, G. R., et al. (1999). Cloning, characterization, and chromosomal location of a novel human K^+-Cl^- cotransporter. *Journal of Biological Chemistry, 274*, 10661–10667.

Hill, W. G., & Zeidel, M. L. (2000). Reconstituting the barrier properties of a water-tight epithelial membrane by design of leaflet-specific liposomes. *Journal of Biological Chemistry, 275*, 30176–30185.

Hinke, J. A. M. (1970). Solvent water for electrolytes in the muscle fiber of the giant barnacle. *The Journal of General Physiology, 56*, 521–541.

Hoffmann, E. K., & Lambert, I. H. (1983). Amino acid transport and cell volume regulation in Ehrlich ascites tumour cells. *Journal of Physiology, 338*, 613–625.

Hoffmann, E. K., & Lambert, I. H. (1994). On the similarity between the small Cl^- channel and the taurine channel activated after cell swelling in Ehrlich ascites tumor cells. *Japanese Journal of Physiology, 44*, S49–S53.

Hoffmann, E. K., Lambert, I. H., & Pedersen, S. F. (2009). Physiology of cell volume regulation in vertebrates. *Physiological Reviews, 89*, 193–277.

Hoffmann, E. K., Lambert, I. H., & Simonsen, L. O. (1986). Separate, Ca^{2+}-activated K^+ and Cl^- transport pathways in Ehrlich ascites tumor cells. *The Journal of Membrane Biology, 91*, 227–244.

Hoffmann, E. K., & Simonsen, E. K. (1989). Membrane mechanisms in volume and pH regulation in vertebrate cells. *Physiological Reviews, 69*, 315–382.

Hoffmann, E. K., Simonsen, L. O., & Lambert, I. H. (1984). Volume-induced increase of K^+ and Cl^- permeabilities in Ehrlich ascites tumor cells. Role of internal Ca^{2+}. *The Journal of Membrane Biology, 78*, 211–222.

Hoffmann, E. K., Simonsen, L. O., & Sjøholm, C. (1979). Membrane potential, chloride exchange, and chloride conductance in Ehrlich mouse ascites tumour cells. *Journal of Physiology, 296*, 61–84.

Hoffmann, E. K., Sjoholm, C., & Simonsen, L. O. (1983). Na^+, Cl^- cotransport in Ehrlich asites tumor cells activated during volume regulation (Regulatory Volume Increase). *The Journal of Membrane Biology, 76*, 269–280.

Hougaard, C., Jørgensen, F., & Hoffmann, E. K. (2001). Modulation of the volume-sensitive K^+ current in Ehrlich ascites tumour cells by pH. *Pflügers Archiv, 442*, 622–633.

Hougaard, C., Klaerke, D. A., Hoffmann, E. K., Olesen, S. P., & Jorgensen, N. K. (2004). Modulation of KCNQ4 channel activity by changes in cell volume. *Biochimica et Biophysica Acta, 1660*, 1–6.

Howard, H. C., Mount, D. B., Rochefort, D., Byun, N., Dupré, N., Lu, J., et al. (2002). Mutations in the K-Cl cotransporter KCC3 cause a severe peripheral neuropathy associated with agenesis of the corpus callosum. *Nature Genetics, 32*, 384–392.

Howard, L. D., & Wondergem, R. (1987). Effects of anisosmotic medium on cell volume, transmembrane potential and intracellular K^+ activity in mouse hepatocytes. *The Journal of Membrane Biology, 100*, 53–61.

Igarashi, P., Vanden Heuvel, G. B., Payne, J. A., & Forbush, B. I. (1995). Cloning, embryonic expression and alternative splicing of a murine kidney specific Na-K-Cl cotransporter. *American Journal of Physiology, 269*, F405–F418.

Ikehara, T., Yamaguschi, H., Hosokawa, K., & Miyamoto, H. (1993). *Kinetic study non the effects*.

Jackson, P. S., Morrison, R., & Strange, K. (1994). The volume-sensitive organic osmolyte-anion channel VSOAC is regulated by nonhydrolytic ATP binding. *American Journal of Physiology, 267*, C1203–C1209.

Jackson, P. S., & Strange, K. (1993). Volume-sensitive anion channels mediate swelling-activated inositol and taurine efflux. *American Journal of Physiology, 265*, C1489–C1500.

Jakab, M., & Ritter, M. (2006). Cell volume regulatory ion transport in the regulation of cell migration. *Contributions To Nephrology, 152*, 151–180.

Jamison, R. L. (1987). Short and long ioop nephrons. *Kidney International, 31*, 597–605.

Jennings, M. L., & al-Rohil, N. (1990). Kinetics of activation and inactivation of swelling-stimulated K^+/Cl^- transport. The volume-sensitive parameter is the rate constant for inactivation. *The Journal of General Physiology, 95*, 1021–1040.

Jennings, M. L., & Schultz, R. K. (1990). Swelling-activated KCl cotransport in rabbit red cells: Flux is determined mainly by cell volume rather than shape. *American Journal of Physiology, 259*, C960–C967.

Johanson, C. E., Duncan, J. A., 3rd, Klinge, P. M., Brinker, T., Stopa, E. G., & Silverberg, G. D. (2008). Multiplicity of cerebrospinal fluid functions: New challenges in health and disease. *Cerebrospinal Fluid Research, 5*, 1–32.

Jorgensen, N. K., Lambert, I. H., & Hoffmann, E. K. (1996). Role of LTD4 in the regulatory volume decrease response in Ehrlich ascites tumor cells. *The Journal of Membrane Biology, 151*, 159–173.

Kahle, K. T., Flores, B., Bharucha-Goebel, D., Zhang, J., Donkervoort, S., Hegde, M., et al. (2016). Peripheral motor neuropathy is associated with defective kinase regulation of the KCC3 cotransporter. *Sci. Signaling, 9*, ra77.

Kaji, D. (1986). Volume-sensitive K transport in human erythrocytes. *The Journal of General Physiology, 88*, 719–738.

Kay, A. R. (2017). How cells can control their size by pumping ions. *Front. Cell Dev. Biol., 5*, 41.

Kelley, D. S., Karson, J. A., Blackman, D. K., Früh-Green, G. L., Butterfield, D. A., Lilley, M. D., et al. (2001). An off-axis hydrothermal vent field near the Mid-Atlantic Ridge at 30 degrees N. *Nature, 412*, 145–149.

Kerr, J. F., Wyllie, A. H., & Currie, A. R. (1972). Apoptosis: A basic biological phenomenon with wide-ranging implications in tissue kinetics. *British Journal of Cancer, 26*, 239–257.

Kim, H. D., Sergeant, S., Forte, L. R., Sohn, D. K., & Im, J. H. (1989). Activation of a Cl-dependent K flux by cAMP in pig red cells. *American Journal of Physiology, 256*, C772–C778.

Kirkegaard, S. S., Lambert, I. H., Gammeltoft, S., & Hoffmann, E. K. (2010). Activation of the TASK-2 channel after cell swelling is dependent on tyrosine phosphorylation. *American Journal of Physiology. Cell Physiology, 299*, C844–C853.

Kirkegaard, S. S., Strøm, P. D., Gammeltoft, S., Hansen, A. J., & Hoffmann, E. K. (2016). The volume activated potassium channel KCNK5 is up-regulated in activated human T cells, but volume regulation is impaired. *Cellular Physiology and Biochemistry: International Journal of Experimental Cellular Physiology, Biochemistry, and Pharmacology, 38*, 883–892.

Klein, T., Cooper, T. G., & Yeung, C. H. (2006). The role of potassium chloride cotransporters in murine and human sperm volume regulation. *Biology of Reproduction, 75*, 853–858.

Klein, J. D., & O'Neill, W. C. (1995). Volume-sensitive myosin phosphorylation in vascular endothelial cells: Correlation with Na-K-2Cl cotransport. *American Journal of Physiology. Cell Physiology, 269*, C1524–C1531.

Koeppe, H. (1895). Ueber den Quellungsgrad der rothen Blutscheiben durch aequimoleculare Salzlösungen und über den osmotischen Druck des Blutplasmas. *Archiv für die Physiologie*, 154–184.

Kopp, C., Linz, P., Dahlmann, A., Hammon, M., Jantsch, J., Müller, D. N., et al. (2013). 23Na magnetic resonance imaging-determined tissue sodium in healthy subjects and hypertensive patients. *Hypertension, 61*, 635–640.

Kregenow, F. M. (1971). The response of duck erythrocytes to nonhemolytic hypotonic media. Evidence for a volume-controlling mechanism. *The Journal of General Physiology, 58*, 372–395.

Krylov, A. V., Pohl, P., Zeidel, M. L., & Hill, W. G. (2001). Water permeability of asymmetric planar lipid bilayers: Leaflets of different composition offer independent and additive resistances to permeation. *The Journal of General Physiology, 118*, 333–340.

Lambert, I. H., & Hoffmann, E. K. (1993). Regulation of taurine transport in Ehrlich ascites tumor cells. *The Journal of Membrane Biology, 131*, 67–69.

Lambert, I. H., & Hoffmann, E. K. (1994). Cell swelling activates separate taurine and chloride channels in Ehrlich mouse ascites tumor cells. *The Journal of Membrane Biology, 142*, 289–298.

Lane, N., & Martin, W. F. (2012). The origin of membrane bioenergetics. *Cell, 151*, 1406–1416.

Lauf, P. K. (1982). Evidence for a chloride-dependent potassium and water transport induced by hypo-osmotic stress in erythrocytes of the marine teleost *Opsanus tau*. *Journal of Comparative Physiology, 146*, 9–16.

Lauf, P. K. (1983). Thiol-dependent passive K/Cl transport in sheep red cells. II. Loss of Cl^- and N-ethylmaleimide sensitivity in maturing high K^+ cells. *The Journal of Membrane Biology, 73*, 247–256.

Lauf, P. K., & Bauer, J. (1987). Direct evidence for chloride-dependent volume reduction in macrocytic sheep reticulocytes. *Biochemical and Biophysical Research Communications, 144*, 849–855.

Lauffenburger, D. A., & Horwitz, A. F. (1996). Cell migration: A physically integrated molecular process. *Cell, 84*, 359–369.

Lauf, P. K., McManus, T. J., Haas, M., Forbush, B., 3rd, Duhm, J., Flatman, P. W., et al. (1987). Physiology and biophysics of chloride and cation cotransport across cell membranes. *Federation Proceedings, 46*, 2377–2394.

Lauf, P. K., & Theg, B. E. (1980). A chloride dependent K^+ flux induced by N-ethylmaleimide in genetically low K^+ sheep and goat erythrocytes. *Biochemical and Biophysical Research Communications, 70*, 221–242.

Law, R. O., & Turner, D. P. (1987). Are ninhydrin-positive substances volume-regulatory osmolytes in rat renal papillary cells? *Journal of Physiology, 386*, 45–61.

Lenart, B., Kintner, D. B., Shull, G. E., & Sun, D. (2004). Na-K-Cl cotransporter-mediated intracellular Na^+ accumulation affects $Ca2^+$ signaling in astrocytes in an in vitro ischemic model. *The Journal of Neuroscience: the Official Journal of the Society for Neuroscience, 24*, 9585–9587.

Lohr, J. W., & Grantham, J. J. (1986). Isovolumetric regulation of isolated S2 proximal tubules in anisosmotic media. *Journal of Clinical Investigation, 78*, 1165–1172.

Lytle, C. (1998). A volume-sensitive protein kinase regulates the Na-K-2Cl cotransporter in duck red blood cells. *American Journal of Physiology. Cell Physiology, 274*, C1002–C1010.

Lytle, C., & Forbush, B. I. (1992). The Na-K-Cl cotransport protein of shark rectal gland. II. regulation by direct phosphorylation. *Journal of Biological Chemistry, 267*, 25438–25443.

MacLeod, R. J., & Hamilton, J. R. (1990). Regulatory volume increase in mammalian jejunal villus cells is due to bumetanide-sensitive NaKCl2 cotransport. *American Journal of Physiology. Gastrointestinal and Liver Physiology, 285*, G665–G674.

Martin, W., & Russell, M. J. (2007). On the origin of biochemistry at an alkaline hydrothermal vent. *Philosophical Transactions of the Royal Society of London - B, 367*, 1887–1925.

Martins, J. R., Faria, D., Kongsuphol, P., Reisch, B., Schreiber, R., & Kunzelmann, K. (2011). Anoctamin 6 is an essential component of the outwardly rectifying chloride channel. *Proceedings of the National Academy of Sciences of the United States of America, 108*, 18168–18172.

Mercado, A., Broumand, V., Zandi-Nejad, K., Enck, A. H., & Mount, D. B. (2006). A C-terminal domain in KCC2 confers constitutive K^+-Cl^- cotransport. *Journal of Biological Chemistry, 281*, 1016–1026.

Merriam, G. R., & Baer, L. (1980). Adrenocorticotropin deficiency: Correction of hyponatremia and hypoaldosteronism with chronic glucocorticoid therapy. *The Journal of Clinical Endocrinology and Metabolism, 50*, 10–14.

Model, M. A. (2014). Possible causes of apoptotic volume decrease: An attempt at quantitative review. *American Journal of Physiology. Cell Physiology, 306*, C417–C424.

Moriguchi, T., Urushiyama, S., Hisamoto, N., Iemura, S. I., Uchida, S., Natsume, T., et al. (2005). WNK1 regulates phosphorylation of cation-chloride-coupled cotransporters via the STE20-related kinases, SPAK and OSR1. *Journal of Biological Chemistry, 280*, 42685–42693.

Mount, D. B., Mercado, A., Song, L., Xu, J., George, J. A. L., Delpire, E., et al. (1999). Cloning and Characterization of KCC3 and KCC4, new members of the cation-chloride cotransporter gene family. *Journal of Biological Chemistry, 274*, 16355–16362.

Nägeli, C. (1855). *Pflanzenphysiologische Untersuchungen von Carl NÄgeli und Carl Cramer* (vol. |Vol. I). Zürich: Friedrich Schulthness.

Nielsen, S., & Knepper, M. A. (1993). Vasopressin activates collecting duct urea transporters and water channels by distinct physical processes. *American Journal of Physiology, 265*, F204–F213.

Niemeyer, M. I., Hougaard, C., Hoffmann, E. K., Jorgensen, F., Stutzin, A., & Sepúlveda, F. V. (2000). Characterisation of a cell swelling-activated K^+-selective conductance of ehrlich mouse ascites tumour cells. *Journal of Physiology, 524*, 757–767.

Nieminen, A. L. (2003). Apoptosis and necrosis in health and disease: Role of mitochondria. *International Review of Cytology, 224*, 29–55.

Nollet, A. (1748). *Lecons de Physique Experimentale*. Paris: Frères Guérin.

Numata, T., Sato, K., Okada, Y., & Wehner, F. (2008). Hypertonicity-induced cation channels rescue cells from staurosporine-elicited apoptosis. *Apoptosis, 13*, 895–903.

O'Donnell, M. E., Martinez, A., & Sun, D. (1995). Endothelial Na-K-Cl cotransport regulation by tonicity and hormones: Phosphorylation of cotransport protein. *American Journal of Physiology. Cell Physiology, 269*, C1513–C1523.

Olson, J., & Holtzman, D. (1982). Respiration and cell volume of primary cultured cerebral astrocytes in media of various osmolarities. *Brain Research, 246*, 273–279.

Orgel, L. E. (2008). The implausibility of metabolic cycles on the prebiotic Earth. *PLoS Biology, 6*, 5–13.

Orlov, S. N., Tremblay, J., & Hamet, P. (1996). Bumetanide-sensitive ion fluxes in vascular smooth muscle cells: Lack of functional Na^+, K^+, 2 Cl^- cotransport. *The Journal of Membrane Biology, 153*, 125–135.

Overton, E. (1907). Uber den Mechanismus der Resorption und der Secretion. In W. Nagel (Ed.), *Handbuch der Physiologie des Menschen* (Volume II, pt. 2). Braunschweig: Vieweg u. Sohn.

Padilla, M. C., Armas-Hernández, M. J., Hernández, R. H., Israili, Z. H., & Valasco, M. (2007). Update of diuretics in the treatment of hypertension. *American Journal of Therapeutics, 14*, 154–160.

Palfrey, H. C., & Pewitt, E. B. (1993). The ATP and Mg^{2+} dependence of Na(+)-K(+)-2Cl^- cotransport reflects a requirement for protein phosphorylation: Studies using calyculin A. *Pflügers Archiv, 425*, 321–328.

Pao, A. C. (2014). Update on the Guytonian view of hypertension. *Current Opinion in Nephrology and Hypertension, 23*, 391–398.

Parker, J. C. (1983). Volume-responsive sodium movements in dog red cells. *American Journal of Physiology, 244*, C324–C330.

Parker, J. C., Colclasure, G. G., & McManus, T. J. (1991). Coordinated regulation of shrinkage-induced Na/H exchange and swelling-induced [K-Cl] cotransport in dog red cells. Further evidence from activation kinetics and phosphatase inhibition. *The Journal of General Physiology, 98*, 869–880.

Paulmichl, M., Li, Y., Wickman, K., Ackerman, M., Peralta, E., & Clapham, D. (1992). New mammalian chloride channel identified by expression cloning. *Nature, 356*, 238–241.

Payne, J. A., & Forbush, B. I. (1994). Alternatively spliced isoforms of the putative renal Na-K-Cl cotransporter are differentially distributed within the rabbit kidney. *Proceedings of the National Academy of Sciences of the United States of America, 91*, 4544–4548.

Payne, J. A., Stevenson, T. J., & Donaldson, L. F. (1996). Molecular characterization of a putative K-Cl cotransporter in rat brain. A neuronal-specific isoform. *Journal of Biological Chemistry, 271*, 16245–16252.

Payne, J. A., Xu, J.-C., Haas, M., Lytle, C. Y., Ward, D., & Forbush, B. I. (1995). Primary structure, functional expression, and chromosome localization of the bumetanide sensitive Na-K-Cl cotransporter in human colon. *Journal of Biological Chemistry, 270*, 17977–17985.

Pfeffer, W. (1877). *Osmotische Untersuchungen – Studien zur Zellmechanik*. Leipzig: Engelmann, W.

Piechotta, K., Garbarini, N. J., England, R., & Delpire, E. (2003). Characterization of the interaction of the stress kinase SPAK with the Na^+-K^+-2Cl^- cotransporter in the nervous system: Evidence for a scaffolding role of the kinase. *Journal of Biological Chemistry, 278*, 52848–52856.

Piechotta, K., Lu, J., & Delpire, E. (2002). Cation-chloride cotransporters interact with the stress-related kinases SPAK and OSR1. *Journal of Biological Chemistry, 277*, 50812–50819.

Plotkin, M. D., Kaplan, M. R., Peterson, L. N., Gullans, S. R., Hebert, S. C., & Delpire, E. (1997). Expression of the Na^+-K^+-2Cl^- cotransporter BSC2 in the nervous system. *American Journal of Physiology. Cell Physiology, 272*, C173–C183.

Ponder, E. (1948). *Hemolysis and related phenomena*. New York: Grune and Stratton.

Prager-Khoutorsky, M. (2017). Mechanosensing in hypothalamic osmosensory neurons. *Seminars in Cell & Developmental Biology, 71*, 13021.

Pringsheim, E. G. (1854). *Untersuchungen über den Bau und die Bildung der Pflanzenzellen (Leipzig)*.

Qiu, Z., Dubin, A. E., Mathur, J., Tu, B., Reddy, K., Miraglia, L. J., et al. (2014). SWELL1, a plasma membrane protein, is an essential component of volume-regulated anion channel. *Cell, 157*, 447–458.

Quarmyne, M. O., Risinger, M., Linkugel, A., Frazier, A., & Joiner, C. (2011). Volume regulation and KCl cotransport in reticulocyte populations of sickle and normal red blood cells. *Blood Cells, Molecules & Diseases, 47*, 95–99.

Race, J. E., Makhlouf, F. N., Logue, P. J., Wilson, F. H., Dunham, P. B., & Holtzman, E. J. (1999). Molecular cloning and functional characterization of KCC3, a new K-Cl cotransporter. *American Journal of Physiology. Cell Physiology, 277*, C1210–C1219.

Rakova, N., Kitada, K., Lerchl, K., Dahlmann, A., Birukov, A., Daub, S., et al. (2017). Increased salt consumption induces body water conservation and decreases fluid intake. *Journal of Clinical Investigation, 127*, 1932–1943.

Ransom, C. B., O'Neal, J. T., & Sontheimer, H. (2001). Volume-activated chloride currents contribute to the resting conductance and invasive migration of human glioma cells. *The Journal of Neuroscience: the Official Journal of the Society for Neuroscience, 21*, 7674–7683.

Rinehart, J., Maksimova, Y. D., Tanis, J. E., Stone, K. L., Hodson, C. A., Zhang, J., et al. (2009). Sites of regulated phosphorylation that control K-Cl cotransporter activity. *Cell, 138*, 525–536.

Ringel, F., & Plesnila, N. (2008). Expression and functional role of potassium-chloride cotransporters (KCC) in astrocytes and C6 glioma cells. *Neuroscience Letters, 442*, 219–223.

Riquelme, G., Sepúlveda, F. V., Jørgensen, F., Pedersen, S., & Hoffmann, E. K. (1998). Swelling-activated potassium currents of Ehrlich ascites tumour cells. *Biochimica et Biophysica Acta, 1371*, 101–106.

Rivera, C., Voipio, J., Payne, J. A., Ruusuvuori, E., Lahtinen, H., Lamsa, K., et al. (1999). The K^+/Cl^- co-transporter KCC2 renders GABA hyperpolarizing during neuronal maturation. *Nature, 397*, 251–255.

Roti Roti, L. W., & Rothstein, A. (1973). Adaptation of mouse leukemic cells (L5178Y) to anisotonic media. I. Cell volume regulation. *Experimental Cell Research, 79*, 295–310.

Rump, K., & Adamzik, M. (2018). Function of aquaporins in sepsis: A systematic review. *Cell & Bioscience, 8*, 10.

Russell, M. J., Daniel, R. M., & Hall, A. (1993). On the emergence of life via catalytic iron-sulphide mem branes. *Terra Nova, 5*, 343–345.

Russell, M. J., Daniel, R. M., Hall, A., & Sherringham, J. A. (1994). A hydrothermally precipitated catalytic iron sulphide membrane as a first step toward life. *Journal of Molecular Evolution, 39*, 231–243.

Russell, M. J., & Martin, W. (2004). The rocky roots of the acetyl-CoA pathway. *Trends in Biochemical Sciences, 29*, 358–363.

Saadoun, S., Papadopoulos, M. C., Hara-Chikuma, M., & Verkman, A. S. (2005). Impairment of angiogenesis and cell migration by targeted aquaporin-1 gene disruption. *Nature, 434*, 786–792.

Sachs, J. R., & Martin, D. W. (1993). The role of ATP in swelling-stimulated K-Cl cotransport in human red cell ghosts. *The Journal of General Physiology, 102*, 551–573.

Sands, J. M. (2004). Renal urea transporters. *Current Opinion in Nephrology and Hypertension, 13*, 525–532.

Schild, L., Lu, Y., Gautschi, I., Schneeberger, E., Lifton, R. P., & Rossier, B. C. (1996). Identification of a PY motif in the epithelial Na channel subunits as a target sequence for mutations causing channel activation found in Liddle syndrome. *The EMBO Journal, 15*, 2381–2387.

Schmidt, W. F., 3rd, & McManus, T. J. (1977). Ouabain-insensitive salt and water movements in duck red cells. I. Kinetics of cation transport under hypertonic conditions. *The Journal of General Physiology, 70*, 59–79.

Schnermann, J., & Briggs, J. P. (2013). Tubular control of renin synthesis and secretion. *Pflügers Archiv, 465*, 39–51.

Schwab, A. (2001). Function and spatial distribution of ion channels and transporters in cell migration. *American Journal of Physiology - Renal Physiology, 280*, F739–F747.

Schwab, A., Wojnowski, L., Gabriel, K., & Oberleithner, H. (1994). Oscillating activity of a Ca(2+)-sensitive K^+ channel. A prerequisite for migration of transformed Madin-Darby canine kidney focus cells. *Journal of Clinical Investigation, 93*, 1631–1636.

Shen, M. R., Chou, C. Y., & Ellory, J. C. (2000). Volume-sensitive KCI cotransport associated with human cervical carcinogenesis. *Pflugers Archiv, 440*, 751–760.

Shen, M. R., Lin, A. C., Hsu, Y. M., Chang, T. J., Tang, M. J., Alper, S. L., et al. (2004). Insulin-like growth factor 1 stimulates KCl cotransport, which is necessary for invasion and proliferation of cervical cancer and ovarian cancer cells. *Journal of Biological Chemistry, 279*, 40017–40025.

Sherbet, G. V., Lakshmi, M. S., & Cajone, F. (1983). Isoelectric characteristics and the secondary structure of some nucleic acids. *Behavioral Sleep Medicine, 10*, 121–128.

Shock, E. L., & Schulte, M. D. (1998). Organic synthesis during fluid mixing in hydrothermal systems. *Journal of Geophysical Research: Planets, 103*, 28513–28527.

Siebens, A. W., & Kregenow, F. M. (1978). Volume regulatory responses of salamander red cells incubated in anisotonic media. Effect of amiloride. *The Physiologist, 21*, 110.

Siebens, A. W., & Kregenow, F. M. (1985). Volume-regulatory responses of *Amphiuma* red cells in anisotonic media. The effect of amiloride. *The Journal of General Physiology, 86*, 527–564.

Simone, L., Gargano, C. D., Pisani, F., Cibelli, A., Mola, M. G., Frigeri, A., et al. (2018). Aquaporin-1 inhibition reduces metastatic formation in a mouse model of melanoma. *Journal of Cellular and Molecular Medicine, 22*, 904–912.

Simon, D. B., Nelson-Williams, C., Johnson Bia, M., Ellison, D., Karet, F. E., Morey Molina, A., et al. (1996). Gitelman's variant of Bartter's syndrome, inherited hypokalaemic alkalosis, is caused by mutations in the thiazide-sensitive Na-Cl cotransporter. *Nature Genetics, 12*, 24–30.

Sleep, N. H., Meibom, A., Fridriksson, T., Coleman, R. G., & Bird, D. K. (2004). H2-rich fluids from serpentinization: Geochemical and biotic implications. *Proceedings of the National Academy of Sciences of the United States of America, 101*, 12818–12823.

Soroceanu, L., Manning, T. J. J., & Sontheimer, H. (1999). Modulation of glioma cell migration and invasion using Cl(-) and K(+) ion channel blockers. *The Journal of Neuroscience: the Official Journal of the Society for Neuroscience, 19*, 5942–5954.

Stokes, J. B., Lee, I., & D'Amico, M. (1984). Sodium chloride absorption by the urinary bladder of the winter flounder: A thiazide-sensitive, electrically neutral transport system. *Journal of Clinical Investigation, 74*, 7–16.

Strange, K., Morrison, R., Shrode, L., & Putnam, R. (1992). Mechanism and regulation of swelling-activated inositol efflux in brain glial cells. *American Journal of Physiology, 265*, C244–C256.

Sung, K.-W., Kirby, M., McDonald, M. P., Lovinger, D. M., & Delpire, E. (2000). Abnormal $GABA_A$-receptor mediated currents in dorsal root ganglion neurons isolated from Na-K-2Cl cotransporter null mice. *The Journal of Neuroscience: the Official Journal of the Society for Neuroscience, 20*, 7531–7538.

Takenaka, M., Preston, A. S., Kwon, H. M., & Handler, J. S. (1994). The tonicity-sensitive element that mediates increased transcription of the betaine transporter gene in response to hypertonic stress. *Journal of Biological Chemistry, 269*, 29379–29381.

Tamura, H., Schild, L., Enomoto, N., Matsui, N., Marumo, F., & Rossier, B. C. (1996). Liddle disease caused by a missense mutation of beta subunit of the epithelial sodium channel gene. *Journal of Clinical Investigation, 97*, 1780–1784.

Tivey, D. R., Simmons, N. L., & Aiton, J. F. (1985). Role of passive potassium fluxes in cell volume regulation in cultured HeLa cells. *The Journal of Membrane Biology, 87*, 93–95.

Tosteson, D. C., & Hoffman, J. F. (1960). Regulation of cell volume by active cation transport in high and low potassium sheep red cells. *The Journal of General Physiology, 44*, 169–194.

Toyomoto, T., Knutsen, D., Soos, G., & Sato, K. (1997). Na-K-2Cl cotransporters are present and regulated in simian eccrine clear cells. *American Journal of Physiology, 273*, R270–R277.

Ueberschär, S., & Bakker-Grunwald, T. (1985). Effects of ATP and cAMP on the (Na^{+}+K^{+}+2Cl^{-})-cotransport system in Turkey erythrocytes. *Biochimica et Biophysica Acta, 818*, 260–266.

Uyanik, G., Elcioglu, N., Penzien, J., Gross, C., Yilmaz, Y., Olmez, A., et al. (2006). Novel truncating and missense mutations of the KCC3 gene associated with Andermann syndrome. *Neurology, 66*, 1044–1048.

Verkman, A. S. (2005). More than just water channels: Unexpected cellular roles of aquaporins. *Journal of Cell Science, 118*, 3225–3232.

Vitari, A. C., Deak, M., Morrice, N. A., & Alessi, D. R. (2005). The WNK1 and WNK4 protein kinases that are mutated in Gordon's hypertension syndrome, phosphorylate and active SPAK and OSR1 protein kinases. *The Biochemical Journal, 391*, 17–24.

Voss, F. K., Ullrich, F., Münch, J., Lazarow, K., Lutter, D., Mah, N., et al. (2014). Identification of LRRC8 heteromers as an essential component of the volume-regulated anion channel VRAC. *Science, 344*, 634–638.

de Vries, H. (1884). *Jahrbücher für Wissenschaftliche Botanik, XIV*, 427.

de Vries, H. (1888). *Zeitschrift für Physikalische Chemie, II*, 415.

Whisenant, N., Zhang, B.-X., Khademazad, M., Loessberg, P., & Muallem, S. (1991). Regulation of Na-K-2Cl cotransport in osteoblasts. *American Journal of Physiology, 261*, C433–C440.

Wilson, F. H., Disse-Nicodeme, S., Choate, K. A., Ishikawa, K., Nelson-Williams, C., Desitter, I., et al. (2001). Human hypertension caused by mutations in WNK kinases. *Science, 293*, 1107–1112.

Wu, Q., Delpire, E., Hebert, S. C., & Strange, K. (1998). Functional demonstration of Na-K-2Cl cotransporter activity in isolated, polarized choroid plexus cells. *American Journal of Physiology, 275*, C1565–C1572.

Xu, J.-C., Lytle, C., Zhu, T. T., Payne, J. A., Benz, E. J., & Forbush, B. I. (1994). Molecular cloning and functional expression of the bumetanide-sensitive Na-K-2Cl cotransporter. *Proceedings of the National Academy of Sciences of the United States of America, 91*, 2201–2205.

Yurinskaya, V. E., Rubashkin, A. A., & Vereninov, A. A. (2011). Balance of unidirectional monovalent ion fluxes in cells undergoing apoptosis: Why does Na^{+}/K^{+} pump suppression not cause cell swelling? *Journal of Physiology, 589*, 2197–2211.

Zhu, L., Lovinger, D., & Delpire, E. (2005). Cortical neurons lacking KCC2 expression show impaired regulation of intracellular chloride. *Journal of Neurophysiology, 93*, 1557–1568.

CHAPTER TWO

Search for Upstream Cell Volume Sensors: The Role of Plasma Membrane and Cytoplasmic Hydrogel

Sergei N. Orlov[*,§,¶,1], **Aleksandra Shiyan**[*], **Francis Boudreault**[||], **Olga Ponomarchuk**[*,||] **and Ryszard Grygorczyk**[||,#]

[*]Faculty of Biology, M.V. Lomonosov Moscow State University, Moscow, Russia
[§]Siberian State Medical University, Tomsk, Russia
[¶]National Research Tomsk State University, Tomsk, Russia
[||]Centre de recherche, Centre hospitalier de l'Université de Montréal (CRCHUM), Montreal, QC, Canada
[#]Department of Medicine, Faculty of Medicine, University of Montreal, Montreal, QC, Canada
[1]Corresponding Author: E-mail: sergeinorlov@yandex.ru

Contents

Abstract

The plasma membrane plays a prominent role in the regulation of cell volume by mediating selective transport of extra- and intracellular osmolytes. Recent studies show that upstream sensors of cell volume changes are mainly located within the cytoplasm that displays properties of a hydrogel and not in the plasma membrane. Cell volume changes occurring in anisosmotic medium as well as in isosmotic environment affect properties of cytoplasmic hydrogel that, in turn, trigger rapid regulatory volume

Current Topics in Membranes, Volume 81
ISSN 1063-5823
https://doi.org/10.1016/bs.ctm.2018.07.001

increase and decrease (RVI and RVD). The downstream signaling pathways include reorganization of 2D cytoskeleton and altered composition of polyphosphoinositides located on the inner surface of the plasma membrane. In addition to its action on physico-chemical properties of cytoplasmic hydrogel, cell volume changes in anisosmotic conditions affect the ionic strength of the cytoplasm and the $[Na^+]_i/[K^+]_i$ ratio. Elevated intracellular ionic strength evoked by long term exposure of cells to hypertonic environment resulted in the activation of TonEBP and augmented expression of genes controlling intracellular organic osmolyte levels. The role of Na^+_i/K^+_i -sensitive, Ca^{2+}_i -mediated and Ca^{2+}_i-independent mechanisms of excitation-transcription coupling in cell volume-adjustment remains unknown.

List of Abbreviations

AFM atomic force microscopy
MAPK mitogen-activated protein kinase
NFAT nuclear factor of activated T cells family
PIP2 phosphatidylinositol 4,5-diphosphate
PIP3 phosphatidylinositol 3,4,5-triphosphate
RVD regulatory volume decrease
RVI regulatory volume increase
SGK1 serum- and glucocorticoid-inducible kinase 1
TonEBP tonicity-responsive enhancer-binding protein
VRAC volume regulated anion channels

1. INTRODUCTION

Changes in cell volume triggered by the altered content of extra- or intracellular osmolytes affect a plethora of cellular functions including macromolecular synthesis and catabolism, cell proliferation, differentiation and death. Because of this, cells regulate their volume with accuracy via accumulation or loss of intracellular osmolytes (so-called regulatory volume increase (RVI) and regulatory volume decrease (RVD), respectively). Activation of RVI and RVD is linked to many cell-type specific signals mediated by intracellular calcium (Ca^{2+}_i), G-proteins, Rho family GTPases, protein kinases, phospholipases, etc., which have been discussed in several comprehensive reviews (Delpire & Austin, 2010; Grygorczyk, Boudreault, Platonova, & Orlov, 2015; Hoffmann & Simonsen, 1989; Hoffmann, Lambert, & Pedersen, 2009; Koivusalo, Kapus, & Grinstein, 2009; Lang, 2009; Lang et al., 1998; Mongin & Orlov, 2001; Pedersen, Kapus, & Hoffmann, 2011; Strange, 2004). In this mini-review, we focus on studies devoted to the identification of up-stream cell volume sensors triggering these signaling pathways with an emphasis on the recently discovered osmosensing properties of cytoplasmic hydrogel and excitation-transcription coupling triggered by elevation of the $[Na+]_i/[K+]_i$ ratio.

2. CELL VOLUME-SENSING BY THE PLASMA MEMBRANE

The control of cell volume has traditionally been modeled as water fluxes driven by solute fluxes across the plasma membrane (Macknight & Leaf, 1977). It is well documented that rapid RVI and RVD responses occurring in several seconds after cell volume changes are caused by the activation of plasma membrane monovalent ion transporters (Hoffmann & Simonsen, 1989; Lang et al., 1998). On the basis of these observations, it might be assumed that the volume-sensing machinery is also located within the plasma membrane. Studies examining this hypothesis are considered in the first section of our mini-review (see also Cytoskeletal Contribution to Cell Stiffness Due to Osmotic Swelling; Extending the Donnan Equilibrium by Pei-Chuan, Sivaselvan, & Sachs, 2018).

2.1 Plasma Membrane Stretch: Role of Intracellular Ca^{2+}

The predicted osmotic pressure across plasma membrane is calculated from the van't Hoff's law as $\Delta\Pi = RT\Delta c$ where Δc is the gradient of osmolyte concentration across the plasma membrane and $RT = 24.4$ atm/M at 25 °C. In accordance with the Laplace's law, the plasma membrane in rounded swollen cells is exposed to tension $\theta = \Delta\Pi r/2$, where r is the cell radius. The model based on these equations predicts that attenuation of extracellular NaCl concentration by 5 mM (~10 mOsM) results in an elevation of surface area $S=4\pi r^2$ by ~5%. Importantly, the lipid bilayer can withstand only 3% increase of the S before collapsing (Kinnunen, 2000). These data suggest that the plasma membrane in swollen cells is stretched and thus leads to conformational transition of membrane-bound proteins involved in cell volume sensing. This hypothesis is consistent with numerous studies demonstrating that mechanical sensitivity of ion channels is a general property, similar to voltage sensitivity (Sachs, 2010).

The electrophysiological properties of mechanosensitive osmolyte transporters were studied in detail using bacterial channels MscL and MscS, the transient receptor potential vanilloid subfamily member TRPV4 functioning as non-selective cation channel and the volume-regulated anion channel (VRAC) encoded by LRRC8 (leucine-rich repeat containing 8) genes and providing non-selective anion and neutral low molecular weight osmolyte fluxes in swollen cells (Cox, Bavi, & Martinac, 2017; Mongin, 2018; Pasantes-Morales, 2016). Importantly, unlike the bacterial channels whose mechanosensation was demonstrated *in vivo* (Martinac, 2011), activation of mammalian channels by stretch has been mainly observed in patch-clamp experiments using excised membrane patches/fragments at room temperature. Moreover, activation of TRPV4 channels assessed by this technique was not detected by the measurement of ion current in whole-cell patch-clamp studies (Strotmann, Harteneck, Nunnenmacher, Schultz, & Plant, 2000).

The canonical model assumes that swollen cells become stiffer because of the plasma membrane stretch (Strange, 2004). However, using atomic force microscopy (AFM), Spagnoli and co-workers demonstrated that during osmotic swelling mammalian cells become softer rather than stiffer, contradicting the assumption that cell membrane being the constraining element in osmotic stress (Spagnoli, Beyder, Besch, & Sachs, 2008). Subsequent study by Zou et al. found a biphasic response of cortical neurons to swelling or to excitotoxic agonists consisting of initial stiffening followed by softening to below-control levels. The initial channel-mediated stiffening was consistent with abrupt elevation of osmotic pressure, while the softening was attributed to the onset of lethal/pathological membrane blebbing subsequent to dissociation of the plasma membrane's bilayer from its spectrin-actomyosin matrix (Zou et al., 2013). Significant increase in membrane stiffness in hypotonically challenged endothelial cells was recently confirmed directly in the AFM study by Ayee, LeMaster, Teng, Lee, and Levitan (2018). They found that cell swelling was not accompanied by any measurable increase in the membrane tension, as estimated by pulling membrane tethers with AFM. Moreover, depolymerization of F-actin did not abrogate but instead enhanced swelling-induced stiffening of the membrane. It was concluded that increased osmotic pressure, rather than changes in membrane tension, could be responsible for activation of volume-sensitive mechanisms in hypotonically challenged cells.

Analyzing this controversy, it should be noted that besides the membrane stretch, volume-sensitive channels may also be activated by heat, oxidative stress, low pH, phospholipase C and phospholipase A2-derived metabolites (Naziroglu & Braidy, 2017). These findings suggest that in swollen cells activity of ion transporters might be affected by the above-listed stimuli triggered by plasma membrane mechanisms. Indeed, in

HeLa cells swelling-dependent activation of TRPV4 was abolished by the depletion of phosphatidylinositol-4,5-biphosphate from the plasma membrane (Garcia-Elias et al., 2013).

Intracellular Ca^{2+} perturbations affect diverse cellular responses (Carafoli, 2002), including monovalent ion transporters involved in cell volume adjustment (Hoffmann et al., 2009; Lang et al., 1998; Mongin & Orlov, 2001). Considering this, it might be proposed that Ca^{2+} influx via stretch-activated cation channels plays a key role in RVD. Indeed, acute and rapid changes in extracellular tonicity resulted in transient elevation of $[Ca^{2+}]_i$ in primary astrocyte cultures (O'Connor & Kimelberg, 1993), cervical cancer cells (Shen, Furla, Chou, & Ellory, 2002), rat lacrimal acinar cells (Speake, Douglas, & Brown, 1998) and Ehrlich ascites tumor cells (Pedersen, Prenen, Droogmans, Hoffmann, & Nilius, 1998). Even though several studies have shown the importance of $[Ca^{2+}]_i$ responses to control RVD (McCarty & O'Neil, 1992; Pasantes-Morales & Morales, 2000), others showed that Ca^{2+} signaling alone is not sufficient to trigger RVD (Doroshenko & Neher, 1992; Jorgensen et al., 1997). Because $[Ca^{2+}]_i$ responses in hypotonic medium differ not only among cell types but also within a single cell population (McCarty & O'Neil, 1992; Boudreault & Grygorczyk, 2004b), we adapted light microscopy-based and non-invasive stereoscopic dual-image surface reconstruction (DISUR) technique (Boudreault & Grygorczyk, 2004a) to simultaneously quantify alteration in volume and $[Ca^{2+}]_i$ within single substrate-adherent cells loaded with fluorescent $Ca^{2+}{}_i$ indicator Fura-2 (Ponomarchuk, Boudreault, Orlov, & Grygorczyk, 2016). Using this unique technique, we found that in 95% of the low density-seeded human lung carcinoma A549 cells, stimulation with 50% hypotonic challenge caused a progressive swelling followed by RVD. Unlike ever-present cell volume perturbations, in 82% of the A549 cells RVD was not accompanied by a measurable increase in intracellular Ca^{2+} concentration (Fig. 1). Importantly, loading cells with BAPTA delayed peak of swelling but did not prevent RVD in 82% of the cells. Additional electrophysiological experiments showed that chelating intracellular calcium with BAPTA did not block swelling-induced outwardly rectifying chloride current mediated by VRAC (Ponomarchuk et al., 2016). These data strongly suggest that intracellular Ca^{2+} signaling is not essential for VRAC activation and subsequent volume restoration in A549 cells.

It is important to note that because of the multiple invaginations of the plasma membrane and ability to increase its surface area by insertion of intracellular membranes in nucleated animal cells, generation of significant

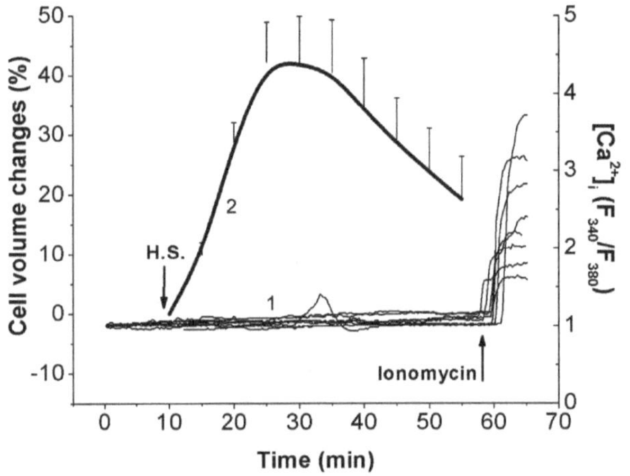

Figure 1 Kinetics of $[Ca^{2+}]_i$ changes in a single cell loaded with Fura-2 (1) and average kinetics of volume changes in A549 cells caused by 50% hypotonic shock. Average kinetics data are presented as mean ± SEM (n = 11). Data plots show relative volume normalized to cell volume at the start of swelling. *Adapted from Ponomarchuk et al. (2016) with permission.*

mechanical forces under swelling or shrinkage seems to be unlikely. Indeed, electrophysiological experiments register no changes in surface-dependent membrane capacitance even upon two-fold increase of cell volume (Okada, 1997). To further explore this issue, we employed DISUR technique for simultaneous measurements of cell height, total surface area and volume. This method allows accurate measurement of these parameters with temporal resolution of ~100 ms in attached cells, whose volume perturbations are mainly caused by alterations of cell height rather than planar area (for more details, see Grygorczyk et al., 2015). We found that nucleated mammalian cells could increase their surface area and volume until membrane rupture by ~3 and 10-fold, respectively (Groulx, Boudreault, Orlov, & Grygorczyk, 2006). Importantly, cells accommodate moderate (2-fold) volume increase by undergoing shape changes and plasma membrane unfolding, while significant endomembrane insertion was observed during extreme swelling. This conclusion is consistent with rupture in modestly hypotonic solutions of mammalian erythrocytes, lacking intracellular membrane reserve and providing ~1.5-fold volume increase exclusively due to discocyte-spherocyte shape transition without any impact on the total surface area (Kageyama, Onoyama, Kogawa, Goto, & Tanabe, 1989). Viewed collectively, these results show that in nucleated cells plasma membrane tension is unlikely to reach levels required for volume-dependent signaling, what is in complete agreement with the AFM study of Ayee et al. (2018) discussed above.

2.2 Signaling Within Space-Limited Membrane Domains: Role of Cortical Cytoskeleton and Polyphosphoinositides

The observations considered above do not exclude the possible generation of cell volume-dependent signals within highly specialized plasma membrane domains such as caveolae. Caveolae, i.e. space-limited plasma membrane invaginations with a diameter of about 50 nm, were detected in all types of animal cells studied so far. They are organized by 4 isoforms of specialized proteins, caveolins, and contain high concentration of cholesterol (Sowa, 2012). Recently it was shown that caveolae act as membrane reserves (Kozera, White, & Calaghan, 2009) responding to mechanical stress by rapid disassembly (Sihna et al., 2011). Trouet and co-workers reported that dominant negative caveolin-1 inhibits VRAC in endothelial cells (Trouet, Hermans, Droogmans, Nilius, & Eggermont, 2001). By contrast, caveolae integrity is not required for cell volume regulation in astrocytes (Mongin & Orlov, 2001) and adipocytes (Eduardsen et al., 2011).

The heterogeneity of plasma membrane is also supported by a cortical two-dimensional (2D) cytoskeleton (membrane scaffold). Nucleus-free mammalian erythrocytes lacking a 3D cytoskeleton are commonly used for the analysis of the structure and functional implications of the cortical cytoskeleton. In these cells 2D cytoskeleton consists of spectrin heterodimers, actin bundles and proteins involved in the regulation of spectrin/actin interaction and anchoring of this network to integral membrane proteins and acid phospholipids (Lux, 2016). In contrast to agents/available for studying the 3D cytoskeleton and cytoskeletal stress, pharmacological tools for identifying cortical cytoskeleton-dependent cellular responses have not yet been developed. Differential scanning microcalorimetry of erythrocyte ghosts indicated that temperature elevation up to 49–50 °C leads to irreversible annealing (melting) of spectrin heterodimers in human (Brandts, Taverna, Salasivan, & Lysko, 1978) and rat erythrocytes (Shnyrov, Orlov, Zhadan, & Pokudin, 1990) (Fig. 2A). We demonstrated that brief preincubation of rat erythrocytes at 49 °C abolishes volume-dependent regulation of $Na^+,K^+,2Cl^-$ and K^+,Cl^- cotransporters (Orlov et al., 1993) (Fig. 2B).

To further explore the role of 2D-membrane scaffolds in volume sensing, we employed atomic force microscopy (Parshina et al., 2013). Using

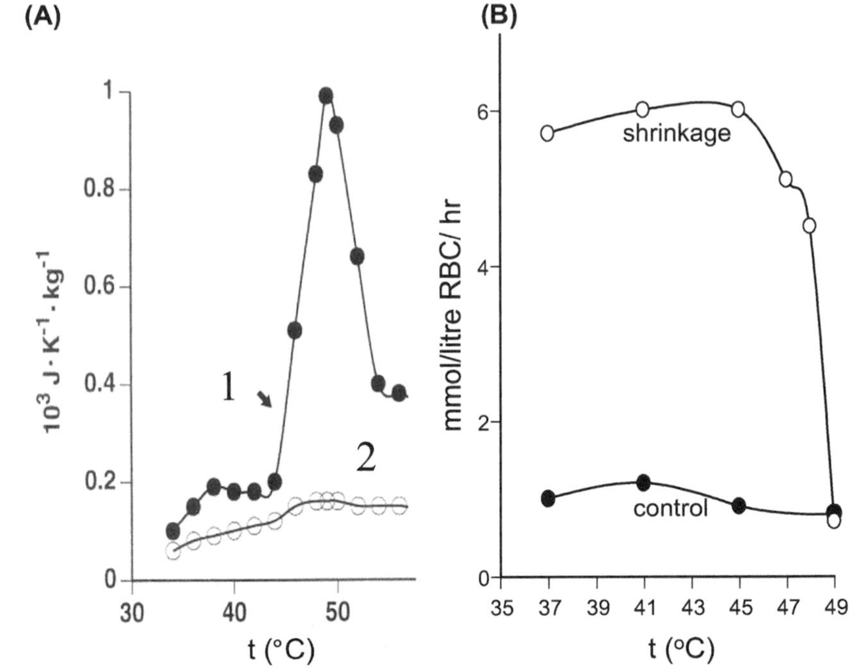

Figure 2 (A) Differential scanning microcalorimetry of rat erythrocyte ghosts. **1** – control; **2** – ghosts preincubated for 10 min at 60 °C. (B) Effect of 10 min preincubation at different temperatures on the activity of Na^+,K^+,$2Cl^-$ cotransport in rat erythrocytes incubated in control (**1**) and hyperosmotic medium (**2**). Note that the activity of ion transporter was measured at 37 °C (Orlov et al., 1993; Shnyrov et al., 1990).

high-pass Fourier filtering, we created images of membrane roughness caused by globular structures with an average wave height of 3–5 nm covered by globular structures with a diameter of 40–50 nm and average height of 1–2 nm (Fig. 3). We did not observe any effect of heat shock on the distance between maximum heights of surface waves, whereas the average diameter of globular structures underlying membrane roughness was decreased from about 50 nm to 30 nm (Fig. 3C). In control erythrocytes, elevation of medium osmolality from 155 to 527 mOsm increased the average height of membrane relief over falls (S_{rel}) from 2.93 ± 0.15 to 4.15 ± 0.22 nm (Fig. 4A). Swelling of mammalian erythrocytes results in the activation of K^+,Cl^- cotransporter (Lauf & Adragna, 2000). In human RBC attenuation of medium osmolality increased K^+,Cl^- cotransport by ~6-fold. Both volume-dependent changes of membrane relief and activity of K^+,Cl^- cotransporter were abolished by 10 min preincubation at 50 °C

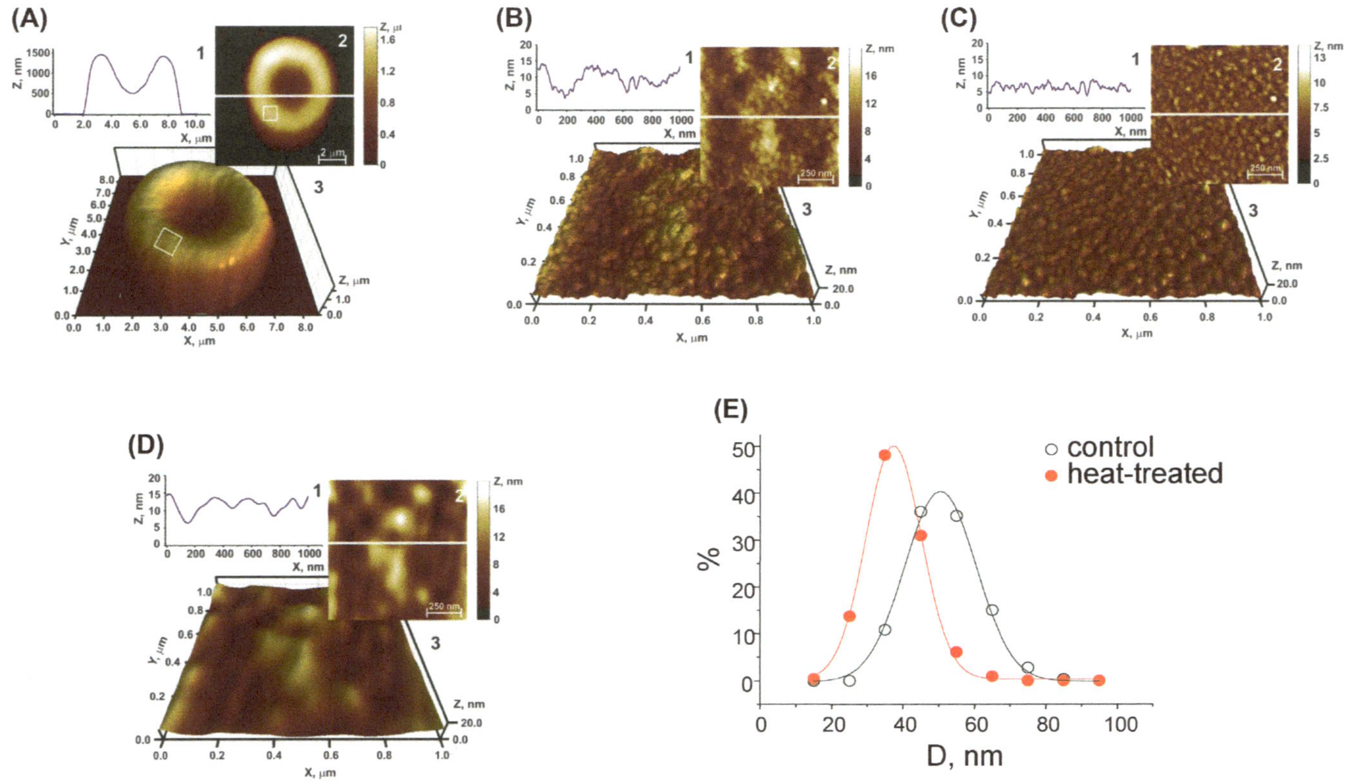

Figure 3 Atomic force microscopy image processing of human erythrocytes. (A) 2D (Section 2) and 3D (Section 3) images of whole human erythrocytes (Sections 2 and 3, respectively); white squares demarcate a 1x1-μm area selected for further analysis. (B) 2D (Section 2) and 3D (Section 3) images of selected membrane area. (C) Quantification of membrane roughness parameters (S_{rough}) in 2D (Section 2) and 3D (Section 3) images. The wavy membrane relief seen in (B) was removed by high-pass Fourier filtration. (D) 2D and 3D images obtained by subtraction of images (C) form images (B) and considered for quantification of geometry of irregular hills-valleys structures within folded membrane surface (parameter S_{rel}). In A–D Section 1 displays the height variation profile across white lines shown in Section 2. (E) The size distribution of globular structures underlying membrane roughness in control and heat-treated (50 °C, 10 min) RBC. *Adapted from Parshina et al. (2013) with permission.*

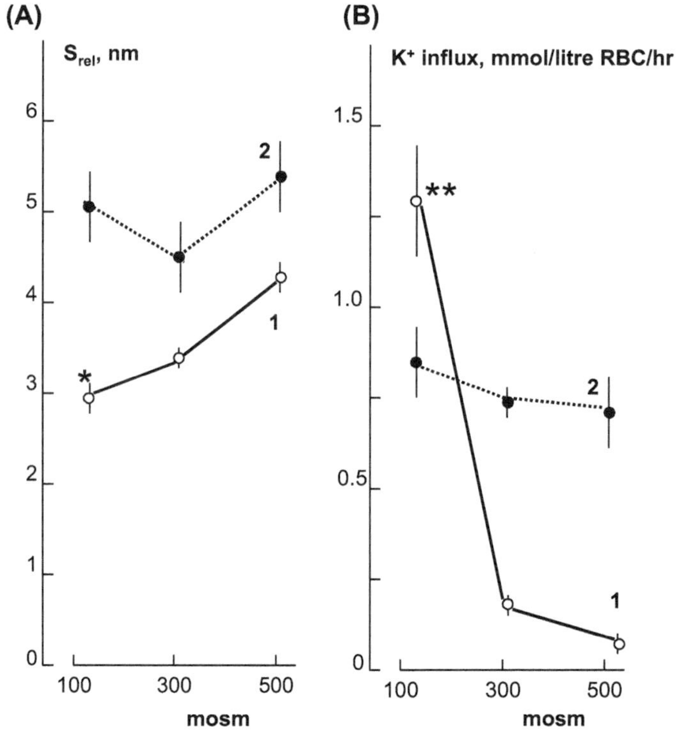

Figure 4 Effect of medium osmolality on membrane relief (S_{rel}, A) and K^+,Cl^- cotransport (B) in control (**1**) and heat-treated (**2**, 50 °C, 10 min) human erythrocytes. *Adapted from Parshina et al. (2013) with permission.*

(Fig. 4B). These results demonstrate that volume-dependent alterations of human erythrocyte membrane relief are caused by reorganization of 2D spectrin-actin network which contributes to the regulation of volume-sensitive ion transporters.

In attached cells, space-limited areas of the plasma membrane abundant with integrin α/β heterodimers might be considered as candidates for shear stress triggered by cell swelling. Indeed, inhibition of integrin/srk kinase signaling blocked RVD in hepatocytes (vom Dahl et al., 2003; Schliess & Haussinger, 2007). In myocytes integrin stretch activates volume-regulated anion channels (Browe & Baumgarten, 2004), whereas in renal epithelial cells it is involved in the expression of tonicity-responsive enhancer-binding protein (TonEBP), triggered by hyperosmotic environment (Moeckel et al., 2013). Importantly, cortical actin microfilaments located close to the plasma membrane interact with dozens of membrane-bound proteins including integrin heterodimers (Ingber, 2003). Numerous studies demonstrated that cortical microfilament regulation affects the activity of all known volume-sensitive ion transporters involved in RVI and RVD (Henson, 1999). Considering this, Cantiello argues that cortical actin microfilaments form a sensory gel, whose volume-dependent physico-chemical properties affect the activity of osmolyte transporters (Cantiello, 1997). The data supporting and contradicting this hypothesis are considered in several comprehensive reviews (Henson, 1999; Pedersen, Hoffmann, & Mills, 2001).

It is well-documented that the phosphorylation of membrane bound-proteins and the content of phosphatidylinositol 4,5-diphosphate (PIP_2), located on the inner surface of the plasma membrane phospholipid bilayer, is a major regulator of actin polymerization and cytoskeleton stability (Janmey & Lindberg, 2004; Pantaleo, De Franceschi, Ferru, Vono, & Turrini, 2010). Two decades ago we demonstrated that unlike activators of protein kinase A (dibutyryl-cAMP), protein kinase G (dibutyryl-cGMP) and protein kinas C (phorbol ester), hyperosmotic shrinkage of rat erythrocytes does not affect phosphorylation of membrane bound proteins (Fig. 5A) but sharply increases the content of PIP_2 (Fig. 5B)

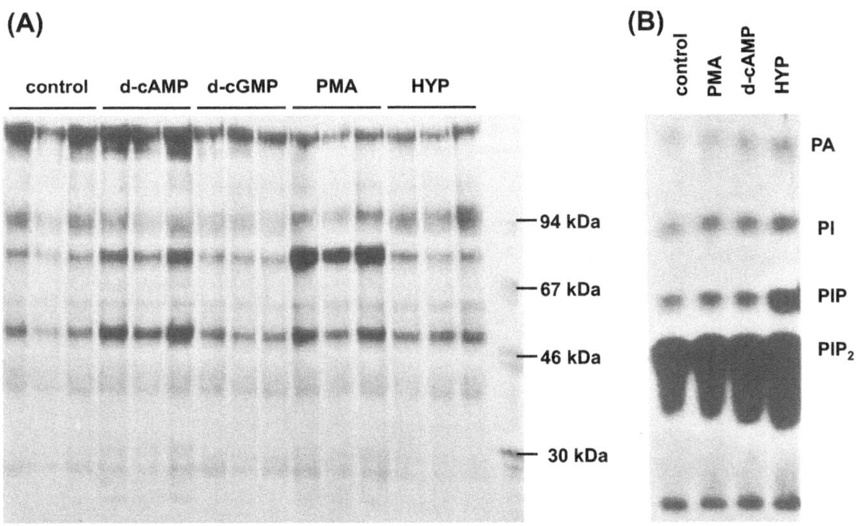

Figure 5 ^{32}P-labeling of membrane-bound proteins (A) and polyphosphoinositides (B) in rat erythrocytes in control conditions and after 30 min incubation in hyperosmotic medium (HYP) or in the presence of phorbol-ester (PMA – 1 μM), dibutyryl-cAMP (d-cAMP - 1 mM) or dibutyryl-cGMP (d-cGMP - 1 mM). PA – phosphatidic acid; PI – phosphatidylinositol; PIP – phosphatidylinositol 4-phosphate; PIP2 – phosphatidylinositol 4,5-diphosphate. *Adapted from Orlov et al. (1989) with permission.*

(Orlov, Pokudin, Kotelevtsev, & Gulak, 1989). Further studies showed elevated content of polyphosphoinositides in hyperosmotically shrunken guinea pig heart (Nasuhoglu et al., 2002), human HeLa cells (Yamamoto et al., 2006) and Ehrlich mouse ascites tumor cells (Nielsen, Jensen, Harbak, Christensen, & Simonsen, 2007). Yamamoto and co-workers reported that hypertonic stress recruited type 1 phosphatidylinositol 4-phosphate 5-kinase β isoform (PIP5KIbeta) to the plasma membrane through its Ser/Thr dephosphorylation, as PIP5KIbeta siRNA abolished both elevation of PIP_2 content and reorganization of actin cytoskeleton in shrunken HeLa cells (Yamamoto et al., 2006). Based on the comparative analysis of hyper- and isosmotic shrinkage, Nielsen and co-workers proposed that accumulation of PIP_2 in shrunken Ehrlich ascites tumor cells is caused by elevated ionic strength rather than by decline of cell volume *per se* (Nielsen et al., 2007). It should be noted that despite the same elevation of ionic strength triggered by hyperosmotic shrinkage, accumulation of PIP_2 has been detected only in rat but not in human erythrocytes, which lack shrinkage-activated $Na^+,K^+,2Cl^-$ cotransporter and Na^+/H^+ exchanger (Fig. 6).

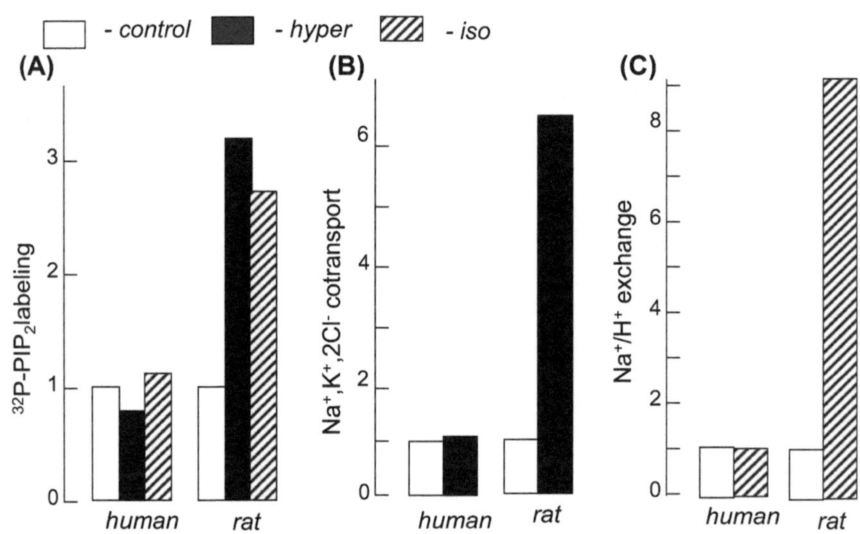

Figure 6 Effect of hyperosmotic (**hyper**) and isosmotic (**iso**) shrinkage on ^{32}P-labeling of phosphatidylinositol 4,5-diphosphate (A), activity of $Na^+,K^+,2Cl^-$ cotransport (B) and Na^+/H^+ exchange (C) in human and rat erythrocytes. Hyperosmotic and isosmotic shrinkage was triggered by the addition of 300 mM sucrose and 2.5 μM valinomycin, respectively. Values obtained in control conditions were normalized as 1 (Orlov et al., 1989, 1993; Parshina et al., 2013).

These species-dependent features strongly support an implication of polyphosphoinositides in the activation of ion transporters providing RVI.

PIP_2 might be further phosphorylated by phosphatidylinositol 3-kinase (PI3K) to produce phosphatidylinositol 3,4,5-triphosphate (PIP_3). Several research teams demonstrated that cell swelling is accompanied by attenuation of PIP_2 (Nielsen et al., 2007) and elevation of PIP_3 content (Feranchak et al., 1999; Krause et al., 1997; Wang et al., 2004). It was also shown that inhibition of PI-3K by wortmannin and LY294002 and application of anti-PIP3 antibodies abolished the activation of volume-sensitive chloride current triggered by cell swelling and/or integrin-dependent stretch (Browe & Baumgarten, 2006; Ichishima, Yamamoto, Iwamoto, & Ehara, 2010; Shi, Barnes, Coca-Prados, & Kelly, 2002; Vessey et al., 2004; Yamamoto, Ichishima, & Ehara, 2008). Does altered content of polyphosphoinositides affect RVI and RVD via direct interaction of PIP2 and PIP3 with the ion transporters or this signaling pathway is mediated by other PIP2/PIP3-sensitive targets such as serum- and glucocorticoid-inducible kinase SGK1 (Wang et al., 2004)? We address this question in the forthcoming studies.

3. CELL VOLUME SENSING BY THE CYTOPLASM

3.1 Three-Dimensional Cytoskeleton

The architecture of nucleated cells is supported by a crosslinked three-dimensional (3D) cytoskeleton located within cytoplasm and formed by actin microfilaments, microtubules and intermediated filaments. The hypothesis implicating 3D cytoskeleton in cell volume sensing is supported by two major observations. *First*, actin microfilaments have been found depolymerized and polymerized in a variety of nonadherent cells subjected to anisosmotic swelling and shrinkage, respectively. The data on cell volume-dependent reorganization of microtubules and intermediate filaments are contradictory and limited to few publications (for more details, see Henson, 1999; Hoffmann et al., 2009). *Second*, in several types of cells addition of F-actin disrupting agents (cytochalasins, latrunculin) and microtubule depolymerization agents (colchicine, vinblastin) suppressed volume-dependent cellular responses including RVI/RVD (for comprehensive reviews, see Papakonstanti, Vardaki, & Stounaras, 2000; Di Ciano-Oliveira, Thirone, Szaszi, & Kapus, 2006; Jakab et al., 2002; Hoffmann et al., 2009).

It should be noted, however, that in primary cultured astrocytes complete actin depolymerization does not prevent swelling-induced activation of K^+ and anion channels and does not affect RVD (Mongin et al., 1996). In our recent experiments we did not detect any significant action of microtubule and microfilament disruption by vinblastine and cytochalasin B, respectively, on RVD in hyposmotically swollen A549 cells (Platonova et al., 2015). Moreover, volume-dependent ion fluxes were detected in mammalian erythrocytes lacking 3D cytoskeleton (Figs. 2, 4, and 6). Collectively, these results indicate that 3D cytoskeleton cannot be considered as a universal sensor of cell volume changes. The role of 3D cytoskeleton in osmotic behavior of cytoplasmic hydrogel is considered below.

3.2 Macromolecular Crowding

In the 1980s to early 1990s, several research groups proposed that high concentration of macromolecules in the cytoplasm (so-called macromolecular crowding) crucially affects molecular and cellular functions (Fulton, 1982; Minton, 1981). Indeed, physical chemistry predicts that at extremely high concentration in cytoplasm thermodynamic activity of the macromolecules is an exponential function of their apparent concentration. This thermodynamic model is consistent with numerous experiments showing drastic impact of high molecular weight solutes on the functional properties of diverse proteins including enzyme activities (Ellis, 2001; Minton, 2001; Zhou, Rivas, & Minton, 2008). These observations allowed researchers to propose that changes in macromolecular crowding may serve as a cell volume sensor (Minton, Colclasure, & Parker, 1992; Zimmerman & Harrison, 1987). The hypothesis was supported by results obtained in resealed ghosts from dog erythrocytes. Indeed, it was shown that volume-dependent regulation of Na^+/H^+ exchange and K^+,Cl^- cotransport occurred when hemoglobin concentration within the resealed ghosts exceeded 300 mg/mL, that corresponds to hemoglobin concentration in intact cells. Importantly, the set points of these carriers were affected by modulation of intracellular macromolecular crowding rather than cell volume *per se* (Colclasure & Parker, 1992; Parker, 1993). Recent studies demonstrated the implication of macromolecular crowding in the regulation of gene transcription (Golkaram, Hellander, Drawert, & Petzold, 2016; Matsuda, Putzel, Backman, & Szleifer, 2014). Role of this phenomenon in transcriptomic changes documented in shrunken cells (see Section 3) has not yet been examined. It should be noted, however, that unlike dog RBC, swelling resealed human erythrocyte ghosts activated K^+,Cl^-

cotransport even though solute concentration was constant and ghosts were virtually free of proteins (Sachs, 1998).

As noted above 10-min preincubation at 49–50 °C completely abolished activation of $Na^+,K^+,2Cl^-$ cotransport and Na^+/H^+ exchange in shrunken rat erythrocytes and K^+,Cl^- cotransport in swollen human RBC (Figs. 2 and 4). Since modest heat treatment does not affect macromolecular crowding, we proposed that this effect is due to inactivation of the osmosensing properties of cytoplasmic hydrogel. This hypothesis has been subjected to experimental verification as discussed in the next section.

3.3 Osmosensing Properties of the Cytoplasmic Hydrogel

The cytoplasm consists of a water-containing fibrous matrix formed by a cytoskeletal network of interconnected polymeric protein fibers that have a net electrical charge and might be better described as an aqueous polyelectrolyte gel (Clegg, 1984; Knull & Minton, 1996; Luby-Phelps, 2000). As such, the cytoplasm may have properties similar to hydrogels, a synthetic or natural water-swollen polymeric network containing chemical or physical crosslinks (Peppas, Huang, Torres-Lugo, Ward, & Zhang, 2000). To verify this hypothesis we employed A549 cells permeabilized with digitonin. We found that single substrate-attached permeabilized cells swell and shrink in hypo- and hypertonic solutions, respectively (Fels, Orlov, & Grygorczyk, 2009). Remarkably, volume changes induced by osmolality alterations were several-fold larger than those observed with intact cells. Indeed, a van't Hoff relationship between volume and inverse osmolality exhibits 3.7-fold steeper dependence in permeabilized cells, compared to their intact counterparts. This observation demonstrates that the cytoplasmic biogel behaves like a highly-sensitive osmometer and suggests that the stable steady-state volume of intact (non-permeabilized) cells is maintained by the plasma membrane, which prevents uncontrolled leakage of osmolytes and reduces intracellular osmotic pressure by energy-dependent ion pumping. Furthermore, since a high proportion of intracellular water is associated with biomolecular interactions and is osmotically unresponsive (Fullerton, Kanal, & Cameron, 2006), such steep dependence indicates that small changes in protein conformation can release or bind significant amounts of water, which may allow rapid modulation of local fluidity, macromolecular crowding and activity of the intracellular environment (Chaplin, 2006; Garlid, 2000).

To further examine the role of cytoplasmic gel in cell volume regulation, we compared the action of short-term heat treatment (10 min, 48 °C) on

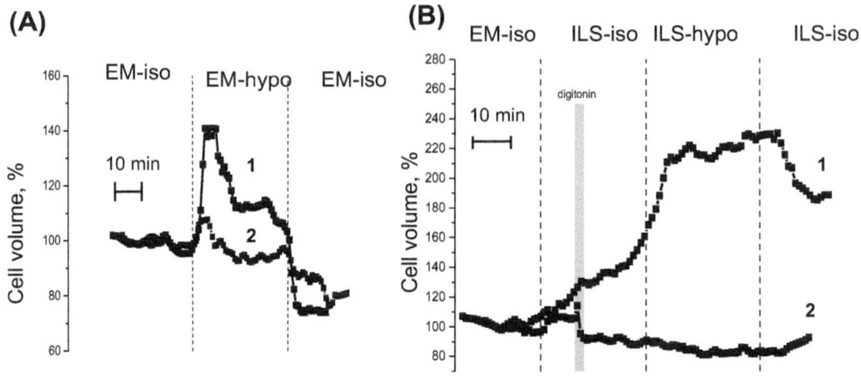

Figure 7 Effect of hyposmotic medium (B-hypo) on volume changes in intact (A) and permeabilized (B) A549 cells. **1** — control cells; **2** — A549 cells were subjected to 10-min preincubation at 48 °C. Note that cell volume was measured at 37 °C. Mean cell volume measured during the first 5 min of incubation was taken as 100% (Platonova et al., 2014).

volume changes triggered by anisosmotic medium in intact and permeabilized A549 cells (Platonova et al., 2014). Fig. 7A shows that hypotonic swelling was accompanied by RVD and almost complete volume normalization in 30 min of incubation (**curve 1**). Consistent with previous observations (Hoffmann et al., 2009; Lang et al., 1998), reperfusion of hypotonically-treated A549 cells with isosmotic medium elicited rapid 30%—40% cell volume decrease (so-called post-RVD/RVI protocol). Ten-min preincubation at 48 °C almost completely abolished hyposmotic swelling of A549 cells as well as RVD (Fig. 7A, **curve 2**). Cell transfer into low-Na^+, high-K^+ intracellular-like solution (ILS) and subsequent plasma membrane permeabilization by digitonin led to elevation of A549 cell volume by ~60% (Fig. 7B, **curve 1**). As noted above, cell permeabilization increased the amplitude of hyposmotic swelling. As predicted, dissipation of osmolyte electrochemical gradients across the plasma membrane in permeabilized cells resulted in RVD suppression. Remarkably, heat treatment completely abolished the swelling of permeabilized cells in hyposomotic medium (Fig. 7B, **curve 2**), demonstrating that temperature elevation suppresses cell osmosensing properties via its effect on cytoplasmic biogel rather than on plasma membrane water permeability.

Ziyadeh, Mills, and Kleinzeller (1992) and Mills, Scwiebert, and Stanton (1994) have shown that treatments which drastically change the structure of plasma membrane-associated actin network are accompanied by massive cell swelling as opposed to modest swelling of intact cells in hypotonic solutions. Keeping in mind the sharp swelling of permeabilized cells (Fig. 7), we compared the cell volume adjustment of intact and permeabilized A549 cells subjected to microtubule and microfilament disruption by vinblastine and cytochalasin B, respectively. Both compounds decreased the maximal amplitude of intact cell swelling in hyposmotic medium and completely abolished the cytoplasmic gel volume increment triggered by hyposmotic medium in permeabilized cells. In contrast to swelling, permeabilization did not significantly affect the amplitude of shrinkage evoked by hyperosmotic medium. In both intact and permeabilized cells the rate of shrinkage was sharply attenuated by cytochalasin B (Platonova et al., 2015). These results demonstrate a key role of the 3D cytoskeleton in the functioning of biogel as an osmosensor. They also suggest that temperature-dependent inhibition of volume-sensitive ion carriers, documented in earlier studies (Orlov et al., 1993), is caused by inactivation of the cytoplasmic biogel osmosensitivity rather than downstream signaling triggered by cell volume changes.

4. DELAYED VOLUME SENSORS: ROLE OF ELEVATED $[NA^+]_i/[K^+]_i$ RATIO

As mentioned above, rapid regulatory responses, triggered in seconds, of cell volume perturbations are mediated by a set of channels and carriers providing the gain (RVI) or the loss (RVD) of monovalent ions. It should be noted, however, that these systems provide cell volume autoregulation via alteration of ionic strength of intracellular environment that, in turn, affects functional properties of proteins and other macromolecules. To escape these side effects caused by long-term elevation of osmolality occurring in distal nephron (Burg, 1995) as well as in cerebrovascular (Orlov & Mongin, 2007) and interstitial (Titze & Machnik, 2010; Titze, 2014) fluids, cells accumulate organic osmolytes via augmented transcription of genes SMIT, BGT1 and TauT, responsible for Na^+-coupled transport of inositol, betaine and taurine or aldoreductase (AR) and esterase (NTE) involved in the synthesis of sorbitol and glycerophospatidyl choline, respectively (Burg, 1995; Kwon, Lim, & Kwon, 2009; Lang & Shumilina, 2017). In hypertonic environment augmented transcription of SMIT, BGT1 and TauT genes is caused by the activation of tonicity responsive enhancer (TonE), consisting of NGGAAAWDHMC(N) tandem repeats, present in the proximal sites of promoter regions of these genes. TonE is activated by interaction with the transcriptional regulator, tonicity-responsive enhancer-binding protein (TonEBP) also known as osmotic-response element-binding protein (OREBP), which is a member of nuclear factor of activated T cells family (NFAT) proteins NFAT5. Molecular mechanisms of NFAT5-mediated transcriptional regulation, including its phosphorylation and nuclear translocation remain poorly understood (for comprehensive review, see Cheung & Ko, 2013; Jeon, Kim, Sheen, & Kwon, 2006; Kim, Jeon, Kwon, Lim, & Kwon, 2007).

Several observations listed below strongly suggest that NFAT5 activation is not the only mechanism of volume-dependent transcriptional regulation. *First*, Cheung and co-workers reported that unlike SMIT, BGT1, TauT and AR, elevated content of NTE mRNA in shrunken cells is not caused by NFAT5-mediated transcriptional activation (Cheung & Ko, 2013). *Second*, NFAT5-mediated transcription might be evoked independently of cell volume changes by signaling via nitric oxide, angiotensin II, TNF-α, TGF-β and other stimuli leading to the activation of Ca^{2+}-dependent and -independent isoforms of protein kinase C (Halterman, Kwon, & Wamhoff, 2012). *Third*, NFAT5-mediated SMIT, BGT1, TauT and AR transcription was observed when medium osmolality was increased by addition of salts but not urea or other highly permeable osmolytes (Woo, Dahl, Handler, & Kwon, 2000), indicating a key role of elevated intracellular ionic strength rather than decreased cell volume *per se* in this process. This conclusion is consistent with the lack of any effect of isosmotic shrinkage on intracellular ionic strength and on the expression of luciferase linked to TonEBP-sensitive promoter (Rodgaard, Schou, Friis, & Hoffmann, 2008). *Fourth*, unlike TonEBP-mediated signaling, enhanced transcription of (serum+glucocorticoid)-sensitive kinase SGK1 was observed both in hyper- and isosmotically shrunken cells (Waldegger, Barth, Raber, & Lang, 1997).

Two decades ago Wehner et al. reported that in rat hepatocytes elevation of osmolality from 300 to 450 mOsm resulted in rapid increase of $[Na^+]_i$ from 15 to 30 mM and $[K^+]_i$ from 80 to 110 mM, that corresponds to an increase in the $[Na^+]_i/[K^+]_i$ ratio by 50% (Wehner & Tinel, 2000). Later on, the same research team demonstrated that this phenomenon is caused by the activation of hypertonicity-induced cation channels (HICCs) permeable to Na^+, K^+, Cs^+ and Li^+ and inhibited by 10 μM amiloride, 100 μM Gd^{3+} and 10 μM flufenamate (Wehner, Lawonn, & Tinel, 2002; Wehner et al., 2006), as well as by siRNA specific for ENaCα-subunit and transient receptor potential cation channel TRPM2 (Bondarova, Li, Endl, & Wehner, 2009; Numata et al., 2012).

Using Affymetrix technology we observed that in human adenocarcinoma HeLa cells, human umbilical vein endothelial cells (HUVECs) and rat aorta vascular smooth muscle cells (RVSMCs) inhibition of the Na^+,K^+-ATPase by ouabain or by K^+-free medium results in altered expression of 684, 737 and 1839 transcripts, respectively (Koltsova, Akimova, Kotelevtsev, Grygorczyk, & Orlov, 2012). Analysing these Na^+_i/K^+_i-sensitive transcripts, we found that 80 ubiquitous genes are affected in all three types of cells. Interleukin 6 (IL-6) was found to be among the ubiquitous Na^+_i/K^+_i-sensitive genes whose content was increased by more than threefold (Koltsova, Tremblay, Hamet, & Orlov, 2015; Koltsova et al., 2012). In RVSMC, the increment of the $[Na^+]_i/[K^+]_i$ ratio triggered by Na^+,K^+-ATPase inhibition was accompanied by up-to 4-fold elevation of SGK1 (Koltsova et al., 2014), water channel AQP1, adhesive factor CD24 and monocyte chemoattractant MCP-1 (manuscript in preparation). Previously, it was shown that expression of these genes might also be enhanced by hyperosmotic shrinkage of epithelial cells from distal tubule, macrophages and hepatocytes (Berga-Bolanos, Drews-Elder, Aramburu, & Lopez-Rodriguez, 2010; Buxade et al., 2012; Kojima et al., 2010; Lanaspa et al., 2010; Waldegger et al., 1997).

Villanutva and co-workers reported that hypoxia in renal epithelial cells leads to NFAT expression and its translocation into the nucleus by a mechanism distinct from the canonical pathway mediated by hypoxia-inducible factor-1α (HIF-1α) (Villanutva et al., 2012). Recently, we observed that HIF-1α-independent transcriptomic changes in hypoxic vascular smooth muscle are mediated by elevation of the $[Na^+]_i/[K^+]_i$ ratio (Koltsova et al., 2014).

The role of mitogen-activated protein kinase (MAPK) in transcriptomic changes in shrunken cells is well-documented (Kojima et al., 2010). We revealed that in spite the same final cell volume changes documented in iso- and hyper-osmotically shrunken C11-MDCK cells, that resembling the intercalated cells from the collecting ducts of Madin-Darby canine kidney collecting ducts (Fig. 8A and B), 5-fold elevation of JNK MAPK phosphorylation is limited to hypertonically shrunken cells (Fig. 8C and D). These features are likely caused by specific effects of cell shrinkage under hyper- and isosmotic conditions on intracellular ionic strength and/or $[Na^+]_i/[K^+]_i$ ratio (Koltsova et al., 2012).

Collectively, the above results allow us to propose that side by side with TonEBP, altered gene expression in shrunken cells is caused by elevation of the $[Na^+]_i/[K^+]_i$ ratio (Fig. 9). This hypothesis assumes that hyperosmotic

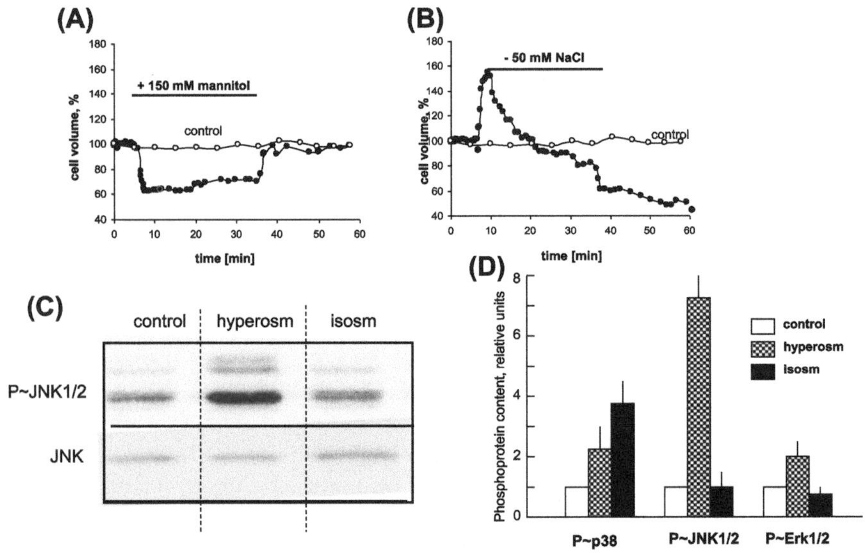

Figure 8 Effect of hyper- (A,C,D) and isoosmotic (B,C,D) shrinkage on cell volume changes (A,B) and JNK MAPK phosphorylation (C,D) in C11-MDCK cells. (A) Time-course of volume changes in cells superfused for 30 min with hyperosmotic medium (+150 mM mannitol) and then transferred to isosmotic medium. (B) Time-course of changes in volume of C11-MDCK cells superfused for 30 min with hyposmotic medium (−50 mM NaCl) and then transferred to isosmotic medium. Initial cell volume values were taken as 100%. (C) Representative Western blots showing the effect of hyperosmotic (hyperosm) and isosmotic (isosm) shrinkage on the total content and phosphorylation of JNK1/2 MAPK. (D) Relative content of phosphorylated MAPKs in cells subjected to hyperosmotic and isosmotic shrinkage. Protein content in control (not subjected to shrinking) cells was considered as 1. *Adapted from Koltsova et al. (2012) with permission.*

shrinkage leads to elevated $[Na^+]_i/[K^+]_i$ ratio, that in turn affects the conformation of intracellular monovalent ion sensors interacting with Na^+/K^+ response elements (Na/KRE) within the promoters of $Na^+{}_i/K^+{}_i$-sensitive genes Ч$_{1,2,3}$. The possible molecular origin of $Na^+{}_i/K^+{}_i$ sensors is considered elsewhere (Orlov & Hamet, 2015). In cells abundant in Na^+/H^+ exchanger and other $Na^+{}_i/K^+{}_i$-sensitive transporters, elevated $[Na^+]_i/[K^+]_i$ ratio leads to increased $[Ca^{2+}]_i$, conformation transition of calmodulin (CaM) and other $Ca^{2+}{}_i$ sensors interacting with diverse CaM-sensitive elements within $Y_{1,2,3}$ gene promoters. The downstream steps of this canonical $Ca^{2+}{}_i$-mediated mechanisms of excitation-transcription coupling are considered in details elsewhere (Alonso & Garcia-Sancho, 2011; Coulon & Blanchard, 2001; Hardingham, Chawla, Johnson, & Bading, 1997;

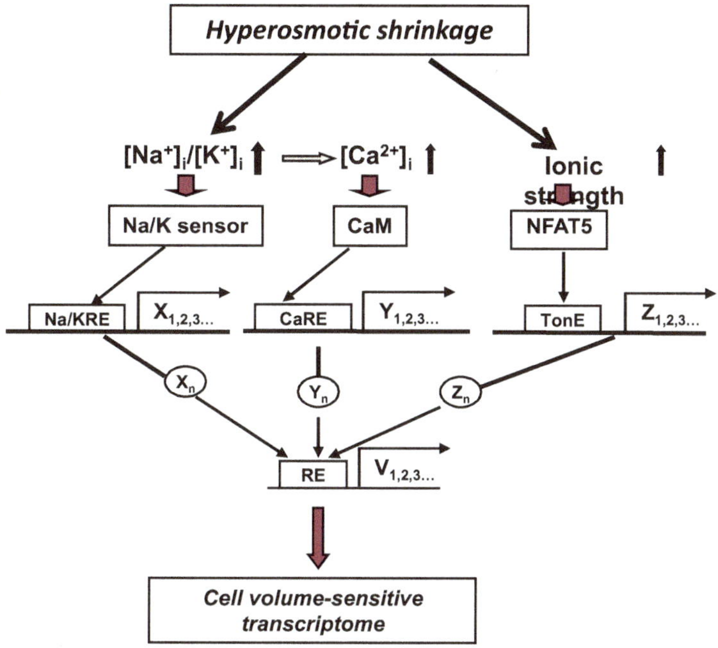

Figure 9 Mechanisms underlying transcriptomic changes in hyperosmotically shrunken cells. Cell shrinkage in hyperosmotic environment results in elevated intracellular ionic strength and $[Na^+]_i/[K^+]_i$ ratio. Increased ionic strength affects transcription by facilitating the interaction of tonicity response element binding protein (TonEBP or NFAT) with tonicity response elements (TonE) in the promoters of $Z_{1,2,3}$ ... genes, whereas elevated $[Na^+]_i/[K^+]_i$ ratio acts on the transcription of $X_{1,2,3}$ ... genes via unidentified monovalent ion sensor(s) (Na/K sensor) and Na/KRE response elements. In addition, dissipation of transmembrane gradient of monovalent cations may increase $[Ca^{2+}]_i$, that in turn affects transcription of $Y_{1,2,3}$ ... genes by binding to calmodulin and other intracellular $Ca^{2+}{}_i$ sensors. $X_{1,2,3}$ ... $Y_{1,2,3}$ ... and $Z_{1,2,3}$ ... symbolize the sets of genes whose expression in shrunken cells is affected by distinct upstream signaling pathways.

Smolyaninova, Koltsova, Sidorenko, & Orlov, 2017). The role of these signaling pathways in volume autoregulation of cells subjected to long-term exposure to hyperosmotic environment is currently studied in our laboratory.

5. CONCLUSIONS AND PERSPECTIVE

The available information discussed above shows that upstream sensors of cell volume changes are mainly located within cytoplasm rather than the plasma membrane (Fig. 10). Indeed, intrinsic cytoplasm osmosensitivity is

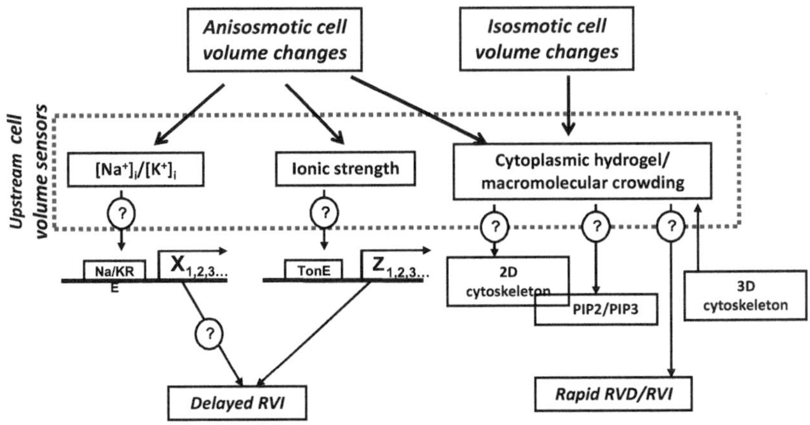

Figure 10 Upstream intermediates of signaling pathways involved in cell volume autoregulation. ? – unknown steps of signal transduction.

3.7-fold greater than in intact cells, demonstrating its ability to absorb and retain large amounts of water. Our experiments (Platonova, Orlov, & Grygorczyk, 2013) and those of others (Spagnoli et al., 2008) provide evidence that in contrast to the tensegrity model mechanical stresses must be distributed within a 3D sponge-like cytoplasmic gel. Cell volume changes occurring in anisosmotic medium as well as in isosmotic environment affect macromolecules-crowded hydrogel that, in turn, triggers rapid RVI and RVD. The downstream signaling pathways include reorganization of 2D cytoskeleton and altered composition of polyphosphoinositides located on the inner surface of the plasma membrane. However, the exact mechanism of the crosstalk between cell volume sensitive properties of cytoplasmic hydrogel and ion transporters involved in RVD and RVI require further investigations.

In addition to its action on the physico-chemical properties of cytoplasmic hydrogel, cell volume changes triggered in anisosmotic but not in isosmotic conditions affect the ionic strength and the $[Na^+]_i/[K^+]_i$ ratio in cytoplasm. Elevation of ionic strength evoked by long term exposure of cells to hypertonic environment resulted in delayed RVI via activation of TonEBP and augmented expression of genes controlling intracellular content of organic osmolytes. Does delayed cell volume autoregulation involve genes whose transcription is under the control of Na^+_i/K^+_i-sensitive, Ca^{2+}_i-mediated and Ca^{2+}_i-independent mechanisms of excitation transcription coupling? What is the molecular origin of intracellular ionic strength and monovalent cation sensors? Does cytoplasmic hydrogel contribute to rapid RVI/RVD via 2D cytoskeleton- and polyphosphoinositide-independent signaling pathways? Clearly this and many other intriguing questions remain unresolved presently and await future studies. Besides, the hypothesis of selective K^+ vs. Na^+ adsorption by the cytoplasm, or contribution and function of different molecular components of the cytoskeleton in cytoplasmic biogel osmosensitivity and water binding capacity, also need to be examined in detail.

ACKNOWLEDGMENTS

This work was supported by grants from the Russian Foundation for Basic Research (SNO #18-04-00063) and Natural Sciences and Engineering Research Council of Canada (RG, 435517-2013).

REFERENCES

Alonso, M. T., & Garcia-Sancho, J. (2011). Nuclear Ca^{2+} signalling. *Cell Calcium, 49*, 280–289.

Ayee, M. A. A., LeMaster, E., Teng, T., Lee, J., & Levitan, I. (2018). Gypotonic challenge of endothelial cells increases membrane stiffness with no effect on tether force. *Biophysical Journal, 114*, 929–938.

Berga-Bolanos, R., Drews-Elder, K., Aramburu, J., & Lopez-Rodriguez, C. (2010). NFAT5 regulates T lymphocyte homeostasis and CD-24-dependent T cell expansion under pathological hypernatremia. *The Journal of Immunology, 185*, 6624–6635.

Bondarova, M., Li, T., Endl, E., & Wehner, F. (2009). α-ENaC is a functional element of the hypertonicity-induced cation channel in HepG2 cells and it mediates proliferation. *Pflügers Archiv - European Journal of Physiology, 458*, 675–687.

Boudreault, F., & Grygorczyk, R. (2004a). Evaluation of rapid volume changes of substrate-adherent cells by conventional microscopy 3D imaging. *Journal of Microscopy, 215*, 302–312.

Boudreault, F., & Grygorczyk, R. (2004b). Cell swelling-induced ATP release is tightly dependent on intracellular calcium elevation. *Journal of Physiology (London), 561*, 499–513.

Brandts, J. F., Taverna, R. D., Salasivan, E., & Lysko, K. A. (1978). Calorimetric studies of structural transitions of human erythrocyte membrane. Studies of B and C transitions. *Biochimica et Biophysica Acta, 512*, 566–578.

Browe, D. M., & Baumgarten, C. M. (2004). Angiotensin II (AT1) receptors and NADH oxidase regulate Cl- current elicited by beta1 integrin stretch in rabbit ventricular myocytes. *The Journal of General Physiology, 124*, 273–287.

Browe, D. M., & Baumgarten, C. M. (2006). EGFR kinase regulates volume-sensitive chloride current elicited by integrin stretch via Pi-3K and NADH oxidaze in ventricular myocytes. *The Journal of General Physiology, 127*, 237–251.

Burg, M. B. (1995). Molecular basis of osmotic regulation. *American Journal of Physiology, 268*, F983–F996.

Buxade, M., Lunazzi, G., Minguillon, J., Iborra, S., Berga-Bolanos, R., Aramburu, J., & López-Rodríguez, C. (2012). Gene expression induced by Toll-like receptors in macrophages reguires the transcription factor NFAT5. *Journal of Experimental Medicine, 13*, 379–393.

Cantiello, H. F. (1997). Role of actin filament organization in cell volume and ion channel regulation. *Journal of Experimental Zoology, 279*, 425–435.

Carafoli, E. (2002). Calcium signaling: A tale for all seasons. *Proceeding of the National Academy of Science of the USA, 99*, 1115–1122.
Chaplin, M. (2006). Do we underestimate the importance of water in cell biology? *Nature Reviews. Molecular Cell Biology, 7*, 861–866.
Cheung, C. K. Y., & Ko, B. C. B. (2013). NFAT5 in cellular adaptation to hypertonic stress - regulations and functional significance. *Journal of Molecular Signaling, 8*, 5.
Clegg, J. S. (1984). Intracellular water and the cytomatrix: Some methods of study and current views. *Journal of Cell Biology, 99*, 167s–171s.
Colclasure, G. C., & Parker, J. C. (1992). Cytosolic protein concentration is the primary volume signal for swelling-induced KCl cotransport in dog red cells. *The Journal of General Physiology, 100*, 1–10.
Coulon, V., & Blanchard, J.-M. (2001). Flux calciques et expression gйnigue. *Medicine Sciences, 17*, 969–978.
Cox, C. D., Bavi, N., & Martinac, B. (2017). Bacterial mechanosensors. *Annual Review of Physiology, 80*, 71–91.
Delpire, E., & Austin, T. M. (2010). Kinase regulation of Na+-K+-2Cl- cotransport in primary neurons. *Journal of Physiology (London), 588*, 3365–3373.
Di Ciano-Oliveira, C., Thirone, A. C. P., Szaszi, K., & Kapus, A. (2006). Osmotic stress and the cytoskeleton: The r(h)ole of Rho GTPases. *Acta Physiologica, 187*, 257–272.
Doroshenko, P., & Neher, E. (1992). Volume-sensitive chloride conductance in bovine chromaffin cell membrane. *Journal of Physiology (London), 449*, 197–218.
Eduardsen, K., Larsen, S. L., Novak, I., Lambert, I. H., Hoffmann, E. K., & Pedersen, S. F. (2011). Cell volume regulation and signaling in 3T3-L1 pre-adipocytes and adipocytes: On the possible role of caveolae, insulin receptors, FAK and ERK1/2. *Cell Physiology and Biochemistry, 28*, 1231–1246.
Ellis, R. J. (2001). Macromolecular crowding: An important but neglected aspect the intracellular environment. *Current Opinion in Structural Biology, 11*, 114–119.
Fels, J., Orlov, S. N., & Grygorczyk, R. (2009). The hydrogel nature of mammalian cytoplasm contributes to osmosensing and extracellular pH sensing. *Biophysical Journal, 96*, 4276–4285.
Feranchak, A. P., Roman, R. M., Doctor, R. B., Salter, K. D., Toker, A., & Fitz, J. G. (1999). The lipid products of phosphoinositide 3-kinase contribute to regulation of cholangiocyte ATP and chloride transport. *Journal of Biological Chemistry, 274*, 30979–30986.
Fullerton, G. D., Kanal, K. M., & Cameron, I. L. (2006). Osmotically unresponsive water fraction on proteins; non-ideal osmotic pressure of bovine serum albumin as a function of pH and salt concentration. *Cell Biology International, 30*, 86–92.
Fulton, A. B. (1982). How crowded is cytoplasm. *Cell, 30*, 345–347.
Garcia-Elias, A., Mrkonjic, S., Pardo-Postor, C., Inada, H., Hellmich, U. A., Rubio-Moscardo, F., Plata, C., Gaudet, R., Vicente, R., & Valverde, M. A. (2013). Phosphatidylinositol-4,5-biphosphate-dependent rearrangement of TRPV4 cytosolic tails enables channel activation by physiological stimuli. *Proceeding of the National Academy of Science of the USA, 110*, 9553–9558.
Garlid, K. D. (2000). The state of water in biological systems. *International Review of Cytology, 192*, 281–302.
Golkaram, M., Hellander, S., Drawert, B., & Petzold, L. R. (2016). Macromolecular crowding regulates gene expression profile by limiting diffusion. *PLoS Computational Biology*. https://doi.org/10.1371/journal.pcbi.1005122.
Groulx, N., Boudreault, F., Orlov, S. N., & Grygorczyk, R. (2006). Membrane reserves and hypotonic cell swelling. *The Journal of Membrane Biology, 214*, 43–56.
Grygorczyk, R., Boudreault, F., Platonova, A., & Orlov, S. N. (2015). Salt and osmosensing: Role of cytoplasmic hydrogel. *Pflügers Archiv - European Journal of Physiology, 467*, 475–487.

Halterman, J. A., Kwon, H. M., & Wamhoff, B. R. (2012). Tonicity-independent regulation of the osmosensitive transcription factor TonEBP (NFAT5). *American Journal of Physiology. Cell Physiology, 302*, C1–C8.
Hardingham, G. E., Chawla, S., Johnson, C. M., & Bading, H. (1997). Distinct functions of nuclear and cytoplasmic calcium in the control of gene expression. *Nature, 385*, 260–265.
Henson, J. H. (1999). Relations between the actin cytoskeleton and cell volume regulation. *Microscopy Research and Technique, 47*, 155–162.
Hoffmann, E. K., Lambert, I. H., & Pedersen, S. F. (2009). Physiology of cell volume regulation in vertebrates. *Physiological Reviews, 89*, 193–277.
Hoffmann, E. K., & Simonsen, L. O. (1989). Membrane mechanisms in volume and pH regulation in vertebrate cells. *Physiological Reviews, 69*, 315–382.
Ichishima, K., Yamamoto, S., Iwamoto, T., & Ehara, T. (2010). α1-Adrenoceptor-mediated depletion of phosphatidylinositol 4,5-biphosphate inhibits activation of volume-regulated anion channels in mouse ventricular myocytes. *British Journal of Pharmacology, 161*, 193–206.
Ingber, D. E. (2003). Tensegrity I. Cell structure and hierarchial systems biology. *Journal of Cell Science, 116*, 1157–1173.
Jakab, M., Furst, J., gschwentner, M., Botta, M., Garavaglia, M.-L., Bazzini, C., Rodighiero, S., Meyer, G., Eichmueller, S., Wöll, E., Chwatal, S., Ritter, M., & Paulmichl, M. (2002). Mechanisms sensing and modulating signals arising from cell swelling. *Cellular Physiology and Biochemistry, 12*, 235–258.
Janmey, P. A., & Lindberg, U. (2004). Cytoskeletal regulation: Rich in lipids. *Nature Reviews. Molecular Cell Biology, 5*, 658–666.
Jeon, U. S., Kim, J. A., Sheen, M. R., & Kwon, H. M. (2006). How tonicity regulates genes: Story of TonEBP transcriptional activator. *Acta Physiologica, 187*, 241–247.
Jorgensen, N. K., Christensen, S., Harbak, H., Brown, A. M., Lambert, I. H., Hoffmann, E. K., & Simonsen, L. O. (1997). On the role of calcium in the regulatory volume decrease (RVD) response in Ehrlich mouse ascites tumor cells. *The Journal of Membrane Biology, 157*, 281–299.
Kageyama, K., Onoyama, Y., Kogawa, H., Goto, E., & Tanabe, K. (1989). The maximum and minimum water content and cell volume of human erythrocytes in vitro. *Biophysical Chemistry, 34*, 79–82.
Kim, J. A., Jeon, U. S., Kwon, M. S., Lim, S. W., & Kwon, H. M. (2007). Transcriptional activator TonE-binding protein in cellular protection and differentiation. *Methods in Enzymology, 428*, 253–267.
Kinnunen, P. K. J. (2000). Lipid bilayers as osmotic response elements. *Cellular Physiology and Biochemistry, 10*, 243–250.
Knull, H., & Minton, A. P. (1996). Structure within eukaryotic cytoplasm and its relationship to glycolytic metabolism. *Cell Biochemistry and Function, 14*, 237–248.
Koivusalo, M., Kapus, A., & Grinstein, S. (2009). Sensors, transducers, and effectors that regulate cell size and shape. *Journal of Biological Chemistry, 284*, 6595–6599.
Kojima, R., Taniguchi, H., Tsuzuki, A., Nakamura, K., Sakakura, Y., & Ito, M. (2010). Hypertonicity-induced expression of monocyte chemoattractant protein-1 through a novel cis-acting element and MAPK-signaling pathways. *The Journal of Immunology, 184*, 5253–5262.
Koltsova, S. V., Akimova, O. A., Kotelevtsev, S. V., Grygorczyk, R., & Orlov, S. N. (2012). Hyperosmotic and isosmotic shrinkage differentially affect protein phosphorylation and ion transport. *Canadian Journal of Physiology and Pharmacology, 90*, 209–217.
Koltsova, S. V., Shilov, B., Burulina, J. G., Akimova, O. A., Haloui, M., Kapilevich, L. V., et al. (2014). Transcriptomic changes triggered by hypoxia: Evidence for HIF-1α -independent, [Na+]i/[K+]i-mediated excitation-transcription coupling. *PLoS One, 9*, e110597.

Koltsova, S. V., Tremblay, J., Hamet, P., & Orlov, S. N. (2015). Transcriptomic changes in Ca2+-depleted cells: Role of elevated intracellular [Na+]/[K+] ratio. *Cell Calcium, 58*, 317–324.

Koltsova, S. V., Trushina, Y., Haloui, M., Akimova, O. A., Tremblay, J., Hamet, P., & Orlov, S. N. (2012). Ubiquitous [Na+]i/[K+]i-sensitive transcriptome in mammalian cells: Evidence for Ca2+i-independent excitation-transcription coupling. *PLoS One, 7*, e38032.

Kozera, L., White, E., & Calaghan, S. (2009). Caveolae act as membrane reserves which limit mechanosensitive I (Cl, swell) channel activation during swelling in the rat ventricular myocyte. *PLoS One, 4*, e8312.

Krause, M., Olsson, T., Law, A. B., Parker, R. A., Lindstrom, D. P., Sundell, H. W., et al. (1997). Effect of volume recruitment on response to surfactant treatment in rabbits with lung injury. *American Journal of Respiratory and Critical Care Medicine, 156*, 862–866.

Kwon, M. S., Lim, S. W., & Kwon, H. M. (2009). Hypertonic stress in the kidney: A necessary evil. *Physiology, 24*, 186–191.

Lanaspa, M. A., Andres-Hernando, A., Li, N., Rivard, C. J., Cirerchi, C., Roncal-Jimenez, C., et al. (2010). The expression of aquaporin-1 in the medulla of the kidney is dependent on the transcription factor associated with hypertonicity, TonEBP. *Journal of Biological Chemistry, 285*, 31694–31703.

Lang, F. (2009). Mechanisms and significance of cell volume regulation. *The Journal of the American College of Nutrition, 26*, 613S–623S.

Lang, F., Busch, G., Ritter, M., Volkl, H., Waldegger, S., Gulbins, E., et al. (1998). Functional significance of cell volume regulatory mechanisms. *Physiological Reviews, 78*, 247–306.

Lang, F., & Shumilina, E. (2017). Regulation of ion channels by the serum- and glucocorticoid-inducible kinase SGK1. *The FASEB Journal, 27*, 3–12.

Lauf, P. K., & Adragna, N. C. (2000). K-Cl cotransport: Properties and molecular mechanism. *Cell Physiology and Biochemistry, 10*, 341–354.

Luby-Phelps, K. (2000). Cytoarchitecture and physical properties of cytoplasm: Volume, viscosity, diffusion, intracellular surface area. *International Review of Cytology, 192*, 189–221.

Lux, S. E., IV (2016). Anatomy of the red cell membrane skeleton: Unanswered questions. *Blood, 127*, 187–199.

Macknight, A. D. C., & Leaf, A. (1977). Regulation of cellular volume. *Physiological Reviews, 57*, 510–573.

Martinac, B. (2011). Bacterial mechanosensitive channels as a paradigm for mechanosensory transduction. *Cellular Physiology and Biochemistry, 28*, 1051–1060.

Matsuda, H., Putzel, G. G., Backman, V., & Szleifer, I. (2014). Macromolecular crowding as a regulator of gene transcription. *Biophysical Journal, 106*, 1801–1810.

McCarty, N. A., & O'Neil, R. G. (1992). Calcium signaling in cell volume regulation. *Physiological Reviews, 72*, 1037–1061.

Mills, J. W., Scwiebert, E. M., & Stanton, B. A. (1994). Evidence for the role of actin filaments in regulating cell swelling. *Journal of Experimental Zoology, 268*, 111–120.

Minton, A. P. (1981). Excluded volume as determinant of macromolecular structure and reactivity. *Biopolymers, 20*, 2093–2120.

Minton, A. P. (2001). The influence of macromolecular crowding and macromolecular confinement on biochemical reactions in physiological media. *Journal of Biological Chemistry, 276*, 10577–10580.

Minton, A. P., Colclasure, G. C., & Parker, J. C. (1992). Model for the role of macromolecular crowding in regulation of cellular volume. *Proceeding of the National Academy of Science of the USA, 89*, 10504–10506.

Moeckel, G. W., Zhang, L., Chen, X., Rossini, M., Zent, R., & Pozzi, A. (2013). Role of integrin $\alpha 1\beta 1$ in the regulation of renal medullary osmolyte concentration. *American Journal of Physiology - Renal Physiology, 290*, F223–F231.

Mongin, A. A. (2018). Volume-regulated anion channel - a frenemy within the brain. *Pflügers Archiv - European Journal of Physiology, 468*, 421–441.
Mongin, A. A., Aksentsev, S. L., Orlov, S. N., Fedulov, A. S., Mezen, N. I., Kvacheva, Z. V., et al. (1996). Swelling-induced activation of Na+,K+,2Cl- cotransport in glial cells: Kinetic properties and intracellular signalling mechanisms. *Biochimica et Biophysica Acta, 1285*, 229–236.
Mongin, A. A., & Orlov, S. N. (2001). Mechanisms of cell volume regulation and possible nature of the cell volume sensor. *Pathophysiology, 8*, 77–88.
Nasuhoglu, C., Feng, S., Mao, Y., Shammat, I., Yamamato, M., Earnest, S., et al. (2002). Modulation of cardiac PIP2 by cardioactive hormones and other physiologically relevant interventions. *American Journal of Physiology. Cell Physiology, 283*, C223–C234.
Naziroglu, M., & Braidy, N. (2017). Thermo-sensitive TRP channels: Novel targets for treating chemotherapy-induced peripheral pain. *Frontiers in Physiology, 8*. article 1040.
Nielsen, D. K., Jensen, A. K., Harbak, H., Christensen, S. C., & Simonsen, L. O. (2007). Cell content of phosphatidylinositol (4,5)biphosphate in Ehrlich mouse ascites tumour cells in response to cell volume perturbations in anisotonic ans isosmotic media. *Journal of Physiology (London), 582*, 1027–1036.
Numata, T., Sato, K., Christmann, J., Marx, R., Mori, Y., Okada, Y., et al. (2012). The ΔC splice-variant of TRPM2 is the hypertonicity-induced cation channel in HeLa cells, and the hypertonicity-induced cation channel in HeLa cells, and the ecto-enzyme CD38 mediates its activation. *Journal of Physiology (London), 590*, 1121–1138.
O'Connor, E. R., & Kimelberg, H. K. (1993). Role of calcium in astrocyte volume regulation and in the release of ions and amino acids. *The Journal of Neuroscience, 13*, 2638–2650.
Okada, Y. (1997). Volume expansion-sensing outward rectifier Cl^- channel: Fresh start to the molecular identity and volume sensor. *American Journal of Physiology, 273*, C755–C789.
Orlov, S. N., & Hamet, P. (2015). Salt and gene expression: Evidence for Na+i,K+i-mediated signaling pathways. *Pflügers Archiv - European Journal of Physiology, 467*, 489–498.
Orlov, S. N., Kolosova, I. A., Cragoe, E. J., Gurlo, T. G., Mongin, A. A., Aksentsev, S. L., et al. (1993). Kinetics and peculiarities of thermal inactivation of volume-dependent Na/H exchange, Na, K,2Cl cotransport and K,Cl cotransport in rat erythrocytes. *Biochimica et Biophysica Acta, 1151*, 186–192.
Orlov, S. N., & Mongin, A. A. (2007). Salt sensing mechanisms in blood pressure regulation and hypertension. *American Journal of Physiology. Heart and Circulatory Physiology, 293*, H2039–H2053.
Orlov, S. N., Pokudin, N. I., Kotelevtsev, Y. V., & Gulak, P. V. (1989). Volume-dependent regulation of ion transport and membrane phosphorylation in human and rat erythrocytes. *Journal of Membrane Biology, 107*, 105–117.
Pantaleo, A., De Franceschi, L., Ferru, E., Vono, R., & Turrini, F. (2010). Current knowledge about the functional roles of phosphorylative changes of membrane proteins in normal and deceased red cells. *Journal of Proteomics, 73*, 445–455.
Papakonstanti, E. A., Vardaki, E. A., & Stounaras, C. (2000). Actin cytoskeleton: A signaling sensor in cell volume regulation. *Cell Physiology and Biochemistry, 10*, 257–264.
Parker, J. C. (1993). In defense of cell volume? *American Journal of Physiology, 265*, C1191–C1200.
Parshina, E. Y., Yusipovich, A. I., Platonova, A. A., Grygorczyk, R., Maksimov, G. V., & Orlov, S. N. (2013). Thermal inactivation of volume-sensitive K+,Cl- cotransport and plasma membrane relief changes in human erythrocytes. *Pflügers Archiv - European Journal of Physiology, 465*, 977–983.
Pasantes-Morales, H. (2016). Channels and volume changes in the life and death of the cell. *Molecular Pharmacology, 90*, 358–370.
Pasantes-Morales, H., & Morales, M. S. (2000). Influence of calcium on regulatory volume decrease: Role of potassium channels. *Nephron, 86*, 414–427.

Pedersen, S. F., Hoffmann, E. K., & Mills, J. W. (2001). The cytoskeleton and cell volume regulation. *Comparative Biochemistry and Physiology — Part A, 130*, 385–390.

Pedersen, S. F., Kapus, A., & Hoffmann, E. K. (2011). Osmosensory mechanisms in cellular and systematic volume regulation. *Journal of the American Society of Nephrology: JASN, 22*, 1587–1597.

Pedersen, S. F., Prenen, J., Droogmans, G., Hoffmann, E. K., & Nilius, B. (1998). Separate selling- and Ca2+-activated anion current in Ehrlich ascites tumor cells. *The Journal of Membrane Biology, 163*, 97–110.

Pei-Chuan, C., Sivaselvan, M., & Sachs, F. (2018). Cytoskeletal contribution to cell stiffness due to osmotic swelling; extending the Donnan equilibrium. In I. Levitan, E. Delpire, & H. Rasgado-Flores (Eds.), *Vol. 81. Cell Volume Regulation and Fluid secretion* (in this volume).

Peppas, N. A., Huang, Y., Torres-Lugo, M., Ward, J. H., & Zhang, J. (2000). Physicochemical foundations and structural design of hydrogels in medicine and biology. *Annual Review of Biomedical Engineering, 2*, 9–29.

Platonova, A., Boudreault, F., Kapilevich, L. V., Maksimov, G. V., Ponomarchuk, O., Grygorczyk, R., et al. (2014). Temperature-induced inactivation of cytoplasmic biogel osmosensing properties is associated with suppression of regulatory volume decrease in A549 cells. *The Journal of Membrane Biology, 247*, 571–579.

Platonova, A. A., Orlov, S. N., & Grygorczyk, R. (2013). Volume changes triggered by anisomotic media in intact and permeabilized cells: Role of cytoskeleton network. *Bull. Sib. Med. 12*, 60.

Platonova, A., Ponomarchuk, O., Boudreault, F., Kapilevich, L. V., Maksimov, G. V., Grygorczyk, R., et al. (2015). Role of cytoskeleton network in anisosmotic volume changes of intact and permeabilized A549 cells. *Biochimica et Biophysica Acta, 1848*, 2337–2343.

Ponomarchuk, O., Boudreault, F., Orlov, S. N., & Grygorczyk, R. (2016). Calcium is not required for triggering volume restoration in hypotonically challenged A549 cells. *Pflügers Archiv - European Journal of Physiology, 468*, 2075–2085.

Rodgaard, T., Schou, K., Friis, M. B., & Hoffmann, E. K. (2008). Does intracellular ionic concentration or cell water content (cell volume) determine the activity of TonEBP in NIH3T3 cells. *American Journal of Physiology. Cell Physiology, 295*, C1528–C1534.

Sachs, J. R. (1998). How do red blood cells know how big they are? In Y. Okada (Ed.), *Cell volume regulation: The molecular mechanism and volume sensing machinery* (pp. 3–13). Tokyo: Elsevier Science.

Sachs, F. (2010). Stretch-activated ion channels: What are they? *Physiology, 25*, 50–56.

Schliess, F., & Haussinger, D. (2007). Osmosensing by integrins in rat liver. *Methods in Enzymology, 428*, 129–144.

Shen, M. R., Furla, P., Chou, C. Y., & Ellory, J. C. (2002). Myosin light chain kinase modulates hypotonicity-induced Ca2+ entry and Cl- channel activity in human cervical cancer cells. *Pflügers Archiv, 444*, 276–285.

Shi, C., Barnes, S., Coca-Prados, M., & Kelly, M. E. (2002). Protein tyrosine kinase and protein phosphatase signaling pathways regulate volume-sensitive chloride currents in a nonpigmented ciliary epithelial cell line. *Investigations in Ophthalmology and Vision Sciences, 43*, 1525–1532.

Shnyrov, V. L., Orlov, S. N., Zhadan, G. G., & Pokudin, N. I. (1990). Thermal inactivation of membrane proteins, volume-dependent Na^{+},K^{+} cotransport, and protein kinase C activator-induced changes of the shape of human and rat erythrocytes. *Biomedica Ret Biochimica Acta, 49*, 445–453.

Sihna, B., Koster, D., Ruez, R., Gonnord, P., Bastiani, M., abankwa, D., et al. (2011). Cells respond to mechanical stress by rapid disassembly of caveolae. *Cell, 144*, 402–413.

Smolyaninova, L. V., Koltsova, S. V., Sidorenko, S. V., & Orlov, S. N. (2017). Augmented gene expression triggered by Na+,K+-ATPase inhibition: Role of Ca2+-mediated and -independent excitation-transcription coupling. *Cell Calcium, 68*, 5–13.

Sowa, G. (2012). Caveolae, caveolins, cavings, and endothelial cell function: New insights. *Frontiers in Physiology, 2*, 120.

Spagnoli, C., Beyder, A., Besch, S., & Sachs, F. (2008). Atomic force microscopy analysis of cell volume regulation. *Physical Review E — Statistical, Nonlinear, and Soft Matter Physics, 78*(3 Pt 1), 031916.

Speake, T., Douglas, I. J., & Brown, P. D. (1998). The role of calcium in the volume regulation of rat lacrimal acinar cells. *The Journal of Membrane Biology, 164*, 283–291.

Strange, K. (2004). Cellular volume homeostasis. *Advances in Physiology Education, 28*, 155–159.

Strotmann, R., Harteneck, C., Nunnenmacher, K., Schultz, G., & Plant, T. D. (2000). OTRPC4, a nonselective cation channel that confers sensitivity to extracellular osmolarity. *Nature Cell Biology, 2*, 695–702.

Titze, J. (2014). Sodium balance is not just a renal affair. *Current Opinion in Nephrology and Hypertension, 23*, 101–105.

Titze, J., & Machnik, A. (2010). Sodium sensing in interstitium and relationship to hypertension. *Current Opinion in Nephrology and Hypertension, 19*, 385–392.

Trouet, D., Hermans, D., Droogmans, G., Nilius, B., & Eggermont, J. (2001). Inhibition of volume-regulated anion channels by dominant-negative caveolin-1. *Biochemical and Biophysical Research Communications, 284*, 461–465.

Vessey, J. P., Shi, C., Jollimore, C. A., Stevens, K. T., Coca-Prados, M., Barnes, S., et al. (2004). Hyposmotic activation of ICl,swell in rabbit nonpigmented ciliary epithelial cells involves increased ClC-2 trafficking to the plasma membrane. *Biochemistry & Cell Biology, 82*, 708–718.

Villanutva, S., Suazo, C., Santapau, D., Perez, F., Quiroz, M., Carreno, J. E., et al. (2012). NFAT5 is activated by hypoxia: Role of ischemia and reperfusion in the rat kidney. *PLoS One, 7*, e39665.

vom Dahl, S., Schliess, F., Reissmann, R., Gorg, B., Weiergraber, O., Kocalkova, M., et al. (2003). Involvement of integrins in osmosensing and signaling toward autophagic proteolysis in rat liver. *Journal of Biological Chemistry, 278*, 27088–27095.

Waldegger, S., Barth, P., Raber, G., & Lang, F. (1997). Cloning and characterization of a putative human serine/threonine protein kinase transcriptionally modified during anisosmotic and isotonic alterations of cell volume. *Proceeding of the National Academy of Science of the USA, 94*, 4440–4445.

Wang, G. X., McCrudden, C., Dai, Y.-P., Horowitz, B., Hume, J. R., & Yamboliev, I. A. (2004). Hyposomotic activation of volume-sensitive outwardly rectifying chloride channels in cultured PASMSc is modulated by SCK. *American Journal of Physiology. Heart and Circulatory Physiology, 287*, H533–H544.

Wehner, F., Bondarova, M., ter Veld, F., Endl, E., Nurnberger, H. R., & Li, T. (2006). Hypertonicity-induced cation channels. *Acta Physiologica, 187*, 21–25.

Wehner, F., Lawonn, P., & Tinel, H. (2002). Ionic mechanisms of regulatory volume increase (RVI) in the human hepatome cell-line HepG2. *Pflügers Archiv - European Journal of Physiology, 443*, 779–790.

Wehner, F., & Tinel, H. (2000). Osmolyte and Na+ transport balances of rat hepatocytes as a function of hypertonic stress. *Pflügers Archiv - European Journal of Physiology, 441*, 12–24.

Woo, S. K., Dahl, S. C., Handler, J. S., & Kwon, H. M. (2000). Bidirectional regulation of tonicity-responsive enhancer binding protein in response to changes in tonicity. *American Journal of Physiology - Renal Physiology, 278*, F1006–F1012.

Yamamoto, M., Chen, M. Z., Wang, Y. J., Sun, H. Q., Wei, Y., Martinez, M., et al. (2006). Hypertonic stress increases phosphatidylinositol 4,5-biphosphate levels by activating PIP5KIbeta. *Journal of Biological Chemistry, 281*, 32630–32638.

Yamamoto, S., Ichishima, K., & Ehara, T. (2008). Regulation of volume-regulated outwardly rectifying anion channels by phosphatidylinositol 3,4,5-triphosphate in mouse ventricular cells. *Biomedical Research, 29*, 307–315.

Zhou, H.-X., Rivas, G., & Minton, A. P. (2008). Macromolecular crowding and confinement: Biochemical, biophysical, and potential physiological consequences. *Annual Review of Biophysics, 37*, 375–397.

Zimmerman, S. B., & Harrison, B. (1987). Macromolecular crowding increases binding of DNA polymerase to DNA; an adaptive effect. *Proceeding of the National Academy of Science of the USA, 84*, 1871–1875.

Ziyadeh, F. N., Mills, J. W., & Kleinzeller, A. (1992). Hypotonicity and volume regulation in shark rectal gland: Role of organic osmolytes and F-actin. *American Journal of Physiology, 262*, F468–F479.

Zou, S., Chisholm, R., Tauskela, J. S., Mealing, G. A., Johnston, L. J., & Morris, C. E. (2013). Force spectroscopy measurements show that cortical neurons exposed to excitotoxic agonists stiffen before showing evidence of bleb damage. *PLoS One, 8*. –e73499.

CHAPTER THREE

Cytoskeletal Contribution to Cell Stiffness Due to Osmotic Swelling; Extending the Donnan Equilibrium

Pei-Chuan Chao*, Mettupalayam Sivaselvan* and Frederick Sachs[§,1]

*Department of Civil, Structural and Environmental Engineering, University at Buffalo, The State University of New York, Buffalo, NY, United States
[§]Department of Physiology and Biophysics, University at Buffalo, The State University of New York, Buffalo, NY, United States
[1]Corresponding Author: E-mail: Sachs@buffalo.edu

Contents

Abstract

Cell volume regulation is commonly analyzed with a model of a closed semipermeable membrane filled with impermeant mobile solutes and the Donnan Equilibrium is used to predict the hydrostatic pressure. This traditional model ignores the fact that most cells are filled with a crosslinked cytoskeleton that is elastic and can be stretched or compressed like a sponge with no obvious need to move mobile solutes. However, calculations show that under osmotic stress, the elastic energy of the cytoskeleton is far greater than the elastic energy of the membrane. Here we expand the traditional Donnan model to include the elasticity of a cytoskeleton with fixed charges and show that cell stiffening happens without a membrane.

Current Topics in Membranes, Volume 81
ISSN 1063-5823
https://doi.org/10.1016/bs.ctm.2018.07.002

1. INTRODUCTION

Volume regulation is vital for cells. Under a variety of osmolarity conditions, animal cells swell or shrink in order to adapt to their environments. Several studies show that as cells swell osmotically, the cell become stiffer (Hoffman & Crocker, 2009; Hoffmann & Simonsen, 1989; Hoffmann, 1992; Zou et al., 2013) while some studies indicate that they become softer (Guilak, Erickson, & Ting-Beall, 2002; Moeendarbary et al., 2013; Spagnoli, Beyder, Besch, & Sachs, 2008; Wang, Irianto, Kazun, Wang, & Knight, 2015). To better understand the observations, we analyzed the basic physics of a charged gel. We adopted what is known as the triphasic theory of a gel developed by Lai, Hou, and Mow (1991). This is based on poroelasticity coupled with ionic equilibrium, to obtain a relation between cell elastic modulus and swelling (volumetric strain) ignoring the membrane since most osmotic stress energy is found in the cytoskeleton (Sachs & Sivaselvan, 2015).

Poroelasticity-ionic transport theory is not often utilized in biology (Sachs & Sivaselvan, 2015), and most cell volume analyses ignore the mechanics of the cytoskeletal gel (Scholz, Phillips, Bucci, Meinzer, & Goldstein, 2011). Poroelastic effects have been used mostly for modeling animal soft tissues including articular cartilage (Gu, Lai, & Mow, 1997; Lai et al., 1991), aorta (Azeloglu, Albro, Thimmappa, Ateshian, & Costa, 2008; Simon, Kaufman, Liu, & Baldwin, 1998) and the intervertebral disc (Lai et al., 1991) as multi-phase mixtures (solid, mobile fluid and ionic phases). Only few studies have modeled single living cells using porous media theory despite their porous gel composition (Iatridis, Laible, & Krag, 2003; Moeendarbary et al., 2013; Nguyen, Oloyede, Singh, & Gu, 2016; Trickey, Baaijens, Laursen, Alexopoulos, & Guilak, 2006; Wang et al., 2015). The common model for animal cell volume regulation treats the cell as an impermeable bag of charged freely diffusing impermeant peptides and ignores the mechanics of the cytoskeleton although under osmotic stress there is much more energy in the cytoskeleton gel than in the cortex (Guilak et al., 2002). We have found that a useful model emphasizing of the role of the cytoskeleton in volume regulation is to think of why common sponges don't lyse when soaked in water. What stops the influx of water? It is the elastic stress in the polymers.

2. MODELING

In this study, following the triphasic theory of Lai et al. (1991), we consider the cell as a mixture of different phases: a solid phase (charged cytoskeletal proteins) and a fluid phase of water as a solvent containing ions as solutes. As for the membrane, the derivation by Sachs and Sivaselvan (2015) shows that, for a given equilibrium pore pressure (pressure in the space between polymers). The ratio of elastic energies in the cytoskeleton to that in the cell membrane is ~500. Most osmotic energy is stored in the cytoplasm. Modeling that ignores the three dimensional elastic cytoskeleton is misleading. In our model, we simplified the cell membrane to simply be an interface, permeable to water and ions but with some elasticity. Within the cell, we have a lattice of crosslinked elastic proteins with a negative fixed charge density (FCD).

The negative FCDs produces an imbalance in the osmolarities of the interstitial fluid and external bathing solution. The osmotic pressure induced by the imbalance is known as the Donnan osmotic pressure. The pressure introduces Donnan osmotic swelling as discussed in the literature (Ateshian, Maas, & Weiss, 2013; Gregor, 1951; Overbeek, 1956). Ateshian et al. (2013) presented a numerical example of how cell volume is affected by fixed charge density and the ion concentration of the bathing solution. Lu, Miller, Chen, Guo, & Mow (2007) also derived a set of explicit formulas to determine the mechanical properties of articular cartilage under Donnan osmotic pressure that is caused by the FCD.

Within poroelastic theory, under an isotropic stress, the total stress is given by

$$\sigma = K\varepsilon - bp \tag{1}$$

where σ, ε and p are respectively the volumetric stress, strain and the excess hydrostatic pore pressure, K is the drained bulk modulus and b is the Biot coefficient which is equal to one when the solid matrix is incompressible (Coussy, 2011). Although linear elastic behavior is assumed for the solid phase, viscoelastic or hyperelastic can be easily applied by replacing K with the desired material model.

3. EQUILIBRIUM OF CHEMICAL POTENTIAL AND OSMOTIC PRESSURE DUE TO FIXED CHARGE

In this section, we treat the equilibrium of chemical potential and osmotic pressure due to fixed charge. For ease of book-keeping, we assume all ions have valence one. The electro-neutrality condition is given by

$$\sum z_\beta c^\beta = c^+ - c^- - c^{\mathrm{F}} = 0 \tag{2}$$

where $z_+ = 1$ and $z_- = -1$. c^- is the internal anion concentration, $c^+ = c^- + c^{\mathrm{F}}$ is the cation concentration and c^{F} is FCD with $z_F = -1$. As for external ion concentration, $c^{+*} = c^{-*} = c^*$ (Donnan, 1924; Gu et al., 1997; Haider, Schugart, Setton, & Guilak, 2006).

Let the chemical potential of water and the ions per mole in the internal and external solution be denoted as $\tilde{\mu}^\alpha$ and $\tilde{\mu}^{\alpha*}$, respectively ($\alpha = w, +, -$). Under isothermal and isobaric conditions, the chemical potentials are given by

Water:

$$\tilde{\mu}^{\mathrm{w}} = \tilde{\mu}_0^{\mathrm{w}} + \frac{1}{\rho_{\mathrm{w}}}\left[p - RT\left(c^{\mathrm{total}}\right)\right] = \tilde{\mu}_0^{\mathrm{w}} + \frac{1}{\rho_{\mathrm{w}}}\left[p - RT\left(2c^- + c^{\mathrm{F}}\right)\right] \tag{3}$$

$$\tilde{\mu}^* = \tilde{\mu}_0^* + \frac{1}{\rho_{\mathrm{w}}}\left[p^* - RT\left(c^{\mathrm{total}*}\right)\right] = \tilde{\mu}_0^* + \frac{1}{\rho_{\mathrm{w}}}\left[p^* - RT(2c^*)\right] \tag{4}$$

Cation:

$$\tilde{\mu}^+ = \tilde{\mu}_0^+ + \frac{RT}{M_+}\ln\left(c^+\right) + \frac{F_c\Psi}{M_+} = \tilde{\mu}_0^+ + \frac{RT}{M_+}\ln\left(c^- + c^{\mathrm{F}}\right) + \frac{F_c\Psi}{M_+} \tag{5}$$

$$\tilde{\mu}^{+^*} = \tilde{\mu}_0^{+*} + \frac{RT}{M_+}\ln\left(c^{+*}\right) + \frac{F_c\Psi^*}{M_+} = \tilde{\mu}_0^{+*} + \frac{RT}{M_+}\ln\left(c^*\right) + \frac{F_c\Psi^*}{M_+} \tag{6}$$

Anion:

$$\tilde{\mu}^- = \tilde{\mu}_0^- + \frac{RT}{M_-}\ln\left(c^-\right) - \frac{F_c\Psi}{M_-} \tag{7}$$

$$\tilde{\mu}^{-*} = \tilde{\mu}_0^{-*} + \frac{RT}{M_-}\ln\left(c^{-*}\right) - \frac{F_c\Psi^*}{M_-} = \tilde{\mu}_0^{-*} + \frac{RT}{M_-}\ln\left(c^*\right) - \frac{F_c\Psi^*}{M_-} \tag{8}$$

where $\tilde{\mu}_0^\alpha$ and $\tilde{\mu}_0^{\alpha*}$ are concentration-independent potentials for the inside and outside of the cell membrane (which are equal). ρ_{w} is the density of water. R is the universal gas constant and T is the absolute temperature. $M_{+,-}$ are molecular weights for the cation and anion. F_c is Faraday's constant and ψ is the electric potential (Haider et al., 2006; Katchalsky & Curran, 1967).

Since the cell membrane is assumed to be permeable to ions and water, the following constraints apply

$$\tilde{\mu}^{\mathrm{w}} = \tilde{\mu}^{*} \tag{9}$$

$$\tilde{\mu}^{+} = \tilde{\mu}^{+*} \tag{10}$$

$$\tilde{\mu}^{-} = \tilde{\mu}^{-*} \tag{11}$$

From Eqs (5–8) and (10–11), we can obtain the relation of the electric potential difference which is negligible:

$$\psi - \psi^{*} = \frac{\mathrm{RT}}{F_c} \ln\left(\frac{c^{-} + c^{\mathrm{F}}}{c^{*}}\right) \tag{12}$$

and ion distribution:

$$\frac{c^{*2}}{c^{-}\left(c^{-} + c^{F}\right)} = 1 \tag{13}$$

or

$$\left(c^{-}\right)^{2} + c^{\mathrm{F}} c^{-} - c^{*2} = 0 \tag{14}$$

The positive solution of Eq. (14) is

$$2c^{-} = -c^{\mathrm{F}} + \sqrt{(2c^{*})^{2} + (c^{\mathrm{F}})^{2}} \tag{15}$$

From Eqs (3), (4) and (9), the fluid pressure can be written as

$$p = RT\left[\left(2c^{-} + c^{\mathrm{F}}\right) - 2c^{*}\right] \tag{16}$$

With Eq. (15), the pressure due to the excess ions becomes

$$p = RT\left[\sqrt{(2c^{*})^{2} + (c^{\mathrm{F}})^{2}} - 2c^{*}\right] \tag{17}$$

4. MECHANICAL EQUILIBRIUM

Fixed charge conservation implies

$$\phi V c^{\mathrm{F}} = \phi_0 V_0 c_0^{F} \tag{18}$$

where ϕ is the porosity of the porous medium and V is the volume of the cell. The subscript zero represents the reference state. Since the skeleton material is assumed to be incompressible, its volume remains constant; in other words

$$(1-\phi_0)V_0 = (1-\phi)V \tag{19}$$

For small deformations $V \simeq (1+\varepsilon)V_0$. Therefore Eq. (19) yields

$$\phi = 1 - \frac{1-\phi_0}{1+\varepsilon} \tag{20}$$

By substituting Eq. (20) into (18), c^{F} can now be presented as a function of ε,

$$c^{\mathrm{F}} = \frac{\phi_0 c_0^{\mathrm{F}}}{\phi_0+\varepsilon} \tag{21}$$

At equilibrium, the osmotic pressure equals the excess pore pressure. By substituting Eq. (21) into (17), we get the pressure as

$$p = RT\left[\sqrt{(2c^*)^2 + \left(\frac{\phi_0 c_0^{\mathrm{F}}}{\phi_0+\varepsilon}\right)^2} - 2c^*\right] \tag{22}$$

The pressure known as the Donnan osmotic pressure is produced by the charge-induced imbalance (Overbeek, 1956). Eq. (1) now can be rewritten as

$$\sigma = K\varepsilon - RT\left[\sqrt{(2c^*)^2 + \left(\frac{\phi_0 c_0^{\mathrm{F}}}{\phi_0+\varepsilon}\right)^2} - 2c^*\right] \tag{23}$$

Now, we derive the effective bulk modulus by differentiating Eq. (23) with respect to ε,

$$K_{eff} = \frac{d\sigma}{d\varepsilon} = K + K_{osmosis} \tag{24}$$

where

$$K_{osmosis} = \frac{RT\left(\phi_0 c_0^{\mathrm{F}}\right)^2}{(\phi_0+\varepsilon)^3}\left[(2c^*)^2 + \left(\frac{\phi_0 c_0^{\mathrm{F}}}{\phi_0+\varepsilon}\right)^2\right]^{-\frac{1}{2}} \tag{25}$$

If no external force is applied ($\sigma = 0$), Eq. (23) can be solved for ε in terms of c^*, it is rewritten in quartic polynomial form

$$\mathrm{A}\varepsilon^4 + \mathrm{B}\varepsilon^3 + \mathrm{C}\varepsilon^2 + \mathrm{D}\varepsilon + \mathrm{E} = 0 \tag{26}$$

where

$$\begin{aligned} \mathrm{A} &= K^2 \\ \mathrm{B} &= 2K^2\phi_0 + 4RTKc^* \\ \mathrm{C} &= K^2\phi_0^2 + 8RTK\phi_0 c^* \\ \mathrm{D} &= 4RTK\phi_0^2 c^* \\ \mathrm{E} &= -\left(RT\phi_o c_0^{\mathrm{F}}\right)^2 \end{aligned} \tag{27}$$

Eq. (26) will always have exactly one sign change through the five coefficients, hence by Descartes' Rule of Signs, there is only one positive real root for the polynomial. By solving Eq. (26) for varying c^*, corresponding ε can be substituted into Eq. (25) and eventually obtain $K_{osmosis}$.

A similar equation to estimate the strain is shown in Azeloglu et al. (2008) and Ateshian et al. (2013). The difference is that we assume the cell as linear elastic material rather than neo-Hookean solid for now. As mentioned, this can be replaced easily in a future study.

5. NUMERICAL RESULTS

In this section, numerical analysis results show the relation of the Young's modulus contributed by osmotic pressure and the external ion concentration, c^*, and hence the volumetric strain, ε. We try to reproduce the mechanical conditions of a homogeneous cell during an indentation test (Spagnoli et al., 2008). The cell is placed in a resting hypotonic environment causing the cell to swell and it is then indented by a probe to estimate its stiffness.

Input parameters for the numerical analysis are in Table 1. In the analysis c^* ranged from 100 to 400 mOsm. Different c_0^{F} show the sensitivity to different input properties. The Young's modulus is obtained from the calculated bulk modulus K with an assumed Poisson's ratio $\upsilon = 0.4$ (Trickey et al., 2006). Fig. 1A shows the volumetric strain increasing with decreasing bathing concentration, c^*, with different initial fixed charge density, c_0^F. Note that ε decreases as c^* increases but never reaches zero. Due to the negative fixed charge in the cell, the Donnan osmotic pressure (22) will always be positive resulting in cell swelling (Ateshian et al., 2013).

Table 1 Input Parameters

Universal gas constant, $R\left(\frac{m^3 \cdot kPa}{K \cdot mol}\right)$	8.314×10^{-3}
Temperature, $T(K)$	294
Poisson's ratio, υ	0.4
Cytoskeleton bulk modulus, K(kPa)	1
Reference porosity, φ_0	0.7
Reference fixed charge density, $c_0^F\left(\frac{mol}{m^3}\right)$	20, 25, 30, 35

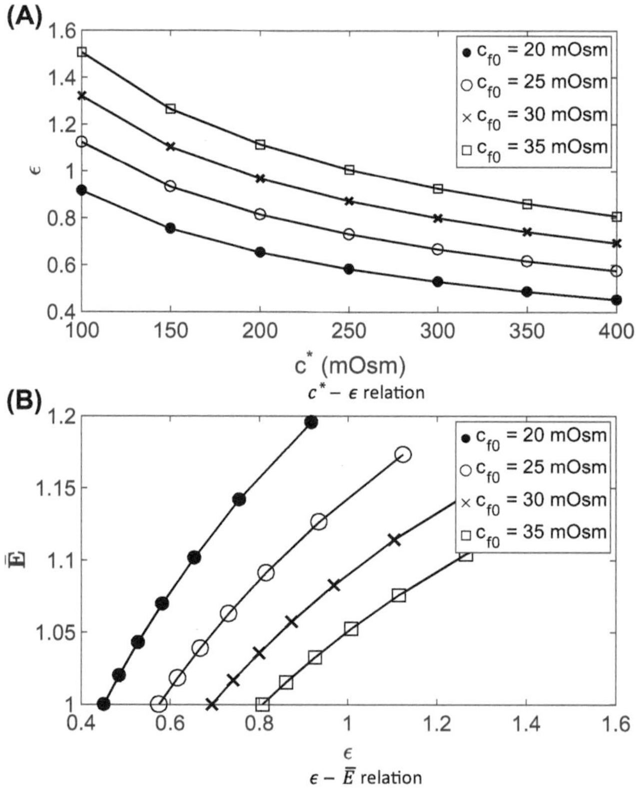

Figure 1 Results of the numerical example. (A) The volumetric strain, ε, versus corresponding bathing concentration, c^*, with different initial fixed charge density, c_0^F. (B) The normalized Young's modulus, $\overline{E}$, of the cell under the corresponding ε with different c_0^F. $\overline{E}$ increases as volumetric strain ε became larger indicates that cell stiffened while swelling under Donnan osmotic pressure.

The $\varepsilon - \overline{E}$ curves are shown in Fig. 1B where $\overline{E}$ is the normalized Young's modulus. All the moduli are normalized by the modulus at $c^* = 400$ mOsm. Although results vary with the parameters, $\overline{E}$ increases up to 20 percent while ε increasing; the cell stiffens with osmotic swelling, and that stiffening does not require a contribution from the cell membrane!

6. SUMMARY

In this study, we constructed a model combining poroelastic and ionic equilibria capable of predicting the changes in cell stiffness due to Donnan osmotic swelling. First, we obtain the Donnan osmotic pressure in Eq. (17) from the equilibrium of chemical potentials, which is substituted into Eq. (1) to calculate the total stress. The bulk modulus under the osmotic pressure is derived by first order differentiation with respect to the volumetric strain.

In numerical examples, we calculated the normalized Young's modulus with bathing concentrations, c^* , ranging from 100 to 400 mOsm as shown in Fig. 1b. The result shows that the cells are stiffened while under hypotonic osmotic challenge. There is no mechanical input from the cell membrane. This observation is consistent with many results from the biological literature that indicate cell stiffening with swelling (Hoffmann, 1992; Hoffman & Crocker, 2009; Hoffmann & Simonsen, 1989; Zou et al., 2013). The derivation above shows that the cell stiffness is dominated by the bulk osmotic pressure of the cytoskeleton, not the membrane (Lu et al., 2007). The input properties for the model in this article are values commonly used in the literature. Although more experimental data are required to validate the hypothesis, the analytic model shows that within certain range of cell properties, the cell stiffness change due to osmotic swelling can be predicted.

7. DISCUSSION

The model shows that basic biophysics predicts that with swelling the mean stiffness of cells will increase and independent of the cell membrane. In practice, cells are mechanically inhomogeneous in space and time so the analysis of "mean cell stiffness" versus external osmotic pressure has many implicit assumptions (Meng, 2008; Meng, Suchyna, & Sachs, 2008).

Referring to a "membrane" as a defined object in mechanics is misleading. Membrane could mean the bilayer, cortical cytoskeleton, cortex, extracellular matrix and the coupling of the cortex to the deeper matrix. Using AFM to measure the forces to indent cells does not relieve the complexity of interpreting the data. Aside from ignoring the potential nonlinearities of the elastic properties of all the contributing components and time dependent changes in material properties, there are measuring artifacts since the cells are thin and do not satisfy the assumptions of Hertzian indentation models (Johnson, 1987). That theory assumes an infinitely thick homogeneous media and that is a bad representation of a cell. As a cell is indented, the bulk strain reaches the substrate (glass or plastic), that is effectively infinitely stiff, and this increases the mean stiffness. This stiffness varies inversely with the fourth power of the thickness (Johnson, 1987). A recent detailed analysis of the substrate effect has been presented by Garcia and Garcia (2018). Cell swelling moves the measuring point (AFM tip) away from the substrate and reduces the contribution of substrate stiffness tending to make the cell look softer.

Steltenkamp, Rommel, Wegener, & Janshoff (2006) measured cell stiffness as a function of osmotic pressure using AFM indentation. They did not address the issue of heterogeneity nor the contribution of substrate stiffness. The model of the cell they used is confusing as they use the terms "cytosol "and "cytoskeleton" as though they were readily separable and distinguishable (Charras, Coughlin, Mitchison, & Mahadevan, 2008; Charras, Mitchison, & Mahadevan, 2009). They refer to the stiffness of the cytoskeleton as cortical actin based whereas later published data shows that actin stress is quite uniformly distributed throughout the cell during anisotonic conditions and is not concentrated in the cortex. Furthermore, the actin tension is time dependent (Guo, Wang, Sachs, & Meng, 2014). The fact that the physical chemistry of a gel predicts the increase in stiffness is not considered.

Zou et al. (2013) examined the effect of osmotic pressure and cation fluxes on neural cell stiffness using an AFM. Much of the data interpretation depended upon a model with where hydrostatic forces acted on a two dimensional membrane and not a three dimensional gel (Sachs & Sivaselvan, 2015). A cation influx through ion channels softened cells and the softening was ascribed to Ca^{+2} activation of proteases that cleaved the cortical cytoskeleton. The influence of charge movement on cell volume (assuming the osmotic stress is in the cortex) is tricky since only a flux of neutral species can significantly alter osmotic strength. A cation flux must be coupled to an anion flux in such a way as to create an osmotic gradient from a net neutral flux.

Ayee, LeMaster, Teng, Lee, and Levitan (2018) did AFM measurements on the effect of osmotic pressure and many of the issues discussed above apply. They tried to separate cortical stresses from deeper stresses using the AFM to assess mean cortical tension and pulling bilayer tethers to assess bilayer tension. They often refer to "the membrane" as the bilayer or as the cortex. As common in cell volume studies, the critical stresses are assigned to the "membrane" (Akinlaja & Sachs, 1998) and ignore the stresses in the three dimensional gel of the cytoplasm (Sachs & Sivaselvan, 2015). The authors assume that by varying the indention force and distance they can separate cortical stress from deeper cytoskeletal stress as though they were mechanically distinct rather than continuous. It is possible that cytoskeletal inhomogeneities allow bilayer blebs to form and act as two dimensional structures, where hydrostatic pressure will make them very stiff (Spagnoli et al., 2008). Again, the presence of heterogeneity of compliance in cells (Guo et al., 2014; Meng & Sachs 2011a; Ye et al, 2014) compromises the interpretation. They found that with actin inhibitors, hypotonic stimuli make the "membrane" stiffer which would fit with the AFM measurement sensing blebs or the cytoskeleton itself (Charras et al., 2008; Spagnoli et al., 2008). The bilayer tether measurements showed no increase in tension with hypotonic swelling suggesting that the osmotic stress is not in the bilayer. The physical models show that swelling of a three dimensional "sponge-like" cytoskeleton, involves much more energy than the surface of the sponge (Sachs & Sivaselvan, 2015) and dominates the response.

Spagnoli et al. (2008) used the AFM to measure cell swelling and stiffness in response to hypotonic stress and remarkably found that cells got softer with swelling. The physical models predict that the cells should get stiffer so that some assumptions of the analysis must be in error. One explanation is the effect of distance from the substrate where swelling moves the indenting tip further from the substrate making the cell appear softer. Another is that the influx of ions, notably calcium, activates proteases decreasing the stiffness of the cytoskeletal gel.

If we approach volume regulation from the point of a design engineer, it might clarify the discussion. Consider a vascular epithelial cell. On the basal side, it is glued to the wall of the vessel. Blood flow over the apical side will try to drag the cell downstream. Since the cell does not wash away, there are normal stresses between the apical and basal sides that are connected by the cytoskeleton. That same internal linking structure will be stressed by changes in volume, allowing the cell to resist gradients of hydrostatic pressure from osmotic sources as well as fluid shear stress. We have exposed many cell types to distilled water and avoided lysis despite a putative 7 Atm gradient (Guo et al., 2014; Meng & Sachs, 2011b, 2012; Verma & et al., 2012).

Another example of the essential role of cytoplasmic elasticity is the freshwater animals. The low osmotic pressure environment imposes drastic gradients on cells that are filled with 300mOsm solutes. How do you maintain equilibrium? You could use powerful ion pumps extracting rare ions from the water, but why bother? Just build the cell with intracellular mechanical reinforcement so that it is in mechanical equilibrium with its normal environment. You can then use some pumps and channels to tune this operating point. A more familiar example is to consider what happens when we drink distilled water. What do our epithelia do with this 7 Atm gradient? A good engineer would just reinforce the cells to withstand that pressure without the need for pumps.

When plant cells, that have a cell wall, mutated to create more mobile animal cells, the rigid wall needed to be removed. However, simple removal would expose the cell to osmotic lysis. Nature, with good engineering, reinforced the cell with a dynamic *interior* skeleton to resist the osmotic pressure in an energy-conserving manner. Since the osmotic stress in the cytoskeleton is the dominant energy term in an osmotic challenge (Sachs & Sivaselvan, 2015), analysis of experiments that do not include cytoskeletal effects will be misleading.

REFERENCES

Akinlaja, J., & Sachs, F. (1998). The breakdown of cell membranes by electrical and mechanical stress. *Biophysical Journal, 75*(1), 247–254.

Ateshian, G. A., Maas, S., & Weiss, J. A. (2013). Multiphasic finite element framework for modeling hydrated mixtures with multiple neutral and charged solutes. *Journal of Biomechanical Engineering, 135*(11), 111001.

Ayee, M. A. A., LeMaster, E., Teng, T., Lee, J., & Levitan, I. (2018). Hypotonic challenge of endothelial cells increases membrane stiffness with No effect on tether force. *Biophysical Journal, 114*(4), 929–938.

Azeloglu, E. U., Albro, M. B., Thimmappa, V. A., Ateshian, G. A., & Costa, K. D. (2008). Heterogeneous transmural proteoglycan distribution provides a mechanism for regulating residual stresses in the aorta. *American Journal of Physiology - heart and Circulatory Physiology, 294*(3), H1197–H1205.

Charras, G. T., Coughlin, M., Mitchison, T. J., & Mahadevan, L. (2008). Life and times of a cellular bleb. *Biophysical Journal, 94*(5), 1836–1853.

Charras, G. T., Mitchison, T. J., & Mahadevan, L. (2009). Animal cell hydraulics. *Journal of Cell Science, 122*(Pt 18), 3233–3241.

Coussy, O. (2011). *Mechanics and physics of porous solids*. John Wiley & Sons.

Donnan, F. G. (1924). The theory of membrane equilibria. *Chemical Reviews, 1*(1), 73–90.

Garcia, P. D., & Garcia, R. (2018). Determination of the elastic moduli of a single cell cultured on a rigid support by force microscopy. *Biophysical Journal, 114*(12), 2923–2932.

Gregor, H. P. (1951). Gibbs-Donnan equilibria in ion exchange resin systems. *Journal of the American Chemical Society, 73*(2), 642–650.

Guilak, F., Erickson, G. R., & Ting-Beall, H. P. (2002). The effects of osmotic stress on the viscoelastic and physical properties of articular chondrocytes. *Biophysical Journal, 82*(2), 720–727.

Gu, W. Y., Lai, W. M., & Mow, V. C. (1997). A triphasic analysis of negative osmotic flows through charged hydrated soft tissues. *Journal of Biomechanics, 30*(1), 71–78.

Guo, J., Wang, Y., Sachs, F., & Meng, F. (2014). Actin stress in cell reprogramming. *Proc Natl Acad Sci U S A, 111*(49), E5252–E5261.

Haider, M. A., Schugart, R. C., Setton, L. A., & Guilak, F. (2006). A mechano-chemical model for the passive swelling response of an isolated chondron under osmotic loading. *Biomechanics and Modeling in Mechanobiology, 5*(2–3), 160.

Hoffmann, E. K. (1992). Cell swelling and volume regulation. *Canadian Journal of Physiology and Pharmacology, 70*(S1), S310–S313.

Hoffman, B. D., & Crocker, J. C. (2009). Cell mechanics: Dissecting the physical responses of cells to force. *Annual Review of Biomedical Engineering, 11*, 259–288.

Hoffmann, E. K., & Simonsen, L. O. (1989). Membrane mechanisms in volume and pH regulation in vertebrate cells. *Physiological Reviews, 69*(2), 315–382.

Iatridis, J. C., Laible, J. P., & Krag, M. H. (2003). Influence of fixed charge density magnitude and distribution on the intervertebral disc: Applications of a poroelastic and chemical electric (PEACE) model. *Journal of Biomechanical Engineering, 125*(1), 12–24.

Johnson, K. L. (1987). *Contact mechanics*. Cambridge university press.

Katchalsky, A., & Curran, P. F. (1967). *Thermodynamics in biophysics*. Cambridge: Harvard University Press. Chaudhuri, O., Parekh, S. H., & Fletcher, D. A. (2007). Reversible stress softening of actin networks. *Nature, 445*(7125), 295–298.

Lai, W. M., Hou, J. S., & Mow, V. C. (1991). A triphasic theory for the swelling and deformation behaviors of articular cartilage. *Journal of Biomechanical Engineering, 113*(3), 245–258.

Lu, X. L., Miller, C., Chen, F. H., Guo, X. E., & Mow, V. C. (2007). The generalized triphasic correspondence principle for simultaneous determination of the mechanical properties and proteoglycan content of articular cartilage by indentation. *Journal of Biomechanics, 40*(11), 2434–2441.

Meng, F. (2008). *stFRET, a novel tool to study molecular force in living cells and animals*. Ph.D. Ann Arbor: State University of New York at Buffalo

Meng, F., & Sachs, F. (2011a). Measuring strain of structural proteins in vivo in real time. In P. Kohl, F. Sachs, & M. R. Franz (Eds.), *Cardiac mechano-electric coupling and Arrhythmia: From pipette to patient* (Vol. 2, pp. 431–434). Oxford University Press, Oxford.

Meng, F., & Sachs, F. (2011b). Visualizing dynamic cytoplasmic forces with a compliance-matched FRET sensor. *Journal of Cell Science, 124*(Pt 2), 261–269.

Meng, F., & Sachs, F. (2012). Orientation-based FRET sensor for real-time imaging of cellular forces. *Journal of Cell Science, 125*(Pt 3), 743–750.

Meng, F., Suchyna, T. M., & Sachs, F. (2008). A fluorescence energy transfer-based mechanical stress sensor for specific proteins in situ. *The FEBS Journal, 275*(12), 3072–3087.

Moeendarbary, E., Valon, L., Fritzsche, M., Harris, A. R., Moulding, D. A., Thrasher, A. J., Stride, E., Mahadevan, L., & Charras, G. T. (2013). The cytoplasm of living cells behaves as a poroelastic material. *Nature Materials, 12*(3), 253–261.

Nguyen, T. D., Oloyede, A., Singh, S., & Gu, Y. (2016). Investigation of the effects of extracellular osmotic pressure on morphology and mechanical properties of individual chondrocyte. *Cell Biochemistry and Biophysics, 74*(2), 229–240.

Overbeek, J. T. (1956). The donnan equilibrium. *Progress in Biophysics and Biophysical Chemistry, 6*(1), 57–84.

Sachs, F., & Sivaselvan, M. V. (2015). Cell volume control in three dimensions: Water movement without solute movement. *The Journal of General Physiology, 145*(5), 373–380.

Scholz, F. G., Phillips, N. G., Bucci, S. J., Meinzer, F. C., & Goldstein, G. (2011). Hydraulic capacitance: Biophysics and functional significance of internal water sources in relation to tree size. In *Size-and age-related changes in tree structure and function* (pp. 341–361). Springer.

Simon, B. R., Kaufman, M. V., Liu, J., & Baldwin, A. L. (1998). Porohyperelastic–transport–swelling theory, material properties and finite element models for large arteries. *International Journal of Solids and Structures, 35*(34–35), 5021–5031.

Spagnoli, C., Beyder, A., Besch, S., & Sachs, F. (2008). Atomic force microscopy analysis of cell volume regulation. *Physical Review E, 78*(3), 031916.

Steltenkamp, S., Rommel, C., Wegener, J., & Janshoff, A. (2006). Membrane stiffness of animal cells challenged by osmotic stress. *Small, 2*(8–9), 1016–1020.

Trickey, W. R., Baaijens, F. P., Laursen, T. A., Alexopoulos, L. G., & Guilak, F. (2006). Determination of the Poisson's ratio of the cell: Recovery properties of chondrocytes after release from complete micropipette aspiration. *Journal of Biomechanics, 39*(1), 78–87.

Verma, D., Ye, N., Meng, F., Sachs, F., Rahimzadeh, J., & Hua, S. Z. (2012). Interplay between cytoskeletal stresses and cell adaptation under chronic flow. *PLoS One, 7*(9), e44167.

Wang, Z., Irianto, J., Kazun, S., Wang, W., & Knight, M. M. (2015). The rate of hypo-osmotic challenge influences regulatory volume decrease (RVD) and mechanical properties of articular chondrocytes. *Osteoarthritis and Cartilage, 23*(2), 289–299.

Ye, N., Verma, D., Meng, F., Davidson, M. W., Suffoletto, K., & Hua, S. Z. (2014). Direct observation of alpha-actinin tension and recruitment at focal adhesions during contact growth. *Experimental Cell Research, 327*(1), 57–67.

Zou, S., Chisholm, R., Tauskela, J. S., Mealing, G. A., Johnston, L. J., & Morris, C. E. (2013). Force spectroscopy measurements show that cortical neurons exposed to excitotoxic agonists stiffen before showing evidence of bleb damage. *PLoS One, 8*(8), e73499.

CHAPTER FOUR

Membrane Stiffening in Osmotic Swelling: Analysis of Membrane Tension and Elastic Modulus

Manuela A.A. Ayee and Irena Levitan[1]
University of Illinois at Chicago, Chicago, IL, United States
[1]Corresponding author: E-mail: levitan@uic.edu

Contents

Abstract

The effects of osmotic swelling on key cellular biomechanical properties are explored in this chapter. We present the governing equations and theoretical backgrounds of the models employed to estimate cell membrane tension and elastic moduli from experimental methods, and provide a summary of the prevailing experimental approaches used to obtain these biomechanical parameters. A detailed analysis of the current evidence of the effects of osmotic swelling on membrane tension and elastic moduli is provided. Briefly, due to the buffering effect of unfolding membrane

Current Topics in Membranes, Volume 81
ISSN 1063-5823
https://doi.org/10.1016/bs.ctm.2018.07.003

reservoirs, mild hypotonic swelling does not change membrane tension or the adhesion of the membrane to the underlying cytoskeleton. Conversely, osmotic swelling causes the cell membrane envelope to stiffen, measured as an increase in the membrane elastic modulus.

1. INTRODUCTION

External mechanical forces and the biomechanical properties of extracellular environments are well recognized today to play major roles in numerous cell functions including proliferation, differentiation, cell—cell interactions and more. Multiple studies showed that cells respond to stretch, shear stress and stiffness of the extracellular matrix by activating mechano-sensitive signaling cascades which couple mechanical signals to cellular responses. Earlier studies traditionally assumed that since cells visibly swell when exposed to hypotonic solution, membrane stretch and increase in membrane tension should play the central role in the ability of cells to respond to hypotonic swelling and induce the regulatory volume decrease (RVD), a process responsible for cell volume homeostasis in hypotonic conditions. The general belief that cell swelling should result in membrane stretch and an increase in membrane tension was also used in multiple studies to determine whether a specific process is stretch-sensitive. However, several considerations led to questioning the assumption that cell swelling necessarily leads to an increase in membrane tension. Instead, accumulating evidence, including studies from our lab, suggest that osmotic swelling leads to stiffening of the cellular envelope with little or no effect on membrane tension. In this chapter, we will review and discuss the current state of knowledge about whether and how osmotic challenge affects membrane tension and elasticity.

1.1 Basic Considerations: Low Extensibility and Limits of Membrane Stretch before Lysis

Mechanically, membrane deformations can be defined by several different parameters: (1) an elastic modulus or an area expansion modulus that corresponds to membrane expansion without shearing or bending (2) a bending modulus corresponding to membrane bending without shearing or expansion; and (3) a shear modulus that corresponds to an elongation of the membrane without changing surface area or bending. For all three parameters, the greater the value of the modulus, the more resistant the membrane is to this particular form of deformation (Hochmuth & Waugh, 1987). Biological membranes were found to have high elastic moduli, indicating resistance to area expansion, low shear moduli, corresponding to the fluid nature of the membrane bilayer, and highly variable bending moduli, which depends on membrane composition and cytoskeleton attachment (Diz-Muñoz, Fletcher, & Weiner, 2013). These parameters suggest that cells have limited ability to increase membrane surface area via stretch in order to accommodate an increase in cell volume during cell swelling.

Consistent with these biomechanical considerations, early studies of the physical properties of the membranes of red blood cells established that these membranes are highly resistant to stretch (Evans & Fung, 1972; Evans & Leblond, 1973; Evans, Waugh, & Melnik, 1976). These studies showed that osmotic swelling of red blood cells resulted in little (up to 7%) or no detectable increase in the surface area. Furthermore, applying well-defined negative pressure to pre-swollen red blood cells, Evans et al. (1976) determined that the maximal area expansion that these cells can achieve before they undergo cell lysis ranges between 2 and 4% of the initial cell area in red blood cells that apparently do not have significant membrane reservoirs to unfold. Since cell swelling may result in a very significant increase in cell volume, the inability of the cell membrane to stretch indicates that swelling should be accompanied by recruiting membrane reservoirs, either from natural membrane unfolding or from fusion of intracellular vesicles (Clark, Wartlick, Salbreux, & Paluch, 2014; Dai & Sheetz, 1999; Sens & Turner, 2006).

1.2 Membrane Unfolding during Cell Swelling

Numerous studies showed that plasma membranes of mammalian cells are highly folded and include different types of invaginations, such as caveolae, clathrin-coated pits, and protrusions, such as microvilli (Vlahakis & Hubmayr, 2000). It is reasonable to assume, therefore, that cell swelling through water intake from the extracellular space during osmotic challenge should result in membrane unfolding, which would allow a significant increase in cell volume as water enters the cell, without stretching and lysing the plasma membrane, and without changing the total membrane surface area. Conceptually, this is a protective mechanism to accommodate acute cell swelling prior to the initiation of the regulatory volume decrease (RVD), a process of cell volume recovery. However, while membrane folds are clearly detected by scanning and TEM microscopy, it is difficult to quantify the unfolding process using microscopy techniques.

In our earlier study, we approached this question, therefore, by comparative analysis of an apparent surface area determined visually or electrically based on measuring cell capacitance (Fig. 1) (Levitan & Garber, 1997, pp. 245–267). The experiment was performed with human myeloma cells whose spherical shape allowed estimating their "visual" surface area by measuring their diameters. The capacitance based estimate of the area is based on the generally accepted assumption that membrane capacitance is directly proportional to the total membrane area and can be calculated using a typical specific capacitance of biological membranes of 1 $\mu F/cm^2$ (Hille, 2001). It is also important to note that membrane area estimated from capacitance is expected to include all membrane folds unless they are electrically uncoupled from the membrane. Therefore, if membrane surface area increases because of the fusion of intracellular vesicles, membrane capacitance increases but if the membrane simply unfolds, capacitance should remain constant. Also, if significant stretching of the membrane were possible without lysis, this would be expected to increase cell capacitance as well. We observed that, as expected, exposing the cells to a hypotonic challenge resulted in a significant increase in the apparent "visual" surface area from below 500 μm^2 to ~1500 ± 290 μm^2, shown

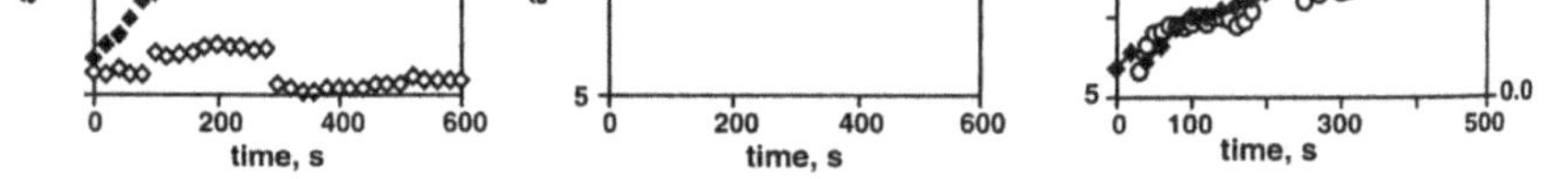

Figure 1 Cell membrane unfolding accompanies volume regulated ion channel (VRAC) activation. (A) The change in average cell surface area over time determined by visual measurement of cell radius using light microscopy with phase optics and a video camera. Data was analyzed using NIHImage software. Measurements made in isotonic solutions are shown with open symbols and those in 75% hypotonic solutions with solid symbols. (B) Average cell capacitance does not change in hypotonically-challenged (filled) or unchallenged (open) cells. For Panels A and B, each point represents n $\geq$ 3. (C) Simultaneous measurement of VRAC current amplitude (open circles) and surface area as determined from visual (solid diamonds) and capacitance (solid squares) measurements is shown for a representative cell in a hypotonic solution. Activation of VRAC correlates with an increase in surface area determined using visual but not electrical, measurement of cell surface area.

in Fig. 1A, corresponding to an increase in the cell diameter from ~30 to ~45 μm. However, the surface area of the cells estimated by measuring cell capacitance did not change during the swelling process (Fig. 1B and C). Furthermore, the "electrical" surface area measured prior to the osmotic challenge was $1590 \pm 70\ \mu m^2$, more than two fold higher than the "visual" area of the same cells. As the swelling developed, the two measurements converged, as the "visual" area increase and the "electrical" area remained constant. These observations, therefore, indicate that visual cell swelling of myeloma cells is a result of membrane unfolding.

Consistent with these observations, we also observed no change in membrane capacitance during cell swelling in endothelial cells suggesting that these cells swell by unfolding membrane invaginations as well (Levitan, Christian, Tulenko, & Rothblat, 2000). These observations are also in agreement with earlier studies showing that while cell swelling in T-lymphocytes is accompanied by an increase in cell capacitance, presumably from the recruitment of intracellular membranes, there is no correlation between the time-courses of cell swelling and capacitance increase, indicating that swelling should involve membrane unfolding (Ross, Garber, & Cahalan, 1994). Similarly, another earlier study by Groulx, Boudreault, Orlov, and Grygorczyk (2006) used a 3D topography imaging technique to determine changes in cell volume and surface area for four mammalian cells types under hypotonic conditions. They found that cells could increase their volume about two fold by drawing on excess membrane from surface reservoirs. Moreover, the volume increased more than 10 fold and the surface area increased 3.6 times when intracellular membrane reserves were recruited, under extreme hypotonic conditions. Taken together, these studies indicate that cell swelling does not necessarily result in an increase in membrane tension because the process of membrane unfolding may accommodate significant cell swelling without any increase in tension. In the next part of this chapter, we will discuss the theoretical basis for estimating membrane tension and the experimental measurements of membrane tension during swelling (*Part II. Membrane Tension*).

Another important biomechanical parameter is the elastic modulus, which gives an indication of the stiffness or deformability of the material. Conceptually, it is important to discriminate between elastic moduli/deformability of the cellular membrane bilayer, the underlying sub-membrane cytoskeleton and the bulk cellular cytoskeleton. It is well-known that the thin lipid bilayer is much more deformable than the underlying cytoskeleton and as such does not contribute significantly to the stiffness of the cells. However, it may still play a significant role in the process of membrane unfolding described above, particularly if the unfolding is accompanied with detachment between the membrane and the sub-membrane cytoskeleton. If membrane unfolding occurs with the sub-membrane cytoskeleton still attached, the elastic properties of the membrane-bilayer/sub-membrane cytoskeleton complex, which together can be termed "cellular envelope," are dominated by the cytoskeleton component. However, different cytoskeleton elements may also have very distinct elastic properties. Clearly, cellular biomechanical properties are highly heterologous and the values of the elastic moduli estimated experimentally may strongly depend on how exactly the deformability was measured. It is not surprising, therefore, that earlier studies produced controversial results for changes in membrane stiffness during osmotic swelling: The study of Zou et al. (2013) showed that cortical neurons undergo transient stiffening upon osmotic swelling, whereas Spagnoli, Beyder, Besch, and Sachs (2008) found significant softening in several cell types. In our recent study, we show that both significant stiffening and mild softening can be observed in the same cells at the same time depending on the depth of the mechanical perturbation used to estimate the stiffness (Ayee, LeMaster, Teng, Lee, & Levitan, 2018). In *Part III* of this chapter, we will discuss the theoretical considerations used to estimate cellular elastic moduli, experimental values for cell stiffening or softening obtained in different studies and their interpretations in terms of the underlying mechanisms (*Part III. Elastic Modulus*).

2. MEMBRANE TENSION

2.1 Theoretical Basis and Experimental Approaches of Measuring Membrane Tension

In the complex environment of a cell membrane, measuring membrane tension is not straightforward and frequently there is a lack of clarity about the interpretations of the measurements. In the next part of this chapter we will discuss the theoretical basis and experimental approaches that are currently used to estimate membrane tension in cell membranes and then discuss the current evidence to determine whether osmotic swelling results in an increase in membrane tension in different cell types and conditions.

The current experimental approaches to estimate membrane tension are based on changes in the free energy of the bilayer upon mechanical perturbation. The lipid bilayer is composed of two monolayers of amphiphilic phospholipid molecules forming a quasi-two-dimensional fluid structure. The thickness of this structure is on the order of a few nanometers (Edidin, 2003), while the lateral size can be several micrometers wide. The mechanical properties of this membrane structure make it highly resistant to lateral stretching or compression, while being very flexible (Lipowsky & Sackmann, 1995; Safran, 2003). These mechanical properties can be estimated by applying the principles of differential geometry and statistical mechanics to represent the lipid bilayer as a mathematical surface (Derényi et al., 2007; Farago & Pincus, 2004; Vaidya, Huang, & Takagi, 2008). Therefore, changes in membrane shape can be represented by associating the free energy of the bilayer with shape deformations caused by bending, surface tension, and changes in pressure. The-free energy of the lipid bilayer describes its energetic stability, with lower free energy values representing more favorable conformations (Canham, 1970). In many experimental studies of cell membrane biomechanics, simplified equations are used to estimate the elastic parameters of the membranes, often with the assumptions underlying the use of particular equations not explicitly stated. In this section, we will discuss the equation for determining membrane tension by beginning with the Helfrich expression for bilayer free energy (Helfrich, 1973). Furthermore, since, as discussed in detail below, membrane tension is estimated experimentally by measuring the force required to pull a membrane tether (nanotube), we discuss how the energy expression is simplified to describe the membrane tension obtained from a tether. We will then discuss the effects of the cytoskeleton on the tension measurements.

Specifically, the estimation of membrane tension is based on the equation that describes the free energy of a bilayer, the Helfrich Equation of Generalized Shape Energy (Bukman, Yao, & Wortis, 1996; Derényi et al., 2002, 2007; Deuling & Helfrich, 1976; Helfrich, 1973; Vaidya et al., 2008). The essential components that constitute the free energy of the bilayer membrane are: (1) surface tension, (2) pressure (3) elastic bending modulus and the spontaneous curvature of the membrane, (4) elastic modulus determining the directionality of bending, called the saddle-splay modulus, and (5) the point force that pulls a membrane tether away from the membrane. The values of these parameters depend on the composition and area density of lipids constituting the membrane and may be positive or negative as shown below in Eq. (1).

$$E_{Helfrich} = [\text{membrane tension}] + [\text{pressure}] + [\text{bending}] + [\text{saddle} - \text{splay}] + [\text{tether}]$$

The various terms in the free energy equation can be defined as follows:

1. The membrane surface tension is defined as the force per unit length acting on a cross-section of membrane (Clark et al., 2014) and is obtained from: $\sigma \int dA$, where σ is the tensile stress (surface tension) and dA a surface area element.
2. The pressure term represents the pressure difference across the membrane (Bukman et al., 1996): $\Delta P \int dV$, where ΔP is the pressure difference between the inner and outer membrane leaflets and dV is a volume element.
3. The bending term represents the energy required to deform the membrane from its intrinsic curvature, while the spontaneous curvature is the preferred curvature that the membrane assumes based on its heterogeneous composition (Dai & Sheetz, 1995; Deuling & Helfrich, 1976): $\frac{\kappa}{2} \int (2H - c_0)^2 dA$, where κ is the bending modulus, H is the mean curvature, and c_0 is the spontaneous curvature.
4. The saddle-splay term represents the Gaussian curvature (Powers, Huber, & Goldstein, 2002) describing the relative rigidity of the membrane with respect to deformations in Gaussian curvature: $\bar{\kappa} \int K dA$, where $\bar{\kappa}$ is the saddle-splay modulus and K is the Gaussian curvature.
5. The tether term, FL, represents the effect of the point force (F) that pulls a tether from the membrane to a tether length of L (Derényi et al., 2002).

Thus, taken together, these terms provide an estimate of the free energy of a bilayer membrane (Bukman et al., 1996; Derényi et al., 2002, 2007; Deuling & Helfrich, 1976; Helfrich, 1973; Vaidya et al., 2008):

$$E_{Helfrich} = \sigma \int dA - \Delta P \int dV + \frac{\kappa}{2} \int (2H - c_0)^2 dA + \bar{\kappa} \int K dA - FL \tag{1}$$

As described briefly above, an experimental approach to estimate membrane tension involves pulling membrane tethers from the bilayer membrane using techniques such as atomic force microscopy (AFM) or laser optical tweezers (Dai & Sheetz, 1999; Dai, Sheetz, Wan, & Morris, 1998;

Diz-Muñoz et al., 2010; Lieber, Yehudai-Resheff, Barnhart, Theriot, & Keren, 2013; Sheetz, 2001; Sinha et al., 2011; Sun et al., 2007). In these techniques, a section of membrane attached to a probe is pulled into the shape of a narrow cylindrical nanotube (tether). The force exerted on the probe by the tether (tether force) is recorded and used to estimate the membrane tension. The relationship between the tether force and membrane tension is based on a simplified version of the free energy equation above (Eq. 1). In the next part of this discussion, we will describe the assumptions that are typically made to eliminate the terms of this equation that are unnecessary for application to membrane tension estimation by tether force.

First, it is possible to eliminate the second term representing the pressure change component of the equation, $\Delta P \int dV$. It is generally assumed that there is no pressure change across the section of membrane constituting the tether (i.e., $\Delta P = 0$) (Powers et al., 2002). It is also possible to eliminate the saddle-splay term, the fourth term of Eq. (1), $\bar{\kappa} \int K dA$. This is based on the definition of the Gaussian curvature, K, in terms the two principal curvatures, c_{max} and c_{min}, which are the maximum and minimum values of the curvature at each given point on the membrane surface. Gaussian curvature is the product of the two principal curvatures at each point on the surface, and is an intrinsic measure of curvature: $K = c_{\mathrm{max}} c_{\mathrm{min}}$. The sign of the Gaussian curvature characterizes the surface shape, with positive Gaussian curvature indicating an elliptic point, negative being a hyperbolic (saddle) point, and a Gaussian curvature of zero representing a parabolic point. Importantly, zero Gaussian curvature is observed in cylinders because one of the principal curvatures would be zero (Fig. 2). Therefore, K is assumed to be zero for membrane tethers. Thus, Eq. (1) can be simplified to give:

$$E = \sigma \int dA + \frac{\kappa}{2} \int (2H - c_0)^2 dA - FL \qquad (2)$$

The first term on the right-hand side of Eq. (2) $\left(\sigma \int dA\right)$ accounts for the surface tension within the membrane, while the second term $\left(\frac{\kappa}{2} \int (2H - c_0)^2 dA\right)$ represents the energetic cost of bending the membrane into a cylindrical tether. The final term (FL) characterizes the contribution of the tether force to the free energy. In other words, three main energy components contribute to the free energy of a tether: as a tether

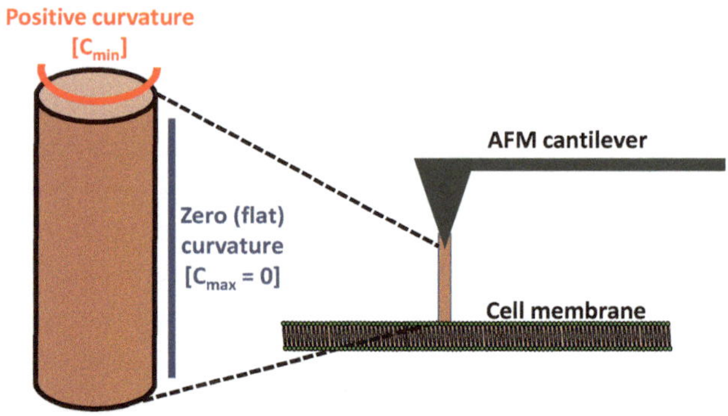

Figure 2 Schematic depicting the formation of a cylindrical membrane tether (nanotube) from a bilayer membrane by an atomic force microscope (AFM) cantilever tip. The cylinder has no Gaussian curvature because one principal curvature is positive while the other is zero (flat). The membrane tether is composed mainly of phospholipids derived from the membrane.

is pulled away from the cell membrane with a point force of F, there is competition between the stretching energy to pull a section of the membrane away from the bulk membrane into an elongated, cylindrical tether (membrane tension), and the bending energy needed to form the membrane into the narrow tubular shape of the tether (Derényi et al., 2007). Integrating Eq. (2) over the surface area (A) of the tether results in:

$$E = \sigma A + \frac{\kappa}{2}(2H - c_0)^2 A - FL \tag{3}$$

For a membrane tether of length L and radius R, the surface area is given as $A = 2\pi RL$, which is the surface area of a cylinder. The mean curvature, H, is the arithmetic mean (average) of the two principal curvatures, and is an extrinsic measure of curvature: $H = \frac{c_{max}+c_{min}}{2}$. Since, as mentioned earlier, one of the principal curvatures for a cylinder is zero, the mean curvature becomes: $H = \frac{c}{2}$, where c is the non-zero curvature. The principal curvature at each point on a curve is defined as the reciprocal of the radius of curvature at that point: $c = \frac{1}{R}$. Thus, substituting $2H = \frac{1}{R}$ and the surface area, $A = 2\pi RL$ into Eq. (3) results in:

$$E = \left(\sigma + \frac{\kappa}{2}\left(\frac{1}{R} - c_0\right)^2\right) 2\pi RL - FL \tag{4}$$

It is often assumed that there is no difference between the two leaflets of the bilayer forming the tether and hence there is no spontaneous curvature (i.e., $c_0 = 0$) (Deuling & Helfrich, 1976; Powers et al., 2002). Therefore the free energy expression in Eq. (4) further simplifies to:

$$E = \left(\sigma R + \frac{\kappa}{2R}\right) 2\pi L - FL \tag{5}$$

An equilibrium radius (R_0) results as the surface tension works towards reducing the tether radius. This equilibrium value is calculated by minimizing Eq. (5) through taking the partial derivative of the free energy with respect to tether radius, R (Derényi et al., 2002):

$$\frac{\partial E}{\partial R} = \left(\sigma - \frac{\kappa}{2R_0^2}\right) 2\pi L = 0 \tag{6}$$

Hence, the equilibrium radius can be obtained as a function of surface tension and bending modulus:

$$R_0 = \sqrt{\frac{\kappa}{2\sigma}} \tag{7}$$

This equation can be rearranged to obtain expressions for bending modulus and surface tension:

$$\kappa = 2\sigma R_0^2 \tag{8}$$

$$\sigma = \frac{\kappa}{2R_0^2} \tag{9}$$

Similarly, the equilibrium tether force is obtained by taking the partial derivative of the free energy with respect to tether length from Eq. (5):

$$\frac{\partial E}{\partial L} = \left(\sigma R + \frac{\kappa}{2R}\right) 2\pi - F_0 = 0 \tag{10}$$

Substituting the expression for equilibrium radius from Eq. (7) for R in Eq. (10) and solving for F_0 results in the equilibrium tether force being a function of surface tension and bending modulus:

$$F_0 = 2\pi\sqrt{2\sigma\kappa} \tag{11}$$

Once again, this equilibrium equation can be rearranged to obtain expressions for the surface tension and bending modulus:

$$\sigma = \frac{1}{2\kappa}\left(\frac{F_0}{2\pi}\right)^2 \tag{12}$$

$$\kappa = \frac{1}{2\sigma}\left(\frac{F_0}{2\pi}\right)^2 \tag{13}$$

Substituting the equations for κ and σ from Eqs. (8) and (9) into Eqs. (12) and (13) provides expressions for the membrane surface tension and bending modulus at the equilibrium condition as functions of only equilibrium tether force and radius (Pontes et al., 2011, 2013). This allows these elastic constants of tension, σ, and bending modulus, κ, to be experimentally estimated by pulling membrane tethers to obtain the steady-state (equilibrium) tether force, F_0 and the tether radius, R_0 as shown in Eqs. (14) and (15):

$$\sigma = \frac{F_0}{4\pi R_0} \tag{14}$$

$$\kappa = \frac{F_0 R_0}{2\pi} \tag{15}$$

These two equations are typically used in experimental studies to estimate membrane tension of lipid vesicles or cellular membranes from measurements of the tether force (F_0) obtained either using atomic force microscopy (AFM) or optical tweezers (Dai & Sheetz, 1999; Dai et al., 1998; Diz-Muñoz et al., 2010, 2013; Hochmuth, Shao, Dai, & Sheetz, 1996; Lieber et al., 2013; Sheetz, 2001; Sinha et al., 2011; Sun et al., 2007). Both of these state-of-the-art techniques are valid to determine the tether force, however, the radius (R_0) of tethers pulled from membranes using laser optical tweezers can be directly determined, whereas, it is not possible to measure the radii of tethers pulled using the AFM experimental technique. This is due to the fact that tethers are pulled in the same direction as the microscope optical axis in AFM measurements and, therefore, cannot be directly visualized. Consequently, with AFM measurements, when comparing different conditions without a change in membrane composition, it can be assumed that the tether radius will remain constant (Ayee et al., 2018). This allows comparison of the relative changes in membrane tension between different conditions.

It is important to note, however, that while these equations (Eq. 14 and 15) are applicable to lipid bilayer membranes such as those of vesicles, they do not take into account the fact that cellular membranes are attached to an underlying cytoskeleton network. This attachment has a major impact on the force required to pull membrane tethers (Dai & Sheetz, 1999; Sheetz, 2001). Indeed, a network of actin filaments forms the cortical cytoskeleton beneath the cell membrane, which is attached to this cortical network by linker proteins such as filamins (Razinia, Mäkelä, Ylänne, & Calderwood, 2012), myosin-1 motors (McConnell & Tyska, 2010; Nambiar, McConnell, & Tyska, 2009), and ezrin-radixin-moesin (ERM) family proteins (Fehon, McClatchey, & Bretscher, 2010). Numerous studies show that disruption of the cytoskeletal networks dramatically decreases the tether force measured in various cell types under different experimental conditions (Ayee et al., 2018; Dai & Sheetz, 1999; Pontes et al., 2011; Raucher et al., 2000; Sun et al., 2005, 2007). These differences in tether force in the presence or absence of the cortical cytoskeleton is attributed to the membrane-cortex adhesion energy (Diz-Muñoz et al., 2010). Consequently, for cells, the equilibrium tether force (Eq. 11) contains an additional term representing the adhesion energy density (γ_0) between the cell membrane and the cytoskeleton (Brochard-Wyart, Borghi, Cuvelier, & Nassoy, 2006; Hochmuth et al., 1996).

$$F_0 = 2\pi\sqrt{2\kappa(\sigma + \gamma_0)} \tag{16}$$

The term $(\sigma + \gamma_0)$, also called σ_m, is the "apparent" membrane tension, which contains terms for both the in-plane bilayer tension and the membrane-cytoskeleton adhesion (Dai & Sheetz, 1999; Sheetz & Dai, 1996). This apparent membrane tension can be calculated from the results of a typical AFM experiment that measures tether force. The close similarity between the values for tether force in lipid vesicles that contain only lipids and in membranes of cells with disrupted cytoskeletons suggests that integral membrane proteins do not contribute significantly to the tether force.

In experimental contexts, tether force values (F_0) have been found to range between approximately 30—40 pN for various cell types. Conversely, F_0 values for the same cell types treated with actin depolymerization agents that significantly disrupt the cytoskeleton (such as Latrunculin-A or cytochalasin-D) are found to be reduced by approximately 50% to about 15 pN in most cases (Ayee et al., 2018; Pontes et al., 2011; Raucher et al., 2000; Sun et al., 2005, 2007). Similarly, the tether forces and membrane-cytoskeleton adhesion energy measured in membrane blebs, which do not contain cytoskeletal structures, have also been found to be significantly lower, by at least 50%, than those of control cells (Dai & Sheetz, 1999; Raucher et al., 2000). Therefore, with respect to the apparent membrane tension defined in Eq. (16), the experimental results above indicate that the attachment of the membrane to the cortical network is as important of a contributor to the tether force as the in-plane bilayer tension. Like-wise, the F_0 values from cells with disrupted cytoskeletons are in close agreement with tether force values obtained from lipid vesicles composed of egg phosphatidylcholine combined with mPEG-DOPE and DSPE-PEG biotin. For these lipid vesicles, a static tether force of 16 pN was obtained using optical tweezers (Cuvelier, Chiaruttini, Bassereau, & Nassoy, 2005). This value is very similar to the tether forces of 15 pN for cytoskeleton disrupted cells discussed above. Hence, experimental measurements of tether forces in cells contain information about both the membrane tension and the energy of adhesion to the cytoskeleton.

In addition to tether pulling experiments, another mechanical method to independently probe cellular membrane properties, including the apparent membrane tension, σ_m, involves indenting cells with an AFM cantilever. In this technique, the cantilever is depressed into a cell and experiences a restoring force that resists this mechanical deformation. The restoring force that results from this indentation is the result of an isotropic tension (σ_{iso}), which is composed of the combined membrane and cortical cytoskeleton tension ($\sigma_0 = \sigma_m + \sigma_c$) as well as the change in membrane surface area $K_A \frac{\Delta A}{A_0}$ resulting from stretching the membrane during indentation (Evans, 1974; Kwok & Evans, 1981; Pietuch & Janshoff, 2013):

$$\sigma_{iso} = \sigma_0 + K_A \frac{\Delta A}{A_0} \tag{17}$$

where K_A is the area compressibility modulus of the membrane connected to the cortical layer and ΔA is the change in surface area as compared to the initial surface area, A_0. The cortical cytoskeleton tension (σ_c) is generated by contractile stress of the actin network (Levayer & Lecuit, 2012; Salbreux, Charras, & Paluch, 2012), while the membrane tension term ($\sigma_m = \sigma + \gamma_0$) is the apparent membrane tension, composed of the in-plane bilayer tension and the membrane-cytoskeleton adhesion, as explained above. Although this indentation method may theoretically be used to obtain an estimate of the apparent membrane tension, it should be used with caution, considering that determination of the surface area of cells is complicated by the fact that the membrane is folded and wrinkled, with excess membrane stored in reservoirs (Clark et al., 2014; Dai & Sheetz, 1999; Sens & Turner, 2006). Hence, a reasonable measurement of

$\frac{\Delta A}{A_0}$, particularly the small change in area upon indentation, would be difficult to obtain (Pietuch & Janshoff, 2013). The indentation approach is most reliable for measuring the cellular elastic modulus, whereas tether pulling is more reliable for estimating apparent membrane tension.

2.2 Osmotic Swelling has Little or No Effect on Membrane Tension

As mentioned previously, the question of the impact of osmotic swelling on membrane tension has been controversial. It was initially believed that an increase in membrane tension would result from hypo-osmotic cell swelling, and that this change in tension would lead to the activation of mechanosensitive ion channels, thus restoring the cellular osmotic equilibrium in the process known as regulatory volume decrease (RVD) (Hoffmann & Dunham, 1995; Okada et al., 2001). However, direct evidence to support the belief that cell swelling would lead to increased cell tension is scarce. A study by Dai et al. of osmotically challenged molluscan neurons observed an increase in the apparent membrane tension measured by membrane tethers pulled using optical tweezers (Dai et al., 1998), while a similar study of mammalian cells by Sinha et al. (2011) saw changes in the apparent tension only when the membrane folds and invaginations were pre-flattened by cholesterol depletion or genetic Caveolin-1 deficiency. As mentioned above, the apparent membrane tension measured in these studies would depend on both the in-plane bilayer tension and the membrane-cytoskeleton adhesion.

We recently undertook a study to simultaneously analyze the impact of a mild osmotic gradient of 20% on the membrane tension and elastic modulus of human aortic endothelial cells (HAECs) by using both AFM nano-indentation and measuring the force required to pull membrane tethers (Ayee et al., 2018). With regards to membrane tension under osmotic challenge, our experiments yielded tether forces in the range of 40 pN for HAECs under both isotonic and hypotonic conditions, while values for cells treated with F-actin disrupting agents were approximately 15 pN under both conditions (Fig. 3). Since there was no difference in the tether force values between the isotonic and hypotonic conditions, we concluded that mild hypotonic swelling did not affect membrane tension or the adhesion energy between the membrane and cortical cytoskeleton.

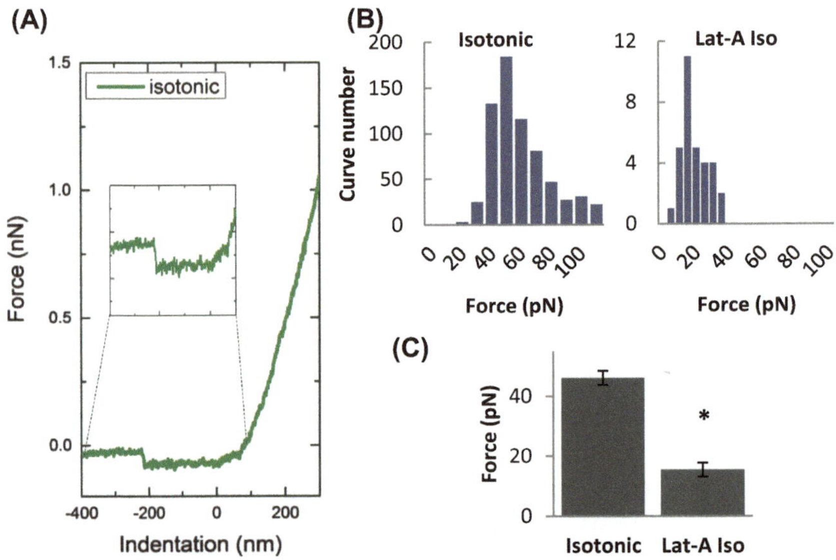

Figure 3 Effect of F-actin depolymerization on the force of membrane tether formation. (A) Representative traces of AFM retraction force curves for cells in isotonic solution with an inset of a representative force discontinuity (used to obtain the tether force). The force discontinuity represents the change in force experienced by the cantilever while retracting from the sample surface upon detachment of the tether. (B) Histograms of membrane tether forces measured in cells exposed to isotonic solution (left) and Latrunculin-A in isotonic medium (right). (C) Mean membrane tether forces of cells with and without exposure to Latrunculin-A (1 μM for 10 min). (Asterisk denotes statistically significant difference using an unpaired samples T-Test ($n = 15-60$ cells; $P < 0.05$) between Latrunculin-A treated and untreated cells).

This conclusion is consistent with the earlier studies both from our group (Levitan & Garber, 1997, pp. 245–267) and other investigators (Dai & Sheetz, 1999; Dai et al., 1997, 1998; Groulx et al., 2006; Raucher & Sheetz, 1999; Sens & Turner, 2006; Sinha et al., 2011), showing that there are significant membrane reservoirs in membrane folds and invaginations, which may accommodate an increase in cell volume and apparent increase in cell surface area without any membrane stretch, thus buffering membrane tension. Indeed, Morris (Morris, 2018) describes the concept of "membrane tension homeostasis" (MTH) as an essential process that works to prevent cellular plasma membranes from lysing under osmotic stress by employing a combination of multiple passive and active stochastic tension modulating steps to resist rupture.

3. ELASTIC PROPERTIES

3.1 Theoretical Basis and Experimental Approaches of Measuring Membrane and Cellular Elastic Modulus

The elastic modulus is another important biomechanical parameter that provides information about cell stiffness. This modulus can be estimated by probing cells with an AFM cantilever to obtain force/distance curves. The curves are obtained by recording the force applied and cantilever deflection at various vertical positions of the cantilever as it approaches, indents, and retracts from each cell (Kuznetsova, Starodubtseva, Yegorenkov, Chizhik, & Zhdanov, 2007). The experimental curves are then fitted to the Hertz model to obtain the Young's elastic modulus of the cells. AFM cantilever tips of various geometries may be used, however, for a pyramidal cantilever tip geometry, the Hertz model gives the force (*F*) applied to the cantilever (loading force) as (Ayee et al., 2018; Lin, Dimitriadis, & Horkay, 2007):

$$F = \frac{E \tan \phi}{2^{1/2}(1 - \upsilon^2)} \delta^2 \tag{18}$$

where E is the Young's elastic modulus, δ is the indentation depth, υ is the cellular Poisson's ratio (assumed to be 0.5 for incompressible biological material, and 2ϕ is the angle of the tip (25°−45°). The exact form of the Hertz model depends on the geometry of the cantilever tip. Although this is the most commonly used model to estimate cellular elastic moduli from AFM experiments (Haase & Pelling, 2015; Schillers et al., 2017; Thomas, Burnham, Camesano, & Wen, 2013), the Hertz model was developed based on several assumptions regarding the material being tested, including sample homogeneity, isotropy, and a linear stress-strain response. These assumptions present a limitation to the model, given that cells are structurally heterogeneous and anisotropic. Therefore, during experiments to estimate the elastic modulus, it is essential to maintain small, uniform indentation depths and probe sizes relative to the sample size, as well as uniform analyses of resulting force/distance curves based on depth of indentation (Dokukin, Guz, & Sokolov, 2013; Lin et al., 2007).

Consequently, the force/distance curves obtained from AFM experiments may be analyzed in different ways to provide elastic modulus estimates that discriminate between the membrane/cortical cytoskeleton region and the whole cell, representing the stiffness of the deeper cytoskeleton

(Askarova, Sun, Sun, Meininger, & Lee, 2013; Ayee et al., 2018). To quantify membrane stiffness and extract values of the elastic modulus, *E*, the force/distance curves can be analyzed from the point at which the AFM cantilever initially touches the cell surface to the point at which it reaches a depth of about 5 nm, representing the approximate thickness of the membrane lipid bilayer (Edidin, 2003). For the elastic modulus of the whole cell, on the other hand, the entire force/distance curve is fitted to the Hertz model. By analyzing force/distance curves in these two ways, the elastic modulus values for the membrane envelope and whole cell are found to be very different (Askarova et al., 2013; Ayee et al., 2018).

3.2 Osmotic Swelling has Differential Effects on Membrane and Cellular Elastic Moduli

Like membrane tension, the issue of the effect of osmotic swelling on cellular elastic moduli has been controversial. Spagnoli et al. (2008) found that osmotic swelling resulted in overall softening of cells measured by AFM using cantilevers with 15 μm polystyrene beads glued to the tips. In this study, the force/distance curves were fitted in their entirety to a version of the Hertz model for hard spheres indenting elastic solid surfaces, thus providing an estimate of the whole cell stiffness. The cell softening observed in this case may be explained in one two ways. First, the cellular cytoskeleton may be thought of as a cross-linked filamentous gel having a poroelastic structure, which has been likened to the structure of a sponge, and has therefore been termed the "sponge-effect" by Sachs et al. (Chao, Sivaselvan, & Sachs, 2018; Charras, Mitchison, & Mahadevan, 2009; Sachs & Sivaselvan, 2015). Another possibility is that the softening could be explained by mild disruption and reorganization of the cytoskeleton as a result of osmotic challenge (Hoffmann, 2000; Hoffmann, Lambert, & Pedersen, 2009; Jorgensen et al., 2003; Lambert, Hoffmann, & Pedersen, 2008; Levitan, Almonte, Mollard, & Garber, 1995; Pasantes-Morales, Cardin, & Tuz, 2000; Pedersen et al., 1999, 2001; Spagnoli et al., 2008; Sun & Levitan, 2003). It is not clear which of these possibilities would account for the observed mild softening of cells as a response to osmotic challenge. In contrast to this observed softening, however, Zou et al. (2013) found that osmotic challenge induced stiffening of cortical neurons. These results were obtained from AFM experiments using 20 μm borosilicate spherical glass beads mounted on the cantilever, and the force/distance curves were fitted to the Hertz model for spherical incompressible tips. Again, in this study, the entire force curves were fitted to the Hertz model to obtain estimates of the elastic moduli. The cortical neurons were briefly exposed to hypotonic shock, resulting in an increase in stiffness, which changed to softening when the solution medium was exchanged. In some cases, softening began even before the medium was changed.

Although both studies described above used whole cell analysis to determine cellular elastic moduli, the differences in their respective experimental methodologies may account for the different effects observed after osmotic challenge. Specifically, the exposure times employed may partly account for the different results obtained. The Spagnoli study exposed various cell types to a hypotonic solution of DMEM with mannitol for at least 10 min. On the other hand, in the Zou study, the cortical neurons were exposed to BPS with reduced NaCl for only 5 min, after which the solution was replaced by an isotonic BPS one. Interestingly, the elastic moduli of the cells in the Spagnoli study initially ranged between about 300 and 400 Pa before treatment, which decreased to about 100 Pa after the hypotonic challenge. Conversely, the elastic moduli of the neurons in the Zou study were initially at approximately 150 Pa, they rose to about 350 Pa within 1−2 min during the 5 min challenge, and then fell back down to the baseline of 150 Pa. Another difference between the two studies was the osmolality of their respective hypotonic solutions. The Spagnoli study used a strong hypotonic solution of 90 mOsm, while the Zou study used a milder osmotic gradient with a solution at 217 mOsm. The use of strong osmotic gradients to study the effects of swelling on cells is not unusual, with most previous studies subjecting cells to osmotic gradients of 50% and above (Dai et al., 1997, 1998; Groulx et al., 2006; Guo, Wang, Sachs, & Meng, 2014; Meng & Sachs, 2010; Pietuch, Brückner, & Janshoff, 2013; Raucher & Sheetz, 1999), some even exposing cells to pure distilled water. Such strong osmotic gradients, however, may not be relevant physiologically. For example, we previously showed that the plasma osmolality decreases by only about 10% in the extreme case of compulsive water drinking (Verghese et al., 1997), due to the tight physiological regulation of the osmolality of the plasma and extracellular fluids. Therefore, osmolality values within the physiological range may be more informative when examining the effects of osmotic swelling on cells. Also, as mentioned above, the manner in which the force/distance curves are analyzed provides information pertinent to different regions of the cell.

In our recent study of hypotonically swollen human aortic endothelial cells (HAECs), we employed a mild 20% osmotic gradient (Ayee et al.,

2018). From our analysis of the two regions of the force/distance curves as described above, we concluded that the elastic modulus of the membrane/cortical cytoskeleton envelope was lower than that of the whole cell (deeper cytoskeleton) with about 10 kPa estimated for the membrane region and 20 kPa for the whole cell. Force curves are continuous; therefore, we distinguished between the membrane region and the whole cell by approximating that the membrane region extended about 5 nm from the cell surface. We found that the osmotic challenge resulted in a significant stiffening of the membrane region in endothelial cells, increasing the elastic modulus twofold (Fig. 4). These values are higher than those observed for cortical neurons and other cell types described above; however, they are in the same range as values for endothelial cells found in our earlier work (Ayee et al., 2017; Shentu et al., 2010, 2012). It is known that the elastic modulus of the cell membrane depends largely on the stiffness of the cortical cytoskeleton underlying the membrane (Fletcher & Mullins, 2010; Pourati et al., 1998; Rotsch & Radmacher, 2000; Sato, Theret, Wheeler, Ohshima, & Nerem, 1990; Wagner, Tharmann, Haase, Fischer, & Bausch, 2006; Wu, Kuhn, & Moy, 1998; Zhang, Long, Wu, & Yu, 2002). As expected, treating our cells with Latrunculin-A caused a significant decrease in the elastic modulus to 1.3 kPa. Unexpectedly, however, the stiffening effect induced by osmotic swelling was enhanced by the disruption of F-actin. Since it

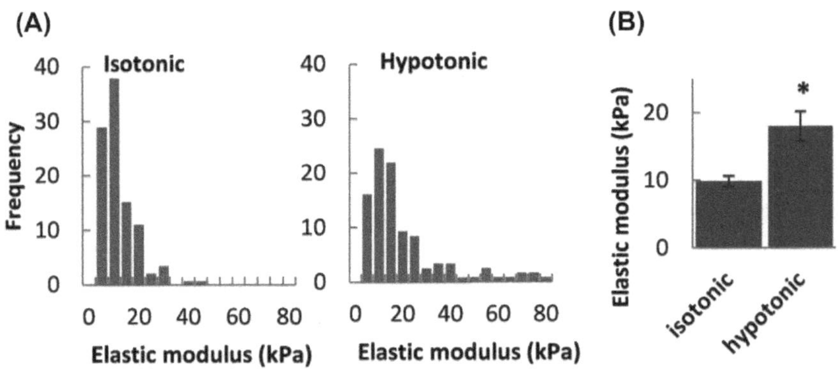

Figure 4 Effect of osmotic challenge on the elastic modulus of endothelial cells. (A) Histograms of elastic moduli measured in the membrane region of cells exposed to isotonic (left) and hypotonic (right) solutions. (B) Mean membrane region elastic moduli of cells in isotonic and hypotonic solutions. (Asterisk denotes significant difference using an unpaired samples T-Test ($n = 15$–60 cells; $P < 0.05$) between moduli measured under each condition).

was shown previously that direct application of hydrostatic pressure to cells with a micropipette results in membrane stiffening (Beyder & Sachs, 2009), we propose that the stiffening of osmotically challenged endothelial cells should be attributed to an increase in internal hydrostatic pressure, originating from the influx of water into the cells due to the osmotic gradient. In contrast to the results obtained for the membrane region, although there was a slight reduction in stiffness of the deeper cytoskeleton, we saw no statistically significant change in the stiffness when cells we exposed to an osmotic challenge, with elastic modulus values remaining between 15 and 20 kPa. Indeed, previous work on Ehrlich ascites tumor cells by Pedersen et al. (1999) showed that osmotic challenge affected only F-actin in the cortical region and not in the deeper cytoskeleton. Consequently, we concluded that osmotic swelling with physiologically relevant hypotonic gradients resulted in depth-dependent changes in the elastic modulus of cells, a fact that should be considered when estimating cell stiffness in experimental settings. Moreover, it is also interesting to note that changes in cell stiffness were shown to affect activation of volume-regulated anion channels (VRAC), one of the major mechanisms of cell volume regulation (Byfield, Hoffman, Romanenko, Fang, Crocker, & Levitan, 2006).

4. CONCLUSIONS

In this chapter we provide a detailed analysis of the methodology used to estimate cell membrane tension and elastic moduli, specifically addressing the theoretical background surrounding the models used in these analyses, the governing equations employed in estimating these biomechanical parameters, and the key assumptions made. We include a summary of the prevalent experimental approaches and describe what is currently understood about how these parameters change during osmotic swelling. From our recent work (Ayee et al., 2018), we find that mild hypotonic swelling does not change membrane tension or the adhesion energy between the membrane and cortical cytoskeleton due to the buffering effect of unfolding membrane reservoirs. These findings challenge the general expectation that cell swelling should result in an increase in membrane tension. We conclude also that the cell membrane envelope stiffens during swelling (Fig. 5), and unexpectedly, stiffening is enhanced when the cytoskeleton was disrupted, indicating that the cytoskeleton has a dampening

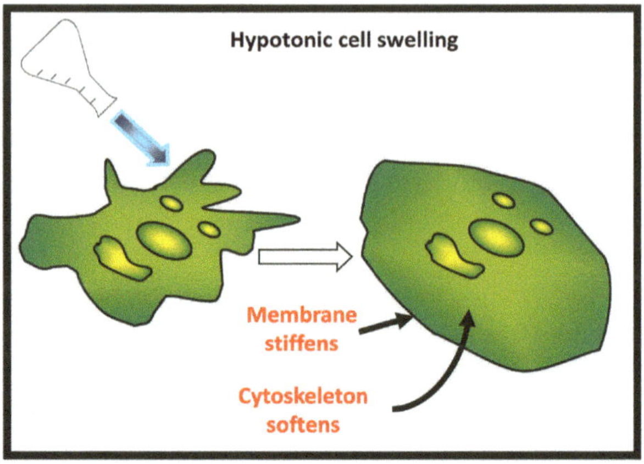

Figure 5 Schematic depicting cell swelling and membrane unfolding as a response to hypotonic challenge resulting in membrane stiffening and mild cytoskeleton softening.

effect on the increase in stiffness. These conclusions have important implications to the field of membrane biomechanics and indicate that hypotonic swelling may not be a suitable method to determine the role of membrane tension in mechanosensitive processes.

ACKNOWLEDGMENTS

We are very grateful to Nicolas Barbera for the critical reading of this manuscript. We also thank Mr. Gregory Kowalsky for his help in designing Fig. 5. The work was supported by the National Institutes of Health (NIH) grants T32 HL-82547 (to M.A.A.A.), HL-083298, and HL-073965 (to I.L.).

REFERENCES

Askarova, S., Sun, Z., Sun, G. Y., Meininger, G. A., & Lee, J. C. (2013). Amyloid-β peptide on sialyl-LewisX-selectin-mediated membrane tether mechanics at the cerebral endothelial cell surface. *PLoS One, 8*. e60972.

Ayee, M. A., LeMaster, E., Shentu, T. P., Singh, D. K., Barbera, N., Soni, D., Tiruppathi, C., Subbaiah, P. V., Berdyshev, E., & Bronova, I. (2017). Molecular-scale biophysical modulation of an endothelial membrane by oxidized phospholipids. *Biophysical Journal, 112*, 325–338.

Ayee, M. A. A., LeMaster, E., Teng, T., Lee, J., & Levitan, I. (2018). Hypotonic challenge of endothelial cells increases membrane stiffness with No effect on tether force. *Biophysical Journal, 114*, 929–938.

Beyder, A., & Sachs, F. (2009). Electromechanical coupling in the membranes of Shaker-transfected HEK cells. *Proceedings of the National Academy of Sciences, 106*, 6626–6631.

Brochard-Wyart, F., Borghi, N., Cuvelier, D., & Nassoy, P. (2006). Hydrodynamic narrowing of tubes extruded from cells. *Proceedings of the National Academy of Sciences, 103*, 7660–7663.

Bukman, D. J., Yao, J. H., & Wortis, M. (1996). Stability of cylindrical vesicles under axial tension. *Physical Review E, 54*, 5463.

Byfield, F. J., Hoffman, B. D., Romanenko, V. G., Fang, Y., Crocker, J. C., & Levitan, I. (2006). Evidence for the role of cell stiffness in the modulation of volume-regulated anion channels. *Acta Physiologica, 187*, 285–294.

Canham, P. B. (1970). The minimum energy of bending as a possible explanation of the biconcave shape of the human red blood cell. *Journal of Theoretical Biology, 26*, 61–81.

Chao, P.-C., Sivaselvan, M., & Sachs, F. (2018). *Cell softening due to osmotic swelling; extending the Donnan equilibrium*. In Press.

Charras, G. T., Mitchison, T. J., & Mahadevan, L. (2009). Animal cell hydraulics. *Journal of Cell Science, 122*, 3233–3241.

Clark, A. G., Wartlick, O., Salbreux, G., & Paluch, E. K. (2014). Stresses at the cell surface during animal cell morphogenesis. *Current Biology, 24*, R484–R494.

Cuvelier, D., Chiaruttini, N., Bassereau, P., & Nassoy, P. (2005). Pulling long tubes from firmly adhered vesicles. *EPL (Europhysics Letters), 71*, 1015.

Dai, J., & Sheetz, M. P. (1995). Mechanical properties of neuronal growth cone membranes studied by tether formation with laser optical tweezers. *Biophysical Journal, 68*, 988–996.

Dai, J., & Sheetz, M. P. (1999). Membrane tether formation from blebbing cells. *Biophysical Journal, 77*, 3363–3370.

Dai, J., Sheetz, M. P., Wan, X., & Morris, C. E. (1998). Membrane tension in swelling and shrinking molluscan neurons. *Journal of Neuroscience, 18*, 6681–6692.

Dai, J., Ting-Beall, H. P., & Sheetz, M. P. (1997). The secretion-coupled endocytosis correlates with membrane tension changes in RBL 2H3 cells. *The Journal of General Physiology, 110*, 1–10.

Derényi, I., Jülicher, F., & Prost, J. (2002). Formation and interaction of membrane tubes. *Physical Review Letters, 88*, 238101.

Derényi, I., Koster, G., Van Duijn, M., Czövek, A., Dogterom, M., & Prost, J. (2007). Membrane nanotubes. In *Controlled nanoscale motion* (pp. 141–159). Springer.

Deuling, H., & Helfrich, W. (1976). The curvature elasticity of fluid membranes: A catalogue of vesicle shapes. *Journal de Physique, 37*, 1335–1345.

Diz-Muñoz, A., Fletcher, D. A., & Weiner, O. D. (2013). Use the force: Membrane tension as an organizer of cell shape and motility. *Transfusion clinique et biologique, 23*, 47–53.

Diz-Muñoz, A., Krieg, M., Bergert, M., Ibarlucea-Benitez, I., Muller, D. J., Paluch, E., & Heisenberg, C.-P. (2010). Control of directed cell migration in vivo by membrane-to-cortex attachment. *PLoS Biology, 8*. e1000544.

Dokukin, M. E., Guz, N. V., & Sokolov, I. (2013). Quantitative study of the elastic modulus of loosely attached cells in AFM indentation experiments. *Biophysical Journal, 104*, 2123–2131.

Edidin, M. (2003). Lipids on the frontier: A century of cell-membrane bilayers. *Nature Reviews Molecular Cell Biology, 4*, 414.

Evans, E. A. (1974). Bending resistance and chemically induced moments in membrane bilayers. *Biophysical Journal, 14*, 923–931.

Evans, E., & Fung, Y.-C. (1972). Improved measurements of the erythrocyte geometry. *Microvascular Research, 4*, 335–347.

Evans, E., & Leblond, P. (1973). Image holograms of single red blood cell discocyte-spherechinocyte transformations. In *Red cell shape* (pp. 131–140). Springer.

Evans, E. A., Waugh, R., & Melnik, L. (1976). Elastic area compressibility modulus of red cell membrane. *Biophysical Journal, 16*, 585–595.

Farago, O., & Pincus, P. (2004). Statistical mechanics of bilayer membrane with a fixed projected area. *The Journal of Chemical Physics, 120*, 2934–2950.

Fehon, R. G., McClatchey, A. I., & Bretscher, A. (2010). Organizing the cell cortex: The role of ERM proteins. *Nature Reviews Molecular Cell Biology, 11*, 276.

Fletcher, D. A., & Mullins, R. D. (2010). Cell mechanics and the cytoskeleton. *Nature, 463*, 485.

Groulx, N., Boudreault, F., Orlov, S. N., & Grygorczyk, R. (2006). Membrane reserves and hypotonic cell swelling. *The Journal of Membrane Biology, 214*, 43–56.
Guo, J., Wang, Y., Sachs, F., & Meng, F. (2014). Actin stress in cell reprogramming. *Proceedings of the National Academy of Sciences, 111*, E5252–E5261.
Haase, K., & Pelling, A. E. (2015). Investigating cell mechanics with atomic force microscopy. *Journal of The Royal Society Interface, 12*, 20140970.
Helfrich, W. (1973). Elastic properties of lipid bilayers: Theory and possible experiments. *Zeitschrift für Naturforschung C, 28*, 693–703.
Hille, B. (2001). *Ion channels of excitable membranes*. MA: Sinauer Sunderland.
Hochmuth, F., Shao, J.-Y., Dai, J., & Sheetz, M. P. (1996). Deformation and flow of membrane into tethers extracted from neuronal growth cones. *Biophysical Journal, 70*, 358–369.
Hochmuth, R., & Waugh, R. (1987). Erythrocyte membrane elasticity and viscosity. *Annual Review of Physiology, 49*, 209–219.
Hoffmann, E. K. (2000). Intracellular signalling involved in volume regulatory decrease. *Cellular Physiology and Biochemistry, 10*, 273–288.
Hoffmann, E. K., & Dunham, P. B. (1995). Membrane mechanisms and intracellular signalling in cell volume regulation. *International Review of Cytology, 161*, 173–262.
Hoffmann, E. K., Lambert, I. H., & Pedersen, S. F. (2009). Physiology of cell volume regulation in vertebrates. *Physiological Reviews, 89*, 193–277.
Jorgensen, N. K., Pedersen, S. F., Rasmussen, H. B., Grunnet, M., Klaerke, D. A., & Olesen, S.-P. (2003). Cell swelling activates cloned Ca^{2+}-activated K^{+} channels: A role for the F-actin cytoskeleton. *Biochimica et Biophysica Acta (BBA)-Biomembranes, 1615*, 115–125.
Kuznetsova, T. G., Starodubtseva, M. N., Yegorenkov, N. I., Chizhik, S. A., & Zhdanov, R. I. (2007). Atomic force microscopy probing of cell elasticity. *Micron: the International Research and Review Journal for Microscopy, 38*, 824–833.
Kwok, R., & Evans, E. (1981). Thermoelasticity of large lecithin bilayer vesicles. *Biophysical Journal, 35*, 637–652.
Lambert, I., Hoffmann, E., & Pedersen, S. (2008). Cell volume regulation: Physiology and pathophysiology. *Acta Physiologica, 194*, 255–282.
Levayer, R., & Lecuit, T. (2012). Biomechanical regulation of contractility: Spatial control and dynamics. *Transfusion clinique et biologique, 22*, 61–81.
Levitan, I., Almonte, C., Mollard, P., & Garber, S. (1995). Modulation of a volume-regulated chloride current by F-actin. *The Journal of Membrane Biology, 147*, 283–294.
Levitan, I., Christian, A. E., Tulenko, T. N., & Rothblat, G. H. (2000). Membrane cholesterol content modulates activation of volume-regulated anion current in bovine endothelial cells. *The Journal of General Physiology, 115*, 405–416.
Levitan, I., & Garber, S. S. (1997). Volume regulated anion channels and cytoskeletal interaction. *From ion channels to cell-to-cell Conversations*. Springer.
Lieber, A. D., Yehudai-Resheff, S., Barnhart, E. L., Theriot, J. A., & Keren, K. (2013). Membrane tension in rapidly moving cells is determined by cytoskeletal forces. *Current Biology, 23*, 1409–1417.
Lin, D. C., Dimitriadis, E. K., & Horkay, F. (2007). Robust strategies for automated AFM force curve analysis-I. Non-adhesive indentation of soft, inhomogeneous materials. *Journal of Biomechanical Engineering, 129*, 430–440.
Lipowsky, R., & Sackmann, E. (1995). *Structure and dynamics of membranes: I. From cells to vesicles/II. generic and specific interactions*. Elsevier.
McConnell, R. E., & Tyska, M. J. (2010). Leveraging the membrane–cytoskeleton interface with myosin-1. *Transfusion clinique et biologique, 20*, 418–426.

Meng, F., & Sachs, F. (2010). Visualizing dynamic cytoplasmic forces with a compliance-matched FRET sensor. *Journal of Cell Science, 124*, 261–269.

Morris, C. E. (2018). Cytotoxic swelling of sick excitable cells - impaired ion homeostasis and membrane tension homeostasis in muscle and neuron. *Current Topics in Membranes, 81*, 457–496.

Nambiar, R., McConnell, R. E., & Tyska, M. J. (2009). Control of cell membrane tension by myosin-I. *Proceedings of the National Academy of Sciences, 106*, 11972–11977.

Okada, Y., Maeno, E., Shimizu, T., Dezaki, K., Wang, J., & Morishima, S. (2001). Receptor-mediated control of regulatory volume decrease (RVD) and apoptotic volume decrease (AVD). *The Journal of Physiology, 532*, 3–16.

Pasantes-Morales, H., Cardin, V., & Tuz, K. (2000). Signaling events during swelling and regulatory volume decrease. *Neurochemical Research, 25*, 1301–1314.

Pedersen, S. F., Hoffmann, E. K., & Mills, J. (2001). The cytoskeleton and cell volume regulation. *Comparative Biochemistry and Physiology Part a: Molecular & Integrative Physiology, 130*, 385–399.

Pedersen, S. F., Mills, J. W., & Hoffmann, E. K. (1999). Role of the F-actin cytoskeleton in the RVD and RVI processes in Ehrlich ascites tumor cells. *Experimental Cell Research, 252*, 63–74.

Pietuch, A., Brückner, B. R., & Janshoff, A. (2013). Membrane tension homeostasis of epithelial cells through surface area regulation in response to osmotic stress. *Biochimica et Biophysica Acta (BBA) - Molecular Cell Research, 1833*, 712–722.

Pietuch, A., & Janshoff, A. (2013). Mechanics of spreading cells probed by atomic force microscopy. *Open biology, 3*, 130084.

Pontes, B., Ayala, Y., Fonseca, A. C. C., Romão, L. F., Amaral, R. F., Salgado, L. T., Lima, F. R., Farina, M., Viana, N. B., & Moura-Neto, V. (2013). Membrane elastic properties and cell function. *PLoS One, 8*. e67708.

Pontes, B., Viana, N., Salgado, L., Farina, M., Neto, V. M., & Nussenzveig, H. (2011). Cell cytoskeleton and tether extraction. *Biophysical Journal, 101*, 43–52.

Pourati, J., Maniotis, A., Spiegel, D., Schaffer, J. L., Butler, J. P., Fredberg, J. J., Ingber, D. E., Stamenovic, D., & Wang, N. (1998). Is cytoskeletal tension a major determinant of cell deformability in adherent endothelial cells? *American Journal of Physiology - cell Physiology, 274*, C1283–C1289.

Powers, T. R., Huber, G., & Goldstein, R. E. (2002). Fluid-membrane tethers: Minimal surfaces and elastic boundary layers. *Physical Review E, 65*, 041901.

Raucher, D., & Sheetz, M. P. (1999). Characteristics of a membrane reservoir buffering membrane tension. *Biophysical Journal*, 77, 1992–2002.

Raucher, D., Stauffer, T., Chen, W., Shen, K., Guo, S., York, J. D., Sheetz, M. P., & Meyer, T. (2000). Phosphatidylinositol 4, 5-bisphosphate functions as a second messenger that regulates cytoskeletonâ€"plasma membrane adhesion. *Cell, 100*, 221–228.

Razinia, Z., Mäkelä, T., Ylänne, J., & Calderwood, D. A. (2012). *Filamins in mechanosensing and signaling.*

Ross, P. E., Garber, S. S., & Cahalan, M. D. (1994). Membrane chloride conductance and capacitance in Jurkat T lymphocytes during osmotic swelling. *Biophysical Journal, 66*, 169–178.

Rotsch, C., & Radmacher, M. (2000). Drug-induced changes of cytoskeletal structure and mechanics in fibroblasts: An atomic force microscopy study. *Biophysical Journal, 78*, 520–535.

Sachs, F., & Sivaselvan, M. V. (2015). Cell volume control in three dimensions: Water movement without solute movement. *The Journal of General Physiology*, 201411297.

Safran, S. (2003). *Statistical thermodynamics of surfaces, interfaces, and membranes*. Boca Raton: CRC Press.
Salbreux, G., Charras, G., & Paluch, E. (2012). Actin cortex mechanics and cellular morphogenesis. *Transfusion clinique et biologique, 22*, 536–545.
Sato, M., Theret, D. P., Wheeler, L. T., Ohshima, N., & Nerem, R. M. (1990). Application of the micropipette technique to the measurement of cultured porcine aortic endothelial cell viscoelastic properties. *Journal of Biomechanical Engineering, 112*, 263–268.
Schillers, H., Rianna, C., Schäpe, J., Luque, T., Doschke, H., Wälte, M., Uriarte, J. J., Campillo, N., Michanetzis, G. P., & Bobrowska, J. (2017). Standardized nanomechanical atomic force microscopy procedure (SNAP) for measuring soft and biological samples. *Scientific Reports, 7*, 5117.
Sens, P., & Turner, M. S. (2006). Budded membrane microdomains as tension regulators. *Physical Review E, 73*, 031918.
Sheetz, M. P. (2001). OPINION: Cell control by membrane-cytoskeleton adhesion. *Nature Reviews Molecular Cell Biology, 2*, 392.
Sheetz, M. P., & Dai, J. (1996). Modulation of membrane dynamics and cell motility by membrane tension. *Trends in cell biology, 6*, 85–89.
Shentu, T. P., Singh, D. K., Oh, M.-J., Sun, S., Sadaat, L., Makino, A., Mazzone, T., Subbaiah, P. V., Cho, M., & Levitan, I. (2012). The role of oxysterols in control of endothelial stiffness. *Journal of Lipid Research, 53*, 1348–1358.
Shentu, T. P., Titushkin, I., Singh, D. K., Gooch, K. J., Subbaiah, P. V., Cho, M., & Levitan, I. (2010). oxLDL-induced decrease in lipid order of membrane domains is inversely correlated with endothelial stiffness and network formation. *American Journal of Physiology - cell Physiology, 299*, C218–C229.
Sinha, B., Köster, D., Ruez, R., Gonnord, P., Bastiani, M., Abankwa, D., Stan, R. V., Butler-Browne, G., Vedie, B., & Johannes, L. (2011). Cells respond to mechanical stress by rapid disassembly of caveolae. *Cell, 144*, 402–413.
Spagnoli, C., Beyder, A., Besch, S., & Sachs, F. (2008). Atomic force microscopy analysis of cell volume regulation. *Physical Review E, 78*, 031916.
Sun, A., & Levitan, I. (2003). Osmotic stress alters the intracellular distribution of non-erythroidal spectrin (Fodrin) in bovine aortic endothelial cells. *Journal of Membrane Biology, 192*, 9–17.
Sun, M., Graham, J. S., HegedÃ¼s, B., Marga, F. o., Zhang, Y., Forgacs, G., & Grandbois, M. (2005). Multiple membrane tethers probed by atomic force microscopy. *Biophysical Journal, 89*, 4320–4329.
Sun, M., Northup, N., Marga, F., Huber, T., Byfield, F. J., Levitan, I., & Forgacs, G. (2007). The effect of cellular cholesterol on membrane-cytoskeleton adhesion. *Journal of Cell Science, 120*, 2223–2231.
Thomas, G., Burnham, N. A., Camesano, T. A., & Wen, Q. (2013). Measuring the mechanical properties of living cells using atomic force microscopy. *Journal of Visualized Experiments*.
Vaidya, N., Huang, H., & Takagi, S. (2008). Correct equilibrium shape equation of axisymmetric vesicles. In *Integral methods in science and engineering* (pp. 267–276). Springer.
Verghese, C., Levitan, I., Nair, C., Abraham, G., Garber, S. S., & Josiassen, R. C. (1997). Impaired lymphocyte volume regulation in schizophrenic patients with polydipsia–hyponatremia. *Biological Psychiatry, 42*, 733–736.
Vlahakis, N. E., & Hubmayr, R. D. (2000). Invited review: Plasma membrane stress failure in alveolar epithelial cells. *Journal of Applied Physiology, 89*, 2490–2496.
Wagner, B., Tharmann, R., Haase, I., Fischer, M., & Bausch, A. R. (2006). Cytoskeletal polymer networks: The molecular structure of cross-linkers determines macroscopic properties. *Proceedings of the National Academy of Sciences, 103*, 13974–13978.

Wu, H., Kuhn, T., & Moy, V. (1998). Mechanical properties of L929 cells measured by atomic force microscopy: Effects of anticytoskeletal drugs and membrane crosslinking. *Scanning, 20*, 389–397.
Zhang, G., Long, M., Wu, Z.-Z., & Yu, W.-Q. (2002). Mechanical properties of hepatocellular carcinoma cells. *World Journal of Gastroenterology, 8*, 243.
Zou, S., Chisholm, R., Tauskela, J. S., Mealing, G. A., Johnston, L. J., & Morris, C. E. (2013). Force spectroscopy measurements show that cortical neurons exposed to excitotoxic agonists stiffen before showing evidence of bleb damage. *PLoS One, 8*. e73499.

CHAPTER FIVE

Molecular Identities and ATP Release Activities of Two Types of Volume-Regulatory Anion Channels, VSOR and Maxi-Cl

Yasunobu Okada*,§,1, Toshiaki Okada§, Md. Rafiqul Islam§ and Ravshan Z. Sabirov§,¶

*Department of Physiology and Systems Bioscience, Kyoto Prefectural University of Medicine, Kyoto, Japan
§Division of Cell Signaling, National Institute for Physiological Sciences (NIPS), Okazaki, Japan
¶Laboratory of Molecular Physiology, Institute of Bioorganic Chemistry, Academy of Sciences of Uzbekistan, Tashkent, Uzbekistan
[1]Corresponding Author: E-mail: okada@nips.ac.jp

Contents

Current Topics in Membranes, Volume 81
ISSN 1063-5823
https://doi.org/10.1016/bs.ctm.2018.07.004

Abstract

An elaborate volume regulation system based on interplay of ion channels and transporters was evolved to cope with constant osmotic challenges caused by intensive metabolism, transport and other physiological/pathophysiological events. In animal cells, two types of anion channels are directly activated by cell swelling and involved in the regulatory volume decrease (RVD): volume-sensitive outwardly rectifying anion channel (VSOR), also called volume-regulated anion channel (VRAC), and Maxi-Cl which is the most major type of maxi-anion channel (MAC). These two channels have very different biophysical profiles and exhibit opposite dependence on intracellular ATP. After several decades of verifying many false-positive candidates for VSOR and Maxi-Cl, LRRC8 family proteins emerged as major VSOR components, and SLCO2A1 protein as a core of Maxi-Cl. Still, neither of these proteins alone can fully reproduce the native channel phenotypes suggesting existence of missing components. Although both VSOR and Maxi-Cl have pores wide enough to accommodate bulky ATP^{4-} and $MgATP^{2-}$ anions, evidence accumulated hitherto, based on pharmacological and gene silencing experiments, suggests that Maxi-Cl, but not VSOR, serves as one of the major pathways for the release of ATP from swollen and ischemic/hypoxic cells. Relations of VSOR and Maxi-Cl with diseases and their selective pharmacology are the topics promoted by recent advance in molecular identification of the two volume-activated, volume-regulatory anion channels.

1. INTRODUCTION

Cell volume regulation is an essential function possessed by animal cells to survive under both physiological and pathophysiological conditions (for Reviews: Hoffmann, Lambert, & Pedersen, 2009; Lang, 1998; Okada, 1998; 2004). Among a variety of anion channel types, two types of anion channels are directly activated by cell swelling and involved in cell volume regulation after cell swelling, the ability of which is dubbed the regulatory volume decrease (RVD), and therefore collectively called volume-regulatory anion channels. One of these volume-regulatory anion channels is the volume-sensitive outwardly rectifying anion channel (VSOR), also called the volume-regulated anion channel (VRAC). Another one is Maxi-Cl which is the most major type of maxi-anion channel (MAC), because 61% of 84 original papers hitherto published exhibited the Maxi-Cl phenotype (see below and Table 1), whereas remaining MACs showed Maxi-Cl-like or non-Maxi-Cl type properties such as a smaller unitary conductance and poor anion/cation selectivity (Sabirov, Merzlyak, Islam, Okada, & Okada, 2016; Sabirov, Merzlyak, Okada, Islam, Uramoto, Mori, Makino, Matsuura, Xie, & Okada, 2017). The molecular identities of the core components for these volume-regulatory anion channels were both recently

Table 1 Comparison of Phenotypical Properties Between VSOR and Maxi-Cl

VSOR	Maxi-Cl
Low electric field anion selectivity: $I^- > Br^- > Cl^- > F^-$	Low electric field anion selectivity: $I^- > Br^- > Cl^- > F^-$
Intermediate unitary conductance: 10–90 pS	Large unitary conductance: 300–500 pS
Mild outward rectification	Linear (ohmic) *I–V* relationship
Inactivation kinetics only at large positive voltages	Inactivation kinetics at small positive and negative voltages
Non-hydrolytic requirement of intracellular ATP	Inhibition by intracellular ATP

discovered (for VSOR or VRAC: Qiu et al., 2014; Voss et al., 2014; for Maxi-Cl: Sabirov et al., 2017) long after the history of false positive candidate molecules. Since ATP exists in anionic forms, $MgATP^{2-}$, ATP^{4-}, $HATP^{3-}$, and $MgHATP^-$, under physiological cytosolic conditions (say, 87.0, 10.7, 2.1, and 0.1%, respectively, in the presence of an equivalent concentration of Mg^{2+} at pH 7.4: Sabirov & Okada, 2005), it is reasonable to conceive that some anion channels serve as a conductive pathway for ATP release from the intracellular compartment to the extracellular space according to a huge outwardly directed electrochemical potential gradient (say, of the order of 10^8 for $MgATP^{2-}$, 10^{10} for ATP^{4-}, 10^9 for $HATP^{3-}$). Actually, possible involvements of VSOR and Maxi-Cl in ATP release activities have so far been intensively explored. Thus, this review article focuses on studies of the molecular identification and the ATP release activities of VSOR and Maxi-Cl.

2. TWO TYPES OF VOLUME-REGULATORY ANION CHANNELS: VSOR AND MAXI-CL

There are numerous types of anion channels in mammalian cells, and most of them are classified into six groups: ligand-gated, voltage-gated, cAMP-activated, Ca^{2+}-activated, volume-activated and acid-activated ones. Among them, VSOR, Maxi-Cl and ClC-2 can be activated by cell swelling, and thereby collectively called volume-activated anion channels (VAACs) (Okada et al., 2009). VSOR and Maxi-Cl are involved in the RVD mechanisms and therefore called volume-regulatory anion channels, whereas ClC-2 is, though activated by cell swelling, not implicated in the RVD process (Bond, Ambikapathy, Mohammad, & Valverde, 1998).

2.1 Roles of VSOR and Maxi-Cl in Cell Volume Regulation

Osmotic cell swelling, or the osmotic volume increase (OVI), is induced by water inflow driven by osmotic gradient across the plasma cell membrane caused by a decrease in the extracellular osmolarity or an increase in the intracellular osmolyte concentration (Fig. 1: OVI). Even when such hypotonic stress persisted, most animal cells can regulate their volume, in the time course within several to several tens of minutes, by the RVD event which is attained by water outflow driven by KCl efflux induced by electroconductive ion channels (Fig. 1: RVD) or electroneutral ion cotransporters (Bazúa-Valenti, Castañeda-Bueno, & Gamba, 2016; Hoffmann & Simonsen, 1989; Hoffmann et al., 2009; Kahle et al., 2015; Okada, 1997, 2004). Under pathological conditions, such as excitotoxicity, lactacidosis, and acidotoxicity, sustained cell swelling is brought about by water inflow driven either by the Donnan effect when Na^+/K^+ pump activity is impaired by cellular ATP depletion and/or by exaggerated NaCl inflow which obligatorily drives water influx (Fig. 1: NVI) (Okada et al., 2001; Okada, Maeno, et al., 2004). When the RVD mechanisms were impaired, cells suffer from persistent severe swelling called the necrotic volume increase

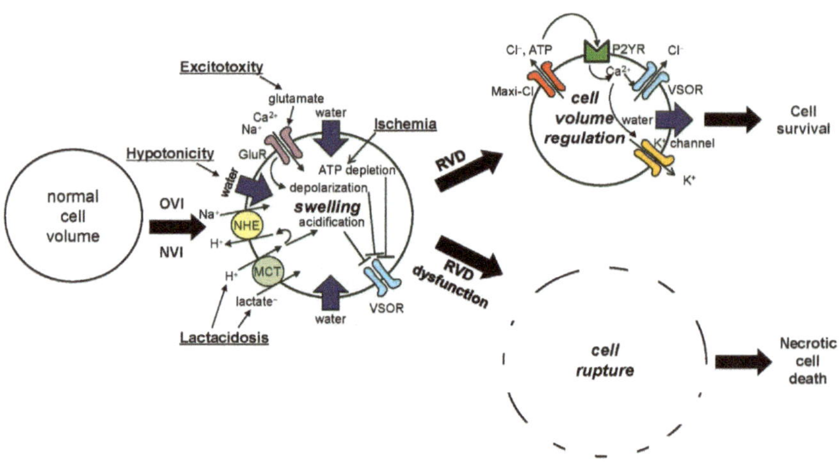

Figure 1 Schematic illustration of ionic mechanisms including roles of VSOR and Maxi-Cl therein for RVD after OVI or NVI. OVI, the osmotic volume increase induced by extracellular hypotonicity (or intracellular hypertonicity); NVI, the necrotic volume increase induced by excitotoxicity, lactacidosis and ischemia (see Okada, Maeno, et al., 2004 for details); RVD, the regulatory volume decrease induced by KCl efflux produced by activation of VSOR and Maxi-Cl channels as well as activation of K^+ channels in parallel (see text for details).

(NVI), thereby ultimately leading to rupture and necrotic cell death (Fig. 1: RVD dysfunction). Under normal conditions, swelling-induced activation of VSOR and Maxi-Cl plays a pivotal role in the RVD event by serving as the most effective pathway for volume-regulatory Cl^- efflux. Maxi-Cl is further involved in the RVD process by releasing anionic forms of ATP which stimulate, in an autocrine fashion, purinergic receptors, thereby leading to intracellular Ca^{2+} mobilization which is dually beneficial for the RVD efficiency because of direct activation of Ca^{2+}-activated K^+ channels (Dezaki, Tsumura, Maeno, & Okada, 2000) and indirect activation of VSOR mediated by Ca^{2+}-nanodomain-stimulated ROS production (Akita & Okada, 2014), as schematically depicted in Fig. 1 (right, top).

2.2 Properties and Activation Mechanisms of VSOR

The functional expression of VSOR (or VRAC) was first discovered in 1988 independently by two groups in human epithelial cells (Hazama & Okada, 1988) and in human lymphoid cells (Cahalan & Lewis, 1988). Succeeding works fully elucidated its phenotypical properties (for Reviews: Akita & Okada, 2014; Nilius et al., 1997; Okada, 1997, 2006; Okada, Kubo, Oiki, Petersen, Tominaga, Hazama, & Morishima, 1994; Okada, Petersen, Kubo, Morishima, & Tominaga, 1994; Strange, Emma, & Jackson, 1996), as summarized in Table 1. Except its low-field strength anion selectivity with Eisenman's type-I sequence ($I^- > Br^- > Cl^- > F^-$), VSOR activity exhibits distinct characteristics including the intermediate single-channel conductance, mild outward rectification of $I-V$ relationship, and voltage-dependent inactivation kinetics only at large positive potentials from those in Maxi-Cl activity (Table 1). Particularly, non-hydrolytic dependence on intracellular ATP of VSOR (Jackson, Morrison, & Strange, 1994; Lewis, Ross, & Cahalan MD, 1993; Oike, Droogmans, & Nilius, 1994; Oiki, Kubo, & Okada, 1994) is the most physiological property distinct from Maxi-Cl.

This non-hydrolytic ATP dependence suggests an essential role of some ATP-binding protein in the VSOR activation mechanism. Actually, our study showed an essential involvement of a cytosolic member of the ATP-binding cassette (ABC) transporter superfamily, ABCF2, therein (Ando-Akatsuka, Shimizu, Numata, & Okada, 2012). Osmotic cell swelling induces provocation of the protein–protein interaction between ABCF2 and an actin-binding protein, α-actinin-4 (ACTN4), thereby preventing ABCF2 from suppressing VSOR activity, presumably by unbinding ABCF2 from the VSOR channel proteins, as schematically depicted in

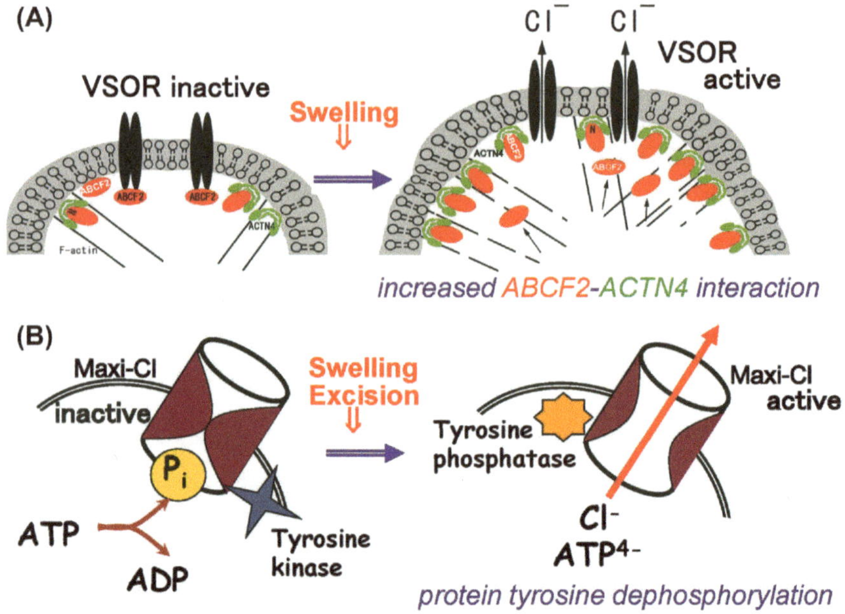

Figure 2 Schematic illustrations of activation mechanisms of VSOR (A) and Maxi-Cl (B) attained by increased interaction between ACTN4 and ABCF2 at its *N*-terminal domain (N) and by dephophorylation of protein tyrosine residues, respectively.

Fig. 2A. Since the sub-membrane actin ring is known to be disrupted by cell swelling (Carton, Hermans, & Eggermont, 2003; Erickson, Northrup, & Guilak, 2003; Levitan, Almonte, Mollard, & Garber, 1995; Morán, Sabanero, Meza, & Pasantes-Morales, 1996; Pedersen, Mills, & Hoffmann, 1999; Pritchard & Guilak, 2004), ACTN4 might be released from actin filaments and become available for the interaction with ABCF2 (Fig. 2A) (Ando-Akatsuka et al., 2012). Supporting this, the disruption of F-actin was found to enhance VSOR activity (Levitan, Almonte, Mollard, & Garber, 1995; Mosishima et al., 2000).

Depletion of plasmalemmal cholesterol was shown to enhance VSOR activity (Byfield et al., 2006; Levitan, Christian, Tulenko, & Rothblat, 2000; Romanenko, Rothblat, & Levitan, 2004), presumably through some mechanism mediated by Rho activity and F-actin (Klausen, Hougaard, Hoffmann, & Pedersen, 2006). Since cholesterol depletion has been shown to release cytoskeletal proteins including α-actinin from the cellular membrane fractions (Harder, Kellner, Parton, & Gruenberg, 1997), the effect of cholesterol depletion on VSOR activity may also be caused by facilitation of the ABCF2-ACTN4 interaction (see Fig. 2A).

VSOR activity was well known to be activated by intracellular application of GTPγS without any visible cell swelling (Doroshenko, Penner, & Neher, 1991; Doroshenko & Neher, 1992; Estevez, Bond, & Strange, 2001; Mitchell, Zhang, Wang, & Jacob, 1997b; Nilius, Oike, Zahradnik, & Droogmans, 1994; Shen, Wu, & Chou, 1996). This effect might be explained by enhancement of the ABCF2-ACTN4 interaction (see Fig. 2A) through the GTPγS effects on small G-proteins, especially Rho family members which regulate actin cytoskeleton dynamics (Pedersen, Hoffmann, & Mills, 2001).

Reduction of intracellular ionic strength was also known to activate VSOR currents (Cannon, Basavappa, & Strange, 1998; Nilius, Prenen, Voets, Eggermont, & Droogmans, 1998; Sabirov, Prenen, Tomita, Droogmans, & Nilius B, 2000; Voets, Droogmans, Raskin; Eggermont, & Nilius, 1999). Since reduced ionic strength increases the surface potential on a number of intracellular multivalent anionic proteins, the protein−protein interaction between ABCF2 and ACTN4 or the VSOR molecule may be much affected. Also, the actin-ACTN4 interaction may be affected by reduced ionic strength, because F-actin is known to be disrupted by reduced ionic strength (Miyata, Kinosita, & Marriott, 1997; Senger & Goldmann, 1995).

2.3 Properties and Activation Mechanisms of Maxi-Cl

In 1983, the anion channel exhibiting a huge unitary conductance, called MAC, was first found in cultured rat muscle cells (Blatz & Magleby, 1983). Most of MACs thereafter reported share with the phenotypical properties, as listed in Table 1, as well as very high anion selectivity with P_{Cl}/P_{Na} or P_{Cl}/P_K of ≥ 6, and therefore they are called Maxi-Cl (for Review: Sabirov et al., 2016). Although Maxi-Cl, like VSOR, is activated by osmotic cell swelling (Sabirov & Merzlyak, 2012; Sabirov & Okada, 2009; Sabirov et al., 2016; Strange et al., 1996), the characteristics of Maxi-Cl currents are distinct from the VSOR currents in the following properties: a very large single-channel conductance (300−500 pS), ohmic $I-V$ relationship, voltage-dependent inactivation kinetics at both positive and negative voltages, and inhibition by intracellular ATP (Table 1).

Maxi-Cl remains in the steadily inactivated state under the normal conditions but becomes activated by a variety of physiological and pathophysiological stimuli, including a hypotonic challenge (Coulombe & Coraboeuf, 1992; Dutta, Okada, & Sabirov, 2002; Dutta, Sabirov, Uramoto, & Okada, 2004, 2008; Falke & Misler, 1989; Jalonen, 1993;

Kurbannazarova, Bessonova, Okada, & Sabirov, 2011; Liu, Tashmukhamedov, Inoue, Okada, & Sabirov, 2006; Sabirov, Dutta, & Okada, 2001), G-protein-coupled receptor simulation (Kawahara & Takuwa, 1991; Sun, Supplisson, & Mayer, 1993; Sun, Supplisson, Torres, Sachs, & Mayer, 1992), salt stress (Bell et al., 2003), and ischemia or hypoxia (Dutta et al., 2004; Liu, Sabirov, & Okada, 2008a; Liu et al., 2006). The most effective stimulus is patch membrane excision in the absence of ATP (Bajnath, Groot, de-Jonge, Kansen, & Bijman, 1993; Bell et al., 2003; Bosma, 1989; Coulombe & Coraboeuf, 1992; Dutta et al., 2002; Dutta et al., 2004; Islam, Uramoto, Okada, Sabirov, & Okada, 2012; Jalonen, 1993; Kanno & Takishima, 1990; Kawahara & Takuwa, 1991; Kurubannazarova et al., 2011; Light et al., 1990; Liu et al., 2006; Mitchell, Wang, & Jacob, 1997a; Sabirov et al., 2006, 2001; Schlichter, Grygorczyk, Pahapill, & Grygorczyk, 1990; Toychiev et al., 2009). Facilitated activation by intracellular ATP deprivation may suggest an involvement of some protein dephosphorylation in the Maxi-Cl activation mechanism. In fact, even without hypotonic stimulation or any cell swelling, application of protein tyrosine kinase (PTK) inhibitors, AG18 and genistein, induced Maxi-Cl activation in the cell-attached mode on mouse mammary C127 cells and mouse adult skin fibroblasts (MAFs) (Toychiev et al., 2009) as well as mouse L929 cells (Islam et al., 2012). Furthermore, excision-induced activation of Maxi-Cl was found to be suppressed not only by several inhibitors of protein tyrosine phosphatases, including an inhibitor of receptor tyrosine phosphatase, in C127 cells and MAFs but also by knockout of a receptor tyrosine phosphatase ς (RPTPς) in MAFs (Toychiev et al., 2009). Thus, as schematically depicted in Fig. 2B, activation of Maxi-Cl is caused by tyrosine dephosphorylation of the Maxi-Cl component mediated by protein tyrosine phosphatases.

3. MOLECULAR IDENTITIES OF VSOR

For more than quarter of a century, since the discovery of VSOR currents, the molecular identity has not been uncovered, although a large number of laboratories devoted a lot of effort to struggle against many, at least five, false-positive candidates so far proposed (for Reviews: Clapham, 1998; Jentsch, Stein, Weinreich, & Zdebik, 2002; Hoffmann et al., 2009; Nilius et al., 1997; Nilius & Droogmans, 2003; Okada, 1997; Pedersen, Klausen, & Nilius, 2015; Pedersen, Okada, & Nilius, 2016). In 2014, one of the core component of VSOR was at last identified (Qiu et al., 2014;

Voss et al., 2014). This breakthrough has, however, opened novel enigmas to be unravelled by future studies.

3.1 False Positive Candidates Proposed Previously for VSOR

First, in 1992, **P-glycoprotein**, a multidrug-resistant ABC transporter, was reported to switch from a drug pump to VSOR upon cell swelling (Valverde et al., 1992). Nevertheless, this interesting 'transporter/channel bi-functional (or switching) hypothesis' was rejected by a variety of subsequent studies (see Reviews: Jentsch et al., 2002; Nilius et al., 1997; Okada, 1997), including our own studies (Miwa, Ueda, & Okada, 1997; Okada et al., 1997; Tominaga, Tominaga, Miwa, & Okada, 1995). Alternatively, we showed that P-glycoprotein modulates VSOR activity by enhancing its sensitivity to volume expansion (Miwa et al., 1997; Okada, 1997; Okada et al., 1997). It can be speculated that some ABC transporter protein, such as P-glycoprotein (ABCB1) and CFTR (ABCC7), may readily affect the VSOR activation mechanism which involves a cytosolic ABC protein, ABCF2 (Fig. 2A).

Second, also in 1992, a cytosolic 235-amino acid protein **pI_{Cln}** identified by expression cloning was reported to function as VSOR by a porin-like insertion mechanism (Paulmichl et al., 1992). The most compelling pieces of evidence for this intriguing 'anchor-insertion channel hypothesis' are the presence of the consensus nucleotide-binding GXGXG motif in the pI_{Cln} sequence and the observation of inhibitory effects of extracellular nucleotides, such as ATP, cAMP and cGMP, on pI_{Cln}-associated chloride currents (Paulmichl et al., 1992). As summarized in previous review articles (Nilius & Droogmans, 2001; Nilius et al., 1997; Okada, 1997; Strange, 1998), however, a variety of data at variance with this pI_{Cln} hypothesis were provided by succeeding studies. Especially, extracellular application of cAMP and cGMP never inhibited native VSOR currents (Tsumura, Oiki; Ueda; Okuma, & Okada, 1996). Also, the pI_{Cln} overexpression-induced Cl^- current was observed to be distinct from the VSOR current (Buyse et al., 1997; Voets et al., 1996). Moreover, pI_{Cln} inserted in lipid bilayers was found to exhibit currents with completely different properties with showing cation selectivity (Li et al., 1998; Strange, 1998). Furthermore, pI_{Cln} was later shown to be a spliceosome component (Pu, Krapivinsky, Krapivinsky, & Clapham, 1999). On the other hand, pI_{Cln} was found to modulate VSOR activity in a number of cell types (Chen, Wang, & Jacob, 1999; Gschwentner et al., 1995; Hubert et al., 2000; Krapivinsly et al., 1994). pI_{Cln} is known to form complexes with soluble actin and other cytosolic proteins (Krapivinsky, Ackerman, Gordon, Krapivinsky, & Clapham, 1994) and also interact with a homolog of the p21-activated protein kinases (PAKs) which is involved in cytoskeletal rearrangements (Krapivinsky, Pu, Wickman, Krapivinsky, & Clapham, 1998). Hence, there remains a possibility that pI_{Cln} may directly or indirectly modulate, under some conditions, the VSOR activity by affecting ABCF2-ACTN4 and ACTN4-actin interactions (Fig. 2A).

Third, in 1997, the hypothesis that the molecular identity of VSOR is **ClC-3** was put forward mainly based on the observations that heterologous overexpression of ClC-3 brought about basally active VSOR-like currents (Duan, Winter, Cowley, Hume, & Horowitz, 1997). A large variety of observations supporting this "ClC-3 = VSOR hypothesis" have subsequently been reported by this group, including inhibition of endogenous VSOR-like currents by some anti-ClC-3 antibodies (Duan et al., 2001; Wang et al., 2003; Yamamoto-Mizuma et al., 2004), by ClC-3 knockdown mediated by antisense oligonucleotides for ClC-3 (Hermoso et al., 2002), and by global knockout (Yamamoto-Mizuma et al., 2004) or cardiac-specific knockout (Xiong et al., 2010) of ClC-3, as well as augmentation of VSOR currents by cardiac-specific overexpression of ClC-3 (Xiong et al., 2009), and the *N*-terminal deletion mutant of ClC-3 yielding constitutively active VSOR-like currents (Rossow et al., 2006). Also, some other groups supported this ClC-3 hypothesis by providing the data showing inhibition of VSOR-like currents by anti-ClC-3 antibodies (Jin et al., 2003; Mao et al., 2011; Zhou, Ren, Qiu, He, & Guan, 2005), by antisense oligonucleotides for ClC-3 (Mao et al., 2008; Zhou, Ren, Qiu, He, & Guan, 2005), by shRNA for ClC-3 (Yin, Tong, Zhu, & Watsky, 2008), and by siRNA for ClC-3 (Liu et al., 2016; Wang et al., 2012; Yang et al., 2011, 2018). In addition, VSOR-like currents activated by TNFα or interleukin 1β were found to be suppressed, but not abolished, in endothelial cells derived from ClC-3 knockout mice (Yang et al., 2012a). On the other hand, a larger number of the data contradictory to the ClC-3 hypothesis have also been accumulated, as follows. (1) Endogenous ClC-3 localizes mainly to endosomes in mammalian cells (Miller et al., 2007; Okada, Akita, Sato-Numata, Islam, & Okada, 2014; Stobrawa et al., 2001; Stauber & Jentsch, 2013; Suzuki et al., 2006; Weylandt et al., 2007) and is involved in endosomal acidification (Hara-Chikuma et al., 2005; Stobrawa et al., 2001) by exerting a $2Cl^-/H^+$ antiporter function (Guzman, Grieschat, Fahlke, & Alekov, 2013; Guzman, Miranda-Laferte, Franzen, & Fahlke, 2015). (2) Transfection of exogenous ClC-3 was unable to express any anionic currents in the experiments performed in several laboratories (Friedrich, Breiderhoff, & Jentsch, 1999; Jentsch, Günther, Pusch, & Schwappach, 1995; Weylandt et al., 2001). (3) Even when a minor fraction of transfected ClC-3 was introduced into the plasma membrane by its overexpression, it was able to express volume-insensitive anion currents, which are distinct from the VSOR current, exhibiting strong outward rectification in the experiments performed in some laboratories (Huang et al., 2001; Li, Shimada, Showalter, & Weinman, 2000; Li, Wang, Zhao, & Weinman, 2002; Ogura et al., 2002; Okada, Akita, Sato-Numata, Islam, & Okada, 2014) with showing dependence on calcium/calmodulin-dependent protein kinase II (CaMKII) activation (Huang et al., 2001) or sensitivity to a heavy metal Cd^{2+} (Okada et al., 2014). (4) Also, endogenous swelling-activated VSOR currents were found to be unaffected by ClC-3 overexpression (Weylandt et al., 2001) and by an anti-ClC-3 antibody (Huang et al., 2001). (5) In addition, swelling-activated VSOR currents could be recorded in rat lacrimal cells which lack expression of endogenous ClC-3 (Majid, Brown, Best, & Park, 2001). (6) Furthermore, a variety of cell types derived from ClC-3 null mice were found to respond to osmotic cell swelling with activation of normal VSOR currents not only at the whole-cell level (Arreola et al., 2002; Behe et al., 2017; Gong et al., 2004; Matsuda, Filali, Moreland, Miller, & Lamb, 2010; Stobrawa et al., 2001) but also at the single-channel level (Wang, Takahashi, Uramoto, & Okada, 2005). (7) Here, we present additional evidence against this "ClC-3 = VSOR hypothesis" in Table 2; that is, the expression level of ClC-3 mRNA (*CLCN3*) in a VSOR-deficient cisplatin-resistant human cell line, KCP-4 (Lee et al., 2007), is not less than but indistinguishable from those of its parental VSOR-rich cisplatin-sensitive cell line, KB, and three other VSOR-rich human cell lines, HEK293T, HeLa and Intestine 407. Among a number of isoforms, say presumably six isoforms (ClC-3a to ClC-3f) in the case of mice, depending on variations in the *N*– and *C*-termini (Fig. 3), previous overexpression and knockdown studies mentioned above were carried out with using or targeting at ClC-3a (Duan et al., 1997; Guzman, Miranda-Laferte, Franzen, & Fahlke, 2015, 2013; Hara-Chikuma et al., 2005; Hermoso et al., 2002; Li, Shimada, Showalter, & Weinman, 2000; Li et al., 2002; Mao et al., 2008, 2011; Matsuda et al., 2010; Okada et al., 2014; Rossow et al., 2006; Weyland et al., 2001; Zhou et al., 2005), ClC-3b (Guzman et al., 2015; Huang et al., 2001; Li et al., 2002; Ogura et al., 2002; Stobrawa et al., 2001; Weyland et al., 2001; Yin, Tong, Zhu, & Watsky, 2008), ClC-3c (Guzman et al., 2015), and ClC-3e (Ogura et al., 2002). We have recently

Table 2 Expression Levels of mRNAs of all the Paralogs for ANO/TMEM16, BEST, and LRRC8 as Well as of ClC-3 Monitored by Microarray Assays in Cisplatin-Resistant KCP-4 and Its Parental KB Cells as Well as in Three Other Human Cell Lines

Gene Symbol	KCP-4 signal	KB signal	HEK293T signal	HeLa signal	Intestine 407 signal
ANO1	29.15	27.97	29.85	34.58	28.21
ANO2	58.85	109.88	57.21	**300.00**	**202.82**
ANO3	27.12	22.39	23.75	26.04	18.16
ANO4	21.29	79.86	34.59	24.26	73.03
ANO5	8.89	8.32	**213.97**	9.60	8.75
ANO6	**1076.77**	**1030.12**	**791.85**	**998.98**	**1271.42**
ANO7	118.21	112.89	104.97	115.86	104.72
ANO8	110.08	119.02	103.42	118.68	146.84
ANO9	99.20	97.01	75.87	88.53	91.26
ANO10	**515.73**	**610.47**	**479.28**	**513.09**	**325.58**
BEST1	46.44	58.07	**232.04**	55.84	63.72
BEST2	44.06	51.89	51.31	53.70	51.31
BEST3	16.46	14.43	13.26	14.11	13.31
BEST4	84.05	76.54	53.57	75.14	64.28
CLCN3	**583.24**	**366.00**	**595.60**	**550.19**	**360.56**
LRRC8A	**554.73**	**443.51**	**304.35**	**523.40**	**517.19**
LRRC8B	**765.92**	**555.64**	**444.46**	**348.98**	186.58
LRRC8C	**432.57**	196.48	**338.63**	171.28	104.46
LRRC8D	**1157.35**	**1241.54**	**1294.62**	**1325.61**	**1595.98**
LRRC8E	99.05	111.85	77.82	97.86	151.35

A signal level less than 100 can be regarded as little or very scanty expression of the given mRNA, and the signal level exceeding 200 is designated with bold font.

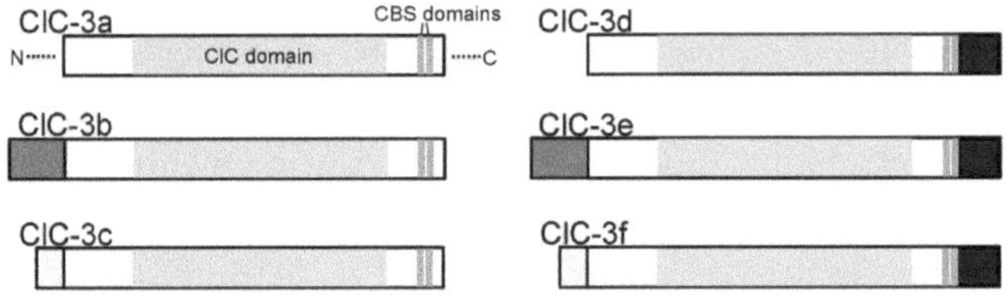

Figure 3 Six splice variants of mouse ClC-3. ClC-3a and -3d are the *N*-terminal short forms (sClC-3) without and with a variant *C*-terminus, respectively. ClC-3b and -3e are the *N*-terminal long forms without and with a variant *C*-terminus, respectively. ClC-3c and -3f are the *N*-terminal middle forms without and with a variant *C*-terminus, respectively.

accomplished a full cloning of ClC-3d (Fig. 3) which has an *N*-terminal amino acid sequence identical to ClC-3a and a *C*-terminal end different from ClC-3a but identical to ClC-3e, and found to exhibit Cd^{2+}-sensitive, volume-insensitive, strongly outward-rectifying anion currents, the properties of which are indistinguishable from those of ClC-3a, when overexpressed in HEK293T cells (Okada et al., 2014). On balance, it is concluded that ClC-3 is primarily expressed in intracellular vesicular components (mainly endosomes) within mammalian cells, and that ClC-3 exhibits volume-insensitive, Cd^{2+}-sensitive, strong outwardly rectifying anion channel currents, even when ClC-3 is forced to be expressed on the plasma membrane by overexpression. Since ClC-3 was reported to stimulate ROS production mediated by NOX under some conditions (Liu et al., 2013; Matsuda et al., 2010; Miller et al., 2007), there is a possibility that endosomal ClC-3 may act as a supporting player in VSOR activation, if any, by elevating ROS production (Shimizu, Numata, & Okada, 2004; Varela, Simon, Riveros, Jørgensen, & Stutzin, 2004) through activation of NOX co-localized with ClC-3 in endosomes, presumably by the charge-neutralizing role of ClC-3 in NOX-mediated electron flow across the endosomal membrane (Miller et al., 2007). ClC-3 is now known to function as an intracellular Cl^-/H^+ antiporter (Guzman, Grieschat;,Fahlke, & Alekov, 2013; Matsuda et al., 2008; Picollo & Pusch, 2005).

Fourth, in *Drosophila* S2 cells, *Drosophila* **bestrophin-1** (dBest1) was identified as one subset of swelling-activated anion channels on the basis of the observations that anionic currents sensitive to dBest1-targeted siRNA were activated by hypoosmotic swelling and inhibited by hyperosmotic shrinkage (Chien & Hartzell, 2007). This finding was later confirmed by unbiased genome-wide RNAi screening and by overexpression of dBest1 in *Drosophila* S2R+ cells (Stotz & Clapham, 2012). Although the properties of dBest1 currents exhibiting little rectification and no inactivation kinetics upon depolarizing pulses were completely different from those of VSOR (Chien & Hartzell, 2007), it was expected that mammalian bestrophins are somehow involved in VSOR activity. This inference looked to be supported by the observations that human bestrophin 1 (hBest1) and mouse bestrophin 2 (mBest2) activities were inhibited by cell shrinkage (Fischmeister & Hartzell, 2005; Xiao, Yu, Cui, & Hartzell, 2009). However, endogenous VSOR currents were found to be preserved intact in macrophages derived from mBest1/2 double knockout mice (Chien & Hartzell, 2008), and overexpression of mBest2 failed to lead to functional expression of any swelling-activated anion channels in the *Drosophila* cells (Stotz & Clapham, 2012). Also, swelling-induced VSOR activation was normally observed in Best1 H178P-expressing neutrophils from a Bestrophinopathy patient (Behe et al., 2017). Recently, on the other hand, a possibility that Best1 is responsible for mammalian VSOR has been reported, based on the observations that swelling-activated currents were increased in *Xenopus* oocytes transfected with hBest1 or mBest1 but not with the R218C or Y227N mutants and that hypotonicity-induced currents sensitive to 4,4'-diisothiocyano-2,2' stilbenedisulfonic acid (DIDS) were observed in retinal pigment epithelial (RPE) cells derived from iPS cells of healthy donor but not of macular dystrophy patients (Milenkovic et al., 2015). However, the swelling-induced currents observed in the oocytes transfected with hBest1 or mBest1 exhibited non-VSOR-typed properties such as ohmic $I-V$ relationship and no inactivation kinetics. Also, little outward rectification of swelling-induced currents was observed in the iPS-derived RPE cells. Moreover, the reversal potential (E_{rev}) of the latter currents was largely different from the E_{Cl} value. Therefore, it is clear that bestrophins are different from the molecular identity of VSOR. In fact, several human cell lines endogenously expressing plenty of VSOR activity were found to virtually lack the molecular expression of all paralogs of Best by our recent gene expression microarray studies, as summarized in Table 2.

Fifth, some members of the family of **TMEM16**, which is also called anoctamin (**ANO**) and forms a family of 10 proteins, have been proposed as the VSOR candidates, based on the observations that functional expression of VSOR in human epithelial cell lines was suppressed by knockdown of TMEM16A/ANO1, TMEM16F/ANO6, TMEM16H/ANO8 and TMEM16J/ANO9 proteins (Almaça et al., 2009). However, VSOR activity is not dependent on TMEM16A, because this mRNA was found to be absent in HL-60 cells which functionally express robust VSOR activity (Hernández-Carballo, De Santiago-Castillo, Rosales-Saavedra, Pérez-Cornejo, & Arreola, 2010). Kunzelmann's group subsequently reported that hTMEM16F is an essential component of outwardly rectifying Cl^- channel activated by apoptotic stimuli such as a Fas ligand (FasL) and STS, in a manner sensitive to siRNA for hTMEM16F, in Jurkat T cells and human airway epithelial cells (Martins et al., 2011). Since VSOR currents are known to be activated by FasL and STS (Okada et al., 2006; Shimizu et al., 2004), the possibility arises that TMEM16F is an essential component of VSOR. This hypothesis seemed consistent with our microarray data showing marked molecular expression only for hTMEM16F and hTMEM16K/hANO10 in several VSOR-rich human cell lines (Table 2). At variance, however, neither overexpression nor siRNA-mediated knockdown of hTMEM16F affected endogenous VSOR currents activated by osmotic swelling or apoptotic stimulation in human epithelial cell lines

(Shimizu et al., 2013). In addition, overexpression and miRNA-mediated knockdown of mTMEM16F failed to increase and suppress, respectively, endogenous VSOR currents in mouse cell lines (Juul et al., 2014). Furthermore, hTMEM16F was found to be a component of CaCC activity characterized by Ca^{2+} sensitivity and strong outward rectification (Grubb et al., 2013; Shimizu et al., 2013; Szteyn et al., 2012) or the Ca^{2+}-activated nonselective cation channel (NSCC) activity associated with phospholipid scramblase (Yang et al., 2012b; Yu et al., 2015). Thus, it is concluded that TMEM16F is unrelated to the VSOR activity that is activated upon cell swelling and apoptotic stimulation. In contrast to this conclusion, Sirianant et al. (Sirianant, Ousingsawat, Wanitchakool, Schreiber, & Kunzelmann, 2016) has recently reported that TMEM16F produces VSOR (or VRAC) activity. However, their TMEM16F-associated currents look to be non-anionic but much contaminated with Ca^{2+}-activated NSCC currents judging from the E_{rev} values. On the other hand, heterologous expression of hTMEM16K, but not its R263H mutant, in *Xenopus* oocytes, HEK293 cells and lymphocytes was reported to augment VSOR currents activated by hypotonic stimulation (Hammer et al., 2015), suggesting ANO10 as a component of VSOR. However, the currents observed in TMEM16K-transfected HEK293 cells are not anion-selective, because E_{rev} values observed before and after reduction of extracellular Cl^- concentration are both markedly different from the E_{Cl} values. When transfected, TMEM16K was moreover observed to be localized primarily in the endoplasmic reticulum (ER) in HEK293 cells, and overexpression of TMEM16K-R263H was found to inhibit the hypotonicity-induced intracellular Ca^{2+} rise, suggesting that TMEM16K is involved in the control of intracellular Ca^{2+} signaling (Hammer et al., 2015). Therefore, it appears that TMEM16K is also unrelated to the molecular identity of VSOR. Nevertheless, the possibility remains that some member(s) of TMEM16 is involved in regulation of VSOR activity by affecting some step(s) in the mechanisms of VSOR activation (Fig. 2A), by taking the following facts into consideration: TMEM16A, TMEM16F and TMEM16K are involved in intracellular Ca^{2+} signaling (Hammer et al., 2015; Singh et al., 2014; Wanitchakool et al., 2017), TMEM16A interacts with the ezrin-radixin-moesin (ERM) complex, which links plasmalemmally associated proteins with the cytoskeleton (Perez-Cornejo et al., 2012), and TMEM16C/ANO3 and TMEM16F regulate the Na^+-activated K^+ channel Slack and the Na^+-Ca^{2+} exchanger 1 (NCX1), respectively (Huang et al., 2013; Ousingsawat et al., 2015).

3.2 Molecules Recently Identified as the Core Component of VSOR

Learning from the fact that every previous failure was caused by making single gene expression or silencing approaches, the fresh start to identify the genuine VSOR molecules has to adopt unbiased genome-wide approaches, because VSOR activity and its activation mechanisms may be susceptible to regulation by multiple molecular factors. At last, actually, both Jentsch's and Patapoutian's groups independently succeeded in identification of the core component of human VSOR, hLRRC8A or SWELL1, by using the same genome-wide siRNA screening methods with an I^--sensitive yellow fluorescent protein and published their papers on the same day in two different journals (Qiu et al., 2014; Voss et al., 2014). Subsequently, mLRRC8A was also demonstrated to be essentially involved in VSOR activity in a number of mouse cells (Behe et al., 2017; Okada, Islam, Tsiferova, Okada, & Sabirov, 2017; Platt et al., 2017; Wang et al., 2017). The specific role of **LRRC8A** in VSOR activity was evidenced by the observations that siRNA-mediated knockdown of LRRC8A prominently inhibited swelling-activated VSOR Cl^- currents but never affected Cl^- currents conducted via the Ca^{2+}-activated Cl^- channel (CaCC), cAMP-activated Cl^- channel (CFTR) channels, Maxi-Cl, and acid-sensitive outwardly rectifying Cl^- channel (ASOR) in mouse C127 cells (Okada et al., 2017).

None of other LRRC8 family members (LRRC8B to 8E) can substitute for LRRC8A, but functional VSOR activity was found to require LRRC8A together with LRRC8C, 8D or 8E in human HCT116 cells (Voss et al., 2014). Consistent with this observation, in human HeLa cells, endogenous VSOR activity was abolished not only by quadruple knockdown of LRRC8B, 8C, 8D and 8E but also by triple knockdown of LRRC8C, 8D and 8E (Sato-Numata, Numata, Inoue, Sabirov, & Okada, 2017). Also, triple disruption of LRRC8C, 8D and 8E abolished endogenous VSOR currents in human HEK293 cells (Lutter, Ullrich, Lueck, Kempa, & Jentsch, 2017). Therefore, it is natural to assume that LRRC8A forms heteromers with LRRC8C, 8D and 8E but presumably not with LRRC8B. Actually, LRRC8D was shown to directly interact with LRRC8A, 8B and 8C by immunoprecipitation studies coupled to mass spectrometry (Lee, Freinkman, Sabatini, & Ploegh, 2014), and the interaction between LRRC8A, 8C and 8E was evidenced by sequential co-immunoprecipitation studies (Lutter, Ullrich, Lueck, Kempa, & Jentsch, 2017).

Since LRRC8 members belong to the pannexin family (Abascal & Zardoya, 2012), it is likely that LRRC8A and LRRC8C/8D/8E form hexameric channels. In fact, recently using cryo-electron microscopy and X-ray crystallography, the hexameric structure of channel formed by purified recombinant LRRC8A alone or together with LRRC8C was verified (Deneka, Sawicka, Lam, Paulino, & Dutzler, 2018).

The following four lines of evidence suggest that LRRC8A *plus* 8C/8D/8E form the pore of VSOR channel or are located very close to it. First, the inactivation kinetics observed at large positive potentials was affected by the combination of LRRC8 members: LRRC8A *plus* 8C heteromers exhibited a slower inactivation time course upon application of depolarizing pulses, whereas LRRC8A *plus* 8E currents inactivated faster in HCT116 cells (Voss et al., 2014; Ullrich, Reincke, Voss, Stauber, & Jentsch, 2016); and the C-terminal part of the first extracellular loop of LRRC8E was found to determine voltage-dependent VSOR inactivation in this cell line (Ullrich, Reincke, Voss, Stauber, & Jentsch, 2016). Second, VSOR permeability to I^- relative to that to Cl^- (P_I/P_{Cl}) was found to be increased by the substitution of threonine-44 of LRRC8A by cysteine (T44C) from 1.3 to 1.6 in HeLa cells (Qiu et al., 2014) and to be slightly decreased by combinatory charge-reversed mutations of K98E of LRRC8A together with K100E of LRRC8C or K91E of LRRC8E from 1.25 to 1.12 of P_I/P_{Cl} in HCT116 cells (Ullrich et al., 2016). Third, the complex of LRRC8A *plus* 8D/8E purified from HeLa cells could form anion channels activated by hypotonic stress or reduced intracellular ionic strength, when reconstituted in lipid bilayers, and its subunit composition affected the single-channel conductance, the degree of outward rectification, and the P_I/P_{Cl} values (Syeda et al., 2016). Fourth, very recently, Deneka and co-authors (Deneka, Sawicka, Lam, Paulino, & Dutzler, 2018) have clearly demonstrated that the LRRC8A hexamer incorporates a pore-like vase-shaped structure with the radius of 0.29 nm at the extracellular subdomain (ESD) and of 3.2 nm at the transmembrane and cytoplasmic-side subdomain (CSD), although this structure may represent the closed-state conformation, because the protein structure was determined in hypertonic, high ionic strength salt solution. Moreover, this study showed that when the total positive charge at the ESD region was decreased by introducing the R103A mutation in LRRC8A, the channel formed by this mutant *plus* LRRC8C exogenously expressed in $LRRC8^{-/-}$ HEK293 cells became more Na^+-permeable and less selective to SO_4^{2-}, the fact suggesting that the ESD region may function as the anion-selective filter.

However, these experimental results cannot as yet provide sufficient proof to convince that LRRC8A *plus* 8C/8D/8E for the VSOR pore *per se*. First, there remains a possibility that LRRC8 members indirectly affect the inactivation kinetics by binding to or interacting with some other regulatory subcomponent serving as a decelerating or accelerating factor. In fact, in HeLa cells, in contrast to HCT116 cells, the inactivation kinetics of VSOR currents was never affected by siRNA-mediated knockdown of LRRC8C and 8E but accelerated rather by the knockdown of LRRC8A *per se*, and moreover the single knockdown of any LRRC8 members failed to retard the inactivation rate (Sato-Numata, Numata, Inoue, Sabirov, & Okada, 2017). Second, the following I^- and Cl^- permeability studies may rather suggest that the T44 and K98 residues of LRRC8A, K100 of LRRC8C, and K91 of LRRC8E do not localize at the anion-selective filter within the VSOR pore: the order of $I^- > Cl^-$ permeability sequence was never affected by a charge-modifying mutation of T44 of LRRC8A (Qiu et al., 2014) and by combinatory charge-modifying mutations of K98 of LRRC8A, K100 of LRRC8C and K91 of LRRC8E (Ullrich et al., 2016), and, furthermore, the anion permeability sequence was not affected by charge-conferring T44E and T44R mutations of LRRC8A (Qiu et al., 2014) and even by charge-reversing K98E, K100E and K91E mutations of LRRC8A, LRRC8C and LRRC8E, respectively (Ullrich et al., 2016). Third, the channels formed by a charge-neutralizing R103A mutant of LRRC8A *plus* LRRC8C still maintained the general preference for anions; and moreover, LRRC8A *plus* a positive charge-conferring L105R mutant of LRRC8C failed to affect anion selectivity (Deneka et al., 2018). Fourth, one cannot exclude a possibility that the channel reconstituted by Syeda et al. (2016) cannot represent the *bona fide* VSOR pore for the following reasons. (1) The reconstituted channel activity was found to be maintained even in the absence of intracellular ATP, the fact is at variance with native VSOR channel activity prerequisitely dependent on intracellular ATP (Jackson et al., 1994; Oiki et al., 1994). (2) The reconstituted LRRC8 channel currents did not show voltage-dependent inactivation kinetics even at +100 mV, although the inactivation was observed in native VSOR currents observed in HeLa cells in this study as well as other studies (Sato-Numata, Numata, Inoue, & Okada, 2016; Sato-Numata et al., 2017). (3) The single-channel conductance (γ) of the reconstituted LRRC8 channels was only around 8 pS and 50–70 pS at +100 mV in 70-mM and 500-mM Cl^- conditions, respectively, and looked considerably smaller than that of native VSOR in HeLa cells (30–80 pS at positive potentials in

~100-mM Cl^- conditions). (4) Also, in this reconstitution system, DCPIB-sensitive hypotonicity-induced single-channel events with γ of ~10 pS at −100 mV (though not at +100 mV) were surprisingly found to be produced by LRRC8A-containing complexes purified from LRRC8(B/C/D/E)$^{-/-}$ HeLa cells in which LRRC8B, 8C, 8D and 8E were quadruply knocked out. Thus, further studies are warranted to determine whether LRRC8A *plus* 8C/8D/8E *per se* represent the native VSOR channel pore which can show a full set of phenotypic properties.

3.3 Missing Components or Subcomponents yet Unidentified for VSOR

Overexpression of LRRC8A alone was found to paradoxically bring about prominent suppression of endogenous VSOR currents in three different human cell lines (Okada et al., 2017; Qiu et al., 2014; Voss et al., 2014), whereas overexpression of LRRC8D or 8E alone never suppressed the endogenously residual small VSOR current in KCP-4 cells (Okada et al., 2017). These results suggest that the excess of LRRC8A expression exhibits a dominant negative effect presumably due to disturbance of subunit stoichiometry leading to formation of pore-disrupting or -closing complex. Actually, LRRC8A-induced suppression was prevented by co-expression with LRRC8C in HEK293 cells (Voss et al., 2014) and with LRRC8D or 8E in KCP-4 cells (Okada et al., 2017). However, such double overexpression of LRRC8A and 8C/8D/8E never increased VSOR currents over the endogenous level (Okada et al., 2017; Voss et al., 2014). This fact strongly suggests that some as-yet-unidentified pore-related component or regulatory subcomponent (X) other than LRRC8 members is required for VSOR activity. Furthermore, the expression levels of mRNAs not only for LRRC8A but also for other LRRC8 isoforms are not much different between VSOR-deficient cisplatin-resistant KCP-4 cells and the parental VSOR-rich cisplatin-sensitive KB cells (Okada et al., 2017; also in Table 2). Moreover, expression levels of LRRC8B and 8C mRNAs in KCP-4 cells look rather higher, though not distinctly, than those in four types of VSOR-rich human cell lines (KB, HEK293T, HeLa and Intestine 407), whereas those of LRRC8A, 8D and 8E mRNAs in the VSOR-deficient cell line are comparable to those in VSOR-rich cells (Table 2). These results provide additional evidence for requirement of X in the VSOR channel complex. In addition, it should be stressed that the channel reconstitution with such a full set of components is to exhibit phenotypical properties including not only anion selectivity and swelling-induced activation but also intracellular ATP requirement, mild outward rectification, voltage-dependent inactivation kinetics, and availability of ABCF2- and DCPIB-binding sites.

More essentially, worth mentioning is that the most compelling evidence for formation of the pore, in which the anion selectivity filter is built in, is still missing in the studies hitherto carried out with LRRC8 members. Thus, a possibility remains that the anion selectivity filter involves the X molecule together with LRRC8 members. In any case, further studies must be done to verify that the charge-modifying mutation(s) of any LRRC8 member(s), especially at the ESD region (Deneka et al., 2018), or of X molecule results in transition of anion selectivity from the anion permeability sequence of Eisenman-I ($I^- > Br^- > Cl^- > F^-$) to another Eisenman's type or conversion of channel property from anion-selective to cation-selective.

[Addendum] The CRISPR-based screening showed that LRRC8A and 8C/8D/8E genes are not essential for survival and proliferation in human cancer cell lines (Wang et al., 2015). Also, HCT116 and HEK293 cells were unexpectedly found to survive and proliferate even after disruption of all five LRRC8 genes (Voss et al., 2014). In addition, a girl carrying a *C*-terminal truncated variant of the LRRC8A gene, from whom the LRRC8A gene was first identified, was able to survive with suffering only from congenital agammaglobulinemia due to inhibition of B cell development (Sawada et al., 2003). Furthermore, *Lrrc8a*$^{-/-}$ mice are not embryonic lethal but appear normal at birth, although they exhibit retarded growth and impaired thymic development and peripheral T cell function as well as postnatal mortality (Kumar et al., 2014). Also, *Lrrc8c*$^{-/-}$ mice can survive without showing marked changes in body weight on a chow diet, although they have significant lower body weight when fed a high-fat diet, indicating that LRRC8C is a key factor in the development of obesity caused by a high-fat diet (Hayashi et al., 2011). These facts are in contrast to known essential roles of VSOR activity in cell volume regulation, proliferation, migration, differentiation, necrosis and apoptosis (for Reviews: Akita & Okada, 2014; Eggermont, Trouet, Carton, & Nilius, 2001; Hoffmann et al., 2009; Okada, Maeno, et al., 2004; Pedersen et al., 2015, 2016). This enigma must be addressed in future studies. It is possible that some other redundant volume regulatory mechanism became activated in the absence of functioning LRRC8 members. Identification of the X molecule might lead to elucidation of this redundant volume regulatory mechanism or *vice versa*.

4. MOLECULAR IDENTITIES OF MAXI-CL

Although Maxi-Cl is known to be functionally expressed with exhibiting well-defined phenotypic properties in a wide variety of cell types, its molecular nature remained undetermined for a long time since the discovery of this channel current in 1983. After dismissing many, at least four, candidates suggested so far (for Reviews: Okada et al., 2009; Sabirov & Okada, 2005, 2009), the core component molecule for Maxi-Cl has been identified in 2017 (Sabirov et al., 2017).

4.1 False Positive Candidates Suggested Previously for Maxi-Cl

The voltage-dependent anion channel (VDAC), also called porin, is normally expressed in the mitochondrial outer membrane and crucial for regulation of apoptosis by forming the complex with the adenine nucleotide translocase (ANT) thereby participating in cytochrome *c* release (Vyssokikh & Brdiczka, 2003) and by interacting with anti-apoptotic proteins (Tsujimoto & Shimizu, 2007). In addition, some VDACs were shown to be expressed also in the plasma membrane (De-Pinto, Messina, Lane, & Lawen, 2010). The plasmalemmal subtype of VDAC (**pl-VDAC**) has long been deemed as the most likely candidate molecule for the Maxi-Cl (Akanda & Elinder, 2006; Buettner, Papoutsoglou, Scemes, Spray, & Dermietzel, 2000; Bahamonde, Fernandez—Fernandez, Guix, Vazquez, & Valverde, 2003; Dermietzel et al., 1994; Elinder et al., 2005; Hardy & Valverde, 1994; Okada, O'Neal, et al., 2004). The first basis for this 'Maxi-Cl = pl-VDAC hypothesis' was similarity of biophysical properties between VDAC and Maxi-Cl, such as a large unitary conductance and voltage-dependent inactivation, thereby exhibiting bell-shaped voltage dependence of P_o (Colombini, 2004; Colombini, Blachly-Dyson, & Forte, 1996; Mannella, 1997; Sabirov et al., 2016) as well as similar pore radii estimated by the non-electrolyte partitioning method (Krasilnikov, Carneiro, Yuldasheva, Campos-de-Carvalho, & Nogueira, 1996; Sabirov & Okada, 2004b) and ATP conductivity with exhibiting open-channel block by ATP (Rostovtseva & Bezrukov, 1998; Rostovtseva & Colombini, 1997; Rostovtseva, Komarov, Bezrukov, & Colombini, 2002; Sabirov et al., 2001). Nevertheless, stricter inspection of biophysical properties revealed crucial differences between VDAC and Maxi-Cl under the same experimental conditions (Sabirov & Merzlyak, 2012; Sabirov et al., 2016). In contrast to the phenotypic properties of Maxi-Cl, VDAC shows poor anion-to-cation selectivity with the P_{Cl}/P_K value of around 1.8 (Blachly-Dyson, Peng, Colombini, & Forte, 1990; Colombini, 1989) and only partial, incomplete voltage-dependent inactivation (Sabirov & Merzlyak, 2012). The second circumstantial evidence supporting the "Maxi-Cl = pl-VDAC hypothesis" was provided by the observation of the inhibitory effect of anti-VDAC1 monoclonal antibody on the MAC activity in rat astrocytes (Dermietzel et al., 1994). Also, anti-VDAC antibodies were found to block Maxi-Cl activity in mouse neuronal hippocampal HT22 cells treated with STS (Elinder et al., 2005). However, it is not easy to rule out a possibility of non-specific effects of antibodies. Moreover, the MAC currents observed by Dermietzel et al. (1994) exhibited poor anion-selectivity with P_{Cl}/P_K of around 1.3, being in stark contrast to strong anion selectivity of Maxi-Cl. The third supporting evidence was the observations that PC12 cells stably transfected with pl-VDAC1 and mitochondrial VDAC1 exhibited MAC-like single-channel events with a bell-shaped voltage dependence of P_o (Buettner et al., 2000). However, their study did not provide any evidence for its anion selectivity. In our laboratory, moreover, overexpression of the pl-VDAC gene in HEK293T cells failed to reproduce phenotypical Maxi-Cl activity and any other MAC activity (Sabirov & Merzlyak, 2012). The fourth basis for the "Maxi-Cl = pl-VDAC hypothesis" was provided by Bahamonde et al. (2003) who reported that antiestrogen-activated MAC-like channel activity was suppressed by transfection with antisense oligonucleotides against the signal peptide of pl-VDAC in C1300 neuroblastoma cells. Nonetheless, the large-conductance anion channel observed in C1300 cells showed slight outward rectification and inactivation gating only at negative voltages, being at variance with the Maxi-Cl phenotype. Moreover, we demonstrated that prominent Maxi-Cl activity was never affected by silencing the VDAC2 gene in VDAC1/VDAC3-double-deficient fibroblasts (Sabirov et al., 2006). Taken together, it is concluded that the molecular identity of Maxi-Cl is distinct from any VDAC isoforms.

In relation to VDAC, ANT or the ADP/ATP carrier (**AAC**), which is coined SLC25A4, plays a central role in the mitochondrial permeability transition (mPT). Although this protein is normally located on the inner mitochondrial membrane, it was also found to be expressed in the neuronal plasmalemma (Gualix, Pintor, & Miras-Portugal, 1999). Also, AAC was shown to form 300—600-pS channels when reconstituted in giant liposomes (Brustovetsky & Klingenberg, 1996), in a manner sensitive to bongkrekic acid (Brustovetsky, Tropschug, Heimpel, Heidkämper, & Klingenberg,

2002). Thus, the possibility arises that AAC is a component of Maxi-Cl. However, this reconstituted large-conductance channel exhibited a cation selectivity with $P_K/P_{Cl} = 4.3$ (Brustovetsky & Klingenberg, 1996). Also, Maxi-Cl activity recorded in C127 cells was found to be insensitive not only to bongkrekate but also to another potent AAC blocker, atractyloside (Sabirov & Okada, 2005). Furthermore, Maxi-Cl channel events were never affected by gene silencing of AAC in C127 cells (Merzlyak PG, Sabirov RZ & Okada Y: unpublished data).

By Suzuki and Mizuno (2004), the human **TTYH1** gene homologous to the gene called *tweety* expressed in the *Drosophila flightless* locus was proposed to form a swelling-activated large-conductance anion channel. In addition, Suzuki (2006) reported that an hTTYH1 splice variant which lacks the *C*-terminal glutamate-rich domain also forms swelling-activated Gd^{3+}-sensitive anion channels. At variance, two splice variants of hTTYH1 failed to functionally express the single-channel activity of Maxi-Cl, when transfected into Maxi-Cl-lacking HEK293 cells (Okada et al., 2009). Also, heterologous expression of hTTYH2 in HEK293 cells failed to reproduce Maxi-Cl activity (Islam, Merzlyak, Okada, Sabirov, & Okada, 2013).

Between Maxi-Cl and pannexin 1 (Panx1), there are functional similarities not only in providing ATP release routes but also in forming plasmalemmal large-conductance ion channels. Thus, the possibility that Panx1 is related to the Maxi-Cl molecule must next be addressed. In the oocyte-expression system, **Panx1** was found to form an ATP-permeable ion channel which exhibits the single-channel conductance of 450−475 pS and is activated by membrane stretch (Bao, Locovei, & Dahl, 2004; Locovei, Bao, & Dahl, 2006), although Panx1 can also form the ATP-impermeable intermediate conductance (~50 pS) ion channel depending on the type of stimulation (Wang et al., 2014). Also, Panx1 was reported to serve as an endogenous large-conductance (530 pS) ion channel with a linear $I-V$ relationship in hippocampal neurons (Thompson, Zhou, & MacVicar, 2006). Despite these several similarities in channel properties, there are remarkable differences between the Panx1-associated channel and Maxi-Cl. For example, in contrast to highly anion-selective Maxi-Cl, Panx1 channels are not selective to anions but permeable to both anions and cations (Bao et al., 2004; Thompson et al., 2006) and did not show inactivation kinetics at large potentials (Locovei et al., 2006; Thompson et al., 2006). In addition, pharmacological properties of both channels are distinct from each other. Panx1 channels were sensitive to a relatively Panx1-specific blocker probenecid (Beckel et al., 2014; Kienitz, Bender, Dermietzel, Pott,

& Zoidl, 2011; Orellana, Montero, & von Bernhardi, 2013; Qiu, Wang, Spray, Scemes, & Dahl, 2011; Sridharan et al., 2010), a hemichannel blocker carbenoxolone (CBX) (Beckel et al., 2014; 2015; Bruzzone, Barbe, Jakob, & Monyer, 2005; Ma, Hui, Pelegrin, & Surprenant, 2009; Kienitz et al., 2011; Pelegrin & Surprenant, 2006; Sridharan et al., 2010; Qiu, Wang, Spray, Scemes, & Dahl, 2011; Orellana et al., 2013; Xia et al., 2012) and a Panx1-specific peptide blocker ^{10}Panx1 (Orellana et al., 2013; Pelegrin & Surprenant, 2006; Sridharan et al., 2010), whereas Maxi-Cl activity was not affected by probenecid, CBX and ^{10}Panx1 (Islam et al., 2012; Sabirov et al., 2016). Also, Panx1 channels were insensitive to a Maxi-Cl blocker Gd^{3+} (Ma et al., 2009). Moreover, Maxi-Cl currents were not inhibited by siRNA-mediated knockdown of Panx1 or Panx2 gene in L929 cells (Islam et al., 2012). In addition, the expression level of Panx1-mRNA in Maxi-Cl-rich C127 cells was not so prominent but rather much lower than those in Maxi-Cl-deficient HEK293T, HeLa and Intestine 407 cells (Table 3).

In relation to pannexins, connexins were also reported to form hemichannels which are involved in ATP release (Bennett et al., 2012; Cotrina et al., 1998). Among connexin family members, connexin 43 (**Cx43**) mRNA (Gja1) was found to be prominently expressed in both Maxi-Cl-rich C127 and L929 cells (Table 3). Moreover, our genome-wide microarray analysis showed that Cx43 is expressed at a markedly lower level in a

Table 3 Expression Levels of mRNAs of Candidate Molecules for ATP-Release Channels Monitored by Microarray Assays in Mouse C127 and L929 Cell Lines as Well as in Three Other Human Cell Lines

Gene Symbol	C127 signal	L929 signal	HEK293T signal	HeLa signal	Intestine 407 signal
Slco2a1	**4283.99**	**855.56**	102.31	70.90	47.95
LRRC8A	**1183.18**	**918.37**	**304.35**	**523.40**	**517.19**
LRRC8D	**530.37**	**602.91**	**1294.62**	**1325.61**	**1595.98**
Gjb2 (CX26)	85.12	72.57	68.72	55.99	54.67
Gjb6 (CX30)	51.59	54.77	42.06	33.96	37.47
Gjb1 (CX32)	101.56	84.41	36.72	39.73	40.25
Gja1 (CX43)	**1704.53**	**1613.41**	**363.14**	59.00	40.33
Gja3 (CX46)	182.01	180.17	123.03	107.06	90.62
Gja8 (CX50)	93.17	93.17	53.81	45.60	57.53
Panx1	139.45	**306.57**	**311.79**	**844.35**	**562.94**
CALHM1	112.38	126.13	72.75	90.20	58.18

A signal level less than 100 can be regarded as little or very scanty expression of the given mRNA, and the signal level exceeding 200 is designated with bold font.

murine cell line which lacks Maxi-Cl activity than that in mouse C127 cells which exhibit robust Maxi-Cl activity (Okada, Islam, Merzlyak, Sabirov, & Okada, 2012). Thus, the possibility arises that Cx43 is a candidate for the Maxi-Cl molecule. However, Maxi-Cl activity in murine L929 cells was not affected not only by a Cx43-specific blocking peptide, Gap27, but also by Cx43-specific siRNA (Islam et al., 2012; Okada et al., 2012). Also, a broad-spectrum connexin inhibitor, octanol-1, failed to inhibit Maxi-Cl activities in cardiomyocytes (Dutta et al., 2004), C127 cells (Sabirov & Okada, 2005) and L929 cells (Islam et al., 2012).

Calcium homeostasis modulator 1 (**CALHM1**) was identified to be a voltage-gated ATP-release channel in mouse taste bud cells (Taruno et al., 2013), mouse nasal epithelial cells (Workman et al., 2017) and porcine urinary bladder cells (Sana-Ur-Rehman, Markus; Moore; Mansfield, & Liu, 2017). Since CALHM1 channels are sensitive to a Maxi-Cl blocker Gd^{3+} (Dreses-Werringloer et al., 2008; Ma et al., 2012; Tanis et al., 2013), a possible involvement of CALHM1 in Maxi-Cl may be deemed. However, the properties of currents conducted by Maxi-Cl are distinct from those by CALHM1 channels which are cation-selective (but not anion-selective) channels exhibiting outward rectification, activation (but not inactivation) kinetics upon depolarization, and small (but not very large) unitary conductance (24 pS) (Dreses-Werringloer et al., 2008; Ma et al., 2012; Tanis et al., 2013). In addition, the expression level of CALHM1 mRNA in Maxi-Cl activity-rich C127 cells was not so prominent (Table 3). Thus, these results provide evidence that the Maxi-Cl molecule is distinct from CALHM1.

4.2 Molecules Recently Identified as the Core Component of Maxi-Cl

By applying unbiased genome-wide approaches, **SLCO2A1** has recently been at last identified as the core component of Maxi-Cl (Sabirov et al., 2017). In this study, the following stepwise procedures were conducted. First, the membrane blebs were isolated from C127 cells, which are known to exhibit the highest Maxi-Cl activity among many cell types so far studied (Sabirov et al., 2001), and confirmed that Maxi-Cl channels are highly accumulated in the constitutively activated state on the bleb membrane. After solubilization, the total membrane proteins were subjected to proteomics analysis by liquid chromatography mass spectrometry (LC-MS/MS), yielding 439 genes including 15 genes encoding proteins with multiple transmembrane-spanning domains. After siRNA and miRNA screening,

SLCO2A1 was selected as the candidate molecule for Maxi-Cl. Genome-wide differential microarray analysis was also, in parallel, performed between Maxi-Cl activity-rich C127 cells and Maxi-Cl activity-deficient C1300 cells, yielding 756 genes. SLCO2A1 gene was found to be top ranked in the list of genes showing large differences. The role of SLCO2A1 as a core component of Maxi-Cl was subsequently verified, as follows. First, CRISPR/Cas9-mediated SLCO2A1 gene disruption was found to eliminate Maxi-Cl activity in C127 cells. Furthermore, heterologous expression of SLCO2A1 was found to produce Maxi-Cl activity in HEK293T cells which endogenously express neither SLCO2A1 gene nor Maxi-Cl activity. In contrast, the disease-causing mutants of SLCO2A1, G222R and P219L, were confirmed to never produce Maxi-Cl activity, despite their successful plasmalemmal expression, when overexpressed in HEK293T cells. Since Maxi-Cl is known to serve as one of ATP-release pathway in osmotically swollen C127 cells (Sabirov & Okada, 2004a; Sabirov et al., 2001) and in ischemic cardiomyocyte (Dutta et al., 2004; 2008), the SLCO2A1 involvement in ATP release was next explored in C127 cells and mouse hearts. Swelling-induced ATP release from C127 cells was found to be suppressed by *Slco2a1* silencing *in vitro*, and ischemia-reperfusion-induced ATP release from Langendorff-perfused mouse hearts was also suppressed when *Slco2a1* expression in the mice was silenced by injection of siRNA *in vivo* via the tail vein in advance. Furthermore, swelling-induced ATP release was found to be augmented by overexpression of SLCO2A1, but not the G222R mutant, in HEK293T cells. Taken together, Sabirov et al. (2017) concluded that SLCO2A1 represents the core component of ATP-release Maxi-Cl channel.

Sabirov et al. (2017) further examined whether SLCO2A1 is involved in the pore formation of Maxi-Cl. First, overexpression of the charge-neutralized K613G mutant of SLCO2A1 was found to remarkably reduce the single-channel conductance and convert ion permeability of the channel from anion-selective to relatively more selective to cations. Then, reconstitution studies were finally performed by incorporating the purified recombinant SLCO2A1 proteins into giant liposomes and found to exhibit Gd^{3+}-sensitive anion-selective Maxi-Cl activity. The charge-neutralized K613G mutant was again found to become more permeable to cations with exhibiting a smaller unitary conductance in the reconstitution system. Thus, it appears that SLCO2A1 constitutes the pore component of Maxi-Cl.

Since SLCO2A1 is known as the molecule responsible for a prostaglandin transporter (PGT), the effects of a PGT substrate, PGE2, and a

PGT blocker, bromsulfophthalein (BSP), were next tested by Sabirov et al. (2017). Then, it was shown that PGE2 partially suppressed the Maxi-Cl currents in wild-type C127, but not in the CRISPR/Cas9-mediated *Slco2a1* knockout C127, cells and that PGE2 again exhibited partial inhibition of the currents observed on proteoliposomes reconstituted with recombinant SLCO2A1 proteins. PGE2 also partially suppressed swelling-induced ATP release from C127 cells. BSP almost completely abolished Maxi-Cl activity in C127 cells and in SLCO2A1-overexpressing HEK293T cells. Also, BSP suppressed swelling-induced ATP release down to the level comparable to that suppressed by Gd^{3+}. These results suggest that SLCO2A1 functions in two modes: as a PGT in the resting state presumably with phosphorylated tyrosine residues (Fig. 2B, left) and as an ATP-conductive Maxi-Cl channel in the activated state presumably with dephosphorylated tyrosine residues (Fig. 2B, right). It is likely that such transitions of SLCO2A1 from a transporter to an anion channel are induced by membrane excision, cell swelling and ischemia(-reperfusion) as well as by tyrosine kinase inhibitors. In this regard, it is noteworthy that the loss of multi-gate barrier in a transporter may often lead to anion channel activity (Minor, 2017).

4.3 Missing Components or Subcomponents yet Unidentified for Maxi-Cl

Maxi-Cl is normally in the non-activated state, and thus the channel event cannot be observed by the on-cell mode patch-clamping without any stimulation. In contrast, the native Maxi-Cl channel in membrane blebs isolated from C127 cells treated with latrunculin B was found to be constitutively activated (Sabirov et al., 2017). In addition, similar constitutively activated channel activities were also found in the channels reconstituted from the proteins extracted from the bleb membranes and recombinant SLCO2A1 proteins (Sabirov et al., 2017). These facts indicate that the whole Maxi–Cl complex natively contains not only SLCO2A1 but also some other component(s), which is involved in the mechanism of channel activation. Since on-cell Maxi-Cl activity was shown to be provoked by application of protein tyrosine kinase blockers (Toychiev et al., 2009) or a Ca^{2+} ionophore A23187 (Bajnath et al., 1993; Kawahara & Takuwa, 1991; Light et al., 1990; Sun et al., 1993), some protein tyrosine kinase/phosphatase and Ca^{2+}-dependent factors may represent the regulatory subcomponents for Maxi-Cl. Furthermore, Maxi-Cl events little show voltage-dependent inactivation kinetics both in the systems of HEK293T heterologous expression and the reconstitution of SLCO2A1, the fact suggesting that there is some missing factor, which is involved in the inactivation gating mechanism and is not endogenously expressed in HEK293T cells, other than SLCO2A1. In addition, Maxi-Cl activity reproduced by reconstituted recombinant SLCO2A1 proteins was totally insensitive to BSP (Sabirov et al., 2017). This fact indicates that some BSP-reactive component is missing in the SLCO2A1-reconstituted channel. Thus, in future reconstitution studies, the full-set phenotypic properties are to be reproduced by incorporating SLCO2A1 and these missing subcomponents.

Maxi-Cl represents the most major type (~60%) of MACs. Thus, there may exist some as-yet-unidentified MAC molecules other than SLCO2A1.

5. ATP RELEASE ACTIVITIES OF VSOR AND MAXI-CL

ATP plays essential roles not only in accumulating and providing energy within the cells but also in exerting as an extracellular ligand for autocrine/paracrine signaling mediated by purinergic receptors once released from the cells (Bodin & Burnstock, 2001; Burnstock, 2007, 2014; Dubyak & el-Moatassim, 1993; Fields & Stevens, 2000). Apart from unregulated leakage from injured cells, there are several pathways for regulated non-lytic release of ATP from stimulated cells. A most well-established regulated pathway for ATP release is exocytosis which occurs in neuronal and secretory cells. Since the ATP concentration within these secretory vesicles was estimated to reach 5 to 200 mM (Bankston & Guidotti, 1996; Bergendorff & Uvnäs, 1973; Van der Kloot, 2003; Zimmermann & Denston, 1976), diffusion of released ATP may effectively raise the extracellular ATP concentration to stimulate purinergic receptors from the resting level (1−10 nM). Recently, accumulating evidence demonstrated that ATP is also released, in a non-lytic regulated manner, via a number of non-exocytotic but conductive pathways in response to a variety of stimuli or stress.

5.1 Channel-Mediated ATP Release

Between the extracellular and intracellular fluids which contain ATP at 1−10 nM and 1−10 mM, respectively, there exists a very steep outward-directed electrochemical potential gradient of the order of 10^8 and 10^{10} for $MgATP^{2-}$ and ATP^{4-}, respectively, which represent the most predominant forms of intracellular ATP, when the intracellular potential is around −60 mV (Fig. 4). Therefore, electro-conductive channels may exert as very efficient ATP-release pathways, provided that the channels have their pores

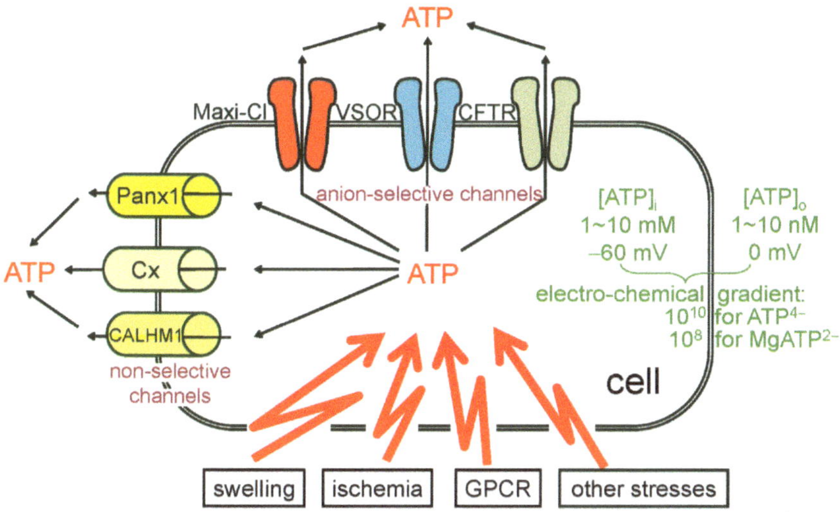

Figure 4 Putative two groups of non-exocytotic conductive pathways (including three types of anion-selective channels and three types of non-selective channels) of regulated ATP release in response to osmotic cell swelling, ischemia/hypoxia, stimulation of G-protein coupled receptors (GPCR) and other types of stress (see text for details).

effectively larger than the size of ATP (~1.15 nm for ATP^{4-} and ~1.20 nm for $MgATP^{2-}$ in diameter: Sabirov & Okada, 2005). Actually, there have been a number of candidates proposed for such conductive pathways for ATP release from mammalian cells (Dubyak, 2012; Lazarowski, 2012; Praetorius & Leipziger, 2009; Sabirov & Okada, 2005; Taruno, 2018). These are classified into two groups (Fig. 4): non-selective or relatively cation-selective channels including connexin (Cx) and Panx1 hemichannels (for Reviews: Bennett et al., 2012; Lohman & Isakson, 2014) and CALHM1 channels (for Review: Taruno, 2018) as well as anion-selective channels including CFTR (for Reviews: Cantiello, 1997; Schwiebert, 1999), VSOR (Braunstein, Zsembery, Tucker, & Schwiebert, 2004; Hisadome et al., 2002) and Maxi-Cl (for Reviews: Sabirov & Okada, 2004a; 2005).

Ample circumstantial evidence has been accumulated for supporting involvements of Cx, Panx1 and CALHM1 in ATP release from a variety of stimulated cells by observing suppression of the ATP release together with channel activity by gene-silencing or -knockout as well as by observing augmentation or induction of them by heterologous expression. As far as ATP release induced by osmotic cell swelling is concerned, these lines of molecular biological evidence have been provided only for Cx30 (Sipos et al., 2009) and Panx1 (Beckel et al., 2014; Islam et al., 2012; Li et al., 2012; Ransford et al., 2009; Seminario-Vidal et al., 2011). However, one cannot explicitly exclude a possibility that these effects of gene-modification on ATP release were caused by eliminating the regulatory actions by these gene products on other ATP-release mechanisms. To directly prove their involvements in conducting ATP-release pathways, the channel-mediated anionic ATP currents *per se* must be detected in future studies. In this regard, attempts to show actual ATP conductance were previously made by observing E_{rev} values under a sharp K_2ATP concentration gradient imposed across the cell membrane of Panx1-expressing oocytes (Bao et al., 2004) and Cx43-expressing C6 astrocytes (Kang et al., 2008). However, under the experimental conditions in these studies, a possibility remains that Cl^- conductance, more or less, contributed to the observed currents, because of contamination of the pipette K_2ATP or Na_2ATP solution by Cl^- which may have come from a Cl^--containing solution back-filled into the very narrow pipette space, from a Cl^--containing salt bridge, or from an Ag—AgCl electrode inserted to the pipette solution.

CFTR was previously believed to conduct anionic forms of ATP. The most compelling evidence was provided by the observations of anionic currents across patch membranes of cells heterologously expressing CFTR using symmetrical ATP solutions (Reisin et al., 1994; Schwiebert et al., 1995). However, a possible contamination of the currents observed during these patch-clamp experiments by Cl^- currents could not again be ruled out, because the shank of the patch pipette was backfilled with Cl^--containing solution which may have been pushed down by hydrostatic pressure or have diffused along a sharp chemical gradient for Cl^- into a Cl^--free ATP-containing tip region. Actually, such ATP currents were not observed in other laboratories (Braunstein et al., 2001; Grygorczyk, Tabcharani, & Hanrahan, 1996; Reddy et al., 1996). Also, Sugita, Yue, and Foskett (1998) reported that even when unitary ATP currents were actually recorded in CFTR-expressing MDCK cells, the ATP currents and CFTR Cl^- currents were found to be distinct from each other in pharmacological properties and sensitivity to the R347 mutant of CFTR, suggesting that CFTR channels and separate ATP-release channels, which are regulated by CFTR, are co-existing. Swelling-induced ATP release was enhanced by overexpression with CFTR in C127 cells (Hazama et al., 2000) and T84 and HeLa cells (Crane, Naeher, Choudhari, & Giroux, 2005). However, the ATP release was insensitive to glibenclamide which blocked CFTR currents but sensitive to Gd^{3+} which failed to affect CFTR Cl^- currents (Hazama et al., 2000). Taken together, it is concluded that CFTR *per se* does not function as an ATP-release channel but instead regulates some other ATP release mechanism. Since VSOR and Maxi-Cl are activated by cell swelling, it is conceived that their involvements in swelling-induced ATP release mechanisms are feasible.

5.2 VSOR-Mediated ATP Release

Nilius and his collaborators (Droogmans, Prenen, Eggermont, Voets, & Niliu, 1998; Droogmans, Maertens, Prenen, & Nilius, 1999) estimated the minimum and maximum of the pore diameter of VSOR of 1.14 and 1.42 nm, respectively, by studying its permeant block by basket-shaped compounds with known sizes. Ternovsky, Okada, and Sabirov (2004) evaluated the cut-off diameter of the VSOR pore to be 1.26 nm by applying the non-electrolyte partitioning method. Since the VSOR pore size is just comparable to the effective size of anionic ATP, ATP can be accommodated by the VSOR pore lumen. In fact, extracellular ATP was found to exhibit a voltage-dependent blocking effect at millimolar concentrations on VSOR currents at positive potentials (Ackerman, Wickman, & Clapham, 1994; Hisadome et al., 2002; Jackson & Strange, 1995; Nilius, Sehrer, & Droogmans, 1994; Poletto Chaves & Varanda, 2008; Tsumura, Oiki, Ueda, Okuma, & Okada, 1996), suggesting a permeant block by ATP plugged into the VSOR pore.

These facts may raise a possibility that ATP can pass through the VSOR pore after unbinding from the plugging site under certain conditions. Supporting this expectation, swelling-induced ATP release was found to be suppressed by known VSOR blockers in bovine aortic endothelial cells (Hisadome et al., 2002), human 1321N1 astrocytoma cells (Blum, Walsh, & Dubyak, 2010), mouse DRG neurons (Fields & Ni, 2010), and mouse RAW 264.7 macrophages (Burow, Klapperstück, & Markwardt F, 2015), although a number of established VSOR blockers including DCPIB, phloretin, glibenclamide and verapamil, in contrast, failed to suppress swelling-induced ATP release from human Intestine 407 cells (Hazama et al., 1999), mouse C127/CFTR cells (Hazama et al., 2000), C127 cells (Sabirov et al., 2001), rat cardiomyocytes (Dutta et al., 2004), mouse embryonic fibroblasts (Sabirov et al., 2006), mouse astrocytes (Liu, Toychiev, Takahashi, Sabirov, & Okada, 2008b), and rat synovial fibroblasts (Hu et al., 2017). Hisadome et al. (2002) then concluded that VSOR can serve as an ATP-release pathway on the basis not only of the above-mentioned pharmacological results but also of following two observations. First, ATP and ADP showed a bell-shaped voltage-dependent inhibition of VSOR Cl^- currents with exhibiting peak suppression at positive potentials of $+30 - +40$ mV and thereafter the relief from this inhibition at larger positive potentials, suggesting that unplugging of the nucleotides and their slippage down to the cytosol were forced by the large positive potentials. However, this event, if any, may represent ATP inflow, but not release, induced under rather unphysiological conditions. Second, in low Cl^- hypotonic conditions, extracellular addition of 10 or 20 mM ADP was found to augment outward currents, suggesting sizable contribution of inward-directed ADP conduction. However, no examination was actually performed whether this augmenting effect was not a consequence of the Cl^- conductance increase induced by purinergic receptor stimulation. Moreover, the E_{rev} values obtained in these whole-cell patch-clamp experiments in the absence and presence of ADP were both around −25 mV, indicating that the nature of observed currents is not anionic but largely cationic, because the intracellular (pipette) solution contained 5 mM Cl^- and 145 mM Cs^+ while the extracellular solution contained 2 mM Cl^- and 5 mM Cs^+. To prove the conductivity to ATP and ADP, in any case, pure ATP or ADP currents must be detected under the conditions in which possible contamination by Cl^- conductance can be safely ruled out.

Recently, hypotonicity-induced ATP release from *Xenopus* oocytes was found to become induced by injection of LRRC8A *plus* 8E in a manner sensitive to carbenoxolone (Gaitán-Peñas et al., 2016). However, carbenoxolone is not specific to VSOR but is a well-known hemichannel blocker. Taken together, it is evident that further studies are necessary to crucially conclude that VSOR exerts as an ATP-release channel under physiological conditions.

5.3 Maxi-Cl-Mediated ATP Release

ATP was found to exert a voltage-dependent open-channel blocking action on Maxi-Cl currents from either the extracellular or intracellular side, exhibiting a weak ATP-binding site (with K_d of 12−13 mM) located approximately in the middle of the channel pore (Sabirov et al., 2001). These results indicate that the Maxi-Cl channel pore provides sufficient room to accommodate ATP. In fact, our pore sizing measurements by the nonelectrolyte exclusion method demonstrated that Maxi-Cl has an average functional pore diameter of around 2.6 nm (Sabirov & Okada, 2004b), which is considerably bigger than the effective diameters of ATP^{4-}

(1.14–1.16 nm), $HATP^{3-}$ (1.12–1.16 nm), $MgATP^{2-}$ (1.18–1.22 nm), ADP^{3-} (1.06–1.12 nm), and UTP^{4-} (1.06–1.08 nm) (Sabirov & Okada, 2005). Thus, Maxi-Cl appears well suited to its function as a conductive pathway for release of these nucleotides.

This inference was actually supported by the following four lines of strong but circumstantial evidence. First, swelling-induced ATP release was shown to be sensitive to a Maxi-Cl-selective blocker, Gd^{3+}, but insensitive to a VSOR and CFTR blocker, glibenclamide, in a large variety of cell types functionally expressing animated Maxi-Cl activity, including mouse mammary C127 and C127/CFTR cells (Hazama et al., 2000; Sabirov et al., 2001), rat cardiomyocytes (Dutta et al., 2004; 2008), mouse embryonic fibroblasts (Sabirov et al., 2006), and mouse astrocytes (Liu et al., 2008b). Second, spatial distribution of Maxi-Cl was shown to coincide with the location of ATP release site from cultured neonatal rat cardiomyocytes and also from freshly isolated adult rat cardiomyocytes (Dutta et al., 2008) by applying both the smart-patch technique, which is the patch-clamp method coupled to the scanning ion conductance microscopy (see Sabirov, Korchev, & Okada, 2012) and the PC12 cell-based ATP biosensor technique (see Hayashi & Tominaga, 2012; Hazama, Hayashi, & Okada, 1998). Third, only a trace of ATP release was induced by a hypoosmotic challenge from HEK293T cells, which endogenously have plenty of VSOR activity but no Maxi-Cl activity, whereas much marked (more than 20 times) ATP release was provoked in response to the hypoosmotic stress from Maxi-Cl-rich C127 cells which also exhibit endogenous VSOR activity as plenty as that in HEK293T cells (Sabirov et al., 2017). Fourth, gene silencing of the core component molecule of Maxi-Cl, SLCO2A1, considerably suppressed swelling-induced ATP release from C127 cells, whereas heterologous expression of SLCO2A1 significantly augmented the swelling-induced ATP release from HEK293T cells (Sabirov et al., 2017).

The direct evidence for actual conductivity of Maxi-Cl to ATP was demonstrated by observing ATP currents and the P_{ATP}/P_{Cl} values evaluated by the E_{rev} values under the conditions, in which Cl^--free ATP-rich solution was filled in the ample bath (but not narrow pipette) space to exclude possible contamination by Cl^-, across the inside-out patch membranes excised from mouse C127 cells (P_{ATP}/P_{Cl} of 0.09–0.10: Dutta, et al., 2002; Sabirov et al., 2001; Sabirov & Okada, 2005), rabbit macula densa cells (P_{ATP}/P_{Cl} of 0.14: Bell et al., 2003), rat cardiomyocytes (P_{ATP}/P_{Cl} of 0.12: Dutta et al., 2004), mouse astrocytes (P_{ATP}/P_{Cl} of 0.11: Liu et al., 2008a), rat cardiac fibroblasts (P_{ATP}/P_{Cl} of 0.14: Okada et al., 2009), and mouse fibroblastic L929 cells (P_{ATP}/P_{Cl} of 0.11: Islam et al., 2012). Maxi-Cl was found to be also permeable to $MgATP^{2-}$ with the even higher P_{MgATP}/P_{Cl} value (around 0.16) in C127 cells (Sabirov & Okada, 2009) and rat cardiomyocytes (Dutta et al., 2004). Maxi-Cl does not discriminate well between the nucleotides with exhibiting P_{ADP}/P_{Cl} of 0.12 and P_{UTP}/P_{Cl} of 0.09 (Sabirov & Okada, 2005). On balance, it is concluded that Maxi-Cl, in fact, provides the conductive pathway for swelling-induced release of ATP and other nucleotides.

6. CONCLUSIONS AND FUTURE DIRECTIONS

The necessity for every living cell to quickly and efficiently respond to the osmotic challenges has been recognized in classic physiological studies. However, only recently, an elaborate volume regulation machinery of animal cells has been deciphered at the functional level, and very recently – at the molecular level after identification of the core components of the VSOR/VRAC and Maxi-Cl channels, which are responsible for the swelling-induced volume-regulatory chloride conductance in cells with normal ATP content and under metabolic starvation, respectively. Swollen cells massively release ATP as well as other nucleotides to stimulate the swollen cells themselves, in an autocrine fashion, to activate their volume regulation mechanisms and to report neighboring cells, in a paracrine fashion, about their physiological/pathophysiological conditions by initiating purinergic signaling cascades. Non-exocytotic conductive pathways for the swelling-induced ATP release play important role in this phenomenon. Although the main contributors to the nucleotide release were suggested mostly based on pharmacological studies, one should keep in mind that the available pharmacological tools remain poorly selective, thereby having caused controversies and confusions in this field. Recent advance in molecular identification efforts will allow to more firmly identify the key components of the physiologically important ATP-release mechanisms. The two new proteins emerged as a molecular basis for VSOR (LRRC8 family members) and Maxi-Cl (SLCO2A1) are expected to be a subject for the extensive investigations of the mechanisms of volume regulation and conductive ATP release, and interrelations between these two processes are to be of great biological significance.

At present, the only means to selectively manipulate VSOR and Maxi-Cl activity is gene silencing, deletion and modification. However, development of small molecules selectively targeting the two channels may accomplish their applicability for therapeutic use. Therefore, search for the selective agonists and antagonists of VSOR and Maxi-Cl, possibly by employing rational drug design methods with LRRC8 family and SLCO2A1 proteins as templates, would become a hot topic in the field. Identification of other missing components and subcomponents of the two channels will be another direction for future studies that will finally attain reproduction of full sets of the phenotypes of native VSOR and Maxi-Cl channels and precise evaluation of their physiological significance and actual involvements in human diseases.

ABBREVIATIONS

ABC	ATP-binding cassette
ACTN4	α-actinin-4
ANO	anoctamin
ANT	adenine nucleotide translocase
ASOR	acid-sensitive outwardly rectifying Cl^- channel
Best	bestrophin
CaCC	Ca^{2+}-activated Cl^- channel
CALHM1	calcium homeostasis modulator 1
CBX	carbenoxolone
Cx	connexin
$\mathbf{E_{rev}}$	reversal potential
MAC	maxi-anion channel
MAF	mouse adult skin fibroblast
NCX	Na^+-Ca^{2+} exchanger
NSCC	nonselective cation channel
NVI	necrotic volume increase
OVI	osmotic volume increase
Panx1	pannexin 1
PGT	prostaglandin transporter
RVD	regulatory volume decrease
VDAC	voltage-dependent anion channel
VRAC	volume-regulated anion channel
VSOR	volume-sensitive outwardly rectifying anion channel

ACKNOWLEDGMENTS

This work was supported by JSPS KAKENHI Grants #26293045 and #17K19517 (to Y.O.) and #1608510 (to T.O.).

REFERENCES

Abascal, F., & Zardoya, R. (2012). LRRC8 proteins share a common ancestor with pannexins, and may form hexameric channels involved in cell-cell communication. *BioEssays: News and Reviews in Molecular, Cellular and Developmental Biology, 34*, 551–560.

Ackerman, M. J., Wickman, K. D., & Clapham, D. E. (1994). Hypotonicity activates a native chloride current in Xenopus oocytes. *The Journal of General Physiology, 103*, 153–179.

Akanda, N., & Elinder, F. (2006). Biophysical properties of the apoptosis-inducing plasma membrane voltage-dependent anion channel. *Biophysical Journal, 90*, 4405–4417.

Akita, T., & Okada, Y. (2014). Characteristics and roles of the volume-sensitive outwardly rectifying (VSOR) anion channel in the central nervous system. *Neuroscience, 275*, 211–231.

Almaça, J., Tian, Y., Aldehni, F., Ousingsawat, J., Kongsuphol, P., Rock, J. R., et al. (2009). TMEM16 proteins produce volume-regulated chloride currents that are reduced in mice lacking TMEM16A. *Journal of Biological Chemistry, 284*, 28571–28578.

Ando-Akatsuka, Y., Shimizu, T., Numata, T., & Okada, Y. (2012). Involvements of the ABC protein ABCF2 and α-actinin-4 in regulation of cell volume and anion channels in human epithelial cells. *Journal of Cellular Physiology, 227*, 3498–3510.

Arreola, J., Begenisich, T., Nehrke, K., Nguyen, H. V., Park, K., Richardson, L., et al. (2002). Secretion and cell volume regulation by salivary acinar cells from mice lacking expression of the Clcn3 Cl– channel gene. *The Journal of Physiology, 545*, 207–216.

Bahamonde, M. I., Fernandez-Fernandez, J. M., Guix, F. X., Vazquez, E., & Valverde, M. A. (2003). Plasma membrane voltage-dependent anion channel mediates antiestrogen-activated maxi Cl^- currents in C1300 neuroblastoma cells. *Journal of Biological Chemistry, 278*, 33284–33289.

Bajnath, R. B., Groot, J. A., de Jonge, H. R., Kansen, M., & Bijman, J. (1993). Calcium ionophore plus excision induce a large conductance chloride channel in membrane patches of human colon carcinoma cells HT-29cl.19A. *Experientia, 49*, 313–316.

Bankston, L. A., & Guidotti, G. (1996). Characterization of ATP transport into chromaffin granule ghosts. Synergy of ATP and serotonin accumulation in chromaffin granule ghosts. *Journal of Biological Chemistry, 271*, 17132–17138.

Bao, L., Locovei, S., & Dahl, G. (2004). Pannexin membrane channels are mechanosensitive conduits for ATP. *FEBS Letters, 572*, 65–68.

Bazúa-Valenti, S., Castañeda-Bueno, M., & Gamba, G. (2016). Physiological role of SLC12 family members in the kidney. *American Journal of Physiology-Renal Physiology, 311*, F131–F144.

Beckel, J. M., Argall, A. J., Lim, J. C., Xia, J., Lu, W., Coffey, E. E., et al. (2014). Mechanosensitive release of adenosine 5'-triphosphate through pannexin channels and mechanosensitive upregulation of pannexin channels in optic nerve head astrocytes: A mechanism for purinergic involvement in chronic strain. *Glia, 62*, 1486–1501.

Beckel, J. M., Daugherty, S. L., Tyagi, P., Wolf-Johnston, A. S., Birder, L. A., Mitchell, C. H., et al. (2015). Pannexin 1 channels mediate the release of ATP into the lumen of the rat urinary bladder. *The Journal of Physiology, 593*, 1857–1871.

Behe, P., Foote, J. R., Levine, A. P., Platt, C. D., Chou, J., Benavides, F., et al. (2017). The LRRC8A mediated "swell activated" chloride conductance is dispensable for vacuolar homeostasis in neutrophils. *Frontiers in Pharmacology, 8*, 262.

Bell, P. D., Lapointe, J. Y., Sabirov, R., Hayashi, S., Peti-Peterdi, J., Manabe, K., et al. (2003). Macula densa cell signaling involves ATP release through a maxi anion channel. *Proceedings of the National Academy of Sciences of the United States of America, 100*, 4322–4327.

Bennett, M. V., Garré, J. M., Orellana, J. A., Bukauskas, F. F., Nedergaard, M., & Sáez, J. C. (2012). Connexin and pannexin hemichannels in inflammatory responses of glia and neurons. *Brain Research, 1487*, 3–15.

Bergendorff, A., & Uvnäs, B. (1973). Storage properties of rat mast cell granules in vitro. *Acta Physiologica Scandinavica, 87*, 213–222.

Blachly-Dyson, E., Peng, S., Colombini, M., & Forte, M. (1990). Selectivity changes in site-directed mutants of the VDAC ion Channel: Structural implications. *Science, 247*, 1233–1236.

Blatz, A. L., & Magleby, K. L. (1983). Single voltage-dependent chloride-selective channels of large conductance in cultured rat muscle. *Biophysical Journal, 43*, 237–241.

Blum, A. E., Walsh, B. C., & Dubyak, G. R. (2010). Extracellular osmolarity modulates G protein-coupled receptor-dependent ATP release from 1321N1 astrocytoma cells. *American Journal of Physiology. Cell Physiology, 298*, C386–C396.

Bodin, P., & Burnstock, G. (2001). Purinergic signalling: ATP release. *Neurochemical Research, 26*, 959–969.

Bond, T. D., Ambikapathy, S., Mohammad, S., & Valverde, M. A. (1998). Osmosensitive C1– currents and their relevance to regulatory volume decrease in human intestinal T84 cells: Outwardly vs. inwardly rectifying currents. *The Journal of Physiology, 511*, 45–54.

Bosma, M. M. (1989). Anion channels with multiple conductance levels in a mouse B lymphocyte cell line. *The Journal of Physiology, 410*, 67–90.

Braunstein, G. M., Roman, R. M., Clancy, J. P., Kudlow, B. A., Taylor, A. L., Shylonsky, V. G., et al. (2001). Cystic fibrosis transmembrane conductance regulator facilitates ATP release by stimulating a separate ATP release channel for autocrine control of cell volume regulation. *Journal of Biological Chemistry, 276*, 6621–6630.

Braunstein, G. M., Zsembery, A., Tucker, T. A., & Schwiebert, E. M. (2004). Purinergic signaling underlies CFTR control of human airway epithelial cell volume. *Journal of Cystic Fibrosis, 3*, 99–117.

Brustovetsky, N., & Klingenberg, M. (1996). Mitochondrial ADP/ATP Carrier can be reversibly converted into a large channel by $Ca2^{+}$. *Biochemistry, 35*, 8483–8488.

Brustovetsky, N., Tropschug, M., Heimpel, S., Heidkämper, D., & Klingenberg, M. (2002). A large $Ca2^{+}$-dependent channel formed by recombinant ADP/ATP Carrier from Neurospora crassa resembles the mitochondrial permeability transition pore. *Biochemistry, 41*, 11804–11811.

Bruzzone, R., Barbe, M. T., Jakob, N. J., & Monyer, H. (2005). Pharmacological properties of homomeric and heteromeric pannexin hemichannels expressed in *Xenopus* oocytes. *Journal of Neurochemistry, 92*, 1033–1043.

Buettner, R., Papoutsoglou, G., Scemes, E., Spray, D. C., & Dermietzel, R. (2000). Evidence for secretory pathway localization of a voltage-dependent anion channel isoform. *Proceedings of the National Academy of Sciences of the United States of America, 97*, 3201–3206.

Burnstock, G. (2007). Physiology and pathophysiology of purinergic neurotransmission. *Physiological Reviews, 87*, 659–797.

Burnstock, G. (2014). Purinergic signalling: From discovery to current developments. *Experimental Physiology, 99*, 16–34.

Burow, P., Klapperstück, M., & Markwardt, F. (2015). Activation of ATP secretion via volume-regulated anion channels by sphingosine-1-phosphate in RAW macrophages. *Pflugers Arch, 467*, 1215–1226.

Buyse, G., Voets, T., Tytgat, J., De Greef, C., Droogmans, G., Nilius, B., et al. (1997). Expression of human pICln and ClC-6 in Xenopus oocytes induces an identical endogenous chloride conductance. *Journal of Biological Chemistry, 272*, 3615–3621.

Byfield, F. J., Hoffman, B. D., Romanenko, V. G., Fang, Y., Crocker, J. C., & Levitan, I. (2006). Evidence for the role of cell stiffness in modulation of volume-regulated anion channels. *Acta Physiologica, 187*, 285–294.

Cahalan, M. D., & Lewis, R. S. (1988). Role of potassium and chloride channels in volume regulation by T lymphocytes. *Society of General Physiologists Series, 43*, 281–301.
Cannon, C. L., Basavappa, S., & Strange, K. (1998). Intracellular ionic strength regulates the volume sensitivity of a swelling-activated anion channel. *American Journal of Physiology, 275*, C416–C422.
Cantiello, H. F. (1997). Nucleotide transport through the cystic fibrosis transmembrane conductance regulator. *Bioscience Reports, 17*, 147–171.
Carton, I., Hermans, D., & Eggermont, J. (2003). Hypotonicity induces membrane protrusions and actin remodeling via activation of small GTPases Rac and Cdc42 in Rat-1 fibroblasts. *American Journal of Physiology. Cell Physiology, 285*, C935–C944.
Chen, L., Wang, L., & Jacob, T. J. (1999). Association of intrinsic pICln with volume-activated Cl^- current and volume regulation in a native epithelial cell. *American Journal of Physiology, 276*, C182–C192.
Chien, L. T., & Hartzell, H. C. (2007). Drosophila bestrophin-1 chloride current is dually regulated by calcium and cell volume. *The Journal of General Physiology, 130*, 513–524.
Chien, L. T., & Hartzell, H. C. (2008). Rescue of volume-regulated anion current by bestrophin mutants with altered charge selectivity. *The Journal of General Physiology, 132*, 537–546.
Clapham, D. E. (1998). The list of potential volume-sensitive chloride currents continues to swell (and shrink). *The Journal of General Physiology, 111*, 623–624.
Colombini, M. (1989). Voltage gating in the mitochondrial channel, VDAC. *Journal of Membrane Biology, 111*, 103–111.
Colombini, M. (2004). Vdac: the channel at the interface between mitochondria and the cytosol. *Molecular and Cellular Biochemistry, 256*, 107–115.
Colombini, M., Blachly-Dyson, E., & Forte, M. (1996). VDAC, a channel in the outer mitochondrial membrane. In T. Narahashi (Ed.), *Ion channels* (pp. 169–202). New York: Plenum Press.
Cotrina, M. L., Lin, J. H., Alves-Rodrigues, A., Liu, S., Li, J., Azmi-Ghadimi, H., et al. (1998). Connexins regulate calcium signaling by controlling ATP release. *Proceedings of the National Academy of Sciences of the United States of America, 95*, 15735–15740.
Coulombe, A., & Coraboeuf, E. (1992). Large-conductance chloride channels of new-born rat cardiac myocytes are activated by hypotonic media. *Pflügers Archiv, 422*, 143–150.
Crane, J. K., Naeher, T. M., Choudhari, S. S., & Giroux, E. M. (2005). Two pathways for ATP release from host cells in enteropathogenic *Escherichia coli* infection. *American Journal of Physiology. Gastrointestinal and Liver Physiology, 289*, G407–G417.
De Pinto, V., Messina, A., Lane, D. J., & Lawen, A. (2010). Voltage-dependent anion-selective channel (VDAC) in the plasma membrane. *FEBS Letters, 584*, 1793–1799.
Deneka, D., Sawicka, M., Lam, A. K. M., Paulino, C., & Dutzler, R. (2018). Structure of a volume-regulated anion channel of the LRRC8 family. *Nature*. https://doi.org/10.1038/s41586-018-0134-y (in press).
Dermietzel, R., Hwang, T. K., Buettner, R., Hofer, A., Dotzler, E., Kremer, M., et al. (1994). Cloning and in situ localization of a brain-derived porin that constitutes a large-conductance anion channel in astrocytic plasma membranes. *Proceedings of the National Academy of Sciences of the United States of America, 91*, 499–503.
Dezaki, K., Tsumura, T., Maeno, E., & Okada, Y. (2000). Receptor-mediated facilitation of cell volume regulation by swelling-induced ATP release in human epithelial cells. *The Japanese Journal of Physiology, 50*, 235–241.
Doroshenko, P., & Neher, E. (1992). Volume-sensitive chloride conductance in bovine chromaffin cell membrane. *The Journal of Physiology, 449*, 197–218.
Doroshenko, P., Penner, R., & Neher, E. (1991). Novel chloride conductance in the membrane of bovine chromaffin cells activated by intracellular GTPγS. *The Journal of Physiology, 436*, 711–724.

Dreses-Werringloer, U., Lambert, J. C., Vingtdeux, V., Zhao, H., Vais, H., Siebert, A., et al. (2008). A polymorphism in CALHM1 influences Ca^{2+} homeostasis, Abeta levels, and Alzheimer's disease risk. *Cell, 133*, 1149–1161.

Droogmans, G., Maertens, C., Prenen, J., & Nilius, B. (1999). Sulphonic acid derivatives as probes of pore properties of volume-regulated anion channels in endothelial cells. *British Journal of Pharmacology, 128*, 35–40.

Droogmans, G., Prenen, J., Eggermont, J., Voets, T., & Nilius, B. (1998). Voltage-dependent block of endothelial volume-regulated anion channels by calix[4]arenes. *American Journal of Physiology, 275*, C646–C652.

Duan, D., Winter, C., Cowley, S., Hume, J. R., & Horowitz, B. (1997). Molecular identification of a volume-regulated chloride channel. *Nature, 390*, 417–421.

Duan, D., Zhong, J., Hermoso, M., Satterwhite, C. M., Rossow, C. F., Hatton, W. J., et al. (2001). Functional inhibition of native volume-sensitive outwardly rectifying anion channels in muscle cells and *Xenopus* oocytes by anti-ClC-3 antibody. *The Journal of Physiology, 531*, 437–444.

Dubyak, G. R. (2012). Function without form: An ongoing search for maxi-anion channel proteins. Focus on "maxi-anion channel and pannexin 1 hemichannel constitute separate pathways for swelling-induced ATP release in murine L929 fibrosarcoma cells". *American Journal of Physiology. Cell Physiology, 303*, C913–C915.

Dubyak, G. R., & el-Moatassim, C. (1993). Signal transduction via P2-purinergic receptors for extracellular ATP and other nucleotides. *American Journal of Physiology, 265*, C577–C606.

Dutta, A. K., Korchev, Y. E., Shevchuk, A. I., Hayashi, S., Okada, Y., & Sabirov, R. Z. (2008). Spatial distribution of maxi-anion channel on cardiomyocytes detected by smart-patch technique. *Biophysical Journal, 94*, 1646–1655.

Dutta, A. K., Okada, Y., & Sabirov, R. Z. (2002). Regulation of an ATP-conductive large-conductance anion channel and swelling-induced ATP release by arachidonic acid. *The Journal of Physiology, 542*, 803–816.

Dutta, A. K., Sabirov, R. Z., Uramoto, H., & Okada, Y. (2004). Role of ATP-conductive anion channel in ATP release from neonatal rat cardiomyocytes in ischaemic or hypoxic conditions. *The Journal of Physiology, 559*, 799–812.

Eggermont, J., Trouet, D., Carton, I., & Nilius, B. (2001). Cellular function and control of volume-regulated anion channels. *Cell Biochemistry and Biophysics, 35*, 263–274.

Elinder, F., Akanda, N., Tofighi, R., Shimizu, S., Tsujimoto, Y., Orrenius, S., et al. (2005). Opening of plasma membrane voltage-dependent anion channels (VDAC) precedes caspase activation in neuronal apoptosis induced by toxic stimuli. *Cell Death and Differentiation, 12*, 1134–1140.

Erickson, G. R., Northrup, D. L., & Guilak, F. (2003). Hypo-osmotic stress induces calcium-dependent actin reorganization in articular chondrocytes. *Osteoarthritis and Cartilage, 11*, 187–197.

Estevez, A. Y., Bond, T., & Strange, K. (2001). Regulation of $I_{Cl,swell}$ in neuroblastoma cells by G protein signaling pathways. *American Journal of Physiology. Cell Physiology, 281*, C89–C98.

Falke, L. C., & Misler, S. (1989). Activity of ion channels during volume regulation by clonal N1E115 neuroblastoma cells. *Proceedings of the National Academy of Sciences of the United States of America, 86*, 3919–3923.

Fields, R. D., & Ni, Y. (2010). Nonsynaptic communication through ATP release from volume-activated anion channels in axons. *Science Signaling [electronic Resource], 3*, ra73.

Fields, R. D., & Stevens, B. (2000). Atp: An extracellular signaling molecule between neurons and glia. *Trends in Neurosciences, 23*, 625–633.

Fischmeister, R., & Hartzell, H. C. (2005). Volume sensitivity of the bestrophin family of chloride channels. *The Journal of Physiology, 562*, 477–491.

Friedrich, T., Breiderhoff, T., & Jentsch, T. J. (1999). Mutational analysis demonstrates that ClC-4 and ClC-5 directly mediate plasma membrane currents. *Journal of Biological Chemistry, 274*, 896–902.

Gaitán-Peñas, H., Gradogna, A., Laparra-Cuervo, L., Solsona, C., Fernández-Dueñas, V., Barrallo-Gimeno, A., et al. (2016). Investigation of LRRC8-mediated volume-regulated anion currents in *Xenopus* oocytes. *Biophysical Journal, 111*, 1429–1443.

Gong, W., Xu, H., Shimizu, T., Morishima, S., Tanabe, S., Tachibe, T., et al. (2004). ClC-3-independent, PKC-dependent activity of volume-sensitive Cl channel in mouse ventricular cardiomyocytes. *Cellular Physiology and Biochemistry: International Journal of Experimental Cellular Physiology, Biochemistry, and Pharmacology, 14*, 213–224.

Grubb, S., Poulsen, K. A., Juul, C. A., Kyed, T., Klausen, T. K., Larsen, E. H., et al. (2013). TMEM16F (Anoctamin 6), an anion channel of delayed Ca^{2+} activation. *The Journal of General Physiology, 141*, 585–600.

Grygorczyk, R., Tabcharani, J. A., & Hanrahan, J. W. (1996). CFTR channels expressed in CHO cells do not have detectable ATP conductance. *The Journal of Membrane Biology, 151*, 139–148.

Gschwentner, M., Nagl, U. O., Wöll, E., Schmarda, A., Ritter, M., & Paulmichl, M. (1995). Antisense oligonucleotides suppress cell-volume-induced activation of chloride channels. *Pflügers Archiv, 430*, 464–470.

Gualix, J., Pintor, J., & Miras-Portugal, M. T. (1999). Characterization of nucleotide transport into rat brain synaptic vesicles. *Journal of Neurochemistry, 73*, 1098–1104.

Guzman, R. E., Grieschat, M., Fahlke, C., & Alekov, A. K. (2013). ClC-3 is an intracellular chloride/proton exchanger with large voltage-dependent nonlinear capacitance. *ACS Chemical Neuroscience, 4*, 994–1003.

Guzman, R. E., Miranda-Laferte, E., Franzen, A., & Fahlke, C. (2015). Neuronal ClC-3 splice variants differ in subcellular localizations, but mediate identical transport functions. *Journal of Biological Chemistry, 290*, 25851–25862.

Hammer, C., Wanitchakool, P., Sirianant, L., Papiol, S., Monnheimer, M., Faria, D., et al. (2015). A coding variant of ANO10, affecting volume regulation of macrophages, is associated with *Borrelia* seropositivity. *Molecular Medicine, 23*, 26–37.

Hara-Chikuma, M., Yang, B., Sonawane, N. D., Sasaki, S., Uchida, S., & Verkman, A. S. (2005). ClC-3 chloride channels facilitate endosomal acidification and chloride accumulation. *Journal of Biological Chemistry, 280*, 1241–1247.

Harder, T., Kellner, R., Parton, R. G., & Gruenberg, J. (1997). Specific release of membrane-bound annexin II and cortical cytoskeletal elements by sequestration of membrane cholesterol. *Molecular Biology of the Cell, 8*, 533–545.

Hardy, S. P., & Valverde, M. A. (1994). Novel plasma membrane action of estrogen and antiestrogens revealed by their regulation of a large conductance chloride channel. *FASEB Journal: Official Publication of the Federation of American Societies for Experimental Biology, 8*, 760–765.

Hayashi, T., Nozaki, Y., Nishizuka, M., Ikawa, M., Osada, S., & Imagawa, M. (2011). Factor for adipocyte differentiation 158 gene disruption prevents the body weight gain and insulin resistance induced by a high-fat diet. *Biological and Pharmaceutical Bulletin, 34*, 1257–1263.

Hayashi, S., & Tominaga, M. (2012). Patch-clamp biosensor method. In Y. Okada (Ed.), *Patch clamp techniques. From biginning to advanced protochols* (pp. 333–342). Tokyo: Springer.

Hazama, A., Fan, H. T., Abdullaev, I., Maeno, E., Tanaka, S., Ando-Akatsuka, Y., et al. (2000). Swelling-activated, cystic fibrosis transmembrane conductance regulator-augmented ATP release and Cl^- conductances in murine C127 cells. *The Journal of Physiology, 523*, 1–11.

Hazama, A., Hayashi, A., & Okada, Y. (1998). Cell surface measurements of ATP release from single pancreatic β cells using a novel biosensor technique. *Pflügers Archiv - European Journal of Physiology, 437*, 31–35.

Hazama, A., & Okada, Y. (1988). Ca^{2+} sensitivity of volume-regulatory K^+ and Cl^- channels in cultured human epithelial cells. *The Journal of Physiology, 402*, 687–702.

Hazama, A., Shimizu, T., Ando-Akatsuka, Y., Hayashi, S., Tanaka, S., Maeno, E., et al. (1999). Swelling-induced, CFTR-independent ATP release from a human epithelial cell line: Lack of correlation with volume-sensitive Cl^- channels. *The Journal of General Physiology, 114*, 525–533.

Hermoso, M., Satterwhite, C. M., Andrade, Y. N., Hidalgo, J., Wilson, S. M., Horowitz, B., et al. (2002). ClC-3 is a fundamental molecular component of volume-sensitive outwardly rectifying Cl^- channels and volume regulation in HeLa cells and *Xenopus* laevis oocytes. *Journal of Biological Chemistry, 277*, 40066–40074.

Hernández-Carballo, C. Y., De Santiago-Castillo, J. A., Rosales-Saavedra, T., Pérez-Cornejo, P., & Arreola, J. (2010). Control of volume-sensitive chloride channel inactivation by the coupled action of intracellular chloride and extracellular protons. *Pflügers Archiv - European Journal of Physiology, 460*, 633–644.

Hisadome, K., Koyama, T., Kimura, C., Droogmans, G., Ito, Y., & Oike, M. (2002). Volume-regulated anion channels serve as an auto/paracrine nucleotide release pathway in aortic endothelial cells. *The Journal of General Physiology, 119*, 511–520.

Hoffmann, E. K., Lambert, I. H., & Pedersen, S. F. (2009). Physiology of cell volume regulation in vertebrates. *Physiological Reviews, 89*, 193–277.

Hoffmann, E. K., & Simonsen, L. O. (1989). Membrane mechanisms in volume and pH regulation in vertebrate cells. *Physiological Reviews, 69*, 315–382.

Hu, F., Hui, Z., Wei, W., Yang, J., Chen, Z., Guo, B., et al. (2017). Hypotonic stress promotes ATP release, reactive oxygen species production and cell proliferation via TRPV4 activation in rheumatoid arthritis rat synovial fibroblasts. *Biochemical and Biophysical Research Communications, 486*, 108–115.

Huang, F., Wang, X., Ostertag, E. M., Nuwal, T., Huang, B., Jan, Y. N., et al. (2013). TMEM16C facilitates Na^+-activated K^+ currents in rat sensory neurons and regulates pain processing. *Nature Neuroscience, 16*, 1284–1290.

Huang, P., Liu, J., Di, A., Robinson, N. C., Musch, M. W., Kaetzel, M. A., et al. (2001). Regulation of human CLC-3 channels by multifunctional Ca^{2+}/calmodulin-dependent protein kinase. *Journal of Biological Chemistry, 276*, 20093–20100.

Hubert, M. D., Levitan, I., Hoffman, M. M., Zraggen, M., Hofreiter, M. E., & Garber, S. S. (2000). Modulation of volume regulated anion current by I(Cln). *Biochimica et Biophysica Acta, 1466*, 105–114.

Islam, M. R., Merzlyak, P., Okada, T., Sabirov, R. Z., & Okada, Y. (2013). Searching for the molecular basis of maxi-anion channel. *The Journal of Physiological Sciences, 63*(Suppl 1), S133.

Islam, M. R., Uramoto, H., Okada, T., Sabirov, R. Z., & Okada, Y. (2012). Maxi-anion channel and pannexin 1 hemichannel constitute separate pathways for swelling-induced ATP release in murine L929 fibrosarcoma cells. *American Journal of Physiology. Cell Physiology, 303*, C924–C935.

Jackson, P. S., Morrison, R., & Strange, K. (1994). The volume-sensitive organic osmolyte-anion channel VSOAC is regulated by nonhydrolytic ATP binding. *American Journal of Physiology. Cell Physiology, 267*, C1203–C1209.

Jackson, P. S., & Strange, K. (1995). Characterization of the voltage-dependent properties of a volume-sensitive anion conductance. *The Journal of General Physiology, 105*, 661–676.

Jalonen, T. (1993). Single-channel characteristics of the large-conductance anion channel in rat cortical astrocytes in primary culture. *Glia, 9*, 227–237.

Jentsch, T. J., Günther, W., Pusch, M., & Schwappach, B. (1995). Properties of voltage-gated chloride channels of the ClC gene family. *The Journal of Physiology, 482*, 19S–25S.

Jentsch, T. J., Stein, V., Weinreich, F., & Zdebik, A. A. (2002). Molecular structure and physiological function of chloride channels. *Physiological Reviews, 82*, 503–568.

Jin, N. G., Kim, J. K., Yang, D. K., Cho, S. J., Kim, J. M., Koh, E. J., et al. (2003). Fundamental role of ClC-3 in volume-sensitive Cl^- channel function and cell volume regulation in AGS cells. *American Journal of Physiology. Gastrointestinal and Liver Physiology, 285*, G938–G948.

Juul, C. A., Grubb, S., Poulsen, K. A., Kyed, T., Hashem, N., Lambert, I. H., et al. (2014). Anoctamin 6 differs from VRAC and VSOAC but is involved in apoptosis and supports volume regulation in the presence of Ca^{2+}. *Pflügers Archiv - European Journal of Physiology, 466*, 1899–1910.

Kahle, K. T., Khanna, A. R., Alper, S. L., Adragna, N. C., Lauf, P. K., Sun, D., et al. (2015). K-Cl cotransporters, cell volume homeostasis, and neurological disease. *Trends in Molecular Medicine, 21*, 513–523.

Kang, J., Kang, N., Lovatt, D., Torres, A., Zhao, Z., Lin, J., et al. (2008). Connexin 43 hemichannels are permeable to ATP. *The Journal of Neuroscience: the Official Journal of the Society for Neuroscience, 28*, 4702–4711.

Kanno, T., & Takishima, T. (1990). Chloride and potassium channels in U937 human monocytes. *Journal of Membrane Biology, 116*, 149–161.

Kawahara, K., & Takuwa, N. (1991). Bombesin activates large-conductance chloride channels in Swiss 3T3 fibroblasts. *Biochemical and Biophysical Research Communications, 177*, 292–298.

Kienitz, M. C., Bender, K., Dermietzel, R., Pott, L., & Zoidl, G. (2011). Pannexin 1 constitutes the large conductance cation channel of cardiac myocytes. *Journal of Biological Chemistry, 286*, 290–298.

Klausen, T. K., Hougaard, C., Hoffmann, E. K., & Pedersen, S. F. (2006). Cholesterol modulates the volume-regulated anion current in Ehrlich-Lettre ascites cells via effects on Rho and F-actin. *American Journal of Physiology. Cell Physiology, 291*, C757–C771.

Krapivinsky, G., Pu, W., Wickman, K., Krapivinsky, L., & Clapham, D. E. (1998). pICln binds to a mammalian homolog of a yeast protein involved in regulation of cell morphology. *Journal of Biological Chemistry, 273*, 10811–10814.

Krapivinsky, G. B., Ackerman, M. J., Gordon, E. A., Krapivinsky, L. D., & Clapham, D. E. (1994). Molecular characterization of a swelling-induced chloride conductance regulatory protein, pICln. *Cell, 76*, 439–448.

Krasilnikov, O. V., Carneiro, C. M., Yuldasheva, L. N., Campos-de-Carvalho, A. C., & Nogueira, R. A. (1996). Diameter of the mammalian porin channel in open and "closed" states: Direct measurement at the single channel level in planar lipid bilayer. *Brazilian Journal of Medical and Biological Research, 29*, 1691–1697.

Kumar, L., Chou, J., Yee, C. S., Borzutzky, A., Vollmann, E. H., von Andrian, U. H., et al. (2014). Leucine-rich repeat containing 8A (LRRC8A) is essential for T lymphocyte development and function. *The Journal of Experimental Medicine, 211*, 929–942.

Kurbannazarova, R. S., Bessonova, S. V., Okada, Y., & Sabirov, Z. (2011). Swelling-activated anion channels are essential for volume regulation of mouse thymocytes. *International Journal of Molecular Sciences, 12*, 9125–9137.

Lang, F. (Ed.). (1998). *Cell volume regulation*. Basel: Karger.

Lazarowski, E. R. (2012). Vesicular and conductive mechanisms of nucleotide release. *Purinergic Signalling, 8*, 359–373.

Lee, C. C., Freinkman, E., Sabatini, D. M., & Ploegh, H. L. (2014). The protein synthesis inhibitor blasticidin S enters mammalian cells via leucine-rich repeat-containing protein 8D. *Journal of Biological Chemistry, 289*, 17124–17131.

Lee, E. L., Shimizu, T., Ise, T., Numata, T., Kohno, K., & Okada, Y. (2007). Impaired activity of volume-sensitive Cl− channel is involved in cisplatin resistance of cancer cells. *Journal of Cellular Physiology, 211*, 513−521.

Levitan, I., Almonte, C., Mollard, P., & Garber, S. S. (1995). Modulation of a volume-regulated chloride current by F-actin. *Journal of Membrane Biology, 147*, 283−294.

Levitan, I., Christian, A. E., Tulenko, T. N., & Rothblat, G. H. (2000). Membrane cholesterol content modulates activation of volume-regulated anion current in bovine endothelial cells. *The Journal of General Physiology, 115*, 405−416.

Lewis, R. S., Ross, P. E., & Cahalan, M. D. (1993). Chloride channels activated by osmotic stress in T lymphocytes. *The Journal of General Physiology, 101*, 801−826.

Li, A., Leung, C. T., Peterson-Yantorno, K., Stamer, W. D., Mitchell, C. H., & Civan, M. M. (2012). Mechanisms of ATP release by human trabecular meshwork cells, the enabling step in purinergic regulation of aqueous humor outflow. *Journal of Cellular Physiology, 227*, 172−182.

Li, C., Breton, S., Morrison, R., Cannon, C. L., Emma, F., Sanchez-Olea, R., et al. (1998). Recombinant pICln forms highly cation-selective channels when reconstituted into artificial and biological membranes. *The Journal of General Physiology, 112*, 727−736.

Li, X., Shimada, K., Showalter, L. A., & Weinman, S. A. (2000). Biophysical properties of ClC-3 differentiate it from swelling-activated chloride channels in Chinese hamster ovary-K1 cells. *Journal of Biological Chemistry, 275*, 35994−35998.

Li, X., Wang, T., Zhao, Z., & Weinman, S. A. (2002). The ClC-3 chloride channel promotes acidification of lysosomes in CHO-K1 and Huh-7 cells. *American Journal of Physiology. Cell Physiology, 282*, C1483−C1491.

Light, D. B., Schwiebert, E. M., Fejes-Toth, G., Naray-Fejes-Toth, A., Karlson, K. H., McCann, F. V., et al. (1990). Chloride channels in the apical membrane of cortical collecting duct cells. *American Journal of Physiology, 258*, F273−F280.

Liu, C. Z., Li, X. Y., Du, R. H., Gao, M., Ma, M. M., Li, F. Y., et al. (2016). Endophilin A2 influences volume-regulated chloride current by mediating ClC-3 trafficking in vascular smooth muscle cells. *Circulation Journal: Official Journal of the Japanese Circulation Society, 80*, 2397−2406.

Liu, H. T., Sabirov, R. Z., & Okada, Y. (2008a). Oxygen-glucose deprivation induces ATP release via maxi-anion channels in astrocytes. *Purinergic Signalling, 4*, 147−154.

Liu, H. T., Tashmukhamedov, B. A., Inoue, H., Okada, Y., & Sabirov, R. Z. (2006). Roles of two types of anion channels in glutamate release from mouse astrocytes under ischemic or osmotic stress. *Glia, 54*, 343−357.

Liu, H. T., Toychiev, A. H., Takahashi, N., Sabirov, R. Z., & Okada, Y. (2008b). Maxi-anion channel as a candidate pathway for osmosensitive ATP release from mouse astrocytes in primary culture. *Cell Research, 18*, 558−565.

Liu, J., Zhang, F. F., Li, L., Yang, J., Liu, J., Guan, Y. Y., et al. (2013). ClC-3 deficiency prevents apoptosis induced by angiotensin II in endothelial progenitor cells via inhibition of NADPH oxidase. *Apoptosis, 18*, 1262−1273.

Locovei, S., Bao, L., & Dahl, G. (2006). Pannexin 1 in erythrocytes: Function without a gap. *Proceedings of the National Academy of Sciences of the United States of America, 103*, 7655−7659.

Lohman, A. W., & Isakson, B. E. (2014). Differentiating connexin hemichannels and pannexin channels in cellular ATP release. *FEBS Letters, 588*, 1379−1388.

Lutter, D., Ullrich, F., Lueck, J. C., Kempa, S., & Jentsch, T. J. (2017). Selective transport of neurotransmitters and modulators by distinct volume-regulated LRRC8 anion channels. *Journal of Cell Science, 130*, 1122−1133.

Ma, W., Hui, H., Pelegrin, P., & Surprenant, A. (2009). Pharmacological characterization of pannexin-1 currents expressed in mammalian cells. *The Journal of Pharmacology and Experimental Therapeutics, 328*, 409−418.

Ma, Z., Siebert, A. P., Cheung, K. H., Lee, R. J., Johnson, B., Cohen, A. S., et al. (2012). Calcium homeostasis modulator 1 (CALHM1) is the pore-forming subunit of an ion channel that mediates extracellular Ca^{2+} regulation of neuronal excitability. *Proceedings of the National Academy of Sciences of the United States of America, 109*, E1963–E1971.

Majid, A., Brown, P. D., Best, L., & Park, K. (2001). Expression of volume-sensitive Cl^- channels and ClC-3 in acinar cells isolated from the rat lacrimal gland and submandibular salivary gland. *The Journal of Physiology, 534*, 409–421.

Mannella, C. A. (1997). Minireview: On the structure and gating mechanism of the mitochondrial channel, VDAC. *Journal of Bioenergetics and Biomembranes, 29*, 525–531.

Mao, J., Chen, L., Xu, B., Wang, L., Li, H., Guo, J., et al. (2008). Suppression of ClC-3 channel expression reduces migration of nasopharyngeal carcinoma cells. *Biochemical Pharmacology, 75*, 1706–1716.

Mao, J., Xu, B., Li, H., Chen, L., Jin, X., Zhu, J., et al. (2011). Lack of association between stretch-activated and volume-activated Cl– currents in hepatocellular carcinoma cells. *Journal of Cellular Physiology, 226*, 1176–1185.

Martins, J. R., Faria, D., Kongsuphol, P., Reisch, B., Schreiber, R., & Kunzelmann, K. (2011). Anoctamin 6 is an essential component of the outwardly rectifying chloride channel. *Proceedings of the National Academy of Sciences of the United States of America, 108*, 18168–18172.

Matsuda, J. J., Filali, M. S., Moreland, J. G., Miller, F. J., & Lamb, F. S. (2010). Activation of swelling-activated chloride current by tumor necrosis factor-alpha requires ClC-3-dependent endosomal reactive oxygen production. *Journal of Biological Chemistry, 285*, 22864–22873.

Matsuda, J. J., Filali, M. S., Volk, K. A., Collins, M. M., Moreland, J. G., & Lamb, F. S. (2008). Overexpression of CLC-3 in HEK293T cells yields novel currents that are pH dependent. *American Journal of Physiology. Cell Physiology, 294*, C251–C262.

Milenkovic, A., Brandl, C., Milenkovic, V. M., Jendryke, T., Sirianant, L., Wanitchakool, P., et al. (2015). Bestrophin 1 is indispensable for volume regulation in human retinal pigment epithelium cells. *Proceedings of the National Academy of Sciences of the United States of America, 112*, E2630–E2639.

Miller, F. J., Jr., Filali, M., Huss, G. J., Stanic, B., Chamseddine, A., Barna, T. J., et al. (2007). Cytokine activation of nuclear factor kappa B in vascular smooth muscle cells requires signaling endosomes containing Nox1 and ClC-3. *Circulation Research, 101*, 663–671.

Minor, D. L., Jr. (2017). Channel surfing uncovers a dual-use transporter. *The EMBO Journal, 36*, 3272–3273.

Mitchell, C. H., Wang, L., & Jacob, T. J. C. (1997a). A large-conductance chloride channel in pigmented ciliary epithelial cells activated by GTPgammaS. *Journal of Membrane Biology, 158*, 167–175.

Mitchell, C. H., Zhang, J. J., Wang, L., & Jacob, T. J. (1997b). Volume-sensitive chloride current in pigmented ciliary epithelial cells: Role of phospholipases. *American Journal of Physiology. Cell Physiology, 272*, C212–C222.

Miwa, A., Ueda, K., & Okada, Y. (1997). Protein kinase C-independent correlation between P-glycoprotein expression and volume sensitivity of Cl^- channel. *Journal of Membrane Biology, 157*, 63–69.

Miyata, H., Kinosita, K., Jr., & Marriott, G. (1997). Cooperative association of actin protomers and crosslinked actin oligomers in filaments at low ionic strength. *Journal of Biochemistry, 121*, 527–533.

Morán, J., Sabanero, M., Meza, I., & Pasantes-Morales, H. (1996). Changes of actin cytoskeleton during swelling and regulatory volume decrease in cultured astrocytes. *American Journal of Physiology, 271*, C1901–C1907.

Morishima, S., Shimizu, T., Kida, H., & Okada, Y. (2000). Volume expansion sensitivity of swelling-activated Cl^- channel in human epithelial cells. *The Japanese Journal of Physiology, 50*, 277–280.

Nilius, B., & Droogmans, G. (2001). Ion channels and their functional role in vascular endothelium. *Physiological Reviews, 81*, 1415–1459.

Nilius, B., & Droogmans, G. (2003). Amazing chloride channels: An overview. *Acta Physiologica Scandinavica, 177*, 119–147.

Nilius, B., Eggermont, J., Voets, T., Buyse, G., Manolopoulos, V., & Droogmans, G. (1997). Properties of volume-regulated anion channels in mammalian cells. *Progress in Biophysics and Molecular Biology, 68*, 69–119.

Nilius, B., Oike, M., Zahradnik, I., & Droogmans, G. (1994). Activation of a Cl^- current by hypotonic volume increase in human endothelial cells. *The Journal of General Physiology, 103*, 787–805.

Nilius, B., Prenen, J., Voets, T., Eggermont, J., & Droogmans, G. (1998). Activation of volume-regulated chloride currents by reduction of intracellular ionic strength in bovine endothelial cells. *The Journal of Physiology, 506*, 353–361.

Nilius, B., Sehrer, J., & Droogmans, G. (1994). Permeation properties and modulation of volume-activated Cl^--currents in human endothelial cells. *British Journal of Pharmacology, 112*, 1049–1056.

Ogura, T., Furukawa, T., Toyozaki, T., Yamada, K., Zheng, Y. J., Katayama, Y., et al. (2002). ClC-3B, a novel ClC-3 splicing variant that interacts with EBP50 and facilitates expression of CFTR-regulated ORCC. *FASEB Journal: Official Publication of the Federation of American Societies for Experimental Biology, 16*, 863–865.

Oike, M., Droogmans, G., & Nilius, B. (1994). The volume-activated chloride current in human endothelial cells depends on intracellular ATP. *Pflugers Arch, 427*, 184–186.

Oiki, S., Kubo, M., & Okada, Y. (1994). Mg^{2+} and ATP-dependence of volume-sensitive Cl^- channels in human epithelial cells. *Jpn The Journal of Physiology, 44*(Suppl. 2), S77–S79.

Okada, S. F., O'Neal, W. K., Huang, P., Nicholas, R. A., Ostrowski, L. E., Craigen, W. J., et al. (2004). Voltage-dependent anion channel-1 (VDAC-1) contributes to ATP release and cell volume regulation in murine cells. *The Journal of General Physiology, 124*, 513–526.

Okada, T., Akita, T., Sato-Numata, K., Islam, M. R., & Okada, Y. (2014). A newly cloned ClC-3 isoform, ClC-3d, as well as ClC-3a mediates Cd-sensitive outwardly rectifying anion currents. *Cellular Physiology and Biochemistry: International Journal of Experimental Cellular Physiology, Biochemistry, and Pharmacology, 33*, 539–556.

Okada, T., Islam, M. R., Merzlyak, P., Sabirov, R. Z., & Okada, Y. (2012). Attempts to identify the gene encoding Maxi-anion channel. *The Journal of Physiological Sciences, 62*(Suppl 1), S85.

Okada, T., Islam, M. R., Tsiferova, N. A., Okada, Y., & Sabirov, R. Z. (2017). Specific and essential but not sufficient roles of LRRC8A in the activity of volume-sensitive outwardly rectifying anion channel (VSOR). *Channels (Austin), 11*, 109–120.

Okada, Y. (1997). Volume expansion-sensing outward rectifier Cl channel: A fresh start to the molecular identity and volume sensor. *American Journal of Physiology. Cell Physiology, 273*, C755–C789.

Okada, Y. (Ed.). (1998). *Cell volume Regulation: The molecular mechanism and volume sensing machinery*. Amsterdam: Elsevier.

Okada, Y. (2004). Ion channels and transporters involved in cell volume regulation and sensor mechanisms. *Cell Biochemistry and Biophysics, 41*, 233–258.

Okada, Y. (2006). Cell volume-sensitive chloride channel: Phenotypic properties and molecular identity. *Contributions To Nephrology, 152*, 9–24.

Okada, Y., Kubo, M., Oiki, S., Petersen, C. C. H., Tominaga, M., Hazama, A., et al. (1994). Properties of volume-sensitive Cl^- channels in a human epithelial cell line. *The Japanese Journal of Physiology, 44*(Suppl. 2), S31–S35.

Okada, Y., Maeno, E., Shimizu, T., Dezaki, K., Wang, J., & Morishima, S. (2001). Receptor-mediated control of regulatory volume decrease (RVD) and apoptotic volume decrease (AVD). *The Journal of Physiology, 532*, 3–16.

Okada, Y., Maeno, E., Shimizu, T., Manabe, K., Mori, S., & Nabekura, T. (2004). Dual roles of plasmalemmal chloride channels in induction of cell death. *Pflügers Archiv, 448*, 287–295.

Okada, Y., Oiki, S., Tominaga, M., Kubo, K., Miwa, A., Tominaga, T., et al. (1997). Volume-sensitive Cl^- channel in human epithelial cells: Regulation by ATP and relation to P-glycoprotein. *The Japanese Journal of Physiology, 47*(Suppl. 1), S19–S20.

Okada, Y., Petersen, C. C. H., Kubo, M., Morishima, S., & Tominaga, M. (1994). Osmotic swelling activates intermediate-conductance Cl^- channels in human intestinal epithelial cells. *The Japanese Journal of Physiology, 44*, 403–409.

Okada, Y., Sato, K., Toychiev, A. H., Suzuki, M., Dutta, A. K., Inoue, H., et al. (2009). The puzzles of volume-activated anion channels. In F. J. Alvarez-Leefmans, & E. Delpire (Eds.), *Physiology and pathology of chloride transporters and channels in the nervous system. From molecules to disease* (pp. 283–306). San Diego: Elsevier.

Okada, Y., Shimizu, T., Maeno, E., Tanabe, S., Wang, X., & Takahashi, N. (2006). Volume-sensitive chloride channels involved in apoptotic volume decrease and cell death. *Journal of Membrane Biology, 209*, 21–29.

Orellana, J. A., Montero, T. D., & von Bernhardi, R. (2013). Astrocytes inhibit nitric oxide-dependent $Ca2^+$ dynamics in activated microglia: Involvement of ATP released via pannexin 1 channels. *Glia, 61*, 2023–2037.

Ousingsawat, J., Wanitchakool, P., Schreiber, R., Wuelling, M., Vortkamp, A., & Kunzelmann, K. (2015). Anoctamin-6 controls bone mineralization by activating the calcium transporter NCX1. *Journal of Biological Chemistry, 290*, 6270–6280.

Paulmichl, M., Li, Y., Wickman, K., Ackerman, M., Peralta, E., & Clapham, D. (1992). New mammalian chloride channel identified by expression cloning. *Nature, 356*, 238–241.

Pedersen, S. F., Hoffmann, E. K., & Mills. (2001). The cytoskeleton and cell volume regulation. *Comparative Biochemistry and Physiology. Part A, Molecular & Integrative Physiology, 130*, 385–399.

Pedersen, S. F., Klausen, T. K., & Nilius, B. (2015). The identification of a volume-regulated anion channel: An amazing odyssey. *Acta Physiologica, 213*, 868–881.

Pedersen, S. F., Mills, J. W., & Hoffmann, E. K. (1999). Role of the F-actin cytoskeleton in the RVD and RVI processes in Ehrlich ascites tumor cells. *Experimental Cell Research, 252*, 63–74.

Pedersen, S. F., Okada, Y., & Nilius, B. (2016). Biophysics and physiology of the volume-regulated anion channel (VRAC)/volume-sensitive outwardly rectifying anion channel (VSOR). *Pflügers Archiv - European Journal of Physiology, 468*, 371–383.

Pelegrin, P., & Surprenant, A. (2006). Pannexin-1 mediates large pore formation and interleukin-1beta release by the ATP-gated P2X7 receptor. *The EMBO Journal, 25*, 5071–5082.

Perez-Cornejo, P., Gokhale, A., Duran, C., Cui, Y., Xiao, Q., Hartzell, H. C., et al. (2012). Anoctamin 1 (Tmem16A) Ca^{2+}-activated chloride channel stoichiometrically interacts with an ezrin-radixin-moesin network. *Proceedings of the National Academy of Sciences of the United States of America, 109*, 10376–10381.

Picollo, A., & Pusch, M. (2005). Chloride/proton antiporter activity of mammalian CLC proteins ClC-4 and ClC-5. *Nature, 436*, 420–423.

Platt, C. D., Chou, J., Houlihan, P., Badran, Y. R., Kumar, L., Bainter, W., et al. (2017). Leucine-rich repeat containing 8A (LRRC8A)-dependent volume-regulated anion channel activity is dispensable for T-cell development and function. *Journal of Allergy and Clinical Immunology, 140*, 1651–1659.

Poletto Chaves, L. A., & Varanda, W. A. (2008). Volume-activated chloride channels in mice Leydig cells. *Pflugers Arch, 457*, 493–504.

Praetorius, H. A., & Leipziger, J. (2009). ATP release from non-excitable cells. *Purinergic Signalling, 5*, 433–446.

Pritchard, S., & Guilak, F. (2004). The role of F-actin in hypo-osmotically induced cell volume change and calcium signaling in anulus fibrosus cells. *Annals of Biomedical Engineering, 32*, 103–111.

Pu, W. T., Krapivinsky, G. B., Krapivinsky, L., & Clapham, D. E. (1999). pICln inhibits snRNP biogenesis by binding core spliceosomal proteins. *Molecular and Cellular Biology, 19*, 4113–4120.

Qiu, F., Wang, J., Spray, D. C., Scemes, E., & Dahl, G. (2011). Two non-vesicular ATP release pathways in the mouse erythrocyte membrane. *FEBS Letters, 585*, 3430–3435.

Qiu, Z., Dubin, A. E., Mathur, J., Tu, B., Reddy, K., Miraglia, L. J., et al. (2014). SWELL1, a plasma membrane protein, is an essential component of volume-regulated anion channel. *Cell, 157*, 447–458.

Ransford, G. A., Fregien, N., Qiu, F., Dahl, G., Conner, G. E., & Salathe, M. (2009). Pannexin 1 contributes to ATP release in airway epithelia. *American Journal of Respiratory Cell and Molecular Biology, 41*, 525–534.

Reddy, M. M., Quinton, P. M., Haws, C., Wine, J. J., Grygorczyk, R., Tabcharani, J. A., et al. (1996). Failure of the cystic fibrosis transmembrane conductance regulator to conduct ATP. *Science, 271*, 1876–1879.

Reisin, I. L., Prat, A. G., Abraham, E. H., Amara, J. F., Gregory, R. J., Ausiello, D. A., et al. (1994). The cystic fibrosis transmembrane conductance regulator is a dual ATP and chloride channel. *Journal of Biological Chemistry, 269*, 20584–20591.

Romanenko, V. G., Rothblat, G. H., & Levitan, I. (2004). Sensitivity of volume-regulated anion current to cholesterol structural analogues. *The Journal of General Physiology, 123*, 77–87.

Rossow, C. F., Duan, D., Hatton, W. J., Britton, F., Hume, J. R., & Horowitz, B. (2006). Functional role of amino terminus in ClC-3 chloride channel regulation by phosphorylation and cell volume. *Acta Physiologica, 187*, 5–19.

Rostovtseva, T., & Colombini, M. (1997). VDAC channels mediate and gate the flow of ATP: Implications for the regulation of mitochondrial function. *Biophysical Journal, 72*, 1954–1962.

Rostovtseva, T. K., & Bezrukov, S. M. (1998). ATP transport through a single mitochondrial channel, VDAC, studied by current fluctuation analysis. *Biophysical Journal, 74*, 2365–2373.

Rostovtseva, T. K., Komarov, A., Bezrukov, S. M., & Colombini, M. (2002). VDAC channels differentiate between natural metabolites and synthetic molecules. *Journal of Membrane Biology, 187*, 147–156.

Sabirov, R. Z., Dutta, A. K., & Okada, Y. (2001). Volume-dependent ATP-conductive large-conductance anion channel as a pathway for swelling-induced ATP release. *The Journal of General Physiology, 118*, 251–266.

Sabirov, R. Z., Korchev, Y. E., & Okada, Y. (2012). Smart-patch technique. In Y. Okada (Ed.), *Patch clamp techniques. From biginning to advanced protochols* (pp. 379–387). Tokyo: Springer.

Sabirov, R. Z., & Merzlyak, P. G. (2012). Plasmalemmal VDAC controversies and maxi-anion channel puzzle. *Biochimica et Biophysica Acta, 1818*, 1570–1580.

Sabirov, R. Z., Merzlyak, P. G., Islam, M. R., Okada, T., & Okada, Y. (2016). The properties, functions, and pathophysiology of maxi-anion channels. *Pflügers Archiv - European Journal of Physiology, 468*, 405–420.

Sabirov, R. Z., Merzlyak, P. G., Okada, T., Islam, M. R., Uramoto, H., Mori, T., et al. (2017). The organic anion transporter SLCO2A1 constitutes the core component of the Maxi-Cl channel. *The EMBO Journal, 36*, 3309–3324.

Sabirov, R. Z., & Okada, Y. (2004a). ATP-conducting maxi-anion channel: A new player in stress-sensory transduction. *The Japanese Journal of Physiology, 54*, 7–14.

Sabirov, R. Z., & Okada, Y. (2004b). Wide nanoscopic pore of maxi-anion channel suits its function as an ATP-conductive pathway. *Biophysical Journal, 87*, 1672–1685.

Sabirov, R. Z., & Okada, Y. (2005). ATP release via anion channels. *Purinergic Signalling, 1*, 311–328.

Sabirov, R. Z., & Okada, Y. (2009). The maxi-anion channel: A classical channel playing novel roles through an unidentified molecular entity. *The Journal of Physiological Sciences, 59*, 3–21.

Sabirov, R. Z., Prenen, J., Tomita, T., Droogmans, G., & Nilius, B. (2000). Reduction of ionic strength activates single volume-regulated anion channels (VRAC) in endothelial cells. *Pflügers Archiv, 439*, 315–320.

Sabirov, R. Z., Sheiko, T., Liu, H., Deng, D., Okada, Y., & Craigen, W. J. (2006). Genetic demonstration that the plasma membrane maxianion channel and voltage-dependent anion channels are unrelated proteins. *Journal of Biological Chemistry, 281*, 1897–1904.

Sana-Ur-Rehman, H., Markus, I., Moore, K. H., Mansfield, K. J., & Liu, L. (2017). Expression and localization of pannexin-1 and CALHM1 in porcine bladder and their involvement in modulating ATP release. *American Journal of Physiology. Regulatory, Integrative and Comparative Physiology, 312*, R763–R772.

Sato-Numata, K., Numata, T., Inoue, R., & Okada, Y. (2016). Distinct pharmacological and molecular properties of the acid-sensitive outwardly rectifying (ASOR) anion channel from those of the volume-sensitive outwardly rectifying (VSOR) anion channel. *Pflügers Archiv - European Journal of Physiology, 468*, 795–803.

Sato-Numata, K., Numata, T., Inoue, R., Sabirov, R. Z., & Okada, Y. (2017). Distinct contributions of LRRC8A and its paralogs to the VSOR anion channel from those of the ASOR anion channel. *Channels (Austin), 11*, 167–172.

Sawada, A., Takihara, Y., Kim, J. Y., Matsuda-Hashii, Y., Tokimasa, S., Fujisaki, H., et al. (2003). A congenital mutation of the novel gene LRRC8 causes agammaglobulinemia in humans. *Journal of Clinical Investigation, 112*, 1707–1713.

Schlichter, L. C., Grygorczyk, R., Pahapill, P. A., & Grygorczyk, C. (1990). A large, multiple-conductance chloride channel in normal human T lymphocytes. *Pflügers Archiv, 416*, 413–421.

Schwiebert, E. M. (1999). ABC transporter-facilitated ATP conductive transport. *American Journal of Physiology, 276*, C1–C8.

Schwiebert, E. M., Egan, M. E., Hwang, T. H., Fulmer, S. B., Allen, S. S., Cutting, G. R., et al. (1995). CFTR regulates outwardly rectifying chloride channels through an autocrine mechanism involving ATP. *Cell, 81*, 1063–1073.

Seminario-Vidal, L., Okada, S. F., Sesma, J. I., Kreda, S. M., van Heusden, C. A., Zhu, Y., et al. (2011). Rho signaling regulates pannexin 1-mediated ATP release from airway epithelia. *Journal of Biological Chemistry, 286*, 26277–26286.

Senger, R., & Goldmann, W. H. (1995). The influence of cations and ionic strength on actin polymerization in the presence/absence of alpha-actinin. *Biochemistry & Molecular Biology International, 35*, 103–109.

Shen, M. R., Wu, S. N., & Chou, C. Y. (1996). Volume-sensitive chloride channels in the primary culture cells of human cervical carcinoma. *Biochimica et Biophysica Acta, 1315*, 138–144.

Shimizu, T., Iehara, T., Sato, K., Fujii, T., Sakai, H., & Okada, Y. (2013). TMEM16F is a component of a Ca^{2+}-activated Cl^- channel but not a volume-sensitive outwardly rectifying Cl^- channel. *American Journal of Physiology. Cell Physiology, 304*, C748–C759.

Shimizu, T., Numata, T., & Okada, Y. (2004). A role of reactive oxygen species in apoptotic activation of volume-sensitive Cl^- channel. *Proceedings of the National Academy of Sciences of the United States of America, 101*, 6770–6773.

Singh, R. D., Gibbons, S. J., Saravanaperumal, S. A., Du, P., Hennig, G. W., Eisenman, S. T., et al. (2014). Ano1, a $Ca2^{+}$-activated Cl^{-} channel, coordinates contractility in mouse intestine by $Ca2^{+}$ transient coordination between interstitial cells of Cajal. *The Journal of Physiology, 592*, 4051–4068.

Sipos, A., Vargas, S. L., Toma, I., Hanner, F., Willecke, K., & Peti-Peterdi, J. (2009). Connexin 30 deficiency impairs renal tubular ATP release and pressure natriuresis. *Journal of the American Society of Nephrology: JASN, 20*, 1724–1732.

Sirianant, L., Ousingsawat, J., Wanitchakool, P., Schreiber, R., & Kunzelmann, K. (2016). Cellular volume regulation by anoctamin 6: $Ca2^{+}$, phospholipase A2 and osmosensing. *Pflügers Archiv - European Journal of Physiology, 468*, 335–349.

Sridharan, M., Adderley, S. P., Bowles, E. A., Egan, T. M., Stephenson, A. H., Ellsworth, M. L., et al. (2010). Pannexin 1 is the conduit for low oxygen tension-induced ATP release from human erythrocytes. *American Journal of Physiology. Heart and Circulatory Physiology, 299*, H1146–H1152.

Stauber, T., & Jentsch, T. J. (2013). Chloride in vesicular trafficking and function. *Annual Review of Physiology, 75*, 453–477.

Stobrawa, S. M., Breiderhoff, T., Takamori, S., Engel, D., Schweizer, M., Zdebik, A. A., et al. (2001). Disruption of ClC-3, a chloride channel expressed on synaptic vesicles, leads to a loss of the hippocampus. *Neuron, 29*, 185–196.

Stotz, S. C., & Clapham, D. E. (2012). Anion-sensitive fluorophore identifies the Drosophila swell-activated chloride channel in a genome-wide RNA interference screen. *PLoS One, 7*, e46865.

Strange, K. (1998). Molecular identity of the outwardly rectifying, swelling-activated anion channel: Time to reevaluate pICln. *The Journal of General Physiology, 111*, 617–622.

Strange, K., Emma, F., & Jackson, P. S. (1996). Cellular and molecular physiology of volume-sensitive anion channels. *American Journal of Physiology. Cell Physiology, 270*, C711–C730.

Sugita, M., Yue, Y., & Foskett, J. K. (1998). CFTR Cl^{-} channel and CFTR-associated ATP channel: Distinct pores regulated by common gates. *The EMBO Journal, 17*, 898–908.

Sun, X. P., Supplisson, S., & Mayer, E. (1993). Chloride channels in myocytes from rabbit colon are regulated by a pertussis toxin-sensitive G protein. *American Journal of Physiology, 264*, G774–G785.

Sun, X. P., Supplisson, S., Torres, R., Sachs, G., & Mayer, E. (1992). Characterization of large-conductance chloride channels in rabbit colonic smooth muscle. *The Journal of Physiology, 448*, 355–382.

Suzuki, M. (2006). The Drosophila tweety family: Molecular candidates for large-conductance $Ca2^{+}$-activated Cl^{-} channels. *Experimental Physiology, 91*, 141–147.

Suzuki, M., & Mizuno, A. (2004). A novel human Cl^{-} channel family related to *Drosophila* flightless locus. *Journal of Biological Chemistry, 279*, 22461–22468.

Suzuki, T., Rai, T., Hayama, A., Sohara, E., Suda, S., Itoh, T., et al. (2006). Intracellular localization of ClC chloride channels and their ability to form hetero-oligomers. *Journal of Cellular Physiology, 206*, 792–798.

Syeda, R., Qiu, Z., Dubin, A. E., Murthy, S. E., Florendo, M. N., Mason, D. E., et al. (2016). LRRC8 proteins form volume-regulated anion channels that sense ionic strength. *Cell, 164*, 499–511.

Szteyn, K., Schmid, E., Nurbaeva, M. K., Yang, W., Münzer, P., Kunzelmann, K., et al. (2012). Expression and functional significance of the $Ca2^{+}$-activated Cl^{-} channel ANO6 in dendritic cells. *Cellular Physiology and Biochemistry: International Journal of Experimental Cellular Physiology, Biochemistry, and Pharmacology, 30*, 1319–1332.

Tanis, J. E., Ma, Z., Krajacic, P., He, L., Foskett, J. K., & Lamitina, T. (2013). CLHM-1 is a functionally conserved and conditionally toxic $Ca2^{+}$-permeable ion channel in *Caenorhabditis elegans. The Journal of Neuroscience: the Official Journal of the Society for Neuroscience, 33*, 12275–12286.
Taruno, A. (2018). ATP release channels. *International Journal of Molecular Sciences, 19*(3), 808.
Taruno, A., Vingtdeux, V., Ohmoto, M., Ma, Z., Dvoryanchikov, G., Li, A., et al. (2013). CALHM1 ion channel mediates purinergic neurotransmission of sweet, bitter and umami tastes. *Nature, 495*, 223–226.
Ternovsky, V. I., Okada, Y., & Sabirov, R. Z. (2004). Nano-sized pore of the volume-sensitive anion channel revealed by differential polymer partitioning. *FEBS Letters, 576*, 433–436.
Thompson, R. J., Zhou, N., & MacVicar, B. A. (2006). Ischemia opens neuronal gap junction hemichannels. *Science, 312*, 924–927.
Tominaga, M., Tominaga, T., Miwa, A., & Okada, Y. (1995). Volume-sensitive chloride channel activity does not depend on endogenous P-glycoprotein. *Journal of Biological Chemistry, 270*, 27887–27893.
Toychiev, A. H., Sabirov, R. Z., Takahashi, N., Ndo-Akatsuka, Y., Liu, H., Shintani, T., et al. (2009). Activation of maxi-anion channel by protein tyrosine dephosphorylation. *American Journal of Physiology. Cell Physiology, 297*, C990–C1000.
Tsujimoto, Y., & Shimizu, S. (2007). Role of the mitochondrial membrane permeability transition in cell death. *Apoptosis: an International Journal on Programmed Cell Death, 12*, 835–840.
Tsumura, T., Oiki, S., Ueda, S., Okuma, M., & Okada, Y. (1996). Sensitivity of volume-sensitive Cl^{-} conductance in human epithelial cells to extracellular nucleotides. *American Journal of Physiology. Cell Physiology, 271*, C1872–C1878.
Ullrich, F., Reincke, S. M., Voss, F. K., Stauber, T., & Jentsch, T. J. (2016). Inactivation and anion selectivity of volume-regulated anion channels (VRACs) depend on C-terminal residues of the first extracellular loop. *Journal of Biological Chemistry, 291*, 17040–17048.
Valverde, M. A., Díaz, M., Sepúlveda, F. V., Gill, D. R., Hyde, S. C., & Higgins, C. F. (1992). Volume-regulated chloride channels associated with the human multidrug-resistance P-glycoprotein. *Nature, 355*, 830–833.
Van der Kloot, W. (2003). Loading and recycling of synaptic vesicles in the Torpedo electric organ and the vertebrate neuromuscular junction. *Progress in Neurobiology, 71*, 269–303.
Varela, D., Simon, F., Riveros, A., Jørgensen, F., & Stutzin, A. (2004). NAD(P)H oxidase-derived H_2O_2 signals chloride channel activation in cell volume regulation and cell proliferation. *Journal of Biological Chemistry, 279*, 13301–13304.
Voets, T., Buyse, G., Tytgat, J., Droogmans, G., Eggermont, J., & Nilius, B. (1996). The chloride current induced by expression of the protein pICln in Xenopus oocytes differs from the endogenous volume-sensitive chloride current. *The Journal of Physiology, 495*, 441–447.
Voets, T., Droogmans, G., Raskin, G., Eggermont, J., & Nilius, B. (1999). Reduced intracellular ionic strength as the initial trigger for activation of endothelial volume-regulated anion channels. *Proceedings of the National Academy of Sciences of the United States of America, 96*, 5298–5303.
Voss, F. K., Ullrich, F., Münch, J., Lazarow, K., Lutter, D., Mah, N., et al. (2014). Identification of LRRC8 heteromers as an essential component of the volume-regulated anion channel VRAC. *Science, 344*, 634–638.
Vyssokikh, M. Y., & Brdiczka, D. (2003). The function of complexes between the outer mitochondrial membrane pore (VDAC) and the adenine nucleotide translocase in regulation of energy metabolism and apoptosis. *Acta Biochimica Polonica, 50*, 389–404.
Wang, G. X., Hatton, W. J., Wang, G. L., Zhong, J., Yamboliev, I., Duan, D., et al. (2003). Functional effects of novel anti-ClC-3 antibodies on native volume-sensitive osmolyte and anion channels in cardiac and smooth muscle cells. *American Journal of Physiology. Heart and Circulatory Physiology, 285*, H1453–H1463.

Wang, J., Ambrosi, C., Qiu, F., Jackson, D. G., Sosinsky, G., & Dahl, G. (2014). The membrane protein Pannexin1 forms two open-channel conformations depending on the mode of activation. *Science Signaling [electronic Resource], 7*, ra69.

Wang, L., Ma, W., Zhu, L., Ye, D., Li, Y., Liu, S., et al. (2012). ClC-3 is a candidate of the channel proteins mediating acid-activated chloride currents in nasopharyngeal carcinoma cells. *American Journal of Physiology. Cell Physiology, 303*, C14–C23.

Wang, R., Lu, Y., Gunasekar, S., Zhang, Y., Benson, C. J., Chapleau, M. W., et al. (2017). The volume-regulated anion channel (LRRC8) in nodose neurons is sensitive to acidic pH. *JCI Insight, 2*, e90632.

Wang, T., Birsoy, K., Hughes, N. W., Krupczak, K. M., Post, Y., Wei, J. J., et al. (2015). Identification and characterization of essential genes in the human genome. *Science, 350*, 1096–1101.

Wang, X., Takahashi, N., Uramoto, H., & Okada, Y. (2005). Chloride channel inhibition prevents ROS-dependent apoptosis induced by ischemia-reperfusion in mouse cardiomyocytes. *Cellular Physiology and Biochemistry: International Journal of Experimental Cellular Physiology, Biochemistry, and Pharmacology, 16*, 147–154.

Wanitchakool, P., Ousingsawat, J., Sirianant, L., Cabrita, I., Faria, D., Schreiber, R., et al. (2017). Cellular defects by deletion of ANO10 are due to deregulated local calcium signaling. *Cellular Signalling, 30*, 41–49.

Weylandt, K. H., Nebrig, M., Jansen-Rosseck, N., Amey, J. S., Carmena, D., Wiedenmann, B., et al. (2007). ClC-3 expression enhances etoposide resistance by increasing acidification of the late endocytic compartment. *Molecular Cancer Therapeutics, 6*, 979–986.

Weylandt, K. H., Valverde, M. A., Nobles, M., Raguz, S., Amey, J. S., Diaz, M., et al. (2001). Human ClC-3 is not the swelling-activated chloride channel involved in cell volume regulation. *Journal of Biological Chemistry, 276*, 17461–17467.

Workman, A. D., Carey, R. M., Chen, B., Saunders, C. J., Marambaud, P., Mitchell, C. H., et al. (2017). CALHM1-mediated ATP release and ciliary beat frequency modulation in nasal epithelial cells. *Scientific Reports, 7*, 6687.

Xia, J., Lim, J. C., Lu, W., Beckel, J. M., Macarak, E. J., Laties, A. M., et al. (2012). Neurones respond directly to mechanical deformation with pannexin-mediated ATP release and autostimulation of P2X7 receptors. *The Journal of Physiology, 590*, 2285–2304.

Xiao, Q., Yu, K., Cui, Y. Y., & Hartzell, H. C. (2009). Dysregulation of human bestrophin-1 by ceramide-induced dephosphorylation. *The Journal of Physiology, 587*, 4379–4391.

Xiong, D., Heyman, N. S., Airey, J., Zhang, M., Singer, C. A., Rawat, S., et al. (2010). Cardiac-specific, inducible ClC-3 gene deletion eliminates native volume-sensitive chloride channels and produces myocardial hypertrophy in adult mice. *Journal of Molecular and Cellular Cardiology, 48*, 211–219.

Xiong, D., Wang, G. X., Burkin, D. J., Yamboliev, I. A., Singer, C. A., Rawat, S., et al. (2009). Cardiac-specific overexpression of the human short CLC-3 chloride channel isoform in mice. *Clinical and Experimental Pharmacology & Physiology, 36*, 386–393.

Yamamoto-Mizuma, S., Wang, G. X., Liu, L. L., Schegg, K., Hatton, W. J., Duan, D., et al. (2004). Altered properties of volume-sensitive osmolyte and anion channels (VSOACs) and membrane protein expression in cardiac and smooth muscle myocytes from $Clcn3^{-/-}$ mice. *The Journal of Physiology, 557*, 439–456.

Yang, H., Huang, L. Y., Zeng, D. Y., Huang, E. W., Liang, S. J., Tang, Y. B., et al. (2012a). Decrease of intracellular chloride concentration promotes endothelial cell inflammation by activating nuclear factor-κB pathway. *Hypertension, 60*, 1287–1293.

Yang, H., Kim, A., David, T., Palmer, D., Jin, T., Tien, J., et al. (2012b). TMEM16F forms a Ca^{2+}-activated cation channel required for lipid scrambling in platelets during blood coagulation. *Cell, 151*, 111–122.

Yang, H., Ma, L., Wang, Y., Zuo, W., Li, B., Yang, Y., et al. (2018). Activation of ClC-3 chloride channel by 17β-estradiol relies on the estrogen receptor α expression in breast cancer. *Journal of Cellular Physiology, 233*, 1071–1081.

Yang, L., Ye, D., Ye, W., Jiao, C., Zhu, L., Mao, J., et al. (2011). ClC-3 is a main component of background chloride channels activated under isotonic conditions by autocrine ATP in nasopharyngeal carcinoma cells. *Journal of Cellular Physiology, 226*, 2516–2526.

Yin, Z., Tong, Y., Zhu, H., & Watsky, M. A. (2008). ClC-3 is required for LPA-activated Cl^- current activity and fibroblast-to-myofibroblast differentiation. *American Journal of Physiology. Cell Physiology, 294*, C535–C542.

Yu, K., Whitlock, J. M., Lee, K., Ortlund, E. A., Cui, Y. Y., & Hartzell, H. C. (2015). Identification of a lipid scrambling domain in ANO6/TMEM16F. *eLife, 4*, e06901.

Zhou, J. G., Ren, J. L., Qiu, Q. Y., He, H., & Guan, Y. Y. (2005). Regulation of intracellular Cl^- concentration through volume-regulated ClC-3 chloride channels in A10 vascular smooth muscle cells. *Journal of Biological Chemistry, 280*, 7301–7308.

Zimmermann, H., & Denston, C. R. (1976). Adenosine triphosphate in cholinergic vesicles isolated from the electric organ of Electrophorus electricus. *Brain Research, 111*, 365–376.

CHAPTER SIX

Molecular Biology and Physiology of Volume-Regulated Anion Channel (VRAC)

James Osei-Owusu*, Junhua Yang*, Maria del Carmen Vitery* and Zhaozhu Qiu*,§,1

*Department of Physiology, Johns Hopkins University School of Medicine, Baltimore, MD, United States
§Solomon H. Snyder Department of Neuroscience, Johns Hopkins University School of Medicine, Baltimore, MD, United States
[1]Corresponding author: E-mail: zhaozhu@jhmi.edu

Contents

Abstract

The Volume-Regulated Anion Channel (VRAC) is activated by cell swelling and plays a key role in cell volume regulation. VRAC is ubiquitously expressed in vertebrate cells and also implicated in many other physiological and cellular processes including fluid secretion, glutamate release, membrane potential regulation, cell proliferation, migration, and apoptosis. Although its biophysical properties have been well characterized,

Current Topics in Membranes, Volume 81
ISSN 1063-5823
https://doi.org/10.1016/bs.ctm.2018.07.005

the molecular identity of VRAC remained a mystery for almost three decades. The field was transformed by recent discoveries showing that the leucine-rich repeat-containing protein 8A (LRRC8A, also named SWELL1) and its four other homologs form heteromeric VRAC channels. The composition of LRRC8 subunits determines channel properties and substrate selectivity of a large variety of different VRACs. Incorporating purified SWELL1-containing protein complexes into lipid bilayers is sufficient to reconstitute channel activities, a finding that supports the decrease in intracellular ionic strength as the mechanism of VRAC activation during cell swelling. Characterization of *Swell1* knockout mice uncovers the important role of VRAC in T cell development, pancreatic β-cell glucose-stimulated insulin secretion, and adipocyte metabolic function. The ability to permeate organic osmolytes and metabolites is a major feature of VRAC. The list of VRAC substrates is expected to grow, now also including some cancer drugs and antibiotics even under non-cell swelling conditions. Therefore, a critical role of VRAC in drug resistance and cell—cell communication is emerging. This review summarizes the exciting recent progress on the structure-function relationship and physiology of VRAC and discusses key future questions to be solved.

1. INTRODUCTION

Because plasma membranes are permeable to water, decreases in extracellular osmolality or increases in intracellular osmolality induces a rapid net influx of water across the cell membrane, resulting in cell swelling. To achieve cell volume homeostasis, cells employ multiple ion channel and transporter mechanisms to mediate the regulatory volume decrease (RVD) through ion (mainly K^+ and Cl^-) efflux followed by release of osmotically obligated water (Hoffmann, Lambert, & Pedersen, 2009; Lang, 2007). A key player in RVD is the Volume-Regulated Anion Channel (VRAC), which is activated by cell swelling and mediates Cl^- efflux (Hoffmann et al., 2009; Jentsch, 2016). The large VRAC current (known as $I_{Cl,\ swell}$) activated over minutes by reducing extracellular osmolality was first described in human lymphocytes and epithelial cells three decades ago (Fig. 1A) (Cahalan & Lewis, 1988; Hazama & Okada, 1988). Subsequently, it has been observed in almost every vertebrate cell type investigated (Nilius et al., 1997; Okada, 1997; Strange, Emma, & Jackson, 1996). The VRAC channel displays characteristic features, such as moderate outward rectification (larger outward currents than inward currents at positive vs. negative membrane potentials, Fig. 1B). Based on this property, VRAC is also referred as VSOR (volume-sensitive outwardly rectifying anion channel). In addition, VRAC currents inactivate at large positive membrane voltages, and have higher permeability to larger halide anions than Cl^- ($I^- > Br^- > Cl^-$).

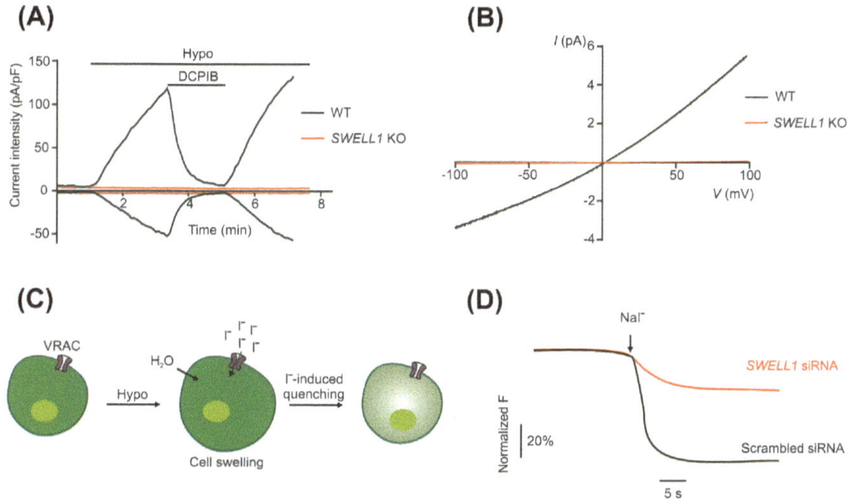

Figure 1 Identification of SWELL1 as an essential component of VRAC by RNAi screening. (A) Typical time course of whole-cell current densities at ± 100 mV induced by hypotonic solution (220 mOsm/kg) for WT and *SWELL1* KO HeLa cells. DCPIB, a potent VRAC blocker, was used as indicated. (B) Representative whole-cell $I_{Cl,\ swell}$ recorded by ramp protocol showing outward rectification. (C) Schematic of the I^--induced YFP quenching as the screening assay for VRAC. (D) Representative traces of I^--induced YFP quenching in HEK-YFP cells transfected with either scrambled siRNA or *SWELL1* siRNA.

Although the electrophysiological characteristics of VRAC have been studied extensively, its molecular identity remained a long-standing mystery (Pedersen, Klausen, & Nilius, 2015). More than half a dozen candidates had been proposed to mediate VRAC, but none of them stood the test of time until the recent discovery of the LRRC8 (leucine-rich repeat-containing protein 8) family proteins as essential components of VRAC. In addition to cell volume regulation, VRAC has been implicated in a wide range of important physiological and cellular functions, such as regulation of membrane potential, salt and fluid secretion, glutamate release, cell proliferation and cell death (Pedersen, Okada, & Nilius, 2016). These physiological and pathological roles assigned to VRAC are mainly based on previous studies with VRAC pharmacological blockers, which often lack specificity and may also inhibit or activate other ion channels and transporters (Bowens, Dohare, Kuo, & Mongin, 2013; Fujii et al., 2015; Minieri et al., 2013). The molecular identification of SWELL1 and its associated LRRC8 homologs as essential pore-forming subunits has transformed the VRAC research field and enabled rapid progress in elucidating the molecular mechanism and physiological role of VRAC.

2. LRRC8 HETEROMERS AS VRAC

2.1 Identification of SWELL1 (LRRC8A) by Genome-Wide RNAi Screen

There are several reasons why it was challenging to uncover the molecular identity of VRAC. First, the popular ion channel expression cloning systems, such as *Xenopus* oocytes and human embryonic kidney cells (HEK), express large endogenous VRAC currents (Ackerman, Wickman, & Clapham, 1994; Helix, Strobaek, Dahl, & Christophersen, 2003). This likely confounded the overexpression results of some previously proposed VRAC candidates and hampered expression cloning, a widely used approach for ion channel identification (Schroeder, Cheng, Jan, & Jan, 2008). Second, VRAC lacks a specific high-affinity channel ligand, such as toxin agonist and antagonist, which hindered direct biochemical purification of the channel protein. Lastly, unlike most cation channels, there are no sequence homologies (for example, conserved pore-lining motif) among known Cl^- channel families identified to date (Duran, Thompson, Xiao, & Hartzell, 2010), including voltage-gated Cl^- channel (CLCs), cAMP-activated Cl^- channel (CFTR), Ca^{2+}-activated Cl^- channels (CaCC, TMEM16/Ano), and ligand-gated anion channels ($GABA_A$ and glycine receptors).

To circumvent these limitations, two research groups independently employed a loss-of-function genetics approach (Qiu et al., 2014; Voss et al., 2014) and performed a genome-wide RNA interference (RNAi) screen using HEK cells stably expressing I^--sensitive yellow fluorescent protein (YFP) (Galietta, Haggie, & Verkman, 2001). Application of hypotonic solutions and NaI triggers a rapid fluorescence quenching due to I^- influx through the endogenous VRAC in HEK-YFP cells (Fig. 1C). This unbiased approach led both groups to the identification of LRRC8A, also called SWELL1 (for its activation by cell swelling), as a key regulator for hypotonicity-induced I^- influx (Fig. 1D). Subsequent RNAi knockdown and CRISPR/Cas9-mediated knockout experiments revealed that SWELL1 is essential for $I_{Cl,\ swell}$ (Fig. 1) and regulates the RVD in various cell types, such as HEK, HeLa and HCT11 cells (Planells-Cases et al., 2015; Qiu et al., 2014; Voss et al., 2014). These two studies highlight the power of loss-of-function genomics screen, which can be applied in the future to other widely expressed Cl^- channels that are well described biophysically, yet without clear molecular identity (Duran et al., 2010). Similar genome-wide RNAi screening based on YFP quenching assay was previously conducted in *Drosophila* S2 cells and identified bestrophin-1 as forming the *Drosophila* $I_{Cl, swell}$ channel (Stotz & Clapham, 2012). Recently, a combination of proteomics and targeted RNAi screen has also been performed to identify the prostaglandin transporter SLCO2A1 as a candidate for the maxi-anion channel (MAC) (Sabirov et al., 2017).

2.2 LRRC8 Heteromers as Essential Components of VRAC

SWELL1 overexpression, although detected on cell surface, didn't enhance but rather decreased the endogenous $I_{Cl, swell}$ (Qiu et al., 2014; Voss et al., 2014), which suggested that SWELL1 could be one of the subunits in a multimeric channel complex. In this scenario, SWELL1 overexpression may result in a stoichiometry that lacks VRAC activity. Indeed, co-immunoprecipitation experiments in transfected HEK cells showed extensive interactions among SWELL1 and its four homologous family members (LRRC8B, 8C, 8D, and 8E) (Lee, Freinkman, Sabatini, & Ploegh, 2014; Voss et al., 2014). Affinity purification of SWELL1 close to the endogenous expression level in HeLa cells revealed only one prominent high molecular weight protein band after native gel electrophoresis (Syeda et al., 2016). Mass spectrometry analysis detected all five LRRC8 proteins, but no other proteins, as specific components within SWELL1-containing protein complexes (Syeda et al., 2016). CRISPR/Cas9-mediated disruption of four LRRC8 homologs (8B-8E) collectively also abolished $I_{Cl, swell}$, indicating SWELL1 and at least one other LRRC8 subunit are required for VRAC activity (Syeda et al., 2016; Voss et al., 2014). Consistently, co-expression of SWELL1 and another LRRC8 member, but not SWELL1 alone reconstituted $I_{Cl, swell}$ in cells lacking all five LRRC8 proteins (Voss et al., 2014). SWELL1 is important for the surface expression of VRAC, as its LRRC8 homologs remained intracellular when transfected alone but were partially trafficked to the plasma membrane when co-transfected with SWELL1 (Voss et al., 2014).

LRRC8 proteins form a conserved protein family in chordates (Abascal & Zardoya, 2012). They are composed of two main parts: N-terminal 4 transmembrane helices (TM) and C-terminal 17 leucine-rich repeats (LRRs) (Fig. 2). The LRRs were initially presumed to face the outside of the cells as potential ligand binding domains (Sawada et al., 2003), however, immunofluorescent staining of non-permeabilized cells indicated that the putative TM1-TM2 and TM3-TM4 loops are extracellular and the N- and C-termini are cytoplasmic (Qiu et al., 2014; Voss et al., 2014).

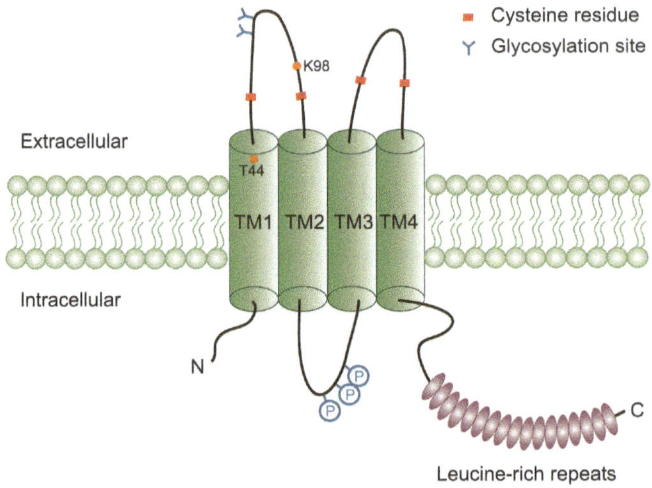

Figure 2 Schematic representation of SWELL1 protein. The diagram contains N-terminal 4 transmembrane helices (TM1-4), C-terminal 17 leucine-rich repeats (ovals), two pairs of conserved cysteine residues in the extracellular loops, and putative N-linked glycosylation sites and phosphorylation sites (P). Threonine (T44) located at the external boundary of TM1 and lysine (K98) within the TM1-TM2 extracellular loop were identifies as potential pore-lining residues.

The cytoplasmic localization of LRR domains of SWELL1 and LRRC8D was also confirmed by protease protection assay (Lee et al., 2014). Based on weak but significant sequence similarities, the TM half of LRRC8 proteins is evolutionarily related to pannexins (Abascal & Zardoya, 2012). Pannexins are the vertebrate homologs of invertebrate gap junction-forming innexins and are structurally related to vertebrate gap junction-forming connexins. Pannexins form hexameric large-pore channels that can release ATP during autocrine and paracrine signaling (Dahl, 2015). Biochemical purification, cross-linking, and single-step photobleaching experiments indicated multi-subunit assemblies of SWELL1-containing protein complexes, probably as hexamers like pannexins (Gaitan-Penas et al., 2016; Syeda et al., 2016). In addition to the similar transmembrane topology, both LRRC8 proteins and pannexins are N-glycosylated and possess two pairs of conserved cysteine residues in their extracellular loops (Fig. 2) (Abascal & Zardoya, 2012; Voss et al., 2014), which likely mediate inter- or intra-molecular disulfide bond formation between different subunits.

The exact stoichiometry and subunit arrangement of functional LRRC8 heteromers are not known. Native gel electrophoresis indicated that the oligomeric state of SWELL1-containing complexes remains the same, despite the deletion of three or all four LRRC8 homologs (Syeda et al., 2016). At least in overexpression systems, sequential co-immunoprecipitation in HEK cells and high-resolution co-localization study in *Xenopus* oocytes suggested that individual VRAC channel may contain SWELL1 and more than one (in theory, up to four) LRRC8 homologs (Gaitan-Penas et al., 2016; Lutter, Ullrich, Lueck, Kempa, & Jentsch, 2017). Most cells and tissues express SWELL1 and all four LRRC8 subunits albeit at variable expression levels (The Human Protein Atlas). Given the potential flexibility in stoichiometry and subunit arrangement within the channel complexes, individual cells may express a large variety of different VRAC channels. Several ligand-gated ion channel families, such as $GABA_A$ receptors, exhibit similar flexibility in subunit composition and their arrangement. Nevertheless, this structural heterogeneity with its consequential functional diversity in channel biophysical properties (discussed in Section 2.4) makes VRAC very unusual among the large ion channel families.

2.3 LRRC8 Heteromers Form the VRAC Channel Pore

Based on the homology with pannexins, LRRC8 proteins are likely pore-forming subunits of VRAC rather than essential auxiliary channel regulators (Abascal & Zardoya, 2012). Both mutagenesis study and *in vitro* reconstitution of purified proteins support this conclusion. Ion selectivity of a channel is typically determined by its pore (Isacoff, Jan, & Minor, 2013). For VRAC, a slight permeability preference of I^- over Cl^- (P_I/P_{Cl}: $\sim$1.2–1.4) is ones of its characteristics. To perform structure–function analysis, mutants of SWELL1 and other LRRC8 proteins were expressed or co-expressed in cells lacking SWELL1, LRRC8B-8E, or all five LRRC8 proteins (Qiu et al., 2014; Voss et al., 2014). To identify SWELL1 mutants that might change this selectivity, cysteine mutations were systematically introduced in individual amino acids of all 4 putative TM domains of SWELL1. This screen led to the identification of threonine at position 44 (T44) (Fig. 2), whose mutations caused significant alterations in anion permeability ratios (Qiu et al., 2014). The T44 residue is also conserved in the other LRRC8 subunits except LRRC8B. Like SWELL1 T44C, mutations at T44 in LRRC8C-8E proteins also increased P_I/P_{Cl} (Syeda et al., 2016), indicating these LRRC8 subunits are also contributing to the VRAC pore. T44 is predicted to be located at the external boundary of TM1, and intriguingly, the similar regions of connexins and pannexins were shown to line part of their channel pores (Maeda et al., 2009; Wang & Dahl, 2010).

In another set of studies of chimeras between LRRC8C and LRRC8E (discussed in the Section 2.4), a conserved lysine residue (K98) within the TM1-TM2 extracellular loop was identified (Fig. 2), which upon charge reversal decreased P_I/P_{Cl} (Ullrich, Reincke, Voss, Stauber, & Jentsch, 2016). The effect on selectivity was further augmented when the same mutation was present in both SWELL1 and LRRC8C or 8E (Ullrich et al., 2016). Collectively, these mutagenesis results support the idea that SWELL1 and its LRRC8 homologs are pore-forming subunits of VRAC. It's unclear why residues from two different regions both play a role in anion selectivity. It has to be noted that although significant, the effects of these mutations on P_I/P_{Cl} are relatively small (decreased to ~1.1 or increased to ~1.6) (Qiu et al., 2014; Ullrich et al., 2016). Furthermore, no effect on cation selectivity was observed (Ullrich et al., 2016). So it's possible that other critical pore-lining residues are yet to be discovered.

In vitro reconstitution provides direct and definitive evidence for pore-forming ability of ion channel candidates. Purified SWELL1-containing protein complexes were reconstituted into droplet interface lipid bilayers assembled from two monolayers (Syeda et al., 2016). Robust Cl^- channel activities, induced by osmotic gradient and sensitive to the VRAC inhibitor DCPIB, were observed (Syeda et al., 2016). This demonstrated that purified SWELL1 and associated LRRC8 proteins are sufficient to recapitulate the key VRAC channel properties. The activation mechanisms of VRAC by cell swelling have long been debated (Hoffmann et al., 2009; Pedersen et al., 2016). Hypotonicity-induced cell swelling evoke numerous changes in cell physiology, which can be generally categorized into three main aspects, including: (1) mechanical changes in the plasma membrane or cytoskeleton; (2) decreased ionic strength or concentrations of specific ions; (3) other signaling pathways downstream of cell swelling. The reconstitution of purified SWELL1 and associated LRRC8 proteins *in vitro* excluded an essential role of other cellular components (such as cytoskeletons) and signaling pathways (such as kinases) in hypotonicity-induced channel activation. Increasing the droplet volume and the subsequent mechanical changes in the lipid bilayers failed to stimulate channel activities. Instead, lowering the ionic strength (from 150 mM KCl to 70 mM KCl) in the absence of an osmotic gradient was sufficient to activate SWELL1-containing protein complexes (Syeda et al., 2016). This is consistent with the seminal finding in endothelial cells that a decrease of intercellular ionic strength rather than an increase in cell volume *per se* is the initial trigger for activation of $I_{Cl,\ swell}$ (Voets, Droogmans, Raskin, Eggermont, & Nilius, 1999). It should

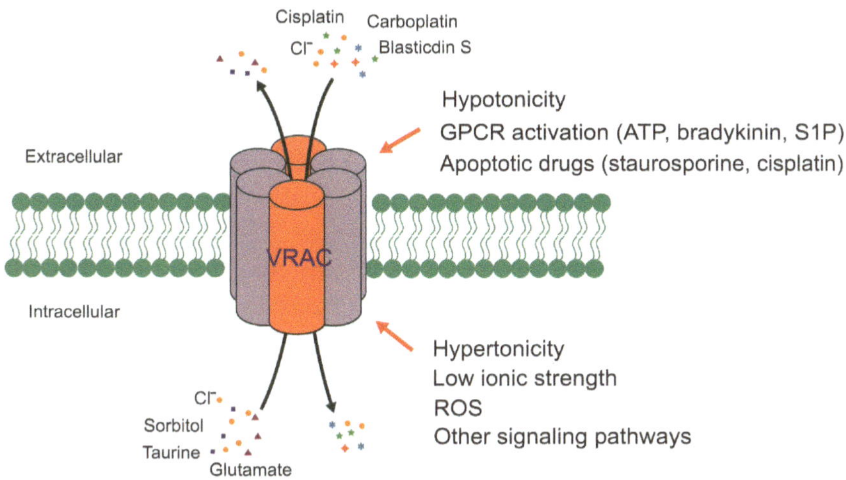

Figure 3 Diverse activation mechanisms and substrates suggesting an important role for VRAC in cell–cell communication. VRAC can be activated by extracellular hypertonicity, GPCR activation, and apoptosis-inducing drugs; and intracellular hypertonicity, reduced ionic strength, and reactive oxygen species (ROS). *S1P*, sphingosine-1-phosphate. In addition to conducting Cl^-, VRAC also permeate many organic osmolytes, metabolites, cancer drugs and antibiotics.

be noted that VRAC can also be activated without cell swelling (Fig. 3, see further discussion in Section 4.1). Their mechanisms of activation are still unclear but unlikely involve ionic strength change. Together, both mutation analysis and *in vitro* reconstitution demonstrated that SWELL1 and associated LRRC8 proteins are pore-forming subunits of VRAC, which is also supported by the contribution of LRRC8 subunits to other VRAC properties discussed in the following two sections.

2.4 LRRC8 Subunit Compositions Determine Key VRAC Properties

Although $I_{Cl,\ swell}$ was widely recorded, some differences of VRAC channels in various tissues and cell types were observed over the years, including: the degree of outward rectification, inactivation kinetics in positive membrane potentials, and even the single channel conductance (Nilius et al., 1997; Okada, 1997; Strange et al., 1996). This suggested potential molecular diversity of VRAC and added to the confusion to its molecular identification. Identification of LRRC8 heteromers as VRAC essential components and appreciation of their biochemical heterogeneity immediately suggested a compelling explanation for the differences in native VRAC currents

(Voss et al., 2014). Indeed, co-expression of SWELL1 and LRRC8C in cells lacking all five LRRC8 proteins generated $I_{Cl,\ swell}$ with slow inactivation kinetics, whereas SWELL1 and LRRC8E yielded fast inactivating currents (Voss et al., 2014). Subsequent chimeric experiments between LRRC8C and LRRC8E identified two amino acid residues in a highly conserved region of TM1-TM2 extracellular loop as the main determinant of VRAC inactivation kinetics (Fig. 2) (Ullrich et al., 2016). There is also a good correlation between the endogenous expression levels of LRRC8C and LRRC8E with differences in $I_{Cl,\ swell}$ inactivation kinetics. Strikingly, lipid bilayer reconstitution of SWELL1-containing protein complexes generated a spectrum of intermediate single-channel conductances ranging from ~10 to 50 pS with various open probabilities (Po) (Syeda et al., 2016). The single-channel conductance range became narrow and non-overlapping when SWELL1 with only one of LRRC8 subunits was incorporated *in vitro*, similar to the observation with cell-attached single-channel recordings on the corresponding LRRC8 triple KO cells (Syeda et al., 2016). Compared to LRRC8C and LRRC8E, LRRC8D-mediated VRAC displayed greater outward rectification and smaller permeability preference of I^- over Cl^- (Syeda et al., 2016), although the change of anion selectivity was relatively small and was not observed in another study (Voss et al., 2014). Finally, the sensitivity to oxidative stress for VRAC heterologously expressed in *Xenopus* oocytes is also dependent on the composition of LRRC8 subunits (Gradogna, Gavazzo, Boccaccio, & Pusch, 2017).

2.5 VRAC as the Conducting Channel for Large Organic Compounds: Amino Acids, Cancer Drugs and Antibiotics

In addition to ions (mainly K^+ and Cl^-), vertebrate cells, in response to cell swelling, also release various organic osmolytes and metabolites to lower the intracellular osmolality, which drives secondary water efflux and facilitates regulatory volume decrease (Hoffmann et al., 2009). These small organic osmolytes and metabolites include sorbitol, myo-inositol, choline, non-essential amino acids and their derivatives (most prominently taurine). The corresponding transport pathway was termed VSOAC (volume-sensitive organic osmolyte anion channel) (Jackson, Morrison, & Strange, 1994). Both VRAC and VSOAC are activated by cell swelling and are inhibited by a similar panel of non-specific anion channel inhibitors. The diameter of the VRAC pore was estimated in the range of 1.2–1.4 nm (Droogmans, Maertens, Prenen, & Nilius, 1999; Ternovsky, Okada, & Sabirov, 2004), large enough to be permeable to larger (and not necessarily negatively charged) molecules, including taurine, glutamate, sugars, and even ATP. Therefore VRAC and VSOAC were thought to share the same molecular identity. However, discrepancies over the time course for their activation and sensitivity to some inhibitors led to the proposal that VSOAC and VRAC were separate entities (Lambert & Hoffmann, 1994; Stutzin et al., 1999). After the successful molecular identification of VRAC, SWELL1 was shown to be essential for VSOAC activities, that is, swelling-induced efflux of a wide range of organic osmolytes, such as taurine, D-aspartate, and myo-inositol (Hyzinski-Garcia, Rudkouskaya, & Mongin, 2014; Lutter et al., 2017; Planells-Cases et al., 2015; Qiu et al., 2014; Schober, Wilson, & Mongin, 2017; Voss et al., 2014). This strongly suggested that VRAC and VSOAC are identical channels or at least share the same essential SWELL1 subunit.

Compositions of the other LRRC8 subunits not only determine VRAC anion-conducting properties, but also significantly contribute to the selectivity of organic osmolytes. Although not absolutely essential, LRRC8D was important for swelling-induced efflux of taurine and myo-inositol in both HEK cells and primary rat astrocytes (Lutter et al., 2017; Planells-Cases et al., 2015; Schober et al., 2017). However, for swelling-induced release of D-aspartate (a non-metabolizable analogue and reporter for glutamate), LRRC8C, 8D and 8E all seem to contribute to some extent depending on the cell types under examination (Lutter et al., 2017; Schober et al., 2017). Heterozygous expression of SWELL1 and LRRC8 subunits in *Xenopus* oocytes also mediated fluxes of amino acids and ATP with somewhat different selectivity compared those in the mammalian cells (Gaitan-Penas et al., 2016). Additionally, two unbiased haploid cell loss-of-function genetic screens surprisingly uncovered an important function of LRRC8D in the uptake of antibiotic blasticidin and cancer drugs (cisplatin and carboplatin) under isotonic culture conditions (Lee et al., 2014; Planells-Cases et al., 2015). These discoveries significantly expanded the role of VRAC channel, especially composed of SWELL1 and LRRC8D, in transporting larger organic compounds in mammalian cells. Importantly, they also suggested alternative activation mechanisms of VRAC under physiological conditions and without cell swelling (Fig. 3).

3. PHYSIOLOGICAL ROLE OF SWELL1 AND VRAC

The discovery of LRRC8 heteromers as VRAC provides tremendous opportunities to evaluate previously proposed physiological role and to discover novel function of this important channel (Fig. 4). *Swell1* knockout

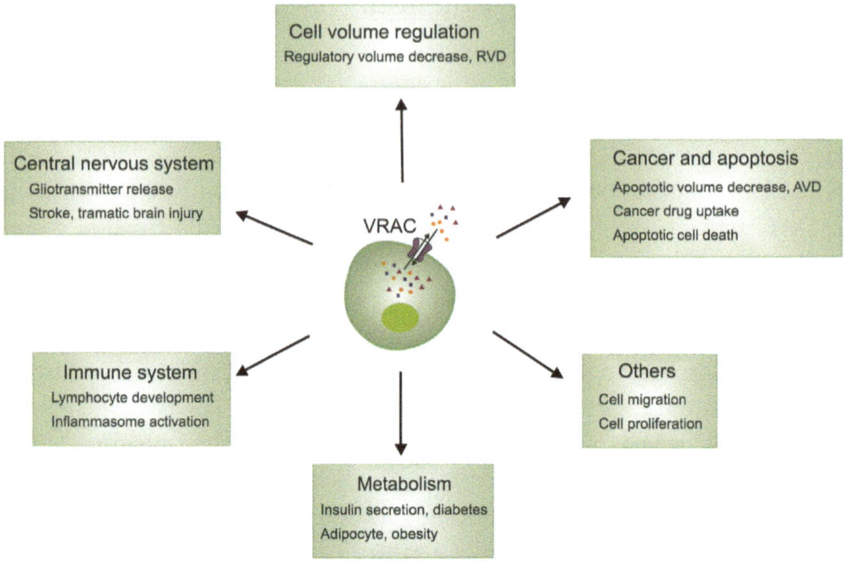

Figure 4 Proposed role of VRAC in cell biology, physiology, and disease.

mice had high embryonic and postnatal lethality, growth retardation, and abnormalities in multiple tissues, including curly hair, thin skeletal muscle bundles, and vacuolated renal tubular cells (Kumar et al., 2014). In zebrafish, Swell1 knockdown caused pericardial edema and defects in trunk elongation and somatogenesis (Yamada, Wondergem, Morrison, Yin, & Strange, 2016). These severe phenotypes highlight the importance of VRAC in animal development and tissue homeostasis. Exciting progress has been made on the role of VRAC in the immune system and metabolism.

3.1 Role of SWELL1 and VRAC in the Immune System

SWELL1 was first cloned from a female patient with congenital agammaglobulinemia and markedly reduced circulating B cells (Sawada et al., 2003). A balanced chromosome translocation was identified which leads to the deletion of the C-terminal two-and-a-half LRRs of SWELL1 (91 aa) and the addition of 35 aa derived from an intronic sequence (SWELL1$^{\Delta 91/+35}$). Bone marrow transplantation in mice with progenitors overexpressing SWELL1$^{\Delta 91/+35}$ resulted in a severe block in B cell development at the pro-B cell to pre-B cell transition and reduced numbers of T cells (Sawada et al., 2003). Thus a potential dominant-negative effect of this heterozygous SWELL1 mutation was proposed as the cause of agammaglobulinemia in this single patient (Sawada et al., 2003). However, reduced number of T cells was not observed in this human case. Additionally, while SWELL1$^{\Delta 91/+35}$ was unable to form functional VRAC, no apparent dominant-negative or gain-of-function effect on $I_{Cl,\ swell}$ was observed in overexpressing cells (Qiu et al., 2014). Therefore the link between the mutant SWELL1 and agammaglobulinemia is not clear at this point.

Swell1 knockout mice exhibited a modest block in B cell development but normal B cell function (Kumar et al., 2014). In contrast, the knockout mice displayed a profound cell-intrinsic block in T cell development at the double-negative stage with decreased proliferation and increased apoptosis, and reduced phosphorylated AKT levels (Kumar et al., 2014). The mechanistic interpretation of this T cell defect was based on a very unusual dual-membrane topology hypothesis that the C-terminal LRR domains of maybe a fraction of SWELL1 is extracellular and interact with a putative SWELL1 ligand, resulting in LCK—ZAP-70—GAB2—PI3K-mediated activation of AKT and T cell survival (Kumar et al., 2014). Recently, a hypomorphic spontaneous mutation in *Swell1*, designated *ébouriffé* (*ebo*), was reported (Platt et al., 2017). This allele was caused by a 2-bp deletion that was predicted to truncate the 15C-terminal LRRs of Swell1. *ebo/ebo* mice shared some similar features as *Swell1* knockout mice, but the overall phenotype is much milder. In contrast to *Swell1* knockout mice, *ebo/ebo* mice had intact T cell development and function, associated, however, with a dramatic reduction of VRAC current in T cells (Platt et al., 2017). It was therefore concluded that the critical role Swell1 plays in T cell development is independent of its VRAC activity, suggesting the involvement of potential channel-independent functions (Platt et al., 2017). However, it should also be noted that as indicated by some residual $I_{Cl,\ swell}$ activity, there might be low levels of intact Swell1 protein expressed in *ebo/ebo* mice (though not detected by western blot). Furthermore, the different genetic backgrounds of *Swell1* knockout (C57B/6) and *ebo* (FVB/N) mice could also account for the differential phenotypes observed. Although the underlying mechanisms still require future investigation, both the human and mouse studies strongly suggested an important role of SWELL1 in the development and function of lymphocytes. In addition to the adaptive immune system, the pharmacological studies also implicated VRAC in the activation of NLRP3 inflammasome in macrophages, which is a critical component of innate immune response and may play a role in the pathogenesis of Alzheimer's disease (Compan et al., 2012; Daniels et al., 2016).

3.2 Role of SWELL1 and VRAC in Metabolism

Most vertebrate animals maintain a stable osmolality in the extracellular fluid (in mammals, close to 300 mOsm/kg of water). Thus extracellular hypotonicity is thought to be unlikely the physiological stimulus to VRAC in most tissues. Instead, VRAC can be activated by increased intracellular osmolality (hypertonicity) resulted from uptake and accumulation of ions or organic osmolytes (Fig. 3). For example, in insulin-secreting pancreatic β-cells, glucose uptake and metabolism, the subsequent intracellular accumulation of glucose metabolites and osmotic entry of water, can induce β-cell swelling which is sufficient to activate VRAC (Best, 2005; Best, Brown, Sener, & Malaisse, 2010; Best, Yates, Decher, Steinmeyer, & Nilius, 2004; Miley, Sheader, Brown, & Best, 1997). In the classic model of glucose-stimulated insulin secretion, glucose metabolism increases the intracellular ATP/ADP ratio, which leads to the closure of the ATP-sensitive K^+ channels (K_{ATP}), thereby inducing membrane depolarization (Ashcroft & Rorsman, 2013). This in turn opens the voltage-gated Ca^{2+} channels, leading to Ca^{2+} entry into the cell, thus triggering insulin release by exocytosis (Ashcroft & Rorsman, 2013). However, pharmacological studies suggested that VRAC may cooperate with K_{ATP} and mediate additional glucose-sensing mechanism (Best et al., 2004). Similar to K_{ATP} closure, glucose-induced VRAC opening was proposed to assist membrane depolarization (due to Cl^- efflux) and thus lead to Ca^{2+} influx and insulin secretion (Best et al., 2010). Consistent with this model, a recent study showed that SWELL1 is required for hypotonicity- and glucose-stimulated $I_{Cl,\ swell}$ in both mouse and human β-cells (Kang et al., 2018). Very strikingly, SWELL1 depletion completely abolished hypotonicity- and glucose-stimulated β-cell membrane depolarization and intracellular calcium signaling (Kang et al., 2018). Furthermore, glucose-stimulated insulin secretion both *in vitro* and *in vivo* was eliminated upon *SWELL1* deletion. Finally, β-cell-specific *Swell1* knockout mice exhibited impaired glucose tolerance (Kang et al., 2018). While future work is warranted to determine the exact relationship and relative contribution of the classical K_{ATP} and VRAC in glucose-stimulated insulin secretion, this study established a critical role for SWELL1 in regulating β-cell excitability and insulin secretion.

The volume of adipocyte, another metabolically active cell type, undergoes considerable expansion in the setting of obesity. Although VRAC is not known to be mechanosensitive, $I_{Cl,\ swell}$-like current was recorded by enlarging adipocytes with the application of positive pressure via the patch pipette (Zhang et al., 2017). Both this mechanically-activated "swell" current and the hypotonicity-activated $I_{Cl,\ swell}$ in adipocytes were shown to be dependent on SWELL1. Interestingly, the hypotonicity-activated $I_{Cl,\ swell}$ activities were augmented in hypertrophic adipocytes of obese mice and humans (Zhang et al., 2017). Functionally, it was shown that SWELL1 positively regulates adipocyte glucose uptake and lipogenesis via insulin-PI3K-AKT2 signaling through LRR domains-mediated interactions with GRB2, a negative regulator of insulin signaling (Zhang et al., 2017). It's still unclear what the physiological stimuli for VRAC activation might be (adipocyte volume expansion or maybe increased intracellular osmolality during lipogenesis) and what the contribution, if any, of VRAC channel function itself is. Importantly, adipocyte-specific *Swell1* knockout reduced adiposity and adipocyte size in obese mice while inducing systemic glucose intolerance and insulin resistance (Zhang et al., 2017). On the other hand, high-fat diet-induced obesity is associated with the upregulation of Swell1 protein expression in adipose, which may be beneficial and contribute to improved systemic glycemia by enhancing insulin sensitivity (Xie et al., 2017).

Intriguingly, SWELL1 expression in human islets is significantly decreased in hyperglycemic individuals based on a global gene expression analysis (Taneera et al., 2012), which may lead to impaired insulin secretion from β-cells. Based on the studies on adipose and liver, it would be important to determine whether VRAC activity or SWELL1 expression is also reduced in these tissues during the development of Type 2 diabetes; if so, this could contribute to insulin resistance. Thus it's possible that enhancing VRAC activity in both insulin-secreting (β-cells) and insulin-sensitive (adipose and liver) tissues may provide a novel therapeutic strategy for Type 2 diabetes and other metabolic disorders.

3.3 Role of VRAC in Apoptosis and Cancer Drug Resistance

Several apoptosis-inducing drugs, such as staurosporine and cisplatin, can slowly activate $I_{Cl,\ swell}$-like currents under isotonic conditions (Fig. 3) (Ise et al., 2005; Maeno, Ishizaki, Kanaseki, Hazama, & Okada, 2000; Poulsen et al., 2010; Shimizu, Numata, & Okada, 2004). Like RVD, these currents may mediate the efflux of intracellular Cl^- and organic osmolytes, which contributes to apoptotic volume decrease (AVD), a critical early step facilitating the progression of apoptosis. VRAC blockers were shown to protect cells from drug-induced apoptosis (Ise et al., 2005; Maeno et al., 2000;

Poulsen et al., 2010). Several drug-resistant cancer cell lines also exhibited decreased cisplatin-induced apoptosis, which is associated with reduced VRAC activity, RVD, and hypotonicity-stimulated taurine release (Lee et al., 2007; Min et al., 2011; Poulsen et al., 2010; Sorensen, Thorsteinsdottir, & Lambert, 2014). Indeed, upon the identification of SWELL1 as an essential VRAC component, *SWELL1* knockout or knockdown significantly diminished cisplatin- or staurosporine-induced VRAC activity, caspase-3 induction and apoptosis (Planells-Cases et al., 2015; Sorensen, Nielsen, Thorsteinsdottir, Hoffmann, & Lambert, 2016). In addition to facilitating AVD-dependent apoptosis, VRAC channels, especially those formed by SWELL1 and LRRC8D, also mediate uptake of platinum-based cancer drugs, including cisplatin and carboplatin (Planells-Cases et al., 2015). The acquirement of cisplatin resistance in cancer cell lines is associated with the downregulation of SWELL1 expression (Sorensen, Dam, Sturup, & Lambert, 2016). Further supporting a clinically relevant role of VRAC in cancer drug sensitivity, low LRRC8D expression was shown to correlate with shorter survival of ovarian cancer patients treated with platinum-based drugs (Planells-Cases et al., 2015).

4. FUTURE STUDIES OF VRAC

Since the first descriptions of $I_{\mathrm{Cl,\ swell}}$ more than three decades ago, a large body of literature has accumulated on various aspects of VRAC in diverse tissues and cell types. This includes its electrophysiological characteristics, pharmacology, activation and regulatory mechanisms, cell biological function, and potential physiological and pathological roles of VRAC, but confusion and unanswered questions persisted due to the lack of a molecular identity. The molecular identification of SWELL1 and its associated LRRC8 homologs as essential VRAC pore-forming subunits represents a breakthrough that has moved the once stagnant field forward. It provides the basis to examine its structure-function relationships and to unambiguously assign cellular and physiological functions to VRAC. It will also likely lead to discoveries of unexpected functions of this important channel family.

4.1 Structural and Molecular Basis of VRAC Permeation, Activation and Regulation

How many SWELL1 and LRRC8 subunits are present in the VRAC channel complex? How their relative positions are arranged in functional VRAC channels? How much flexibility of subunit stoichiometry (composition and position) it exists in the native channel complexes? To address these questions, a combination of single-molecule and structural studies (cryo-EM in particular) on both overexpressed and endogenous channel complexes is required. The structure information and subsequent mutagenesis studies will also provide molecular insights in many fundamental questions about the conducting and gating mechanisms of VRAC. First, what are the pore-lining permeation pathways for both Cl^- and organic compounds? What is the molecular basis for anion and substrate selectivity of different LRRC8 subunit compositions? Second, lowering ionic strength has been shown to activate VRAC both *in vitro* constitution and in whole-cell recording. Does a decrease in ionic strength cause conformational change and open the VRAC channel pore? Are the LRR domains the potential ionic strength sensor? Third, how is VRAC activity modulated by pharmacological inhibitors (such as DCPIB and DIDS), ATP and lipids? The interaction between ATP and VRAC is particularly interesting. Intracellular ATP but not its hydrolysis is required for maintaining VRAC activity (Jackson et al., 1994; Oike, Droogmans, & Nilius, 1994). This ATP dependence was postulated as a self-protection mechanism preventing cells from losing too much organic osmolytes during RVD. However, it's unclear whether it's due to direct ATP binding to VRAC. ATP can also permeate the VRAC pore, but a high concentration of extracellular ATP acts as a voltage-dependent VRAC blocker, reminiscent of an open-pore blocker (Jackson & Strange, 1995; Nilius, Oike, Zahradnik, & Droogmans, 1994; Tsumura, Oiki, Ueda, Okuma, & Okada, 1996). Specific and potent chemical antagonists and agonists have been indispensable in elucidating the mechanism of action and probing physiological and pathological function of ion channels. Such reagents are still not available for VRAC. High-throughput YFP quenching assay and the now known molecular identity of VRAC will undoubtedly facilitate the identification of better and more useful chemical probes for VRAC channels.

In addition to cell swelling, VRAC can also be activated by other physiologically relevant stimuli, such as apoptosis-inducing drugs (Maeno et al., 2000; Shimizu et al., 2004) and G protein-coupled receptor (GPCR) activation downstream of ATP, bradykinin, or sphingosine-1-phosphate (Burow, Klapperstuck, & Markwardt, 2015; Liu, Akita, Shimizu, Sabirov, & Okada, 2009; Takano et al., 2005) (Fig. 3). Unlike the classical hypotonicity-activated $I_{Cl,\ swell}$, these VRAC currents typically take longer time to develop and are smaller in amplitude. These features indicate alternative mode of VRAC activation with potential involvement of second messenger systems. Indeed, reactive oxygen species (ROS) and various protein kinases were proposed to either activate or modulate VRAC activity (Browe & Baumgarten, 2006; Pedersen et al., 2016; Shimizu et al., 2004; Varela, Simon, Riveros, Jorgensen, & Stutzin, 2004). Future biochemical study, especially on the post-translational modification, of SWELL1 and LRRC8 subunits may provide clues on the mechanisms and signaling cascades leading to VRAC activation under non-swelling conditions.

Up until now, co-expression of SWELL1 and LRRC8 homologs in mammalian cells failed to significantly increase $I_{Cl,\ swell}$ (Voss et al., 2014), although purified LRRC8 heteromers were sufficient to reconstitute VRAC channel activity *in vitro* (Syeda et al., 2016). It is therefore possible that there is yet another factor or maybe cellular mechanism to limit the amplitude of $I_{Cl,\ swell}$. This is also supported by a recent study that a pair of closely related cells with very similar LRRC8 gene expression nevertheless had a big difference in their VRAC activity (Okada, Islam, Tsiferova, Okada, & Sabirov, 2017). Alternatively, it's still a possibility that overexpression of mixed LRRC8 proteins may fail to generate large amounts of correctly assembled functional heteromers due to a stricter stoichiometry than expected. Regardless, it's important to understand the basis for this observation and examine what cellular factor or mechanism may modulate or limit the VRAC amplitude.

4.2 Role of SWELL1 and VRAC in the Brain Physiology and Disease

Robust VRAC activity is ubiquitously observed and future studies will surely uncover many interesting physiological functions in different tissues and organs (Catalan et al., 2015). Its significance in the brain is particularly intriguing (see two excellent recent reviews on this topic (Akita & Okada, 2014; Mongin, 2016)). Astrocytes undergo dramatic and sustained swelling in a number of pathological conditions, such as ischemic stroke, traumatic brain injury, hyponatremia, brain tumor, and epilepsy (Kimelberg, 2005). Due to VRAC's unusual ability to support the permeation of organic compounds, activation of VRAC in swelling astrocytes is proposed to be a major pathway for the release of glutamate, the primary excitatory neurotransmitter (Feustel, Jin, & Kimelberg, 2004; Kimelberg, 2005; Kimelberg, Goderie, Higman, Pang, & Waniewski, 1990; Liu, Tashmukhamedov, Inoue, Okada, & Sabirov, 2006). Increase in extracellular glutamate and subsequent excessive activation of neuronal glutamate receptors, known as excitotoxicity, has emerged as an important mechanism of neuronal cell damage in brain diseases (Lai, Zhang, & Wang, 2014). In addition to its role in astrocytic glutamate release, VRAC activation in neurons may mediate Cl^- influx and further contribute to cell damage by promoting neuronal swelling (Inoue & Okada, 2007), although a recent study suggested the involvement of a new type of Cl^- channel (Rungta et al., 2015). Consistently, VRAC blockers inhibited the release of glutamate and dramatically reduced brain damage in rodent models of ischemic stroke (Alibrahim et al., 2013; Zhang, Zhang, Feustel, & Kimelberg, 2008). With the establishment of the essential role of SWELL1 in swelling-induced neurotransmitter release *in vitro* (Hyzinski-Garcia et al., 2014; Lutter et al., 2017; Schober et al., 2017), brain-specific *Swell1* knockout mouse models will be useful to evaluate its importance in many neurological diseases associated with astrocyte swelling.

One of the paradigm shifts in neuroscience over the last two decades is the appreciation of gliotransmission, the active modulation of neuronal synaptic transmission by astrocytes (Gundersen, Storm-Mathisen, & Bergersen, 2015). However, the mechanism of astrocytic release of gliotransmitters (such as glutamate, D-serine, ATP) has been heavily debated (Fiacco & McCarthy, 2018; Hamilton & Attwell, 2010; Savtchouk & Volterra, 2018). Whereas astrocyte release of neurotransmitters by vesicular exocytosis is controversial, there is strong evidence for the involvement of ion channel-mediated release, such as VRAC, which can be activated by physiological stimuli, including GPCR activation (Liu et al., 2009; Takano et al., 2005). Future studies will elucidate the contribution of SWELL1 and other LRRC8 subunits to the communication between neurons and astrocytes and their physiological and behavioral significance.

4.3 Contribution of LRRC8 Subunits to VRAC Physiology and Beyond

So far, most physiological studies have been focused on the essential VRAC component SWELL1 except for the previous reports on the link between LRRC8C and adipose metabolism (Hayashi et al., 2011; Tominaga et al., 2004). Although the contribution of the other LRRC8 subunits to VRAC diversity is well established, what is the physiological relevance of the electrophysiological and substrate selectivity differences? In the future, thorough analyses on their tissue expression profiles and knockout mouse models will help answer this question. In addition to forming cell surface VRAC channels, it's conceivable that SWELL1 and its LRRC8 homologs may also form volume-regulated channels on intracellular organelles, such as the endoplasmic reticulum (ER), and that LRRC8B-8E may even have VRAC-independent functions. Interestingly, LRRC8B, which has a minimal role in various aspects of VRAC activity, was recently shown to mediate ER Ca^{2+} leak in HEK cells (Ghosh, Khandelwal, Kumar, & Bera, 2017). Whether it functions directly as an ER Ca^{2+} channel or regulates ER Ca^{2+} levels by indirect mechanisms still requires further investigation. Three groups recently reported the cryo-EM structures of homomeric mouse and human SWELL1 channels, revealing the overall hexameric architecture and ion-conducting pore of VRAC (Deneka et al., 2018; Kefauver et al., 2018; Kasuya et al., 2018). These structures provide a solid foundation for further dissecting the heterogeneity and mechanisms of activation and regulation of VRAC.

ACKNOWLEDGMENTS

We thank the reviewers for their many insightful comments and suggestions. The work in the senior author's (ZQ) laboratory is supported by a grant from the NIH (R35 GM124824).

REFERENCES

Abascal, F., & Zardoya, R. (2012). LRRC8 proteins share a common ancestor with pannexins, and may form hexameric channels involved in cell-cell communication. *BioEssays: News and Reviews in Molecular, Cellular and Developmental Biology, 34*(7), 551–560. https://doi.org/10.1002/bies.201100173.

Ackerman, M. J., Wickman, K. D., & Clapham, D. E. (1994). Hypotonicity activates a native chloride current in Xenopus oocytes. *The Journal of General Physiology, 103*(2), 153–179.

Akita, T., & Okada, Y. (2014). Characteristics and roles of the volume-sensitive outwardly rectifying (VSOR) anion channel in the central nervous system. *Neuroscience, 275,* 211–231. https://doi.org/10.1016/j.neuroscience.2014.06.015.

Alibrahim, A., Zhao, L. Y., Bae, C. Y., Barszczyk, A., Sun, C. L., Wang, G. L., & Sun, H. S. (2013). Neuroprotective effects of volume-regulated anion channel blocker DCPIB on neonatal hypoxic-ischemic injury. *Acta Pharmacologica Sinica, 34*(1), 113–118. https://doi.org/10.1038/aps.2012.148.

Ashcroft, F. M., & Rorsman, P. (2013). K(ATP) channels and islet hormone secretion: New insights and controversies. *Nature Reviews. Endocrinology, 9*(11), 660–669. https://doi.org/10.1038/nrendo.2013.166.

Best, L. (2005). Glucose-induced electrical activity in rat pancreatic beta-cells: Dependence on intracellular chloride concentration. *The Journal of Physiology, 568*(Pt 1), 137–144. https://doi.org/10.1113/jphysiol.2005.093740.

Best, L., Brown, P. D., Sener, A., & Malaisse, W. J. (2010). Electrical activity in pancreatic islet cells: The VRAC hypothesis. *Islets, 2*(2), 59–64. https://doi.org/10.4161/isl.2.2.11171.

Best, L., Yates, A. P., Decher, N., Steinmeyer, K., & Nilius, B. (2004). Inhibition of glucose-induced electrical activity in rat pancreatic beta-cells by DCPIB, a selective inhibitor of volume-sensitive anion currents. *European Journal of Pharmacology, 489*(1–2), 13–19. https://doi.org/10.1016/j.ejphar.2004.02.030.

Bowens, N. H., Dohare, P., Kuo, Y. H., & Mongin, A. A. (2013). DCPIB, the proposed selective blocker of volume-regulated anion channels, inhibits several glutamate transport pathways in glial cells. *Molecular Pharmacology, 83*(1), 22–32. https://doi.org/10.1124/mol.112.080457.

Browe, D. M., & Baumgarten, C. M. (2006). EGFR kinase regulates volume-sensitive chloride current elicited by integrin stretch via PI-3K and NADPH oxidase in ventricular myocytes. *The Journal of General Physiology, 127*(3), 237–251. https://doi.org/10.1085/jgp.200509366.

Burow, P., Klapperstuck, M., & Markwardt, F. (2015). Activation of ATP secretion via volume-regulated anion channels by sphingosine-1-phosphate in RAW macrophages. *Pflügers Archiv, 467*(6), 1215–1226. https://doi.org/10.1007/s00424-014-1561-8.

Cahalan, M. D., & Lewis, R. S. (1988). Role of potassium and chloride channels in volume regulation by T lymphocytes. *Society of General Physiologists Series, 43*, 281–301.

Catalan, M. A., Kondo, Y., Pena-Munzenmayer, G., Jaramillo, Y., Liu, F., Choi, S., & Melvin, J. E. (2015). A fluid secretion pathway unmasked by acinar-specific Tmem16A gene ablation in the adult mouse salivary gland. *Proceedings of the National Academy of Sciences of the United States of America, 112*(7), 2263–2268. https://doi.org/10.1073/pnas.1415739112.

Compan, V., Baroja-Mazo, A., Lopez-Castejon, G., Gomez, A. I., Martinez, C. M., Angosto, D., & Pelegrin, P. (2012). Cell volume regulation modulates NLRP3 inflammasome activation. *Immunity, 37*(3), 487–500. https://doi.org/10.1016/j.immuni.2012.06.013.

Dahl, G. (2015). ATP release through pannexon channels. *Philosophical Transactions of the Royal Society of London B Biological Sciences, 370*(1672). https://doi.org/10.1098/rstb.2014.0191.

Daniels, M. J., Rivers-Auty, J., Schilling, T., Spencer, N. G., Watremez, W., Fasolino, V., & Brough, D. (2016). Fenamate NSAIDs inhibit the NLRP3 inflammasome and protect against Alzheimer's disease in rodent models. *Nature Communications, 7*, 12504. https://doi.org/10.1038/ncomms12504.

Deneka, D., Sawicka, M., Lam, A. K. M., Paulino, C., & Dutzle, R. (2018). Structure of a volume-regulated anion channel of the LRRC8 family. *Nature, 558*, 254–259. https://doi.org/10.1038/s41586-018-0134-y.

Droogmans, G., Maertens, C., Prenen, J., & Nilius, B. (1999). Sulphonic acid derivatives as probes of pore properties of volume-regulated anion channels in endothelial cells. *British Journal of Pharmacology, 128*(1), 35–40. https://doi.org/10.1038/sj.bjp.0702770.

Duran, C., Thompson, C. H., Xiao, Q., & Hartzell, H. C. (2010). Chloride channels: Often enigmatic, rarely predictable. *Annual Review of Physiology, 72*, 95–121. https://doi.org/10.1146/annurev-physiol-021909-135811.

Feustel, P. J., Jin, Y., & Kimelberg, H. K. (2004). Volume-regulated anion channels are the predominant contributors to release of excitatory amino acids in the ischemic cortical penumbra. *Stroke, 35*(5), 1164–1168. https://doi.org/10.1161/01.STR.0000124127.57946.a1.

Fiacco, T. A., & McCarthy, K. D. (2018). Multiple lines of evidence indicate that gliotransmission does not occur under physiological conditions. *The Journal of Neuroscience: The Official Journal of the Society for Neuroscience, 38*(1), 3–13. https://doi.org/10.1523/JNEUROSCI.0016-17.2017.

Fujii, T., Takahashi, Y., Takeshima, H., Saitoh, C., Shimizu, T., Takeguchi, N., & Sakai, H. (2015). Inhibition of gastric H+,K+-ATPase by 4-(2-butyl-6,7-dichloro-2-cyclopentylindan-1-on-5-yl)oxybutyric acid (DCPIB), an inhibitor of volume-regulated anion channel. *European Journal of Pharmacology, 765*, 34–41. https://doi.org/10.1016/j.ejphar.2015.08.011.

Gaitan-Penas, H., Gradogna, A., Laparra-Cuervo, L., Solsona, C., Fernandez-Duenas, V., Barrallo-Gimeno, A., & Estevez, R. (2016). Investigation of LRRC8-mediated volume-regulated anion currents in Xenopus oocytes. *Biophysical Journal, 111*(7), 1429–1443. https://doi.org/10.1016/j.bpj.2016.08.030.

Galietta, L. J., Haggie, P. M., & Verkman, A. S. (2001). Green fluorescent protein-based halide indicators with improved chloride and iodide affinities. *FEBS Letters, 499*(3), 220–224.

Ghosh, A., Khandelwal, N., Kumar, A., & Bera, A. K. (2017). Leucine-rich repeat-containing 8B protein is associated with the endoplasmic reticulum Ca(2+) leak in HEK293 cells. *Journal of Cell Science, 130*(22), 3818–3828. https://doi.org/10.1242/jcs.203646.

Gradogna, A., Gavazzo, P., Boccaccio, A., & Pusch, M. (2017). Subunit-dependent oxidative stress sensitivity of LRRC8 volume-regulated anion channels. *The Journal of Physiology, 595*(21), 6719–6733. https://doi.org/10.1113/JP274795.

Gundersen, V., Storm-Mathisen, J., & Bergersen, L. H. (2015). Neuroglial transmission. *Physiological Reviews, 95*(3), 695–726. https://doi.org/10.1152/physrev.00024.2014.

Hamilton, N. B., & Attwell, D. (2010). Do astrocytes really exocytose neurotransmitters? *Nature Reviews. Neuroscience, 11*(4), 227–238. https://doi.org/10.1038/nrn2803.

Hayashi, T., Nozaki, Y., Nishizuka, M., Ikawa, M., Osada, S., & Imagawa, M. (2011). Factor for adipocyte differentiation 158 gene disruption prevents the body weight gain and insulin resistance induced by a high-fat diet. *Biological and Pharmaceutical Bulletin, 34*(8), 1257–1263.

Hazama, A., & Okada, Y. (1988). Ca^{2+} sensitivity of volume-regulatory K^+ and Cl^- channels in cultured human epithelial cells. *The Journal of Physiology, 402*, 687–702.

Helix, N., Strobaek, D., Dahl, B. H., & Christophersen, P. (2003). Inhibition of the endogenous volume-regulated anion channel (VRAC) in HEK293 cells by acidic di-aryl-ureas. *The Journal of Membrane Biology, 196*(2), 83–94. https://doi.org/10.1007/s00232-003-0627-x.

Hoffmann, E. K., Lambert, I. H., & Pedersen, S. F. (2009). Physiology of cell volume regulation in vertebrates. *Physiological Reviews, 89*(1), 193–277. https://doi.org/10.1152/physrev.00037.2007.

Hyzinski-Garcia, M. C., Rudkouskaya, A., & Mongin, A. A. (2014). LRRC8A protein is indispensable for swelling-activated and ATP-induced release of excitatory amino acids in rat astrocytes. *The Journal of Physiology, 592*(22), 4855–4862. https://doi.org/10.1113/jphysiol.2014.278887.

Inoue, H., & Okada, Y. (2007). Roles of volume-sensitive chloride channel in excitotoxic neuronal injury. *The Journal of Neuroscience: The Official Journal of the Society for Neuroscience, 27*(6), 1445–1455. https://doi.org/10.1523/JNEUROSCI.4694-06.2007.

Isacoff, E. Y., Jan, L. Y., & Minor, D. L., Jr. (2013). Conduits of life's spark: A perspective on ion channel research since the birth of neuron. *Neuron, 80*(3), 658–674. https://doi.org/10.1016/j.neuron.2013.10.040.

Ise, T., Shimizu, T., Lee, E. L., Inoue, H., Kohno, K., & Okada, Y. (2005). Roles of volume-sensitive Cl^- channel in cisplatin-induced apoptosis in human epidermoid cancer cells. *The Journal of Membrane Biology, 205*(3), 139–145. https://doi.org/10.1007/s00232-005-0779-y.

Jackson, P. S., Morrison, R., & Strange, K. (1994). The volume-sensitive organic osmolyte-anion channel VSOAC is regulated by nonhydrolytic ATP binding. *American Journal of Physiology, 267*(5 Pt 1), C1203–C1209. https://doi.org/10.1152/ajpcell.1994.267.5.C1203.

Jackson, P. S., & Strange, K. (1995). Characterization of the voltage-dependent properties of a volume-sensitive anion conductance. *The Journal of General Physiology, 105*(5), 661–676.

Jentsch, T. J. (2016). VRACs and other ion channels and transporters in the regulation of cell volume and beyond. *Nature Reviews. Molecular Cell Biology, 17*(5), 293–307. https://doi.org/10.1038/nrm.2016.29.

Kang, C., Xie, L., Gunasekar, S. K., Mishra, A., Zhang, Y., Pai, S., ... Sah, R. (2018). SWELL1 is a glucose sensor regulating beta-cell excitability and systemic glycaemia. *Nature Communications, 9*(1), 367. https://doi.org/10.1038/s41467-017-02664-0.

Kasuya, G., Nakane, T., Yokoyama, T., Jia, Y., Inoue, M., Watanabe, K., Nakamura, R., Nishizawa, T., Kusakizako, T., Tsutsumi, A., Yanagisawa, H., Dohmae, N., Hattori, M., Ichijo, H., Yan, Z., Kikkawa, M., Shirouzu, M., Ishitani, R., & Nureki, O. (2018). Cryo-EM structure of the volume-regulated anion channel LRRC8. *bioRxiv*. https://doi.org/10.1101/331207.

Kefauver, J. M., Saotome, K., Dubin, A. E., Pallesen, J., Cottrell, C. A., Hong, G., Cahalan, S. M., Qiu, Z., Crowley, C. S., Whitwam, T., Lee, W.-H., Ward, A. B., & Patapoutian, A. (2018). Structure of the Human Volume Regulated Anion Channel. *bioRxiv*. https://doi.org/10.1101/323584.

Kimelberg, H. K. (2005). Astrocytic swelling in cerebral ischemia as a possible cause of injury and target for therapy. *Glia, 50*(4), 389–397. https://doi.org/10.1002/glia.20174.

Kimelberg, H. K., Goderie, S. K., Higman, S., Pang, S., & Waniewski, R. A. (1990). Swelling-induced release of glutamate, aspartate, and taurine from astrocyte cultures. *The Journal of Neuroscience: The Official Journal of the Society for Neuroscience, 10*(5), 1583–1591.

Kumar, L., Chou, J., Yee, C. S., Borzutzky, A., Vollmann, E. H., von Andrian, U. H., & Geha, R. S. (2014). Leucine-rich repeat containing 8A (LRRC8A) is essential for T lymphocyte development and function. *The Journal of Experimental Medicine, 211*(5), 929–942. https://doi.org/10.1084/jem.20131379.

Lai, T. W., Zhang, S., & Wang, Y. T. (2014). Excitotoxicity and stroke: Identifying novel targets for neuroprotection. *Progress in Neurobiology, 115*, 157–188. https://doi.org/10.1016/j.pneurobio.2013.11.006.

Lambert, I. H., & Hoffmann, E. K. (1994). Cell swelling activates separate taurine and chloride channels in Ehrlich mouse ascites tumor cells. *The Journal of Membrane Biology, 142*(3), 289–298.

Lang, F. (2007). Mechanisms and significance of cell volume regulation. *Journal of the American College of Nutrition, 26*(5 Suppl), 613S–623S.

Lee, C. C., Freinkman, E., Sabatini, D. M., & Ploegh, H. L. (2014). The protein synthesis inhibitor blasticidin s enters mammalian cells via leucine-rich repeat-containing protein 8D. *Journal of Biological Chemistry, 289*(24), 17124–17131. https://doi.org/10.1074/jbc.M114.571257.

Lee, E. L., Shimizu, T., Ise, T., Numata, T., Kohno, K., & Okada, Y. (2007). Impaired activity of volume-sensitive Cl^- channel is involved in cisplatin resistance of cancer cells. *Journal of Cellular Physiology, 211*(2), 513–521. https://doi.org/10.1002/jcp.20961.

Liu, H. T., Akita, T., Shimizu, T., Sabirov, R. Z., & Okada, Y. (2009). Bradykinin-induced astrocyte-neuron signalling: Glutamate release is mediated by ROS-activated volume-sensitive outwardly rectifying anion channels. *The Journal of Physiology, 587*(Pt 10), 2197–2209. https://doi.org/10.1113/jphysiol.2008.165084.

Liu, H. T., Tashmukhamedov, B. A., Inoue, H., Okada, Y., & Sabirov, R. Z. (2006). Roles of two types of anion channels in glutamate release from mouse astrocytes under ischemic or osmotic stress. *Glia, 54*(5), 343–357. https://doi.org/10.1002/glia.20400.

Lutter, D., Ullrich, F., Lueck, J. C., Kempa, S., & Jentsch, T. J. (2017). Selective transport of neurotransmitters and modulators by distinct volume-regulated LRRC8 anion channels. *Journal of Cell Science, 130*(6), 1122–1133. https://doi.org/10.1242/jcs.196253.

Maeda, S., Nakagawa, S., Suga, M., Yamashita, E., Oshima, A., Fujiyoshi, Y., & Tsukihara, T. (2009). Structure of the connexin 26 gap junction channel at 3.5 A resolution. *Nature, 458*(7238), 597–602. https://doi.org/10.1038/nature07869.

Maeno, E., Ishizaki, Y., Kanaseki, T., Hazama, A., & Okada, Y. (2000). Normotonic cell shrinkage because of disordered volume regulation is an early prerequisite to apoptosis. *Proceedings of the National Academy of Sciences of the United States of America, 97*(17), 9487–9492. https://doi.org/10.1073/pnas.140216197.

Miley, H. E., Sheader, E. A., Brown, P. D., & Best, L. (1997). Glucose-induced swelling in rat pancreatic beta-cells. *The Journal of Physiology, 504*(Pt 1), 191–198.

Minieri, L., Pivonkova, H., Caprini, M., Harantova, L., Anderova, M., & Ferroni, S. (2013). The inhibitor of volume-regulated anion channels DCPIB activates TREK potassium channels in cultured astrocytes. *British Journal of Pharmacology, 168*(5), 1240–1254. https://doi.org/10.1111/bph.12011.

Min, X. J., Li, H., Hou, S. C., He, W., Liu, J., Hu, B., & Wang, J. (2011). Dysfunction of volume-sensitive chloride channels contributes to cisplatin resistance in human lung adenocarcinoma cells. *Experimental Biology and Medicine (Maywood), 236*(4), 483–491. https://doi.org/10.1258/ebm.2011.010297.

Mongin, A. A. (2016). Volume-regulated anion channel — a frenemy within the brain. *Pflügers Archiv, 468*(3), 421–441. https://doi.org/10.1007/s00424-015-1765-6.

Nilius, B., Eggermont, J., Voets, T., Buyse, G., Manolopoulos, V., & Droogmans, G. (1997). Properties of volume-regulated anion channels in mammalian cells. *Progress in Biophysics and Molecular Biology, 68*(1), 69–119.

Nilius, B., Oike, M., Zahradnik, I., & Droogmans, G. (1994). Activation of a Cl^- current by hypotonic volume increase in human endothelial cells. *The Journal of General Physiology, 103*(5), 787–805.

Oike, M., Droogmans, G., & Nilius, B. (1994). The volume-activated chloride current in human endothelial cells depends on intracellular ATP. *Pflügers Archiv, 427*(1–2), 184–186.

Okada, Y. (1997). Volume expansion-sensing outward-rectifier Cl^- channel: Fresh start to the molecular identity and volume sensor. *American Journal of Physiology, 273*(3 Pt 1), C755–C789. https://doi.org/10.1152/ajpcell.1997.273.3.C755.

Okada, T., Islam, M. R., Tsiferova, N. A., Okada, Y., & Sabirov, R. Z. (2017). Specific and essential but not sufficient roles of LRRC8A in the activity of volume-sensitive outwardly rectifying anion channel (VSOR). *Channels, 11*(2), 109–120. https://doi.org/10.1080/19336950.2016.1247133.

Pedersen, S. F., Klausen, T. K., & Nilius, B. (2015). The identification of a volume-regulated anion channel: An amazing Odyssey. *Acta Physiologica, 213*(4), 868–881. https://doi.org/10.1111/apha.12450.

Pedersen, S. F., Okada, Y., & Nilius, B. (2016). Biophysics and physiology of the volume-regulated anion channel (VRAC)/Volume-Sensitive outwardly rectifying anion channel (VSOR). *Pflügers Archiv, 468*(3), 371–383. https://doi.org/10.1007/s00424-015-1781-6.

Planells-Cases, R., Lutter, D., Guyader, C., Gerhards, N. M., Ullrich, F., Elger, D. A., … Jentsch, T. J. (2015). Subunit composition of VRAC channels determines substrate specificity and cellular resistance to Pt-based anti-cancer drugs. *The EMBO Journal, 34*(24), 2993–3008. https://doi.org/10.15252/embj.201592409.

Platt, C. D., Chou, J., Houlihan, P., Badran, Y. R., Kumar, L., Bainter, W., … Geha, R. S. (2017). Leucine-rich repeat containing 8A (LRRC8A)-dependent volume-regulated anion channel activity is dispensable for T-cell development and function. *The Journal of Allergy and Clinical Immunology, 140*(6), 1651–1659. https://doi.org/10.1016/j.jaci.2016.12.974. e1651.

Poulsen, K. A., Andersen, E. C., Hansen, C. F., Klausen, T. K., Hougaard, C., Lambert, I. H., & Hoffmann, E. K. (2010). Deregulation of apoptotic volume decrease and ionic movements in multidrug-resistant tumor cells: Role of chloride channels. *American Journal of Physiology. Cell Physiology, 298*(1), C14–C25. https://doi.org/10.1152/ajpcell.00654.2008.

Qiu, Z., Dubin, A. E., Mathur, J., Tu, B., Reddy, K., Miraglia, L. J., & Patapoutian, A. (2014). SWELL1, a plasma membrane protein, is an essential component of volume-regulated anion channel. *Cell, 157*(2), 447–458. https://doi.org/10.1016/j.cell.2014.03.024.

Rungta, R. L., Choi, H. B., Tyson, J. R., Malik, A., Dissing-Olesen, L., Lin, P. J. C., & MacVicar, B. A. (2015). The cellular mechanisms of neuronal swelling underlying cytotoxic edema. *Cell, 161*(3), 610–621. https://doi.org/10.1016/j.cell.2015.03.029.

Sabirov, R. Z., Merzlyak, P. G., Okada, T., Islam, M. R., Uramoto, H., Mori, T., & Okada, Y. (2017). The organic anion transporter SLCO2A1 constitutes the core component of the maxi-Cl channel. *The EMBO Journal, 36*(22), 3309–3324. https://doi.org/10.15252/embj.201796685.

Savtchouk, I., & Volterra, A. (2018). Gliotransmission: Beyond black-and-white. *The Journal of Neuroscience: The Official Journal of the Society for Neuroscience, 38*(1), 14–25. https://doi.org/10.1523/JNEUROSCI.0017-17.2017.

Sawada, A., Takihara, Y., Kim, J. Y., Matsuda-Hashii, Y., Tokimasa, S., Fujisaki, H., … Hara, J. (2003). A congenital mutation of the novel gene LRRC8 causes agammaglobulinemia in humans. *Journal of Clinical Investigation, 112*(11), 1707–1713. https://doi.org/10.1172/JCI18937.

Schober, A. L., Wilson, C. S., & Mongin, A. A. (2017). Molecular composition and heterogeneity of the LRRC8-containing swelling-activated osmolyte channels in primary rat astrocytes. *The Journal of Physiology, 595*(22), 6939–6951. https://doi.org/10.1113/JP275053.

Schroeder, B. C., Cheng, T., Jan, Y. N., & Jan, L. Y. (2008). Expression cloning of TMEM16A as a calcium-activated chloride channel subunit. *Cell, 134*(6), 1019–1029. https://doi.org/10.1016/j.cell.2008.09.003.

Shimizu, T., Numata, T., & Okada, Y. (2004). A role of reactive oxygen species in apoptotic activation of volume-sensitive Cl(−) channel. *Proceedings of the National Academy of Sciences of the United States of America, 101*(17), 6770–6773. https://doi.org/10.1073/pnas.0401604101.

Sorensen, B. H., Dam, C. S., Sturup, S., & Lambert, I. H. (2016). Dual role of LRRC8A-containing transporters on cisplatin resistance in human ovarian cancer cells. *Journal of Inorganic Biochemistry, 160*, 287–295. https://doi.org/10.1016/j.jinorgbio.2016.04.004.

Sorensen, B. H., Nielsen, D., Thorsteinsdottir, U. A., Hoffmann, E. K., & Lambert, I. H. (2016). Downregulation of LRRC8A protects human ovarian and alveolar carcinoma cells against Cisplatin-induced expression of p53, MDM2, p21Waf1/Cip1, and Caspase-9/-3 activation. *American Journal of Physiology. Cell Physiology, 310*(11), C857–C873. https://doi.org/10.1152/ajpcell.00256.2015.

Sorensen, B. H., Thorsteinsdottir, U. A., & Lambert, I. H. (2014). Acquired cisplatin resistance in human ovarian A2780 cancer cells correlates with shift in taurine homeostasis and ability to volume regulate. *American Journal of Physiology. Cell Physiology, 307*(12), C1071–C1080. https://doi.org/10.1152/ajpcell.00274.2014.

Stotz, S. C., & Clapham, D. E. (2012). Anion-sensitive fluorophore identifies the Drosophila swell-activated chloride channel in a genome-wide RNA interference screen. *PLoS One, 7*(10), e46865. https://doi.org/10.1371/journal.pone.0046865.

Strange, K., Emma, F., & Jackson, P. S. (1996). Cellular and molecular physiology of volume-sensitive anion channels. *American Journal of Physiology, 270*(3 Pt 1), C711–C730. https://doi.org/10.1152/ajpcell.1996.270.3.C711.

Stutzin, A., Torres, R., Oporto, M., Pacheco, P., Eguiguren, A. L., Cid, L. P., & Sepulveda, F. V. (1999). Separate taurine and chloride efflux pathways activated during regulatory volume decrease. *American Journal of Physiology, 277*(3 Pt 1), C392–C402.

Syeda, R., Qiu, Z., Dubin, A. E., Murthy, S. E., Florendo, M. N., Mason, D. E., … Patapoutian, A. (2016). LRRC8 proteins form volume-regulated anion channels that sense ionic strength. *Cell, 164*(3), 499–511. https://doi.org/10.1016/j.cell.2015.12.031.

Takano, T., Kang, J., Jaiswal, J. K., Simon, S. M., Lin, J. H., Yu, Y., & Nedergaard, M. (2005). Receptor-mediated glutamate release from volume sensitive channels in astrocytes. *Proceedings of the National Academy of Sciences of the United States of America, 102*(45), 16466–16471. https://doi.org/10.1073/pnas.0506382102.

Taneera, J., Lang, S., Sharma, A., Fadista, J., Zhou, Y., Ahlqvist, E., & Groop, L. (2012). A systems genetics approach identifies genes and pathways for type 2 diabetes in human islets. *Cell Metabolism, 16*(1), 122–134. https://doi.org/10.1016/j.cmet.2012.06.006.

Ternovsky, V. I., Okada, Y., & Sabirov, R. Z. (2004). Sizing the pore of the volume-sensitive anion channel by differential polymer partitioning. *FEBS Letters, 576*(3), 433–436. https://doi.org/10.1016/j.febslet.2004.09.051.

Tominaga, K., Kondo, C., Kagata, T., Hishida, T., Nishizuka, M., & Imagawa, M. (2004). The novel gene fad158, having a transmembrane domain and leucine-rich repeat, stimulates adipocyte differentiation. *Journal of Biological Chemistry, 279*(33), 34840–34848. https://doi.org/10.1074/jbc.M312927200.

Tsumura, T., Oiki, S., Ueda, S., Okuma, M., & Okada, Y. (1996). Sensitivity of volume-sensitive Cl^- conductance in human epithelial cells to extracellular nucleotides. *American Journal of Physiology, 271*(6 Pt 1), C1872–C1878. https://doi.org/10.1152/ajpcell.1996.271.6.C1872.

Ullrich, F., Reincke, S. M., Voss, F. K., Stauber, T., & Jentsch, T. J. (2016). Inactivation and anion selectivity of volume-regulated anion channels (VRACs) depend on C-terminal residues of the first extracellular loop. *Journal of Biological Chemistry, 291*(33), 17040–17048. https://doi.org/10.1074/jbc.M116.739342.

Varela, D., Simon, F., Riveros, A., Jorgensen, F., & Stutzin, A. (2004). NAD(P)H oxidase-derived H(2)O(2) signals chloride channel activation in cell volume regulation and cell proliferation. *Journal of Biological Chemistry, 279*(14), 13301–13304. https://doi.org/10.1074/jbc.C400020200.

Voets, T., Droogmans, G., Raskin, G., Eggermont, J., & Nilius, B. (1999). Reduced intracellular ionic strength as the initial trigger for activation of endothelial volume-regulated anion channels. *Proceedings of the National Academy of Sciences of the United States of America, 96*(9), 5298–5303.

Voss, F. K., Ullrich, F., Munch, J., Lazarow, K., Lutter, D., Mah, N., & Jentsch, T. J. (2014). Identification of LRRC8 heteromers as an essential component of the volume-regulated anion channel VRAC. *Science, 344*(6184), 634–638. https://doi.org/10.1126/science.1252826.

Wang, J., & Dahl, G. (2010). SCAM analysis of Panx1 suggests a peculiar pore structure. *The Journal of General Physiology, 136*(5), 515–527. https://doi.org/10.1085/jgp.201010440.

Xie, L., Zhang, Y., Gunasekar, S. K., Mishra, A., Cao, L., & Sah, R. (2017). Induction of adipose and hepatic SWELL1 expression is required for maintaining systemic insulin-sensitivity in obesity. *Channels, 11*(6), 673–677. https://doi.org/10.1080/19336950.2017.1373225.

Yamada, T., Wondergem, R., Morrison, R., Yin, V. P., & Strange, K. (2016). Leucine-rich repeat containing protein LRRC8A is essential for swelling-activated Cl^- currents and embryonic development in zebrafish. *Physiological Reports, 4*(19). https://doi.org/10.14814/phy2.12940.

Zhang, Y., Xie, L., Gunasekar, S. K., Tong, D., Mishra, A., Gibson, W. J., & Sah, R. (2017). SWELL1 is a regulator of adipocyte size, insulin signalling and glucose homeostasis. *Nature Cell Biology, 19*(5), 504–517. https://doi.org/10.1038/ncb3514.

Zhang, Y., Zhang, H., Feustel, P. J., & Kimelberg, H. K. (2008). DCPIB, a specific inhibitor of volume regulated anion channels (VRACs), reduces infarct size in MCAo and the release of glutamate in the ischemic cortical penumbra. *Experimental Neurology, 210*(2), 514–520. https://doi.org/10.1016/j.expneurol.2007.11.027.

SECTION 2

Regulation of Intracellular Chloride and Water Homeostasis

CHAPTER SEVEN

Role of WNK Kinases in the Modulation of Cell Volume

Paola de los Heros*, Diana Pacheco-Alvarez§ and Gerardo Gamba¶,||,1

*División de Investigación, Facultad de Medicina, Universidad Nacional Autónoma de México, Coyoacán, Mexico City, Mexico
§Escuela de Medicina, Universidad Panamericana, Mexico City, Mexico
¶Molecular Physiology Unit, Instituto de Investigaciones Biomédicas, Universidad Nacional Autónoma de México, Tlalpan, Mexico City, Mexico
||Department of Nephrology and Mineral Metabolism, Instituto Nacional de Ciencias Médicas y Nutrición Salvador Zubirán, Tlalpan, Mexico City, Mexico
[1]Corresponding author: E-mail: gerardo.gambaa@incmnsz.mx

Contents

Abstract

Ion Transport across the cell membrane is required to maintain cell volume homeostasis. In response to changes in extracellular osmolarity, most cells activate specific metabolic or membrane-transport pathways to respond to cell swelling or shrinkage and return their volume to its normal resting state. This process involves the rapid adjustment of the activities of channels and transporters that mediate flux of K^+, Na^+, Cl^-,

Current Topics in Membranes, Volume 81
ISSN 1063-5823
https://doi.org/10.1016/bs.ctm.2018.08.002

and small organic osmolytes. Cation chloride cotransporters (CCCs) NKCCs and KCCs are a family of membrane proteins modulated by changes in cell volume and/or in the intracellular chloride concentration ($[Cl^-]_i$). Cell swelling triggers regulatory volume decrease (RVD), promoting solute and water efflux to restore normal cell volume. Swelling-activated KCCs mediate RVD in most cell types. In contrast, cell shrinkage triggers regulatory volume increase (RVI), which involves the activation of the NKCC1 cotransporter of the CCC family. Regulation of the CCCs during RVI and RVD by protein phosphorylation is a well-characterized mechanism, where WNK kinases and their downstream kinase substrates, SPAK and OSR1 constitute the essential phospho-regulators. WNKs-SPAK/OSR1-CCCs complex is required to regulate cell shrinkage-induced RVI or cell swelling-induced RVD via activating or inhibitory phosphorylation of NKCCs or KCCs, respectively. WNK1 and WNK4 kinases have been established as $[Cl^-]_i$ sensors/regulators, while a role for WNK3 kinase as a cell volume-sensing kinase has emerged and is proposed in this chapter.

1. INTRODUCTION

Animal cells are constantly challenged to a variety of stimuli. In response to an extracellular or intracellular osmotic perturbation, cells must maintain a constant cell volume to survive and enable normal cell function. In response to changes in extracellular osmolarity, most cells activate specific metabolic or membrane-transport pathways to respond to swelling or shrinkage and return cell volume to its normal resting state. Among the mediators and signaling pathways responding to cell volume modification are lipid and lipid modifying enzymes, intracellular Ca^{2+}, protein phosphorylation and Rho family GTPases (Koivusalo, Kapus, & Grinstein, 2009). Several transport membrane proteins are involved in cell volume regulation, including a diversity of channels and cotransporters. This chapter will focus on the Cation Chloride Cotransporters (CCCs) family that play a role in cell volume regulation and how the volume-regulatory responses of CCCs are controlled by the signaling kinases WNKs (With no Lysine Kinases) to maintain intracellular ionic homeostasis and proper cellular function. A potential role for WNK3, as the kinase translating the cell volume changes to the CCCs to modulate their response to cell swelling or shrinkage is proposed.

2. PRINCIPLES OF CELL VOLUME REGULATION

Differences in extracellular and intracellular solute concentrations create a transmembrane osmotic gradient. This gradient results in the flow of water into or out the cell freely across the membrane, causing cell swelling or shrinkage until osmotic equilibrium is restored. The processes by which swollen or shrunken cells return to their normal volume are called Regulatory Volume Decrease (RVD) and Regulatory Volume Increase (RVI), respectively (Lauf & Adragna, 2000; McManus, Churchwell, & Strange, 1995; Strange, 2004) (Fig. 1). During these processes transport systems on the membrane are activated within seconds of volume deviation and in response to cell volume changes as small as 3%–5%. Volume regulatory loss and gain of electrolytes is accomplished through membrane transport proteins that regulate movement of mainly inorganic anions, such as sodium (Na^+), potassium (K^+) and chloride (Cl^-) or small organic molecules called osmolytes. During RVI, cell shrinkage is regulated by gain of potassium (K–Cl) and sodium chloride (Na–Cl) and organic solutes with the accompanying osmotically driven influx of water molecules (Koivusalo et al., 2009). On the other hand, during RVD in swollen

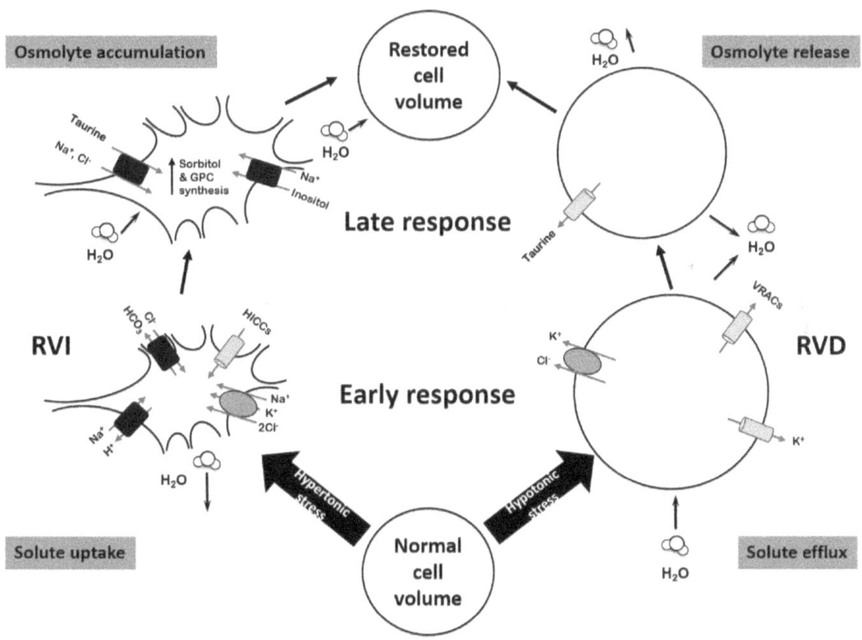

Figure 1 ***Cell volume regulation mechanisms.*** Differences in extracellular and intracellular solute concentrations causes cell swelling or shrinkage until normal cell volume is restored by RVI or RVD. During RVI, the cell activates solute influx which drives influx of water osmotically; during RVD, decrease in cell volume is accomplished by K–Cl and organic osmolytes efflux followed by loss of water. Principal RVI, RVD and late response solute transport pathways are shown.

cells, the decrease in cell volume is accomplished by K–Cl and organic osmolyte efflux carried out by a different set of proteins, followed by loss of water (Okada, 2004).

2.1 RVI

Regulatory Volume Increase is mediated by several mechanisms depending on the cell type. The main effectors during the response are the $Na^+ - H^+$ exchanger (NHE), the $Cl^- - HCO_3^-$ anion exchanger (AE) and the $Na^+ - K^+ - 2Cl^-$ cotransporter (NKCC). These proteins exploit the electrochemical gradient generated by the $Na^+ - K^+$- ATPase and the negative membrane potential to drive Na^+ cations into the cells. In some cell types, like rat hepatocytes, Na^+-permeable channels play a significant role in the RVI response.

Na^+-H^+ exchanger: The sodium-hydrogen exchanger (NHE) protein family transports Na^+ ions in exchange of intracellular H^+. Besides regulation of cell volume they are responsible for maintaining the intracellular pH balance of many cells. In addition, in the kidney NHE are important for the whole body acid-base homeostasis (Schultheis et al., 1998). Several isoforms have been described. Cell shrinkage activates NHE1, NHE2 and NHE4, but not NHE3 and NHE5 (Alexander & Grinstein, 2006). Activation of the NHE exchanger in response to cell shrinkage depends on intracellular ATP, but it does not involve a change in its phosphorylation level. Nevertheless, auxiliary proteins that controls its activity might be the target of phosphorylation during RVI. It has been shown that osmotic activation of NHE1 results from a shift in the pH of its "set point" and from enhanced affinity for extracellular Na^+ but the biochemical reactions underlying these events remains unclear (Grinstein, Woodside, Sardet, Pouyssegur, & Rotin, 1992).

$Cl^- - HCO_3^-$ anion exchanger (AE): These membrane proteins transport bicarbonate (HCO_3^-) in exchange for Cl^- and are the principal regulators of pH in animal cells and acid/base secretion. $Cl^- - HCO_3^-$ anion exchangers are expressed in many cell types and can be found in two different gene families: SLC4A and SLC26A (Bonar & Casey, 2008). They operate in parallel with the NHE to respond to RVI. It has been shown that shrinkage-induced activation of the AE is secondary to intracellular alkalinisation resulting from NHE activation (Humphreys, Jiang, Chernova, & Alper, 1995). Transport of HCO_3^- out of the cell in exchange for Cl^- by the AE, along with the transport of protons out of the cell and Na^+ into the

cell by NHE1, contributes to the Na−Cl gain necessary to counteract cell shrinkage without a change in pH. Accumulation of Na−Cl inside the cell will cause the osmotic movement of water and the recovery to normal cell volume.

Na − K−2Cl cotransporter: NKCC participates in RVI contributing to Na^+ influx coupled to K^+ and Cl^-. It belongs to the cation-chloride cotransporter family, along with the Na−Cl cotransporter (NCC) and four K−Cl cotransporters (KCCs) (Arroyo, Kahle, & Gamba, 2013). There are two Na−K−2Cl cotransporters: NKCC1 and NKCC2, and both are stimulated by cell shrinkage. NKCC1 is expressed in the basolateral membrane of almost all cell types, while NKCC2 expression is restricted to the apical membrane of the thick ascending loop of Henle in the kidney (TAL), where it plays an important role in salt reabsorption and the whole body homeostasis of salt and water (Gamba, 2005a). Activation of NKCCs in response to cell volume decrease takes place upon direct phosphorylation of the cotransporters by Serine (Ser)/Threonine (Thr) kinases of the STE20 and the WNK [With No Lysine (K)] kinases family. Further in the chapter it will be discussed how this is achieved (Delpire & Gagnon, 2008; Kahle, Rinehart, et al., 2008).

Hypertonicity Activated Cation Channels (HICCs): Channel-mediated ion transport constitute a rapid and effective mechanism of RVI in several cell types. Once they become active, they allow the indiscriminate passage of monovalent cations, mainly import of Na^+ and K^+ at the same time. Following their electrochemical gradient, a considerable higher number of cations will enter the cell via hypertonicity activated cation channels, accompanied by parallel conductive entry of Cl^-, increasing intracellular osmolarity (Wehner et al., 2006). The molecular identity of HICCs is not completely know. However, previous studies have identified α-ENaC as a functional element of the hepatocytes HICCs (Soleimani et al., 2012) and the TRPM2 channel as the molecular correlate of the HICC in HeLa cells (Numata et al., 2012). Likewise, during a small interfering RNA (siRNA) screen in hepatocytes cell line HepG2, channels δ-ENaC, TRPM2 and TRPM5 were identified as HICCs contributing to RVI, cell proliferation and apoptosis (Koos et al., 2018). On the other hand, based on their pharmacological properties three main HICCs classes can be distinguished: the first one, is either inhibited by the diuretic amiloride, but insensitive to Gadolinium (Gd^{3+}) and the anti-inflammatory compound flufenamate, the second class are efficiently blocked by Gd^{3+} and flufenamate, but insensitive to amiloride and the third class, both sensitive to

Gd^{3+} and amiloride (Pedersen, Owsianik, & Nilius, 2005). In addition to their role in cell homeostasis, HICCs have been shown to be involved in gene expression, cell locomotion, cell proliferation and apoptosis (Wehner & Olsen, 2004).

Finally, although the transport systems described above rapidly restore the cell size, a secondary or late response system of RVI is activated by the cell to withstand the acute uptake of organic ions (Burg, Ferraris, & Dmitrieva, 2007). An increased ionic intracellular concentration is harmful to cellular metabolism, and to folding, association, and function of macromolecules including proteins. To counteract it, over a period of hours, the cell starts accumulating osmolytes. These organic osmolytes are "non-perturbing" solutes, that have unique biochemical and biophysical properties that allow the cell to accumulate them in high concentrations, without altering cell function and structure (Strange, 2004) (Fig. 1). This accumulation permits the cells to release the excess ions, restore intracellular ion concentration and reduce ionic strength, but most important, maintain cell volume in the presence of a continued hypertonic stress.

Osmolytes: The main organic osmolytes employed by the cell are taurine (2-aminoethylsulfonic acid), betaine, inositol, sorbitol and glycerol-phosphocholine. Accumulation of taurine, betaine, and inositol after a hypertonic challenge occurs by uptake from the extracellular media by the taurine, betaine and sodium myo-inositol cotransporters which couple their transport to Na^+ and Cl^- ions (Koivusalo et al., 2009). Accumulation of sorbitol and glycerol-phosphocholine occurs after stimulation of transcription of the transporters or enzymes involved in their synthesis by the transcription factor osmotic response element-binding protein, which is activated by several volume-activated kinases (Burg et al., 2007).

2.2 RVD

During Regulatory Volume Decrease, K—Cl efflux is mainly involved. The electrochemical gradient of K^+ drives it out of the cell, leading to net efflux of water and electrolytes and reduction of cell volume. K^+ transport is accompanied by Cl^- anions efflux achieved by the main effectors of RVD, the electroneutral K—Cl cotransporters (KCCs) and separate K^+ and Cl^- channels. In animal cells, in addition to K—Cl efflux, the RVD mechanism involves channel- or transporter-mediated efflux of amino acids and osmolytes (mainly taurine and sorbitol).

K—Cl cotransporter: Electroneutral coupled K—Cl transport by the KCCs was first described in volume regulation in red blood cells

(Lauf & Adragna, 2000). These membrane proteins mediate K^+ and Cl^- exit during RVD. The four KCCs described to date: KCC1, KCC2, KCC3 and KCC4 are expressed in many different cell types and all respond to cell swelling (Adragna, Fulvio, & Lauf, 2004). KCCs are regulated by phosphorylation/dephosphorylation cycles. Upon cell swelling KCCs become active by direct dephosphorylation of the cotransporter, which drives K—Cl outside the cell and decreases intracellular osmolarity. In contrast with the NKCCs, phosphorylation of KCCs by members of the Ste-20 and WNK kinases in response to cell shrinkage inhibits their activity. Because of this phosphorylation/dephosphorylation regulation, a volume-sensitive kinase/phosphatase pair was suggested to be responsible for KCCs/NKCCs RVD and RVI response, respectively (Dunham, Klimczak, & Logue, 1993; Kaji & Tsukitani, 1991). Regulation and response of KCCs by WNK kinases, will be addressed later in this chapter.

K^+ channels: Multiple well-characterized K^+ channels become activated upon cell swelling. These channels conduct K^+ efflux as a result of the outwardly directed K^+ electrochemical gradient. In the course of RVD, different types of K^+ channels can become active in the same cell at the same time (Wehner et al., 2006), but their individual contribution to the decrease of cell volume is not well established. A parallel anion efflux accompanies the conductive K^+ efflux to accomplish the necessary loss of solutes so RVD can take place.

Anion channels: In the face of cell swelling, most types of cells force out Cl^- anions through many different anion channels including the Volume Regulated Anion Channels or VRACs which are the most associated to RVD. The preferred substrate of VRACs are Cl^- anions, but they may also be the transport pathway for hypotonicity-induced osmolyte efflux (Stutzin & Hoffmann, 2006). The molecular entity of this channel was unknown until very recently, when studies by several laboratories demonstrated conclusively that VRACs are encoded by members of the leucine-rich repeat containing 8 (*Lrrc8*) gene family which comprises five members, termed *Lrrc8a-e* (Jentsch, Lutter, Planells-Cases, Ullrich, & Voss, 2016). In addition, cell volume VRAC's have been proposed to be critically involved in cell proliferation and migration, apoptosis, *trans*-epithelial transport, swelling-induced exocytosis and intercellular communication; and they might also play a role in pathological states like cancer and ischemia (Stauber, 2015).

Amino acids release: Just as in RVI, after a prolonged cell swelling state, the cell activates a late secondary volume regulatory response, where common amino acid osmolytes including taurine, aspartate and glutamate are released from the cell. Large amounts of taurine leave the cell after osmotic swelling through a pathway distinct from the Na^+/taurine transporter mentioned above. Taurine efflux may occur via the VRACs, but the precise molecular identity of the pathway involved is unknown (Voss, Pedersen, Christensen, & Lambert, 2004).

2.3 The SLC12 Family of Cation-Chloride Cotransporters

As described above, while there are many pathways involved in RVI and RVD, the electroneutral cation-chloride cotransporters (CCCs) play a central role in the modulation of cell volume by WNKs and will be the focus of the rest of the chapter. The CCCs are encoded in the solute carrier family 12 (SLC12) genes consisting of nine members (Gamba, 2005a). The common functional characteristics of all proteins in this family are the following: (1) the coupled transport of one cation (Na^+ and/or K^+) per each transported anion, hence the name of electroneutral; (2) the transported anion is always chloride; (3) they are all modulated by changes in cell volume; (4) they are all modulated by changes in the intracellular chloride concentration ($[Cl^-]_i$); and (5) regulation of CCCs activity is achieved by phosphorylation and dephosphorylation events.

Two main branches are easily distinguishable in the SLC12 family by structural and functional characteristics. One is the Na^+-coupled chloride cotransporters, collectively named NKCCs, that include the Na–K–2Cl cotransporters 1 and 2 (NKCC1 and NKCC2) and the Na–Cl cotransporter (NCC). The degree of identity between them is about 50%. The other branch is the K^+-coupled chloride cotransporters known as KCCs that encompasses four members known as KCC1, KCC2, KCC3 and KCC4. The percentage of identity among KCCs is 70% and with the NKCCs about 25%. Following the Na^+ or K^+ gradients, NKCCs move chloride into the cell, while KCCs move chloride outside the cell, respectively. Thus, NKCCs promote an increase of $[Cl^-]_i$ participating in the pathways for RVI as mentioned above, while the KCCs promote decrease in the $[Cl^-]_i$ as one of the RVD mechanisms.

CCCs are involved in many different and important cellular functions like cell volume regulation, *trans*-epithelial ion transport, and maintenance of $[Cl^-]_i$. Multiple members of this family and their upstream regulators are mutated in human Mendelian disorders featuring brain or renal phenotypes resulting from impaired ion homeostasis (Table 1). For an in-depth review of the extensive amount of information for this family, readers are

Table 1 The SLC12 Family of Cation-Chloride Cotransporters

Gene/Protein	Tissue Expression	Physiological Function	Genetic Disease
SLC12A1/ NKCC2	Kidney (TAL)	NaCl reabsorption in the TAL; regulation of Ca^{2+} excretion; urine concentration	Bartter syndrome
SLC12A2/ NKCC1	Ubiquitous	Cell volume regulation (RVI); provide ions for secretion	Potential role in human schizophrenia Multi-organ system failure
SLC12A3/NCC	Kidney (DCT)	NaCl reabsorption in the DCT; regulation of Ca^{2+} and K* renal excretion;	Gitelman syndrome
SLC12A4/KCC1	Ubiquitous	Cell volume regulation (RVD), KCl epithelial transport	ND
SLC12A5/KCC2	CNS	Intraneuronal Cl^- concentration regulation	Idiopathic generalized epilepsy
SLC12A6/KCC3	Lung, Heart, Kidney, Brain	Volume regulation in the brain; K^+ recycling in the kidney	Andermann syndrome
SLC12A7/KCC4	Lung, Heart, Kidney, Brain	Participates in acid excretion in α-IC cells of CD	ND

CD, collecting duct; CNS, central nervous system; DCT, distal convoluted tubule; ND, no data; TAL, Thick ascending loop of Henle; α-IC, alpha intercalated cell. For references go to Section 2.3.

referred to recent reviews on the subject (Ares, Caceres, & Ortiz, 2011; Blaesse, Airaksinen, Rivera, & Kaila, 2009; Delpire & Austin, 2010; Gamba, 2009; Hebert, Mount, & Gamba, 2004) (Kahle, Staley, et al., 2008; Pacheco-Alvarez & Gamba, 2011; Russell, 2000).

Function of the NKCCs and the KCCs is reciprocally regulated: volume-sensitive KCCs (KCC1, 3, and 4) are largely inactive in isotonic conditions, but rapidly activated by hypotonicity-induced cell swelling, while NKCCs become active under hypertonic-induced cell shrinkage. Swelling-induced KCCs activation are associated with inhibition of protein kinase phosphorylation and the activation of protein phosphatases (PPs). Inhibition of protein phosphatases abolishes KCCs RVD response, demonstrating an essential regulatory role for PPs in this process. Conversely, activation of NKCCs is associated with protein kinase phosphorylation and inhibition by dephosphorylation.

Intracellular chloride concentration constitutes another important stimulus for CCCs regulation (Kahle et al., 2006). NKCCs are activated when $[Cl^-]_i$ is reduced, while the KCCs are activated when the $[Cl^-]_i$ is increased. These however seem to conflict with expected changes in cell volume, because it is expected that cell shrinkage is associated with $[Cl^-]_i$ increase and cell swelling with $[Cl^-]_i$ decrease. Thus, as will be further discussed in this chapter, it is likely that kinases responsible for translating the cell volume or $[Cl^-]_i$ changes to the CCCs are distinct.

2.4 Regulation of Cation Chloride Cotransporters During RVI and RVD by Protein Phosphorylation

Activated Ser/Thr kinases can control osmolarity by different pathways like: direct regulation of channels and cotransporters, induction of genes encoding osmolyte-transporting systems, and reorganization of the cell cytoskeleton. Regulation of the cation chloride cotransporters during RVI and RVD by protein phosphorylation is a very well-characterized mechanism, where much evidence has shown the WNK Serine/Threonine kinases and their downstream kinase substrates, the Ste-20 kinases SPAK/OSR1 (Ste 20/SPS1-related proline/Alanine-rich kinase) and OSR1 (oxidative stress-responsive kinase 1) are the essential phospho-regulators of the CCCs family activity (Gagnon & Delpire, 2012; Shekarabi et al., 2017).

When activated by osmotic stress or low Cl^- intracellular concentration (discussed further in the chapter), WNK kinases undergo autophosphorylation of a conserved Ser residue leading to activation and subsequent phosphorylation of their downstream effectors. WNKs phosphorylate a key Thr residue of the T loop in the catalytic domain of SPAK/OSR1 (Thr 233 in human SPAK and Thr 185 in human OSR1) and a Ser residue in the non-catalytic carboxy S motif (Vitari, Deak, Morrice, & Alessi, 2005). In turn, once phosphorylated and activated, SPAK/OSR1 will induce NKCC1, NKCC2 and NCC transport by direct phosphorylation of a series of conserved Thr residues located in the amino terminal region of the cotransporters (Darman & Forbush, 2002; Flemmer, Gimenez, Dowd, Darman, & Forbush, 2002; Pacheco-Alvarez et al., 2006; Ponce-Coria et al., 2008; Richardson et al., 2008, 2011). Interaction between SPAK/OSR1 and the CCCs occurs via a conserved carboxy terminal region named the CCT domain located in the kinases that recognizes a specific binding motif in the cotransporters. This motif is formed by the sequence of amino acids RFXV/I, it is present and conserved in many of the CCCs family members and is generally referred as the "SPAK binding site" (Delpire & Gagnon, 2007). Remarkably the RFXV/I motif is also present in the WNKs kinases and is required for interaction with the SPAK/OSR1 CCT domain (Fig. 2) (Delpire & Gagnon, 2007; Piechotta, Lu, & Delpire, 2002; Richardson et al., 2008; Thastrup et al., 2012; Villa et al., 2007; Vitari et al., 2005; Zagorska et al., 2007). In this way upon cell shrinkage, RVD response will trigger WNK and SPAK/OSR1 activity. As a result, influx of Na^+, Cl^- and K^+ through NKCCs will promote osmotically driven entry of water, supporting the recovery of cell volume. On the other hand, during cell swelling and RVI, these regulatory kinases will become inactive and unable to phosphorylate CCCs, PP dephosphorylated KCCs will become active, transporting K^+ coupled to Cl^- outside the cell, followed by water and eventually recovery of normal cell size.

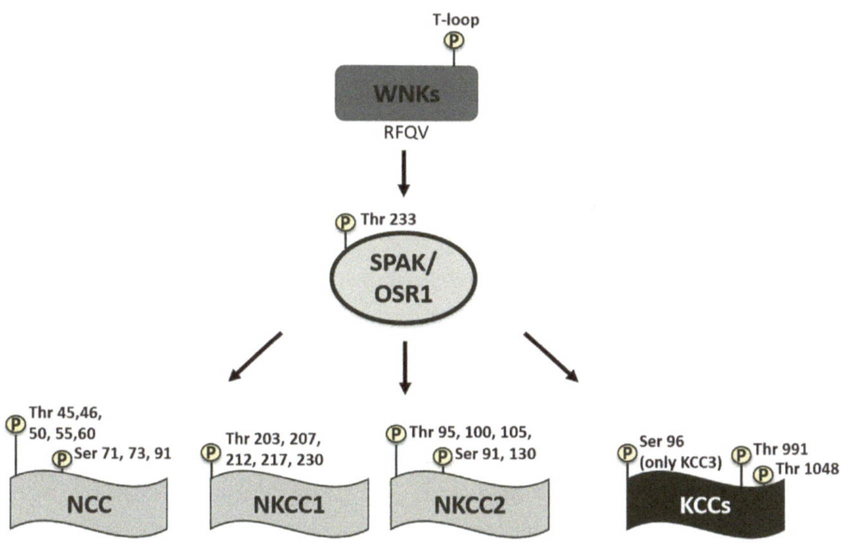

Figure 2 ***WNKs-SPAK/OSR1-CCCs signaling pathway.*** Activation of WNKs kinases in response to osmotic stress or low Cl^- intracellular concentration leads to SPAK/OSR1 phosphorylation and subsequent phosphorylation of amino terminal domain residues of downstream effectors NCC, NKCC1, NKKC2, or amino and carboxy residues in KCCs. Phosphorylation sites important for regulation of the WNK-SPAK/OSR1-CCCs signaling pathway are shown.

The first binding partner and substrate for SPAK and OSR1 to be identified was NKCC1. As mentioned before the kinases directly phosphorylate a cluster of conserved Thr residues in the amino terminal cytosolic domain, corresponding to Thr203, Thr207, Thr212 and Thr217 on human NKCC1. Mutation of Thr212 or Thr217 inhibited activation of NKCC1 by SPAK in response to hyperosmotic stress or hypotonic low Cl^- conditions (Dowd & Forbush, 2003; Piechotta, Garbarini, England, & Delpire, 2003; Piechotta et al., 2002). Sequence alignments of NCC and NKCC2 human and mouse amino acid sequences showed that the NKCC1 SPAK/OSR1 phosphorylation sites are conserved, suggesting that this signaling pathway could also activate them. Indeed, interaction and phosphorylation of NCC and NKCC2 with SPAK/OSR1 was demonstrated (Pacheco-Alvarez et al., 2006; Richardson & Alessi, 2008). NCC residues Thr46, Thr55 and Thr60, equivalent to human NKCC1 residues Thr203, Thr212, and Thr217, are directly phosphorylated by SPAK/OSR1 (Fig. 2). Mutation of Thr60 to Alanine (Ala), prevented activation of the cotransporter in response to hypotonic low Cl^- conditions. Moreover, Thr60 mutation prevented phosphorylation of the other sites, indicating that Thr60 phosphorylation is essential for enabling and/or maintaining the phosphorylation of the other residues. Likewise, hypertonic conditions were shown to stimulate human NKCC2 activity by phosphorylation of corresponding Thr100 and Thr105 residues (Gimenez & Forbush, 2005; Rinehart et al., 2005). However, the effect of tonicity is higher in NKCC1 than in NKCC2 (Gamba et al., 1994).

Inhibition of the KCCs by the WNK activated SPAK/OSR1 kinases is also promoted by direct phosphorylation of the cotransporters in one of two KCC phospho-sites identified and described as critical in controlling KCCs activity (Rinehart et al., 2009). These sites correspond to residues Thr991 and Thr1048 in KCC3 amino acid sequence and are conserved in all KCCs. Recent findings have shown that SPAK/OSR1 directly phosphorylate Thr1048, but not Thr991, in all the KCCs, inhibiting their activity (de los Heros et al., 2014) (Fig. 2). Single or double Threonine to Alanine mutations in KCC3 rendered the cotransporter constitutively active. However, activation of the mutant cotransporter can be further increased by hypotonicity, suggesting the existence of additional inhibitory phosphorylation sites. Indeed, in KCC3 there is a third site located at the amino terminal domain, Serine 96, that needs to be dephosphorylated for the full activation of this cotransporter (Melo et al., 2013).

3. WITH NO LYSINE KINASES (WNKS)

The WNK kinases are a small family of proteins characterized by having a Cysteine in place of a highly conserved Lysine in the subdomain II of their kinase activation domain. In most Ser/Thr kinases this Lysine residue usually mediates ATP binding and the catalysis of phosphoryl transfer. However, in the WNK kinases, this Lysine is in an atypical position in the subdomain I of the catalytic domain (Verissimo & Jordan, 2001; Xu et al., 2000). Four members of the WNK kinase family have been identified in the human: WNK1, WNK2, WNK3 and WNK4. In 2001, it was shown that mutations in *WNK1* and *WNK4* cause the human monogenic disease Familiar Hyperkalemic Hypertension (FHHt) (also known as pseudohypoaldosteronism type 2 or Gordon syndrome) (Wilson et al., 2001). Because this disease phenotype is associated with renal handling and balance of Na^+, Cl^- and K^+ electrolytes, the study of this signaling pathway uncovered an important role for WNKs in the regulation of solute transport and arterial pressure.

WNK family members exhibit only about 40% amino acid identity but nevertheless all share a similar topology (Fig. 3). Their structure features a highly-conserved kinase domain divided in 12 subdomains characteristic of the Ser/Thr kinases, located in the amino terminal side of the protein;

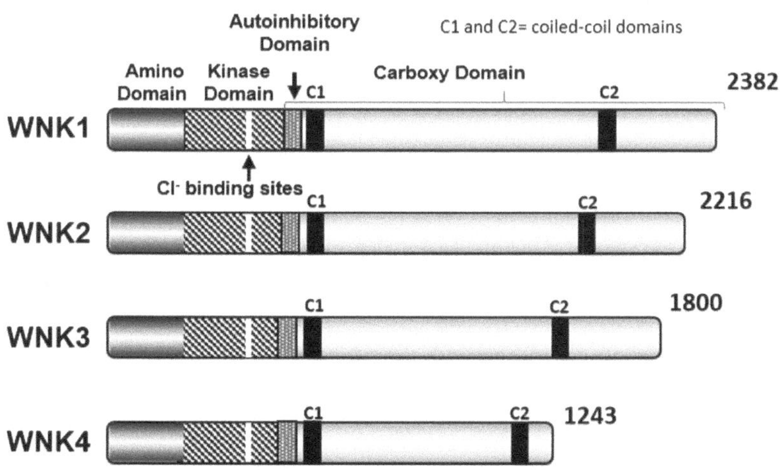

Figure 3 ***WNK kinases domain structure.*** All four human WNK kinases share a common topology. Kinase and autoinhibitory domain are located towards the amino terminal domain of the protein, and coiled coil domains placed in the carboxy terminal domain of the kinase, also known as the regulatory domain.

a conserved auto-inhibitory domain; and a carboxy-terminal regulatory domain (Gamba, 2005b). WNKs, like most kinases are regulated by their auto-inhibitory domain, which ablates their activity until an activation signal releases it. In response to an osmotic stress condition WNK1 undergoes autophosphorylation at two Serine residues located in the kinase domain (Ser378/382), stimulating its activity. Autoinhibition of WNK1 is dependent of two highly conserved Phenylalanine residues (Phe522/524), since mutation of these residues to Alanine ablates it (Xu et al., 2002).

WNK1: Of the WNK kinases, WNK1 is the most extensively studied. It was the first WNK to be identified in a search for new MAP kinases members. The human *WNK1* gene is located on chromosome 12, and its transcript encodes a 28 exon mRNA translated to a protein of approximately 2382 amino acids, known as L-WNK1 for Long-WNK1 (B. Xu et al., 2000). A second transcription site in exon 4a produces a slightly shorter transcript only present in the kidney and thus known as kidney-specific or KS-WNK1. KS-WNK1 lacks a kinase domain and is specifically expressed in the distal convoluted tubule (DCT) and connecting tubule (CNT) of the nephron at high levels (Delaloy et al., 2008; O'Reilly, Marshall, Speirs, & Brown, 2003; Vidal-Petiot et al., 2012). Several other isoforms are generated through alternative splicing of exons 9, 11, 12 and 26, but their precise tissue distribution and functions are still not known. In addition to being mutated in FHHt, WNK1 mutations are also associated with the hereditary disease sensory neuropathy type II, a form of congenital insensitivity to pain. The majority of the WNK1 studies conducted to date have focused on its role in the kidney (Kahle, Rinehart, et al., 2008), where they have placed WNK1 as a key regulator of blood pressure and serum potassium by controlling the transport of Na−Cl through the CCCs including NCC and the epithelial Na^+ channel (ENaC), as well as the renal outer medullary K^+ channel (ROMK). Recent studies have started to analyze WNK1 extrarenal functions, and these include regulating vasoconstriction in response to α_1-adrenergic receptors in cardiovascular tissues (Bergaya et al., 2011). Also, WNK1 is involved in cell migration, division, and differentiation, somatic WNK1 mutations have been identified in several types of cancer by large-scale sequencing (Greenman et al., 2007; Sjoblom et al., 2006).

WNK2: Is the least characterized of all WNK kinases. The human *WNK2* gene is located on chromosome 9, position q22.31 (Verissimo & Jordan, 2001), and encodes a 2216 amino protein. Alternative splicing produces multiple transcript variants. WNK2 is highly expressed in brain neurons of the developing and mature brain, where it is expressed in a regional and temporally differential way during development (Rinehart et al., 2011). Unlike other WNKs, WNK2 is not expressed in the kidney. Studies have shown WNK2 functions as an antiproliferative tumor suppressor, inhibiting the proliferation pathway of MEK1/ERK1 in cancer cells (Moniz et al., 2007). WNK2 levels are significantly reduced in human glioma and meningioma due to epigenetic silencing of the *WNK2* promoter. Hypermethylation of the *WNK2* promoter increases tumor cell proliferation, migration, and motility (Moniz et al., 2013). Apart from its tumor-suppressing role in the brain, WNK2 regulates cell volume and/or GABAergic signaling in the central nervous system by promoting Cl^- entry through activating NKCC1 and inhibiting KCC2 in a kinase-dependent way (Rinehart et al., 2011). It is unknown if WNK2 regulation of neuronal CCCs is linked to its role in cell growth and proliferation. Nevertheless, it has been shown that NKCC1 constitutes the main Cl^- entry in glioma cells and that its genetic or pharmacological inhibition results in reduced glioma cells invasion *in vivo* (Garzon-Muvdi et al., 2012).

WNK3: In the human genome, *WNK3* is located on chromosome Xp11.22 and encodes a 1800 amino acid residue protein, translated from a 24 exon transcript. Alternative splicing of exons 18 and 22 produces 2 variants (Holden, Cox, & Raymond, 2004). Like WNK1 and WNK4, WNK3 is expressed in many secreting epithelial tissues. It is expressed abundantly in almost every tissue in the human, with higher expression in brain and kidney. Unlike WNK1 and 4, WNK3 can be found all along the nephron localized specifically in the tight junctions (Rinehart et al., 2005). Like WNK2, WNK3 brain expression is regionally and temporally regulated during embryonic development. WNK3 has been shown to function as a positive regulator of the transcellular Ca^{2+} transport pathway, and it also plays a role in the increase of cell survival by delaying the onset of apoptosis in a caspase-3-dependent pathway (Verissimo, Silva, Morris, Pepperkok, & Jordan, 2006). WNK3 regulates CCCs activity in a kinase-dependent manner, bypassing the tonicity requirements for their activation or inhibition. This mechanism will be further discussed in the chapter. Finally, recent findings have implicated WNK3 in CNS injury during ischemia (Begum et al., 2015). WNK3 knockout mice showed reduced infarct volume, cerebral edema, and axonal demyelination after a stroke episode, compared to wild type mice. Focal ischemia appeared to stimulate NKCC1 phosphorylation most likely via the WNK and SPAK/OSR1 pathway. Loss of WNK3 significantly reduced the ischemia-associated brain damage associated with NKCC1 activation, suggesting a neuroprotective rol.

WNK4: The human *WNK4* gene is located on chromosome 17q21.2a. The 19 exon transcript encodes a 1243 amino acid protein (Wilson et al., 2001). There are no spliced variants identified to date. Studies in wild type and knock out mice, have shown, that is expressed in cytoplasmic and apical membrane in the kidney, specifically in TAL, DCT, CNT and CD (Collecting Duct) (Ohno et al., 2011). It is also enriched in secretory epithelium from brain, colon, pancreas, biliary ducts, and epididymis. WNK4 regulates the balance between Na—Cl reabsorption and K^+ secretion by inhibiting and activating several types of ion channels, like the Cl^- channel CFTR (Cystic Fibrosis Transmembrane Conductance Regulator), cotransporters like NCC, NKCC1 and KCC4, and exchangers involved in electrolyte flux like the electrogenic Na^+/HCO_3, NBCe1 (Hadchouel, Ellison, & Gamba, 2016; Mendes et al., 2011; Shekarabi et al., 2017; Welling, 2016). As previously mentioned, mutations in *WNK4* cause the human monogenic disease FHHt.

4. THE WNK KINASES AS INTRACELLULAR CHLORIDE SENSORS

How the cell signals size and how this is transduced into volume-regulatory responses is still not fully understood. It is clear, however, that these mechanisms are extremely sensitive and tightly regulated. Recent data suggest that responses to cell swelling or shrinkage are not as simple as they appear. The cell volume detector and effector mechanisms appear to respond selectively to both the magnitude and the nature of the volume perturbation, allowing the cell to control both intracellular ionic composition and volume. A major challenge to the field is to identify the specific signals linking the upstream sensors and their corresponding downstream effectors. For WNKs kinases, changes in osmolarity and/or intracellular chloride concentration seem to be separate signals that regulate their activities and effectors in a differential way.

4.1 L-WNK1 Kinase as an Intracellular Chloride Sensor

Evidence that WNK1 constitutes a primary regulator of intracellular Cl^- changes in renal and extrarenal tissues was revealed by the findings that WNK1 can bind Cl^- ions, and that WNK1 and WNK4 activity is modulated by $[Cl^-]_i$ (Bazua-Valenti et al., 2015; Hadchouel et al., 2016; Piala et al., 2014). By means of crystallographic studies, Piala and coworkers identified a chloride-binding site in the long WNK1 isoform (L-WNK1) situated on the catalytic domain at the amino terminal of the kinase. The Cl^- ion binds directly to two leucine residues, Leu369 and Leu371 as determined by site-directed mutagenesis. Leu369 confers sensitivity to chloride to a greater extent than Leu371. Cl^- ions also form hydrophobic interactions with surrounding residues Phe283 and Leu299. Direct binding between chloride and L-WNK1 kinase domain has an inhibitory effect, preventing kinase autophosphorylation and activation. The fact that the motif is located in the kinase domain likely contributes to the rapid cellular response to changes in the chloride concentration (Piala et al., 2014). Chloride was found to stabilize L-WNK1 in the inactive conformation, inhibiting its autophosphorylation with an inhibition coefficient (IC_{50}) of approximately 20 mM, which is slightly above the chloride concentration found in neurons (10–15 Mm) and slightly below that observed in the kidney (50 mM). Therefore, this IC_{50} is consistent with the intracellular chloride concentrations found in different cell types, meaning Cl^- is regulating L-WNK1 within a concentration appropriate to the functions of various tissues (Piala et al., 2014).

4.2 WNK4 Kinase as an Intracellular Chloride Sensor

At the same time that the L-WNK1 chloride-binding site was described, Bazúa-Valenti and colleagues demonstrated that WNK4 kinase is also chloride-sensitive (Bazua-Valenti & Gamba, 2015; Bazua-Valenti et al., 2015). The substitution of Leu322, the residue homologous to the chloride binding site in L-WNK1, to phenylalanine (Leu322Phe), converts WNK4 to a constitutively phosphorylated kinase that consequently activates the NCC cotransporter without requiring changes in $[Cl^-]_i$. Leu 324 in WNK4 kinase, equivalent to leucine residue 371 in L-WNK1, is not necessary for NCC activation (Fig. 4).

As previously described the Cl− IC_{50} for L-WNK1 is 20 mM (Piala et al., 2014), while in *Xenopus laevis* oocytes and the HEK293 cell line the $[Cl^-]_i$ is around 40–50mM, thus allowing the partial activity of L-WNK1 and completely inhibiting the activity of WNK4 (Chang et al., 2009; Cruz-Rangel, Gamba, Ramos-Mandujano, & Pasantes-Morales, 2012; Keicher & Meech, 1994). This data is consistent with the observations that the basal activity of NCC depends on endogenous L-WNK1 (Bazua-Valenti & Gamba, 2015). The increase in NCC activity induced by low chloride hypotonic conditions in *X. laevis* oocytes overexpressing the NCC cotransporter is possibly caused by the endogenous activity of L-WNK1. Interestingly, injection of *X. laevis* oocytes with the anion

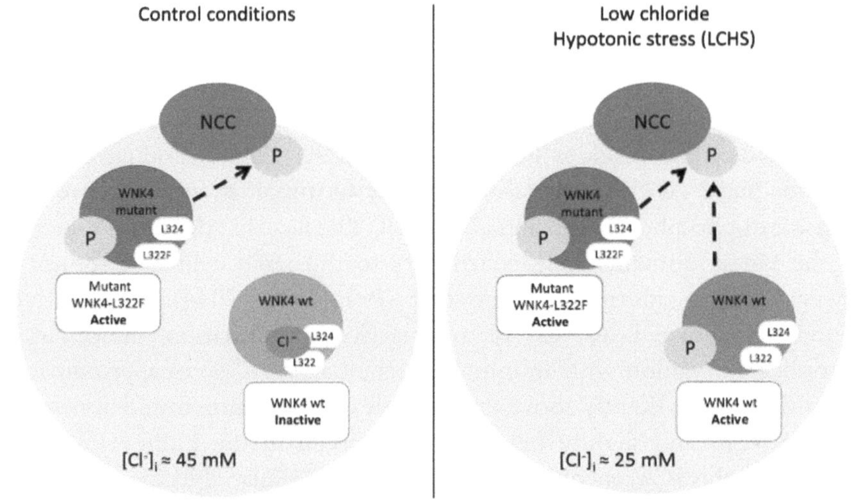

Figure 4 ***The activity of WNK4 towards SPAK-NCC is modulated by intracellular chloride concentration (Bazua-Valenti et al., 2015).*** In control conditions in oocytes, a high intracellular chloride concentration (45 mM) inactivates wild type (wt) WNK4 wt, so NCC activity is not stimulated. Under low chloride conditions (25 mM), wt WNK4 is phosphorylated and now activates NCC. Elimination of the chloride-binding site in WNK4 (Mutant WNK4-L322F), turns WNK4 into a constitutively-active kinase that is able to activate NCC, regardless of the intracellular chloride concentration.

exchanger SLC26A9 reduced the $[Cl^-]_i$ to approximately 30 mM, giving rise to the activation of co-injected NCC, without need for the direct effect of low hypotonic chloride condition (Bazua-Valenti & Gamba, 2015).

4.3 WNK3 Kinase Is Not Sensitive to Intracellular Chloride Concentration

WNK3 kinase has been described as a chloride-sensitive kinase since its relevant function in the modification of the $[Cl^-]_i$, through the activation/inhibition of the SLC12 cotransporters via SPAK kinase phosphorylation was shown (Cruz-Rangel et al., 2012; Pacheco-Alvarez et al., 2012; Zhang et al., 2016). WNK3 kinase promotes the entry of Cl^-, activating NCC, NKCC1, and NKCC2, and prevents its exit from the cell by inhibiting the KCCs. In this manner, WNK3 kinase is the only member of the WNK family that has the peculiar property of regulating the two branches of the CCC family in opposite ways (Kahle et al., 2006). Initially these data led to the proposal that WNK3 kinase is chloride-sensitive (Pacheco-Alvarez & Gamba, 2011). However, more recent evidence demonstrates WNK3 has a low sensitivity to Cl^- compared to the other WNK proteins.

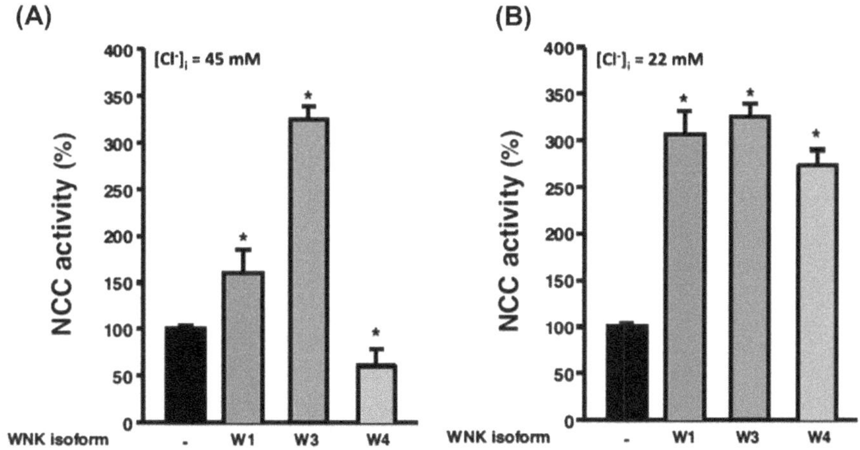

Figure 5 ***Sensitivity of WNK kinases to intracellular chloride, as revealed by their effects towards NCC (Bazua-Valenti et al., 2015).*** The effect of WNK kinases on NCC activity in *Xenopus laevis* oocytes microinjected with NCC cRNA alone or together with each of the WNKs cRNA. (A) uptake in control conditions in which intracellular chloride concentration $[Cl^-]_i$ is 45 mM (*$P > 0.05$ vs. black bar). In this condition WNK1 has a slight activating effect, WNK3 a strong activating effect and WNK4 inhibits NCC. (B) *Xenopus* oocytes from the same experiments incubated in a low-chloride hypotonic stress medium for 16 h in which $[Cl^-]_i$ is reduced to 22 mM (*$P > 0.05$ vs. black bar). In this condition, all three WNKs activate NCC. Thus, the apparent sensitivity to chloride is WNK4 > WNK1 > WNK3.

Fig. 5 shows that in the functional expression model of *X. laevis* oocytes, under basal conditions in which the average intracellular chloride concentration is around 45 mM, the most potent activator of NCC is WNK3 (about 300% activation), followed by WNK1 (about 50% activation), whereas WNK4 has no effect on the basal activity of the cotransporter or can even induce a negative effect of NCC, as some studies suggested originally (Wilson et al., 2003; Yang, Angell, Mitchell, & Ellison, 2003). After exposing oocytes to low chloride hypotonic stress that reduced the intracellular chloride concentration to 22 mM, all three kinases activated NCC by about 300% (Bazua-Valenti & Gamba, 2015). As Fig. 5B shows, the reduction of intracellular chloride had no further effect on the WNK3-induced activation of NCC compared to control conditions (Fig. 5A), meaning WNK3 regulation of NCC activity is independent of changes in the $[Cl^-]_i$ and that WNK1 and WNK4 regulation is clearly dependent in $[Cl^-]_i$ changes. In this respect WNK4 seems to be the most sensitive member of the WNK family to chloride as shown by Terker and colleagues, who assessed WNKs-induced SPAK phosphorylation under increasing $[Cl^-]_i$ *in vitro* (Terker et al., 2016). It is worth mentioning that the study by Piala and colleagues (Piala et al., 2014) measured an Cl− IC_{50} of 20 mM for L-WNK1 autophosphorylation, while the study by Terker et al. (2016) measured a much higher value, around 60−150 mM, using SPAK phosphorylation as the readout. Based on the studies, it is therefore possible that L-WNK1 could either be partially or fully active in HEK cells and oocytes at baseline. Another interesting observation is that while L-WNK1 and WNK4 require an $[Cl^-]_i$ decrease for their activation by phosphorylation, WNK3 kinase auto-phosphorylation is independent of the $[Cl^-]_i$. Thus, although WNK3 kinase is capable of modifying $[Cl^-]_i$, through the regulation of the SLC12 cotransporters, WNK3 sensitivity to chloride is not the main stimulus inducing its activity (Fig. 6).

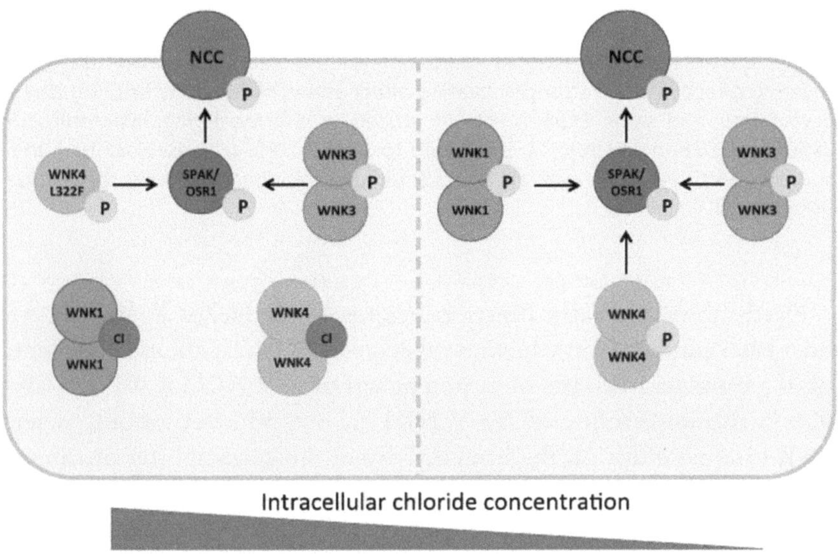

Figure 6 ***Proposed model of WNKś regulation of NCC in response to intracellular chloride concentration*** $[Cl^-]_i$ ***(Bazua-Valenti et al., 2015).*** In a high intracellular chloride concentration (left), chloride binds to WNK1 and WNK4, preventing kinase autophosphorylation, which reduces or inhibits NCC activation. In contrast, WNK3 autophosphorylation seems to be insensitive to chloride and it thus activates SPAK and NCC at any chloride concentration. In low chloride conditions (right) WNK1 and WNK4 autophosphorylation is no longer inhibited by chloride, resulting in SPAK and NCC activation. Elimination of the chloride binding site in WNK4 (WNK4-L322F) permits constitutive WNK4 autophosphorylation even in high intracellular chloride conditions.

5. WNKS AND CELL VOLUME REGULATION

5.1 WNK1 Kinase Is Regulated by Hyperosmotic Stress

WNK1 kinase has been extensively studied for its role in FHHt but data suggest it could be involved in modulation of cell volume. Studies in renal epithelial, colon, and breast cancer cells revealed an increase in WNK1 activity (phosphorylation) after 30s of incubation in conditions of hypertonic stress induced by glucose, sucrose, mannitol or sorbitol. However, evidence was also obtained that hypotonic stress conditions also induced WNK1 activation, but to a lesser degree (Lenertz et al., 2005; Xu et al., 2002; Zagorska et al., 2007). Changes in cellular tonicity induce phosphorylation of human WNK1 at Ser382, a residue conserved in all species (Lenertz et al., 2005; Xu et al., 2002; Zagorska et al., 2007). In addition, using a WNK1 knockout cell line obtained by Crisper/Cas mediated genome editing, L-WNK1 was shown to be essential for bumetanide-sensitive RVI in mammalian cells following hypertonic stress, despite an increase in WNK3 protein expression in these cells (Roy et al., 2015). Thus, evidence suggest a role of WNK1 in cell volume regulation that requires further investigation.

5.2 WNK3 Kinase and Cell Volume Regulation

Among WNKs kinases, WNK3 has shown unique biochemical and functional properties that suggest it could be the kinase transducing cell volume changes into activation or inhibition of CCCs (Rinehart et al., 2005).

Data published since 2005 have shown that WNK3 kinase and its catalytically inactive version WNK3-D294A have no requirement for changes in cell volume to activate or inhibit the CCCs cotransporters. A summary of the results obtained through the functional expression model in *X. laevis* oocytes is shown in Fig. 7 (de los Heros et al., 2006; Kahle et al., 2005; Rinehart et al., 2005) demonstrating that the NKCCs, which are typically inhibited under hypotonic conditions, become active under this inhibitory condition when co-expressed with WNK3 (Fig. 7A). In contrast, in a hypertonic state, in which the NKCCs are maximally activated, the co-expression with the catalytically inactive WNK3-D294A results in their complete inhibition (Fig. 7B). Thus, in the presence of wild type WNK3 or inactive WNK3, changes in cell volume are no longer required to activate or inhibit NCC, NKCC1 and NKCC2. A similar, but opposite situation

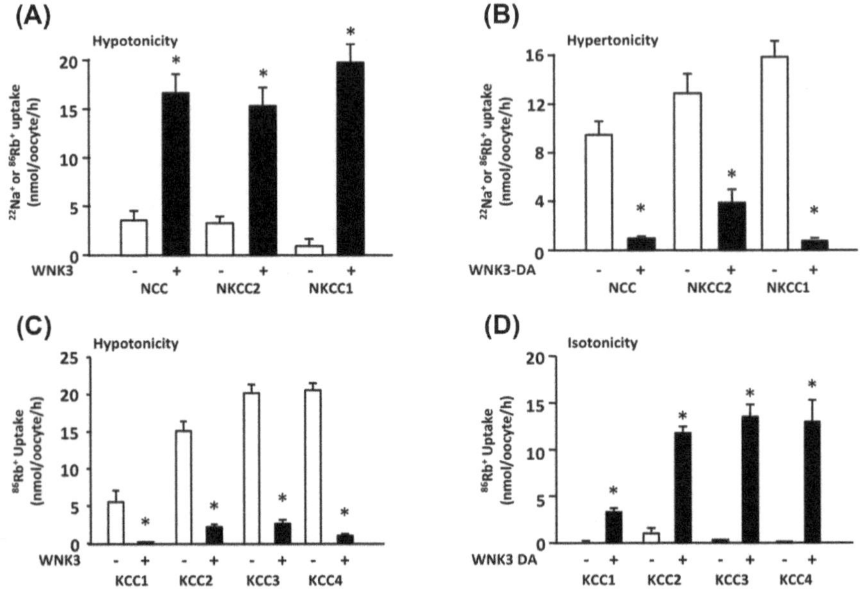

Figure 7 ***WNK3 wild type and the catalytically inactive WNK3-DA bypasses the tonicity requirements for CCCś activity regulation (Pacheco-Alvarez et al., 2012).*** Summary of functional expression assays in *Xenopus laevis* oocytes microinjected with appropriate CCCs alone or together with wild type WNK3 or the catalytically inactive WNK3-DA cRNA, by measuring $^{86}Rb^+$ uptake for NKCC1 and NKCC2 (*Panel A and B*) and KCCs (*Panel C and D*), and $^{22}Na^+$ measuring for NCC activity (*Panel A and B*). Functional assay experiments were performed under isotonic, hypertonic and hypotonic conditions as stated.

occurred for KCCs. Cell swelling-induced hypotonic conditions are associated with KCC activation due to their dephosphorylation. In the presence of wild type WNK3, however, KCCs remain inactive despite the hypotonic stimuli (Fig. 7C). Additionally, in isotonic conditions where KCCs are usually inactive, when co-expressed with WNK3-D294A, all four KCCs are activated without any change in extracellular tonicity (Fig. 7D). Thus, the presence of WNK3 or WNK3-D294A bypasses the tonicity requirements for KCCs activation or inhibition. These observations using *X. laevis* oocytes were later confirmed in mammalian cells (Cruz-Rangel et al., 2012) where it was shown that WNK3 has a direct effect on the RVD and RVI volume compensatory mechanisms.

Modulation of the CCCs by WNK3 kinase has physiological important implications in many tissues. It was recently established that WNK3 is involved in a critical node of volume-sensitive cation-Cl cotransporter regulation in the mammalian brain (Zhang et al., 2016). *In vitro* and *in vivo* data found the WNK3/SPAK complex as an essential regulator of NKCC1 and KCC3 phosphorylation mediated in part by an interaction between SPAK CCT domain and the RFXV/I domains. By treating HEK293 cells transfected either with wild type or kinase-dead WNK3 with the SPAK/OSR1 CCT domain inhibitor STOCK1S-50699, it was demonstrated that WNK3/SPAK inhibition prevents acute cell swelling in response to osmotic stress by stimulating KCC3 activity via reduced phosphorylation. These observations again place WNK3 kinase as a sensor simultaneously signaling to the RVI effector, NKCC1 and the RVD effector KCC3.

5.3 Perspective

Since WNK3 is the non-sensitive kinase of the family to $[Cl^-]_i$ and that expression of wild type or catalytically inactive WNK3 with the CCCs bypasses the tonicity requirement for activation or inactivation of CCCs, it is reasonable to propose that WNK3 could be the cell volume sensitive kinase translating cell volume changes to the CCCs. This may explain the unsolved question about how $[Cl^-]_i$ and cell volume can modulate the activity of CCCs in opposite directions. That is, during cell shrinkage the NKCCs are activated due to RVI, but during cell shrinkage the $[Cl^-]_i$ is increased, which would counter NKCCs activation. On the other hand, during cell swelling the KCCs cotransporters are activated, but during cell swelling dilution decreases $[Cl^-]_i$, which would prevent KCC activation. Given that WNK1 and most importantly WNK4 are sensitive to $[Cl^-]_i$, while WNK3 is not, it is possible that WNK1 and WNK4 are the chloride-sensitive kinases while WNK3 is the cell volume-sensitive kinase. These are testable hypotheses that can be resolved in the years to come.

REFERENCES

Adragna, N. C., Fulvio, M. D., & Lauf, P. K. (2004). Regulation of K-Cl cotransport: From function to genes. *The Journal of Membrane Biology, 201*(3), 109–137.

Alexander, R. T., & Grinstein, S. (2006). Na^+/H^+ exchangers and the regulation of volume. *Acta Physiologica (Oxford), 187*(1–2), 159–167. https://doi.org/10.1111/j.1748-1716.2006.01558.x.

Ares, G. R., Caceres, P. S., & Ortiz, P. A. (2011). Molecular regulation of NKCC2 in the thick ascending limb. *American Journal of Physiology - Renal Physiology, 301*(6), F1143–F1159.

Arroyo, J. P., Kahle, K. T., & Gamba, G. (2013). The SLC12 family of electroneutral cation-coupled chloride cotransporters. *Molecular Aspects of Medicine, 34*(2–3), 288–298. https://doi.org/10.1016/j.mam.2012.05.002.

Bazua-Valenti, S., Chavez-Canales, M., Rojas-Vega, L., Gonzalez-Rodriguez, X., Vazquez, N., Rodriguez-Gama, A., et al. (2015). The effect of WNK4 on the Na^+-Cl- cotransporter is modulated by intracellular chloride. *Journal of the American Society of Nephrology: JASN, 26*, 1781–1786. https://doi.org/10.1681/ASN.2014050470.

Bazua-Valenti, S., & Gamba, G. (2015). Revisiting the NaCl cotransporter regulation by with-no-lysine kinases. *American Journal of Physiology. Cell Physiology, 308*(10), C779–C791. https://doi.org/10.1152/ajpcell.00065.2015.

Begum, G., Yuan, H., Kahle, K. T., Li, L., Wang, S., Shi, Y., et al. (2015). Inhibition of WNK3 kinase signaling reduces brain damage and accelerates neurological recovery after stroke. *Stroke, 46*(7), 1956–1965. https://doi.org/10.1161/STROKEAHA.115.008939.

Bergaya, S., Faure, S., Baudrie, V., Rio, M., Escoubet, B., Bonnin, P., et al. (2011). WNK1 regulates vasoconstriction and blood pressure response to alpha 1-adrenergic stimulation in mice. *Hypertension, 58*(3), 439–445. https://doi.org/10.1161/HYPERTENSIONAHA.111.172429.

Blaesse, P., Airaksinen, M. S., Rivera, C., & Kaila, K. (2009). Cation-chloride cotransporters and neuronal function. *Neuron, 61*(6), 820–838.

Bonar, P. T., & Casey, J. R. (2008). Plasma membrane Cl(-)/HCO(3)(-) exchangers: Structure, mechanism and physiology. *Channels (Austin), 2*(5), 337–345.

Burg, M. B., Ferraris, J. D., & Dmitrieva, N. I. (2007). Cellular response to hyperosmotic stresses. *Physiological Reviews, 87*(4), 1441–1474. https://doi.org/10.1152/physrev.00056.2006.

Chang, M. H., Plata, C., Zandi-Nejad, K., Sindic, A., Sussman, C. R., Mercado, A., et al. (2009). Slc26a9–anion exchanger, channel and Na^+ transporter. *The Journal of Membrane Biology, 228*(3), 125–140. https://doi.org/10.1007/s00232-009-9165-5.

Cruz-Rangel, S., Gamba, G., Ramos-Mandujano, G., & Pasantes-Morales, H. (2012). Influence of WNK3 on intracellular chloride concentration and volume regulation in HEK293 cells. *Pflügers Archiv, 464*(3), 317–330. https://doi.org/10.1007/s00424-012-1137-4.

Darman, R. B., & Forbush, B. (2002). A regulatory locus of phosphorylation in the N terminus of the Na-K-Cl cotransporter, NKCC1. *Journal of Biological Chemistry, 277*(40), 37542–37550.

de los Heros, P., Alessi, D. R., Gourlay, R., Campbell, D. G., Deak, M., Macartney, T. J., et al. (2014). The WNK-regulated SPAK/OSR1 kinases directly phosphorylate and inhibit the K^+-Cl- co-transporters. *The Biochemical Journal, 458*(3), 559–573. https://doi.org/10.1042/BJ20131478.

de los Heros, P., Kahle, K. T., Rinehart, J., Bobadilla, N. A., Vazquez, N., San Cristobal, P., et al. (2006). WNK3 bypasses the tonicity requirement for K-Cl cotransporter activation via a phosphatase-dependent pathway. *Proceedings of the National Academy of Sciences of the United States of America, 103*, 1976–1981.

Delaloy, C., Elvira-Matelot, E., Clemessy, M., Zhou, X. O., Imbert-Teboul, M., Houot, A. M., et al. (2008). Deletion of WNK1 first intron results in misregulation of both isoforms in renal and extrarenal tissues. *Hypertension, 52*(6), 1149–1159.

Delpire, E., & Austin, T. M. (2010). Kinase regulation of Na^+-K+-2Cl- cotransport in primary afferent neurons. *Journal of Physiology, 588*(Pt 18), 3365–3373.

Delpire, E., & Gagnon, K. B. (2007). Genome-wide analysis of SPAK/OSR1 binding motifs. *Physiological Genomics, 28*(2), 223–231.

Delpire, E., & Gagnon, K. B. (2008). SPAK and OSR1: STE20 kinases involved in the regulation of ion homoeostasis and volume control in mammalian cells. *The Biochemical Journal, 409*(2), 321–331.

Dowd, B. F., & Forbush, B. (2003). PASK (proline-alanine-rich STE20-related kinase), a regulatory kinase of the Na-K-Cl cotransporter (NKCC1). *Journal of Biological Chemistry, 278*(30), 27347–27353.

Dunham, P. B., Klimczak, J., & Logue, P. J. (1993). Swelling activation of K-Cl cotransport in LK sheep erythrocytes: A three-state process. *The Journal of General Physiology, 101*(5), 733–765.

Flemmer, A. W., Gimenez, I., Dowd, B. F., Darman, R. B., & Forbush, B. (2002). Activation of the Na-K-Cl otransporter NKCC1 detected with a phospho-specific antibody. *Journal of Biological Chemistry, 277*(40), 37551–37558.

Gagnon, K. B., & Delpire, E. (2012). Molecular physiology of SPAK and OSR1: Two Ste20-related protein kinases regulating ion transport. *Physiological Reviews, 92*(4), 1577–1617. https://doi.org/10.1152/physrev.00009.2012.

Gamba, G. (2005a). Molecular physiology and pathophysiology of the electroneutral cation-chloride cotransporters. *Physiological Reviews, 85*(2), 423–493.

Gamba, G. (2005b). Role of WNK kinases in regulating tubular salt and potassium transport and in the development of hypertension. *American Journal of Physiology - Renal Physiology, 288*(2), F245–F252.

Gamba, G. (2009). The thiazide-sensitive Na^+-Cl- cotransporter: Molecular biology, functional properties, and regulation by WNKs. *American Journal of Physiology - Renal Physiology, 297*(4), F838–F848.

Gamba, G., Miyanoshita, A., Lombardi, M., Lytton, J., Lee, W. S., Hediger, M. A., et al. (1994). Molecular cloning, primary structure and characterization of two members of the mammalian electroneutral sodium-(potassium)-chloride cotransporter family expressed in kidney. *Journal of Biological Chemistry, 269*(26), 17713–17722.

Garzon-Muvdi, T., Schiapparelli, P., Ap Rhys, C., Guerrero-Cazares, H., Smith, C., Kim, D. H., et al. (2012). Regulation of brain tumor dispersal by NKCC1 through a novel role in focal adhesion regulation. *PLos Biology, 10*(5). https://doi.org/10.1371/journal.pbio.1001320. e1001320.

Gimenez, I., & Forbush, B. (2005). Regulatory phosphorylation sites in the N-terminus of the renal Na-K-Cl cotransporter (NKCC2). *American Journal of Physiology - Renal Physiology, 289*(6), F1341–F1345.

Greenman, C., Stephens, P., Smith, R., Dalgliesh, G. L., Hunter, C., Bignell, G., et al. (2007). Patterns of somatic mutation in human cancer genomes. *Nature, 446*(7132), 153–158.

Grinstein, S., Woodside, M., Sardet, C., Pouyssegur, J., & Rotin, D. (1992). Activation of the Na^+/H^+ antiporter during cell volume regulation. Evidence for a phosphorylation-independent mechanism. *Journal of Biological Chemistry, 267*(33), 23823–23828.

Hadchouel, J., Ellison, D. H., & Gamba, G. (2016). Regulation of renal electrolyte transport by WNK and SPAK-OSR1 kinases. *Annual Review of Physiology, 78*, 367–389. https://doi.org/10.1146/annurev-physiol-021115-105431.

Hebert, S. C., Mount, D. B., & Gamba, G. (2004). Molecular physiology of cation-coupled Cl(-) cotransport: The SLC12 family. *Pflugers Archives, 447*(5), 580–593.

Holden, S., Cox, J., & Raymond, F. L. (2004). Cloning, genomic organization, alternative splicing and expression analysis of the human gene WNK3 (PRKWNK3). *Gene, 335*, 109–119.

Humphreys, B. D., Jiang, L., Chernova, M. N., & Alper, S. L. (1995). Hypertonic activation of AE2 anion exchanger in Xenopus oocytes via NHE-mediated intracellular alkalinization. *American Journal of Physiology. Cell Physiology, 268*, C201–C209.

Jentsch, T. J., Lutter, D., Planells-Cases, R., Ullrich, F., & Voss, F. K. (2016). VRAC: Molecular identification as LRRC8 heteromers with differential functions. *Pflügers Archiv, 468*(3), 385–393. https://doi.org/10.1007/s00424-015-1766-5.

Kahle, K. T., Rinehart, J., De Los Heros, P., Louvi, A., Meade, P., Vazquez, N., et al. (2005). WNK3 modulates transport of Cl- in and out of cells: Implications for control of cell volume and neuronal excitability. *Proceedings of the National Academy of Sciences of the United States of America, 102*(46), 16783–16788.

Kahle, K. T., Rinehart, J., Giebisch, G., Gamba, G., Hebert, S. C., & Lifton, R. P. (2008). A novel protein kinase signaling pathway essential for blood pressure regulation in humans. *Trends in Endocrinology and Metabolism, 19*(3), 91–95.

Kahle, K. T., Rinehart, J., Ring, A., Gimenez, I., Gamba, G., Hebert, S. C., et al. (2006). WNK protein kinases modulate cellular Cl- flux by altering the phosphorylation state of the Na-K-Cl and K-Cl cotransporters. *Physiology, 21*, 326–335.

Kahle, K. T., Staley, K. J., Nahed, B. V., Gamba, G., Hebert, S. C., Lifton, R. P., et al. (2008). Roles of the cation-chloride cotransporters in neurological disease. *Nature Clinical Practice Neurology, 4*(9), 490–503.

Kaji, D. M., & Tsukitani, Y. (1991). Role of protein phosphatase in activation of KCl cotransport in human erithrocytes. *American Journal of Physiology. Cell Physiology, 260*, C178–C182.

Keicher, E., & Meech, R. (1994). Endogenous Na^{+}-K+ (or NH_4^{+})-2Cl- cotransport in Rana oocytes; anomalous effect of external NH_4^{+} on PHi. *Journal of Physiology, Paris, 475*, 45–57.

Koivusalo, M., Kapus, A., & Grinstein, S. (2009). Sensors, transducers, and effectors that regulate cell size and shape. *Journal of Biological Chemistry, 284*(11), 6595–6599. https://doi.org/10.1074/jbc.R800049200.

Koos, B., Christmann, J., Plettenberg, S., Kading, D., Becker, J., Keteku, M., et al. (2018). Hypertonicity-induced cation channels in HepG2 cells: Architecture and role in proliferation vs. apoptosis. *Journal of Physiology, 596*(7), 1227–1241. https://doi.org/10.1113/JP275827.

Lauf, P. K., & Adragna, N. C. (2000). K-Cl cotransport: Properties and molecular mechanism. *Cellular Physiology and Biochemistry: International Journal of Experimental Cellular Physiology, Biochemistry, and Pharmacology, 10*(5–6), 341–354.

Lenertz, L. Y., Lee, B. H., Min, X., Xu, B. E., Wedin, K., Earnest, S., et al. (2005). Properties of WNK1 and implications for other family members. *Journal of Biological Chemistry, 280*(29), 26653–26658.

McManus, M. L., Churchwell, K. B., & Strange, K. (1995). Regulation of cell volume in health and disease. *The New England Journal of Medicine, 333*(19), 1260–1266.

Melo, Z., de los Heros, P., Cruz-Rangel, S., Vazquez, N., Bobadilla, N. A., Pasantes-Morales, H., et al. (2013). N-terminal serine dephosphorylation is required for KCC3 full activation by cell swelling. *Journal of Biological Chemistry, 288*(44), 31468–31476. https://doi.org/10.1074/jbc.M113.475574.

Mendes, A. I., Matos, P., Moniz, S., Luz, S., Amaral, M. D., Farinha, C. M., et al. (2011). Antagonistic regulation of cystic Fibrosis transmembrane conductance regulator cell surface expression by protein kinases WNK4 and spleen tyrosine kinase. *Molecular and Cellular Biology, 31*(19), 4076–4086.

Moniz, S., Martinho, O., Pinto, F., Sousa, B., Loureiro, C., Oliveira, M. J., et al. (2013). Loss of WNK2 expression by promoter gene methylation occurs in adult gliomas and triggers Rac1-mediated tumour cell invasiveness. *Human Molecular Genetics, 22*(1), 84–95. https://doi.org/10.1093/hmg/dds405.

Moniz, S., Verissimo, F., Matos, P., Brazao, R., Silva, E., Kotevelets, L., et al. (2007). Protein kinase WNK2 inhibits cell proliferation by negatively modulating the activation of MEK1/ERK1/2. *Oncogene, 26*(41), 6071–6081. https://doi.org/10.1038/sj.onc.1210706.

Numata, T., Sato, K., Christmann, J., Marx, R., Mori, Y., Okada, Y., et al. (2012). The Delta C splice-variant of TRPM2 is the hypertonicity-induced cation channel in HeLa cells, and the ecto-enzyme CD38 mediates its activation. *Journal of Physiology, 590*(5), 1121–1138. https://doi.org/10.1113/jphysiol.2011.220947.

O'Reilly, M., Marshall, E., Speirs, H. J., & Brown, R. W. (2003). WNK1, a gene within a novel blood pressure control pathway, tissue-specifically generates radically different isoforms with and without a kinase domain. *Journal of the American Society of Nephrology, 14*(10), 2447–2456.

Ohno, M., Uchida, K., Ohashi, T., Nitta, K., Ohta, A., Chiga, M., et al. (2011). Immunolocalization of WNK4 in mouse kidney. *Histochemistry and Cell Biology, 136*(1), 25–35.

Okada, Y. (2004). Ion channels and transporters involved in cell volume regulation and sensor mechanisms. *Cell Biochemistry and Biophysics, 41*(2), 233–258. https://doi.org/10.1385/CBB:41:2:233.

Pacheco-Alvarez, D., & Gamba, G. (2011). WNK3 is a putative chloride-sensing kinase. *Cellular Physiology and Biochemistry: International Journal of Experimental Cellular Physiology, Biochemistry, and Pharmacology, 28*(6), 1123–1134.

Pacheco-Alvarez, D., San Cristobal, P., Meade, P., Moreno, E., Vazquez, N., Munoz, E., et al. (2006). The Na-Cl cotransporter is activated and phosphorylated at the amino terminal domain upon intracellular chloride depletion. *Journal of Biological Chemistry, 281*(39), 28755–28763.

Pacheco-Alvarez, D., Vazquez, N., Castaneda-Bueno, M., de los Heros, P., Cortes-Gonzalez, C., Moreno, E., et al. (2012). WNK3-SPAK interaction is required for the modulation of NCC and other members of the SLC12 family. *Cellular Physiology and Biochemistry: International Journal of Experimental Cellular Physiology, Biochemistry, and Pharmacology, 29*(1–2), 291–302.

Pedersen, S. F., Owsianik, G., & Nilius, B. (2005). TRP channels: An overview. *Cell Calcium, 38*(3–4), 233–252. https://doi.org/10.1016/j.ceca.2005.06.028.

Piala, A. T., Moon, T. M., Akella, R., He, H., Cobb, M. H., & Goldsmith, E. J. (2014). Chloride sensing by WNK1 involves inhibition of autophosphorylation. *Science Signaling [electronic Resource], 7*(324), ra41. https://doi.org/10.1126/scisignal.2005050.

Piechotta, K., Garbarini, N., England, R., & Delpire, E. (2003). Characterization of the interaction of the stress kinase SPAK with the Na^+-K+-2Cl- cotransporter in the nervous system: Evidence for a scaffolding role of the kinase. *Journal of Biological Chemistry, 278*(52), 52848–52856.

Piechotta, K., Lu, J., & Delpire, E. (2002). Cation chloride cotransporters interact with the stress-related kinases Ste20-related proline-alanine-rich kinase (SPAK) and oxidative stress response 1 (OSR1). *Journal of Biological Chemistry, 277*(52), 50812–50819.

Ponce-Coria, J., San Cristobal, P., Kahle, K. T., Vazquez, N., Pacheco-Alvarez, D., de los Heros, P., et al. (2008). Regulation of NKCC2 by a chloride-sensing mechanism involving the WNK3 and SPAK kinases. *Proceedings of the National Academy of Sciences of the United States of America, 105*(24), 8458–8463.

Richardson, C., & Alessi, D. R. (2008). The regulation of salt transport and blood pressure by the WNK-SPAK/OSR1 signalling pathway. *Journal of Cell Sciences, 121*(Pt 20), 3293–3304.

Richardson, C., Rafiqi, F. H., Karlsson, H. K., Moleleki, N., Vandewalle, A., Campbell, D. G., et al. (2008). Activation of the thiazide-sensitive Na^+-Cl- cotransporter by the WNK-regulated kinases SPAK and OSR1. *Journal of Cell Sciences, 121*(Pt 5), 675–684.

Richardson, C., Sakamoto, K., De Los Heros, P., Deak, M., Campbell, D. G., Prescott, A. R., et al. (2011). Regulation of the NKCC2 ion cotransporter by SPAK-OSR1-dependent and -independent pathways. *Journal of Cell Sciences, 124*(Pt 5), 789–800.

Rinehart, J., Kahle, K. T., De Los Heros, P., Vazquez, N., Meade, P., Wilson, F. H., et al. (2005). WNK3 kinase is a positive regulator of NKCC2 and NCC, renal cation-Cl-cotransporters required for normal blood pressure homeostasis. *Proceedings of the National Academy of Sciences of the United States of America, 102*(46), 16777–16782.

Rinehart, J., Maksimova, Y. D., Tanis, J. E., Stone, K. L., Hodson, C. A., Zhang, J., et al. (2009). Sites of regulated phosphorylation that control K-Cl cotransporter activity. *Cell, 138*(3), 525–536.

Rinehart, J., Vazquez, N., Kahle, K. T., Hodson, C. A., Ring, A. M., Gulcicek, E. E., et al. (2011). WNK2 is a novel regulator of essential neuronal cation-chloride cotransporters. *Journal of Biological Chemistry, 286*(34), 30171–30180.

Roy, A., Goodman, J. H., Begum, G., Donnelly, B. F., Pittman, G., Weinman, E. J., et al. (2015). Generation of WNK1 knockout cell lines by CRISPR/Cas-mediated genome editing. *American Journal of Physiology - Renal Physiology, 308*(4), F366–F376. https://doi.org/10.1152/ajprenal.00612.2014.

Russell, J. M. (2000). Sodium-potassium-chloride cotransport. *Physiological Reviews, 80*(1), 211–276.

Schultheis, P. J., Clarke, L. L., Meneton, P., Miller, M. L., Soleimani, M., Gawenis, L. R., et al. (1998). Renal and intestinal absorptive defects in mice lacking the NHE3 Na^+/H^+ exchanger. *Nature Genetics, 19*(3), 282–285.

Shekarabi, M., Zhang, J., Khanna, A. R., Ellison, D. H., Delpire, E., & Kahle, K. T. (2017). WNK kinase signaling in ion homeostasis and human disease. *Cell Metabolism, 25*(2), 285–299. https://doi.org/10.1016/j.cmet.2017.01.007.

Sjoblom, T., Jones, S., Wood, L. D., Parsons, D. W., Lin, J., Barber, T. D., et al. (2006). The consensus coding sequences of human breast and colorectal cancers. *Science, 314*(5797), 268–274.

Soleimani, M., Barone, S., Xu, J., Shull, G. E., Siddiqui, F., Zahedi, K., et al. (2012). Double knockout of pendrin and Na-Cl cotransporter (NCC) causes severe salt wasting, volume depletion, and renal failure. *Proceedings of the National Academy of Sciences of the United States of America, 109*(33), 13368–13373. https://doi.org/10.1073/pnas.1202671109.

Stauber, T. (2015). The volume-regulated anion channel is formed by LRRC8 heteromers - molecular identification and roles in membrane transport and physiology. *Biological Chemistry, 396*(9–10), 975–990. https://doi.org/10.1515/hsz-2015-0127.

Strange, K. (2004). Cellular volume homeostasis. *Advances in Physiology Education, 28*(1–4), 155–159. https://doi.org/10.1152/advan.00034.2004.

Stutzin, A., & Hoffmann, E. K. (2006). Swelling-activated ion channels: Functional regulation in cell-swelling, proliferation and apoptosis. *Acta Physiologica (Oxford), 187*(1–2), 27–42. https://doi.org/10.1111/j.1748-1716.2006.01537.x.

Terker, A. S., Zhang, C., Erspamer, K. J., Gamba, G., Yang, C. L., & Ellison, D. H. (2016). Unique chloride-sensing properties of WNK4 permit the distal nephron to modulate potassium homeostasis. *Kidney International, 89*, 127–134. https://doi.org/10.1038/ki.2015.289.

Thastrup, J. O., Rafiqi, F. H., Vitari, A. C., Pozo-Guisado, E., Deak, M., Mehellou, Y., et al. (2012). SPAK/OSR1 regulate NKCC1 and WNK activity: Analysis of WNK isoform interactions and activation by t-loop trans-autophosphorylation. *The Biochemical Journal, 441*(1), 325–337.

Verissimo, F., & Jordan, P. (2001). WNK kinases, a novel protein kinase subfamily in multicellular organisms. *Oncogene, 20*(39), 5562–5569.

Verissimo, F., Silva, E., Morris, J. D., Pepperkok, R., & Jordan, P. (2006). Protein kinase WNK3 increases cell survival in a caspase-3-dependent pathway. *Oncogene, 25*(30), 4172–4182. https://doi.org/10.1038/sj.onc.1209449.

Vidal-Petiot, E., Cheval, L., Faugeroux, J., Malard, T., Doucet, A., Jeunemaitre, X., et al. (2012). A new methodology for quantification of alternatively spliced exons reveals a highly tissue-specific expression pattern of WNK1 isoforms. *PLoS One, 7*(5). https://doi.org/10.1371/journal.pone.0037751. e37751.

Villa, F., Goebel, J., Rafiqi, F. H., Deak, M., Thastrup, J., Alessi, D. R., et al. (2007). Structural insights into the recognition of substrates and activators by the OSR1 kinase. *EMBO Reports, 8*(9), 839–845.

Vitari, A. C., Deak, M., Morrice, N. A., & Alessi, D. R. (2005). The WNK1 and WNK4 protein kinases that are mutated in Gordon's hypertension syndrome, phosphorylate and active SPAK and OSR1 protein kinases. *The Biochemical Journal, 391*(Pt. 1), 17–24.

Voss, J. W., Pedersen, S. F., Christensen, S. T., & Lambert, I. H. (2004). Regulation of the expression and subcellular localization of the taurine transporter TauT in mouse NIH3T3 fibroblasts. *European Journal of Biochemistry, 271*(23–24), 4646–4658. https://doi.org/10.1111/j.1432-1033.2004.04420.x.

Wehner, F., Bondarava, M., ter Veld, F., Endl, E., Nurnberger, H. R., & Li, T. (2006). Hypertonicity-induced cation channels. *Acta Physiologica (Oxford), 187*(1–2), 21–25. https://doi.org/10.1111/j.1748-1716.2006.01561.x.

Wehner, F., & Olsen, H. (2004). Hypertonicity-induced cation channels in rat hepatocytes and their intracellular regulation. *Advances in Experimental Medicine and Biology, 559*, 253–261.

Welling, P. A. (2016). Roles and regulation of renal K channels. *Annual Review of Physiology, 78*, 415–435. https://doi.org/10.1146/annurev-physiol-021115-105423.

Wilson, F. H., Disse-Nicodeme, S., Choate, K. A., Ishikawa, K., Nelson-Williams, C., Desitter, I., et al. (2001). Human hypertension caused by mutations in WNK kinases. *Science, 293*(5532), 1107–1112.

Wilson, F. H., Kahle, K. T., Sabath, E., Lalioti, M. D., Rapson, A. K., Hoover, R. S., et al. (2003). Molecular pathogenesis of inherited hypertension with hyperkalemia: The Na-Cl cotransporter is inhibited by wild-type but not mutant WNK4. *Proceedings of the National Academy of Sciences of the United States of America, 100*(2), 680–684.

Xu, B., English, J. M., Wilsbacher, J. L., Stippec, S., Goldsmith, E. J., & Cobb, M. H. (2000). WNK1, a novel mammalian serine/threonine protein kinase lacking the catalytic lysine in subdomain II. *Journal of Biological Chemistry, 275*(22), 16795–16801.

Xu, B. E., Min, X., Stippec, S., Lee, B. H., Goldsmith, E. J., & Cobb, M. H. (2002). Regulation of WNK1 by an autoinhibitory domain and autophosphorylation. *Journal of Biological Chemistry, 277*(50), 48456–48462.

Yang, C. L., Angell, J., Mitchell, R., & Ellison, D. H. (2003). WNK kinases regulate thiazide-sensitive Na-Cl cotransport. *Journal of Clinical Investigation, 111*(7), 1039–1045.

Zagorska, A., Pozo-Guisado, E., Boudeau, J., Vitari, A. C., Rafiqi, F. H., Thastrup, J., et al. (2007). Regulation of activity and localization of the WNK1 protein kinase by hyperosmotic stress. *Journal of Cell and Biology, 176*(1), 89–100.

Zhang, J., Gao, G., Begum, G., Wang, J., Khanna, A. R., Shmukler, B. E., et al. (2016). Functional kinomics establishes a critical node of volume-sensitive cation-Cl- cotransporter regulation in the mammalian brain. *Scientific Reports, 6*, 35986. https://doi.org/10.1038/srep35986.

CHAPTER EIGHT

Intracellular Macromolecules in Cell Volume Control and Methods of Their Quantification

Michael A. Model*[,1] and Jonathan C. Petruccelli[§]
*Kent State University, Kent, OH, United States
[§]University at Albany, Albany, NY, United States
[1]Corresponding author: E-mail: mmodel@kent.edu

Contents

Abstract

The traditional theories of cell volume regulation focus on monovalent ions and small organic osmolytes. The main subject of this review is macromolecular content of the cell and its role in cell volume. We start by reviewing general information about cellular macromolecules and present some quantitative relationships. Next, we review a wide range of methods for measuring intracellular macromolecular concentration and related

Current Topics in Membranes, Volume 81
ISSN 1063-5823
https://doi.org/10.1016/bs.ctm.2018.06.002

parameters; in particular, a large section is devoted to the so-called quantitative phase imaging methods based on transmission light microscopy. In the last part, we discuss three specific biological examples where quantitative analysis of macromolecular concentrations is expected to generate valuable insights into biological processes: the biology of organelles, long-term cell volume maintenance and apoptotic volume decrease.

1. INTRODUCTION

Much of the research on cell volume regulation has been focusing on the relationships between the total cell volume (V_{cell}) and osmotically most significant ions: potassium, sodium, and chloride. Small organic molecules (A_S) with molecular weights in the range 100−200 are the other recognized player in osmotic balance: most of them are membrane-impermeant under normal conditions, but some, such as urea, taurine, betaine or inositol, can sometimes cross the membrane and contribute to volume maintenance. In this review, we take a closer look at a less discussed factor in cell volume regulation − large macromolecules, such as proteins and nucleic acids (A_L). Their direct contribution to intracellular osmolarity is modest, but they provide the bulk of "fixed" electric charges; because these charges are mostly negative at biological pH, they must be balanced by an excess of intracellular cations over anions, thus indirectly affecting V_{cell}. In addition, macromolecules occupy a significant part of the cell volume, leaving less room for water. We will keep referring, more or less interchangeably, to three quantities: the total organic matter A_T; its macromolecular portion A_L, and intracellular water that occupies volume V_W. A_T and A_L are similar in terms of mass because the cumulative mass of A_S is relatively small (see below); V_W can be viewed as the difference between V_{cell} and the volume occupied by A_T (or A_L). The ratio V_W/V_{cell} will be designated as cell water content W. We will discuss the proven, as well as anticipated, benefits of quantifying A_L (A_T, V_W) and describe the available techniques for their measurement.

This paper is not intended to serve as a comprehensive review but partly reflects the authors' personal interests and expertise. For example, a large section is devoted to optical (in particular, quantitative phase) methods, but magnetic resonance techniques are omitted. While we discuss apoptotic shrinkage at considerable length, we skip non-mammalian cells and organisms. Likewise, we do not delve into molecular signaling, which has been addressed in numerous and extensive reviews by other authors. Certain opinions and statements may appear controversial and speculative to some readers, but part of our goal is to call attention to hitherto neglected possibilities.

The amount of intracellular water V_W is, in many cases, the main determinant of the total cell volume V_{cell}. The importance, as well as measurement methods of V_{cell}, have been reviewed before (Hoffmann, Lambert, & Pedersen, 2009; Lang, 2007; Model, 2018; Orlov, Platonova, Hamet, & Grygorczyk, 2013). V_{cell} would be a natural parameter, for example, for describing cell growth (unless one is interested in possible variations in cell water content during the cell cycle). It is equally useful in measurements of membrane water permeability following an abrupt anisosmotic shock, when rapid volume changes occur almost entirely by water redistribution. There are situations, however, where V_W and V_{cell} have distinct roles and should be considered separately. Below are some examples.

(1) The apoptotic volume decrease (AVD) is a highly reproducible feature of apoptotic cell death and, arguably, the most essential difference between inflammation-associated necrosis and often asymptomatic apoptosis (Bortner & Cidlowski, 2002). The AVD has two aspects of very different nature: dehydration and fragmentation. In the vast majority of publications concerning AVD, the entire volume loss has been attributed to dehydration. This is not necessarily true in all cases, and to distinguish between these two routes (which may involve different cellular processes), intracellular dry matter must be quantified along with V_{cell}. (2) Cell volume recovery after an initial shrinkage or swelling in an anisosmotic solution is a fundamental property of many cell types (Lang et al., 1998). However, it has been studied mostly in short-term experiments, presumably to avoid complications resulting from cell growth. Quantification of A_L would remove this difficulty and may reveal new aspects of osmotic adaptation (Mudrak, Rana, & Model, 2018). (3) Some parts of the cell and organelles may have different refractive indices than the cytosol, suggesting different protein concentrations and different water content. Because stable water gradients within a cell are unlikely, this creates a situation that calls for an explanation. These three examples will be discussed in more detail later in this chapter.

We only briefly mention some other applications where the ability to quantify A_L could be beneficial. (4) Some pathological conditions are characterized by abnormal cell volume. Some of them result from disruption of water balance, such as edema, and others from abnormal amount of protein, such as hepatomegaly (Baraona, Leo, Borowsky, & Lieber, 1975). It is quite possible, however, that when a condition persists, both water and protein may be affected, and a single parameter of cell volume would fail to pinpoint the origin of the pathological change. (5) Significant changes in refractive index have been observed in various types of cancer and infection (Liu et al., 2016). These observations hold promise for diagnostics and for revealing new aspects of the disease. (6) Autophagy results in cell volume decrease due to protein digestion (Hosokawa, Hara, & Mizushima, 2007);

however, the cell size and the amount of protein in the cell are regulated separately (Miettinen & Björklund, 2015), and measurements of water alone may only tell part of the story. (7) Macromolecular crowding is believed to affect enzymatic activity and cellular metabolism (Minton, 2001; Ovádi & Saks, 2004), but its most defining parameter, the protein concentration, cannot be obtained without appropriate tools. (8) Removal of cellular water is critical for survival of cryopreserved cells, and cell water behavior is an important direction in cryopreservation research (Baust, Gao, & Baust, 2009). It is quite possible that measurements of A_L can supply additional information to what can be learned from V_{cell}. (9) Water regulation in plants is obviously the subject of enormous practical importance (Kramer, 1988). The potential of optical techniques for water determination in plant physiology has yet to be explored. (10) Even in cases when volume measurements might be adequate to address the problem at hand, the mass/volume concentration of $\mathbf{A_L}$ varies within a much narrower range, and analysis of only 10–20 cells may generate statistically significant data (Grover et al., 2011; Model & Schonbrun, 2013; Bryan et al., 2014; Mudrak et al., 2018; Model et al., 2018).

2. THE COMPONENTS OF CELL VOLUME AND THE MAIN RELATIONSHIPS

In considering the osmotic balance between the cell and environment, it is customary to present the organic matter as a single entity with total molecular concentration $[A_T]$:

$$[K] + [Na] + [Cl] + [A_T] = P \quad (1)$$

where P stands for the osmolarity of extracellular medium (we are making no distinction between osmolarity and osmolality). Here, and for the rest of the text, square brackets denote concentrations, bold font is used when concentrations are expressed in mass/volume units, and the same symbols, when used in equations without brackets, denote the total mass of the substance. The excess of P ($\sim$290 mM in extracellular fluid) over the sum $[K] + [Na] + [Cl]$ is usually greater than 100 mM, providing a rough estimate for $[A_T]$. In prokaryotes, the concentrations of ions and metabolites are substantially higher than in mammalian cells (Bennett et al., 2009; Theillet et al., 2014).

A word of caution would be in place here. The more rigorous form of Eq. (1) would include activities rather than molar concentrations, and ion activities can be strongly affected by the presence of concentrated protein (Ocon, Acerete, & Reboiras, 1987; Pfister & Pauly, 1972). Because this effect is difficult to account for quantitatively, we follow the tradition of presenting Eq. (1) in a simplified form. In our defense, there are numerous examples of successful modeling of the cell osmotic behavior based on such simplified theory (e.g., Armstrong, 2003; Fraser & Huang, 2004; Lew & Bookchin, 1986; Terashima et al., 2006; Vereninov et al., 2007).

The complementary condition of electric neutrality assigns an average negative charge z_{av} to organic molecules, producing the condition

$$[K] + [Na] = [Cl] + z_{av}[A_T] \tag{2}$$

It follows from Eqs. (1) and (2) that z_{av} has to be close or slightly larger than 1.

Although all intracellular organic matter has been lumped into a single quantity A_T, this does not accurately reflect the reality. About 60–75% of the dry weight of a typical mammalian cell is due to proteins, and nucleic acids contribute to it 5%–10% (Feijó Delgado et al., 2013; Canepa et al., 2002; https://www.thermofisher.com/us/en/home/references/ambion-tech-support/rna-tools-and-calculators/macromolecular-components-of-e.HTML). About a half of the protein is present in solution, the other half being structural (Luby-Phelps, 2000). Low-molecular compounds A_S can be cell-specific. For example, the prominent osmolytes of the rat muscle are phosphocreatine (37 mM) and anserine (30 mM) (Conway, 1957). Molecular concentrations of amino acids range from 4 mM in the red blood cells to 15 mM in the rat muscle (Burton, 1983; Conway, 1957) to 35 mM in granulocytes or in the brain (Canepa et al., 2002; Law, 1991). Intracellular lactate can reach 20–40 mM in cultures cells (Schmid & Blanch, 1992). Among nucleotides, ATP is the most abundant, with concentration in the low mM range (Burton, 1983; Traut, 1994).

If we take the average protein size as 40 KDa (Brocchieri & Karlin, 2005) and its intracellular concentration as 0.2 g/mL, the molecular concentration of protein turns out to be only a few mM. To obtain osmolarity, this number may have to be multiplied several fold due to non-ideal behavior of proteins at concentrations approaching intracellular (Maughan & Godt, 2001; Minton, 2008; Vilker, Colton, & Smith, 1981). On the other hand, many proteins, such as those forming cytoskeleton, are present in complexes, which may reduce their effective osmolarity. In any event, it is clear that most of the osmotic contribution from organic compounds comes from small molecules.

The situation with electric charges is a little more complicated. In the osmotic and electric balance compiled by Conway (1957), only about 50% of intracellular negative charge in the rat muscle is associated with proteins at pH 7; much of the remaining charge comes from phosphocreatine and anserine. Many other cell types contain less potassium and more chloride than does the muscle, so it is possible that in those cells, the contribution of large macromolecules A_L (proteins and nucleic acids) to the overall negative charge is greater than in the muscle. Thus, we can slightly modify the above equations by separating the most osmotically important and electrically important organic compounds:

$$[K] + [Na] + [Cl] + [A_S] \approx P \tag{3}$$

$$[K] + [Na] \approx [Cl] + z_L[A_L] \tag{4}$$

By substituting [K] + [Na] from Eq. (3) into Eq. (4), we obtain an equation for a balanced cell state:

$$P - 2[Cl] \approx [A_S] + z_L[A_L] \tag{5}$$

(Alternatively, we could have replaced chloride with the sum of sodium and potassium). Unlike z_{av} in Eq. (2), z_L can be closer to 20–30, and therefore, both terms on the right side of the equation are comparable and must be kept. By replacing the concentrations of organic osmolytes through their total molar amounts A_S and A_L divided by the volume of intracellular water, an expression for V_w follows:

$$V_w = \frac{A_S + z_L A_L}{P - 2[Cl]} = \frac{A_S - z_L A_L}{P - 2([K] + [Na])} \tag{6}$$

Eq. (6) is slightly different from the standard formula

$$V_{cell} = \frac{(1 + z_{av})A_T}{P - 2[Cl]} = \frac{(1 - z_{av})A_T}{P - 2([K] + [Na])} \tag{7}$$

which deals with averaged characteristics of the cell organic material (Fraser & Huang, 2007; Jakobsson, 1980; Kay, 2017). While Eq. (7) is more inclusive in that it does not neglect any components, Eq. (6) emphasizes the different roles played by small and large organic molecules, which are measured separately and by different techniques. The other difference is the use of V_w rather than V_{cell}; because of the aforementioned difficulty with interpreting intracellular activities, this difference is probably inessential, and the choice between V_{cell} and V_w is dictated by what constitutes the main interest.

Although the main subject of this chapter is organic matter, we also mention the role of small inorganic ions because this will be important for understanding the problem of the AVD. It follows from Eqs. (5) and (6) that, as long as organic mass and intracellular pH remain constant, either chloride or the sum of cations uniquely determine cell volume at a balanced state. Because z and A are usually unknown, only relative changes in ion concentrations carry an easily interpretable information:

$$\frac{V_{w2}}{V_{w1}} = \frac{W_2(1 - W_1)}{W_1(1 - W_2)} = \frac{P - 2[Cl]_1}{P - 2[Cl]_2} = \frac{P - 2([K]_1 + [Na]_1)}{P - 2([K]_2 + [Na]_2)} \tag{8}$$

This equation sets an upper limit on the possible relative decrease in V_W (and therefore V_{cell}) caused by ion redistribution alone: it expresses constancy of osmolarity with the additional requirement that changes in chloride must be matched by cations. If the initial chloride concentration is $[Cl]_1$, then its complete depletion would shrink the cell by the factor of

$$\frac{V_2}{V_1} = 1 - \frac{2[Cl]_1}{P} \tag{9}$$

For example, a cell with normal concentration of intracellular chloride of 30 mM cannot shrink by more than 10% in a medium with 300 mOsm/L due to an efflux of inorganic ions; a greater relative volume decrease would be a sign that other mechanisms are at play. There is no similar limit on maximal swelling.

As one would expect, V_{cell} is proportional to the amount of dry matter A (in molar units), given everything else the same. Thus, phosphocreatine and lactate metabolism may have a significant impact on the amount of cell water through A_S (Kemp, 2007). It is less clear whether processes like protein degradation (which is supposed to increase the number of molecules) lead in practice to measurable swelling; such events might be more readily observable on the organelle level. Some research has been done on the effect of the average macromolecular charge z. This parameter is a strong function of pH: as pH shifts from 7.0 to 6.0, z can be reduced by a half (Conway, 1957; Vilker et al., 1981). Its effect on the cell volume has been demonstrated experimentally by Fraser, Middlebrook, Usher-Smith, Schwiening, and Huang (2005) and Usher-Smith, Fraser, Bailey, Griffin, and Huang (2006) and further investigated theoretically by Kemp (2007). However, other authors did not come to the same conclusions (de Paoli, Overgaard, Pedersen, & Nielsen, 2007). This is not surprising because, depending on the available membrane transporters, volume reduction through a decrease in z may be masked by the opposite effect of acidification on the Na/H exchange, causing cell swelling (Grinstein & Foskett, 1990; Kimelberg, 2005), or, conversely, by stimulation of the K/Cl transport, favoring cell shrinkage (Lew, Freeman, Ortiz, & Bookchin, 1991).

If one wishes to convert dry mass concentration into the units of cell water content W, this can be done be presenting cell volume as a sum of the volumes occupied by water (V_W) and dry mass (V_A):

$$V_{cell} = V_w + V_A = V_w + \frac{[\mathbf{A_T}]V_{cell}}{d_A} \tag{10}$$

Here d_A is the density of dry mass (in solution) and the concentration is expressed in mass, rather than in molar, units. Assuming a typical value $d_A \approx$ 1.4 g/mL, an expression for W results:

$$W = V_W/V_{cell} = 1 - [\mathbf{A_T}]/d_A \approx 1 - 0.7[\mathbf{A_T}] \tag{11}$$

Although dry density is not constant but depends on the DNA/protein ratio and on the cell metabolic and proliferative state (Feijó Delgado et al., 2013), the resulting uncertainty in W should not exceed 1%–2%.

The other frequently used parameters are the ratio of water to total mass and the average cell density. The bulk of information on the amount of intracellular water has been provided by drying and weighing experiments. The quantity directly measurable by this method, the mass fraction of water W_m, is slightly different from its volume fraction W: they are related as

$$\begin{aligned} W_m &= \frac{\text{Mass of water}}{\text{Mass of cell}} = \frac{V_W d_W}{V_W d_W + (V_{cell} - V_W) d_A} \\ &= \frac{W}{W - (1 - W) d_A} = \frac{W}{1.4 - 0.4W} \end{aligned} \tag{12}$$

This could be the reason why commonly reported values W_m for cell water content are slightly lower than those calculated from Eq. (11).

The denominator of Eq. (12) expresses cell density. Thus, all four parameters, $[\mathbf{A_T}]$, W, W_m, and d_{cell}, are interconvertible and carry nearly identical information:

$$d_{cell} = 1.4 - 0.4W = 1 + 0.28[\mathbf{A_T}] = \frac{1.4}{1 + 0.4W_m} \tag{13}$$

3. METHODS FOR MEASUREMENT OF W, A_T, AND A_S (NON-QPI)

3.1 Non-optical Methods

Drying and weighing. The drying method for determination of W_m is rather straightforward when applied to tissues; the practical problem with applying it to cell suspensions is the need to correct for the amount of solvent trapped between the cells. This can be done by adding a radioactive membrane-impermeant probe to the medium; the experimental details can be found elsewhere (Bratbak & Dundas, 1984; Kubitschek & Friske, 1986, Savitz, Sidel, & Solomon, 1964). Similar methods can be used for cell monolayers (Vasquez, Ishibashi, & Howard, 1982).

Buoyant density. Cell separation by density centrifugation is an old biological technique (Sharpe, 1988); in particular, it has been used to investigate changes in cell hydration during apoptosis (Cohen, Sun, Snowden, Dinsdale, & Skilleter, 1992; Patterson, Grossman, D'Andrea, & Latter, 1995; Wyllie & Morris, 1982; Yamada & Ohyama, 1988; Yurinskaya et al., 2005). It has been estimated that a 10% change in the water/protein ratio results in an easily measurable separation in gradient tubes (Yurinskaya et al., 2005). Microfluidic systems for measuring cell density, volume and mass have been constructed in the laboratory of S. Manalis (Bryan et al., 2014; Feijó Delgado et al., 2013; Grover et al., 2011). Cell are passed in a microfluidic channel over a vibrating cantilever, whose resonant frequency changes according to the cell's buoyant mass. Next, the same cell is measured second time in a liquid with a different density. These two measurements allow calculation of the cell density, volume, and mass. Instruments for mass measurement of particles operating on this principle are available from Affinity Biosensors (Santa Barbara, CA).

Chemical assays for small osmolytes. The diverse group of low-molecular osmolytes A_S can be quantified by mass spectrometry and related techniques (e.g., Cocuron, Tsogtbaatar, & Alonso, 2017; Schneider et al., 2009). We are not aware of any systematic application of modern metabolomics to problems related to cell volume regulation. On the other hand, the much simpler ninhydrin colorimetric reaction for amines, such as in amino acids and peptides (Friedman, 2004), has been frequently used to investigate cell responses to osmotic challenges. Ninhydrin assay is performed on extracts of cells or whole tissues. There have been numerous reports of the use of ninhydrin test to study see animals or bacteria (e.g., Darling, Burrows,

& Blum, 1990; Shumway & Gabbott, 1977), but not so much for cells of mammalian origin (Alfieri et al., 2002; Fumarola, Zerbini, & Guidotti, 2001; Hendil & Hoffmann, 1974; Law & Turner, 1987; Poulsen et al., 2010). Despite a semiquantitative nature of the test (different amino compounds respond differently), ninhydrin staining has produced consistent data supporting the active role of small osmolytes in cell volume maintenance.

We are not qualified to discuss nuclear magnetic resonance (Hoffman, Kumar, Spitzer, & Gupta, 1986; Kreis, Ernst, & Ross, 1993; Riqing & Nanming, 1997; Shinar & Navon, 1985), electron spin resonance (Hagedorn et al., 1997), or electron microscopy (Terryn, Michel, Kilian, Bonhomme, & Balossier, 2000), and refer the reader to the original literature.

3.2 Optical Methods: Fluorescence

Protein staining. Relative changes in A_T can be determined by covalent staining of intracellular amine groups. CellTracker Probes (ThermoFisher Scientific) have the advantage of low pH sensitivity in the neutral range, unlike the popular carboxyfluorescein succinimidyl ester (CFSE).

Ion concentration. As follows from Eq. (8), relative changes in W due to water redistribution under balanced conditions (i.e., *not* in osmotic shock experiments) are related to concentrations of either chloride or of the sum of cations. Chloride has indeed been used as a water permeability reporter (Baumgart, Rossi, & Verkman, 2012; Esteva-Font, Phuan, Anderson, & Verkman, 2013). Our recent results with potassium probe Asante Potassium Green-2 (TEFLabs, Austin, TX) indicate that it does not discriminate between potassium and sodium and acts as a cation sensor in its intracellular form (Rana et al., Submitted).[a] Its utility for measuring changes in W has yet to be investigated.

Quenching of fluorophores by proteins. Calcein is a popular reagent for measuring relative volume changes; however, its sensitivity to volume comes apparently from quenching by cytosolic proteins (Davidson & Higgins, 2013; Solenov, Watanabe, Manley, & Verkman, 2004). Likewise, the yellow fluorescent protein is quenched by high density of protein in solution (Morikawa et al., 2016). These agents can be used for detecting relative changes in intracellular protein density.

[a] A similar dye, Asante Potassium Green-4, is selective for potassium.

Crowding sensors. Some large molecules tend to assume a more compact shape in a crowded environment; if distant parts of such a molecule carry two fluorophores which form a FRET pair, their proximity can be estimated by FRET analysis. One such sensor was designed in the form of polyethylene glycol labeled with Atto 488 and Atto 565 on the opposite ends; it had to be put into cells by injection (Gnutt et al, 2015, 2017). Although FRET was found to increase during osmotically induced cell compression, it was also sensitive to the nature of the crowding agent (e.g., cytosol vs. DNA vs. BSA *in vitro*), and not just to its concentration.

Crowding sensors based on fluorescent proteins composed of CFP and YFP parts have also been constructed and tested in bacteria and mammalian cells (Boersma, Zuhorn, & Poolman, 2015; Germond, Fujita, Ichimura, & Watanabe, 2016; Morikawa et al., 2016). Their responses often correlated with data obtained by other methods. Some FRET images showed large variations between cells (Morikawa et al., 2016), as well as within individual cells (Machiyama, Morikawa, Okamoto, Watanabe, & Fujita, 2017). It is not immediately clear if any additional information can be gathered from heterogeneity of FRET.

Diffusion. Measurements of diffusion provide indirect assessment of protein concentration. Densely packed proteins impede long-range movements of molecules and also increase the viscosity of intracellular water. The first of these effects can be detected by measuring the translational diffusion rate of fluorescent probes by fluorescence recovery after photobleaching (FRAP) or by fluorescence correlation spectroscopy (FCS), and the second — from rotational diffusion using time-resolved fluorescence anisotropy (Luby-Phelps, 2000; Puchkov, 2013; Soleimaninejad, Chen, Lou, Smith, & Hong, 2017; Verkman, 2002). Fluorescent probes with viscosity-sensitive spectral shifts have been synthesized as well (Kuimova et al., 2009; Liu et al., 2013).

Experiments have confirmed the strong effect of osmotic volume changes on the diffusion of fluorescent probes (Guigas, Kalla, & Weiss, 2007; Kao, Abney, & Verkman, 1993; Miermont et al., 2013; Soleimaninejad et al., 2017; Swaminathan, Hoang, & Verkman, 1997) and showed an increase in viscosity during apoptotic shrinkage (Kuimova et al., 2009). However, as a tool for *measuring* protein density, these methods are not quantitative enough: there is no simple and universal relationship between diffusion coefficients and protein concentration, and intracellular viscosity can be affected by interactions between protein molecules (Yadav, Shire, & Kalonia, 2011).

Fluorescence and refractive index. The presence of organic compounds increases the refractive index of solution in a linear manner:

$$n = n_0 + \gamma c \tag{14}$$

where n_0 is the refractive index of the solvent, c is the mass/volume concentration of the solute and γ is the refractive index increment, specific for each chemical. It has been found that, for most biological molecules, including proteins and nucleic acids, the values of γ are close to 0.19 mL/g (Theisen, Johann, Deacon, & Harding, 2000), and the linearity is maintained for a wide range of concentrations (Barer & Tkaczyk, 1954; Zangle & Teitell, 2014; Zhao, Brown, & Schuck, 2011). Therefore, refractive index measurement is a rather direct approach to determination of $[\mathbf{A_T}]$.

In fluorescence lifetime imaging microscopy (FLIM), fluorescence is excited either with a very brief pulse of light or by a high-frequency periodic illumination. Such methods of excitation permit measurements of fluorescence lifetime - the average time τ that a molecule remains in the excited state before emitting a photon and returning to the ground state. It follows from theory that τ is proportional to the inverse square of refractive index and can, in principle, be used as a refractive index sensor. Not every fluorophore would be useful in this capacity because fluorescence lifetime is also affected by fluorescence quenchers and by viscosity of the medium (Suhling et al., 2015). But in the green fluorescent protein, the fluorescence moiety is shielded from the environment, making refractive index the only external parameter affecting τ (Suhling, Davis, & Phillips, 2002). The spatial resolution of refractive index mapping by FLIM may be moderate because τ senses not only the immediate surrounding of the fluorophore but also the refractive index of the medium several microns away (Jones & Suhling, 2006). Nevertheless, it has been possible to observe differences in the refractive index between the cytosol and the membrane (van Manen et al., 2008).

Another large group of methods for measuring refractive index is discussed in Section 4.

3.3 Other Optical Methods

Reflection. Reflection and optical scattering are caused by differences in the refractive index between the sample and the environment, and there have been attempts to utilize both of these phenomena for quantitative characterization of biological cells. However, direct reflection from the cell-coverglass boundary is contaminated by reflections from other surfaces and by optical interference between them (Bereiter-Hahn, Fox, & Thorell, 1979; Limozin & Sengupta, 2009). The use of the total internal reflection (TIR) obviates some of these difficulties because evanescent wave usually does not reach the apical side of the cell. Cells often recede from the cover-glass between focal adhesion points, and these gaps may affect the reflected intensity; correction for this factor can be made by varying the illumination angle in the TIR experiment (Bohannon, Holz, & Axelrod, 2017). The other complicating factor is scattering by cellular inhomogeneities, which reduces the amount of light returning to the detector (Axelrod & Axelrod, 2018).

Surface plasmon resonance (SPR) also senses the refractive index of the contact medium through observation of TIR intensity. While this is a standard technique for characterization of molecular interactions, its cellular applications run into difficulties. Claims that SPR can detect changes in the refractive index of cells (Moh, Yuan, Bu, Zhu, & Gao, 2008; Yanase et al., 2010) need qualifications because the reflected intensity is also sensitive to various small perturbations just above the surface, such as small amounts of material deposited onto the metal film. Thus, biological inferences from SPR investigation of cells can be uncertain (Kosaihira & Ona, 2008; Yanase et al., 2007).

Scatter. Light scatter in flow cytometers is used to discriminate between different cell populations based on their size and refractive index. However, scattering intensity depends not only on the refractive index of the cytosol, but also on cell shape, size, inclusions with different refractive indices and the angle at which the scattered light is collected (Mourant et al., 2000). The multitude of factors that influence the scattered intensity makes it very difficult to extract the refractive index from the signal.

UV transmission. The common method of protein and DNA quantification in solutions utilizes their differential absorption in the UV. The peak of DNA spectrum is located at 260 nm, and proteins exhibit two peaks: one at 280 nm, due to aromatic amino acids, and the other at 220 nm, due to peptide bonds. By taking absorption measurements at two wavelengths, 260/280 nm or 260/220 nm, the amount of protein and DNA can be determined.

A similar idea has been applied to measurements on whole cells. Transmission images under UV illumination show a contrast that quantitatively reflect their DNA and protein content (Cheung et al, 2011, 2013; Zeskind et al., 2007). Transmission imaging under 300 nm requires custom setup because some optical elements of standard microscopes are not transparent enough in this spectral range. It may be possible to combine

UV transmission with thickness/volume measurements using dye exclusion transmission imaging at 630 nm (Gregg, McGuire, Focht, & Model, 2010) because the dye Acid Blue 9 used in that method is relatively transparent at 260–280 nm. That would additionally convert the total protein and DNA content measured by UV transmission into intracellular concentrations.

Raman imaging. Raman microscopy is a powerful tool for locating and quantifying intracellular macromolecules. Carbon-hydrogen bonds in proteins produce a distinct peak at 2880–3000 cm^{-1}, which is well-separated from the water peak at 3350–3550 cm^{-1}. Intracellular water concentration or protein/water ratio can be found from spectral intensities (Bauer et al., 1998; Caspers, Lucassen, Carter, Bruining, & Puppels, 2001; Egawa, Hirao, & Takahashi, 2007; Huizinga, Bot, de Mul, Vrensen, & Greve, 1989). The depth discrimination of a confocal Raman microscope can be around 2–3 μm (Adar, Lee, Mamedov, & Whitley, 2010; Uzunbajakava et al., 2003), which should allow such measurements even on many types of cultured cells.

Table 1 makes an attempt to summarize the above methods.

Table 1 Comparison and Characteristics of Discussed Methods

Method	**Most Directly Measurable Parameter**	**Accuracy**
Drying	W_m	+++
Buoyant density	d_{cell}	+++
FRAP, FCS, spectral shift, fluorescence anisotropy	Diffusion coefficient or viscosity	+
FLIM of GFP	n	++
Reflection, TIR, SPR	n	++
Scatter	n	+
UV transmission	$\mathbf{A}_T$	+++
Fluorescent labeling of proteins	$\mathbf{A}_T$	++
Raman microscopy	W, [$\mathbf{A}_T$]	+++
QPI	$\mathbf{A}_T$	+++
Quenching of fluorophores, crowding sensors	[$\mathbf{A}_T$]	++
Mass spectrometry	$\mathbf{A}_S$	+++
Ninhydrin test	$\mathbf{A}_S$	++
pH	z (related to V_W)	++
Monovalent ions	Ion concentrations (related to V_W)	++

The ranking of accuracy is very approximate and may depend on various conditions.

4. QUANTITATIVE PHASE IMAGING (QPI)

4.1 The Principle of QPI

As discussed earlier in this chapter, intracellular organic matter present in the cells produces an increase in their refractive index (Eq. 14). Differences in refractive index between a cell and its environment are imprinted upon light as different delays to the phase of light oscillations. Because frequencies of visible light are very high (400–800 THz), its oscillations cannot be directly recorded by imaging sensors. To become detectable, phase delay must be converted into measurable intensity. This is done in standard phase contrast and differential interference contrast microscopes, but those are typically used only to improve sample visibility and not for quantitative purposes. Quantification of the phase is accomplished by a group of methods known as quantitative phase imaging (QPI).

Despite intense research and numerous publications in optics journals, QPI remains little known to biologists. The capabilities offered by QPI are directly relevant to the task of measuring intracellular organic matter. We limit our discussion to the basic ideas behind various implementations of QPI and their relative merits. Furthermore, due to the breadth of QPI methods and their variations, we will only attempt to review the most common and promising new techniques with an emphasis on those most compatible with conventional microscopes.

QPI is based on transmitted illumination that crosses a refractive sample (Fig. 1). When the cell's refractive index varies both across the sample and in

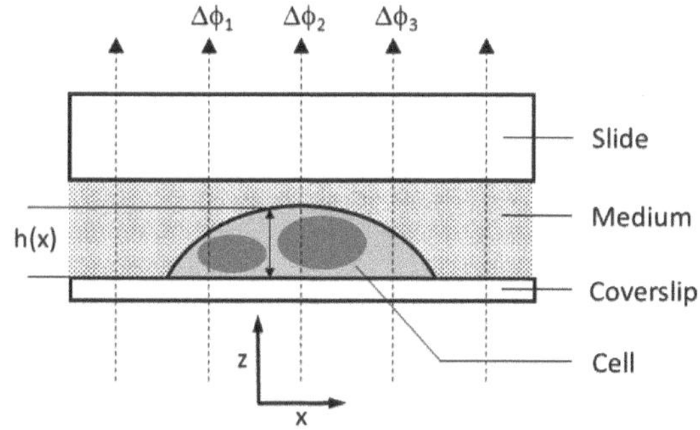

Figure 1 A diagram showing accumulation of phase delay $\Delta\phi$ of light passing through a cell. The cell is shown having a higher refractive index than the surrounding medium and, additionally, it has two inclusions with still higher n. Because both cell thickness h(**x**) and refractive index contribute to $\Delta\phi$, it varies in the order $\Delta\phi_2 > \Delta\phi_1 > \Delta\phi_3$.

depth, the relative phase accrued by light passing through the cell at horizontal position $\mathbf{x}$ is

$$\Delta\phi(\mathbf{x}) = \frac{2\pi}{\lambda}\int_{0}^{h(\mathbf{x})}[n_{cell}(\mathbf{x}, z) - n_0]dz \tag{15}$$

where z denotes depth within the sample, $h(\mathbf{x})$ is the cell thickness profile, n_{cell} is the cell's refractive index (which may vary depending on the position), n_0 is a constant refractive index of the medium, and λ is the wavelength of light. Scattering is assumed insignificant in this model. Sometimes, spatial variability of refractive index is disregarded, and Eq. (15) takes the much simpler form:

$$\Delta\phi(\mathbf{x}) = (2\pi/\lambda)[n_{cell} - n_0]h(\mathbf{x}) \tag{16}$$

QPI can be used for the following applications:

(1) Volume/profile measurements. Simpler models based on Eq. (16) can be employed, but finding cell volume or cell thickness profile from the phase data requires *a priori* knowledge of refractive index.

(2) Refractive index measurements. If the cell's height profile or its volume are known from independent measurements (Model, 2018), Eq. (16) can be used to reconstruct the average refractive index.

(3) The total mass of dry matter. This quantity is directly related to phase and requires no additional measurements. According to Eq. (14), $n_{cell} - n_0 = \gamma[\mathbf{A_T}]$, and the total organic matter contained in the cell is

$$\mathbf{A_T} = \frac{\lambda}{2\pi\gamma}\sum\phi \tag{17}$$

where summation is carried out over the entire cell area.

(4) Both cell shape and refractive index. For spatially varying refractive index, this can be done by adopting techniques from computed tomography (CT): the cell is illuminated from different angles while performing QPI measurements at each angle (Charrière et al., 2016; Choi et al., 2007; Popescu et al., 2008; Kim et al., 2014a). These techniques achieve high-resolution 3D mapping of refractive index, enabling the identification of organelles and variations in concentration of intracellular matter. Such 3D methods known as diffraction tomography are described in detail in Section 4.5.

QPI methods can be compared by several characteristics. One of them is transverse resolution—the smallest resolvable phase feature in the image. While the Abbe diffraction limit adequately quantifies the resolution of a conventional intensity image, QPI relies on a computational reconstruction of phase from an intensity image or images. Defining an adequate measure of resolution applicable to all QPI methods is not straightforward. However, some rules of thumb do exist and, in what follows, the relative performance of each method will be discussed, with high-resolution methods reaching or exceeding the diffraction-limited performance of the imaging optics. Generally, methods employing partially coherent illumination (illumination from a larger range of angles) can produce higher transverse resolution. Methods that require multiple images also typically have higher transverse resolution and can sometimes exceed the diffraction-limited performance of single-snapshot imaging.

A second important metric of a QPI method is the image acquisition rate, of importance when imaging dynamic processes. Since acquiring multiple images improves transverse resolution, there is commonly a trade-off between transverse resolution and acquisition rate.

A third metric is the accuracy of the phase reconstruction. Like transverse resolution, it is difficult to quantify because of the computational step in the reconstruction. For example, algorithms can be fine-tuned to accurately reconstruct particular phase features but may completely misrepresent others. Additionally, the phase imaging methods themselves may produce artifacts for particular classes of phase objects (for example, large, slowly-varying phase objects are difficult for non-interferometric methods to accurately reproduce). One way to quantify resolution and accuracy is to use phase targets, such as plastic beads or specially prepared objects with known profile and refractive index. These can be used to validate a particular phase imaging technique for similar classes of objects. Replacing the medium with one that has a different refractive index allows to perform calibration on live cells (Mudrak et al., 2018; Rappaz et al, 2005, 2009). Additionally, some broad rules of thumb can be established. Techniques based on coherent (e.g., laser) sources are more susceptible to errors due to speckle noise and diffraction artifacts that appear in images. Sources of lower spatial and temporal coherence (e.g. LED and white-light sources) typically improves phase accuracy. In addition, methods utilizing many images can improve phase accuracy by acquiring redundant measurements. Another reason QPI may produce inaccurate phase reconstructions is that the light's phase can be changed by mechanical vibrations in optical elements or even by air currents in the system. This can be significant for vibrations that change the positions of the optics by fractions of a wavelength (changes on the scale of 100 nm or less depending on the required accuracy). Systems that utilize interferometers with different paths for the sample and reference beam are most subject to this type of noise, since changes in one path relative to the other can introduce substantial phase errors, and such systems often require significant vibration isolation. Techniques utilizing a single optical path have significantly reduced sensitivity to vibrations.

Finally, it may be desirable to use methods compatible with conventional bright-field or fluorescence microscopy without significant alterations, particularly when QPI is used to measure refractive index while cell volume is quantified by other optical methods.

Table 2 summarizes common QPI methods with regard to these metrics. These methods are described in more detail in the following sections. However, QPI is a vibrant field and the following is not intended as a comprehensive review.

4.2 Interferometric Methods

Classical interferometry. Classical interference microscopy is the first and the best-established quantitative phase technique (Davis, Wilkins, Chayen, & La Cour, 1954). It relies on the fact that when two light waves of different phase delays overlap, the total intensity is modulated by an interference term that is proportional to the cosine of the phase delay between them. Nowadays, it is used mostly for materials applications. It is unlikely that Jamin-Lebedeff or Mach-Zehnder microscopes would make a comeback to biology in their original form, and we refer the interested reader to older literature (Tolansky, 1968).

Digital holographic microscopy. Digital holographic microscopy (DHM) is a well-developed QPI technique that has been extensively reviewed (Kim, 2010, 2011). Though many variations exist, they typically utilize a highly coherent, narrow-band light source (commonly a laser, although highly monochromatic LED sources have also been used). The light from the source is collected and allowed to pass through an interferometer which splits the light into two beams traveling along different paths: an object beam, which is passed through the sample, and a reference beam. The two beams are then recombined at an image sensor. The interference pattern is recorded in an out-of-focus plane, which allows computational back-

Table 2 A Comparison of Common Metrics for the Discussed QPI Methods

QPI Method	High Transverse Resolution	Fast Acquisition	Low Sensitivity to Vibrations	Low Sensitivity to Coherent Noise	Compatibility with Conventional Microscope
On-axis DHM	No	Yes	No	No	Requires significant modifications
Off-axis DHM	Yes	No	No	No	Requires significant modifications
Hilbert Phase Microscopy	No	Yes	No	No	Requires significant modifications
Fourier Phase Microscopy	Yes	No	Yes	No	Compatible, but requires additional optics before the camera
Diffraction Phase Microscopy	No	Yes	Yes	No	Compatible, but requires additional optics before the camera
SLIM	Yes	No	Yes	Yes	Compatible, but requires additional optics before the camera
iSLIM	No	Yes	Yes	Yes	Compatible, but requires additional optics before the camera
Linear propagation-based methods (TIE, CTF, WOTF)	Yes	Yes	Yes	Yes	Compatible
Fourier Ptychography and Ptychography	Yes	No	Yes	Yes	Compatible, but requires significant alteration of illumination.

propagation methods to be employed to isolate the phase of the sample from the measured intensity at the sensor.

There is a tradeoff in DHM between speed and spatial resolution. If the two interfering beams propagate parallel to each other on their way to the sensor, computational backpropagation will produce artifacts which reduce the accuracy of the phase reconstruction. By tilting the reference beam (a method known as off-axis DHM) this can be avoided, but off-axis methods come with a reduction of the transverse resolution of the reconstructed phase. An alternative method is temporal phase shifting which relies on multiple recorded holograms to avoid artifacts in the reconstruction while preserving resolution. However, the acquisition of multiple images places a limit on the temporal resolution of the method, as several frames must be acquired without significant changes in the sample.

One additional consideration is that because the interference pattern depends on the cosine of the phase difference between the beams, the reconstructed phase from backpropagation is unique only to the addition of any integer multiple of 2π radians. This problem can be addressed either by a computational process known as phase unwrapping or by acquiring additional images while illuminating the sample with more than one wavelength of light (Gass, Dakoff, & Kim, 2003).

DHM has been used successfully on a wide range of cell types including neurons (Marquet et al., 2005; Marguet, Depeursinge, & Magistretti, 2014), fibroblasts (Mann et al, 2006), ovarian cancer cells (Khmaladze, Kim, & Lo, 2008; Mann, Yu, Lo, & Kim, 2005) as well as imaging cell dynamics (Kemper et al., 2006; Rappaz et al., 2005). Although DHM is a well-developed technique supported by numerous commercially available systems, the interferometer requires specialized hardware and is therefore not immediately compatible with a conventional microscope.

Hilbert phase microscopy. Hilbert phase microscopy (HPM) utilizes similar hardware to DHM to produce an interference pattern between an object and reference beam. However, the interference pattern is recorded in-focus, so backpropagation cannot be employed to reconstruct the sample phase. Instead, a computational process based on the Hilbert transform is applied to reconstruct the phase (Ikeda, Popescu, Dasari, & Feld, 2005; Lue et al., 2006; Popescu et al., 2006a). The advantage of HPM is that since it is a single-shot technique performed in-focus, it is capable of high-speed quantitative phase imaging and has been used to image erythrocyte dynamics (Popescu et al., 2006a; Xue, 2011). However, it has lower transverse resolution than methods utilizing multiple images. Moreover, the reconstruction of rapidly varying phase features may be limited due to the nature of the reconstruction algorithm, although techniques have been developed to address these limitations (Trusiak, Mico, Garcia, & Patorski, 2016).

4.3 Common-Path Methods

While interferometry as described previously applies to systems in which the illumination is split into two beams prior to the sample, common path QPI methods implement a considerably more compact interferometer after the sample. In this geometry, the two interfering light beams propagate through many of the same optical components, increasing robustness to vibrations and misalignments of the system as well as generally reducing the system's footprint. Because they can be placed after the sample, common-path systems can also be easily implemented at the output port of a microscope.

There are a variety of common-path methods available (Popescu, 2011). All rely on the fact that phase variations introduced by the sample will scatter light out of the incident beam. In a conventional microscope, the incident beam follows a path through the optics that is defined by the condenser aperture (or annulus in the case of a phase contrast microscope). The aperture can be re-imaged after the exit port of the microscope providing easy access to the light that was not scattered from the incident beam. Phase imparted by the specimen causes light to scatter out of the incident beam as it propagates, forming a halo around the image of the condenser aperture. In common-path methods, the unscattered light is used as a reference beam and modulated to interfere with the scattered light. The primary differences between the various common-path methods are in how they treat the scattered and unscattered light and what reconstruction algorithm is applied.

Fourier phase microscopy (FPM) utilizes spatially coherent, monochromatic illumination. The method takes its name from the fact that modulation of the reference beam occurs in the Fourier plane of the imaging system, which is where the incident beam would focus to a (diffraction-limited) point in the absence of the sample. The scattered light, carrying the sample's phase information forms a halo about this point. A spatial light modulator is placed in the Fourier plane and used to introduce a phase shift between the scattered and unscattered light. All the light is then allowed to propagate to the image plane. Four images are acquired in succession with different phases imprinted on the unscattered light. Phase is reconstructed from combinations of these four images. Fourier phase microscopy has been demonstrated to recover refractive index in live cells (Lue et al., 2007; Popescu et al., 2004, 2006b) and has recently been extended to work with white light (Bhaduri, Tangella, & Popescu, 2013).

Diffraction phase microscopy (DPM) utilizes monochromatic, coherent illumination along with a diffraction grating to separate the light at an exit port of a microscope into two spatially separated beams which proceed through an imaging system to an image sensor. In the Fourier plane of the imaging system, one beam is allowed to pass unperturbed, while the other is passed through a pinhole so that only the unscattered illumination proceeds through the system. When these beams are recombined at the camera, the system acts as a very compact interferometer that produces images similar to HPM, making this a single-shot technique. These images are processed via the same algorithms as HPM. Diffraction phase microscopy has been demonstrated on blood cells (Popescu et al., 2006b) and kidney cells (Park, Popescu, Badizadegan, Dasari, & Feld, 2006). DPM has recently been extended to utilize white light (Bhaduri, Pham, Mir, & Popescu, 2012) which reduces speckle artifacts in phase reconstructions.

Spatial light interference microscopy (SLIM) (Wang et al., 2011) and *instantaneous spatial light interference microscopy* (iSLIM) (Ding & Popescu, 2010) are two methods that implement common-path methods in a Zernike phase contrast microscope utilizing white light illumination. Unlike the coherent, point-source illumination used in FPM and DPM, the phase contrast microscope illuminates the sample through an annulus. As a result, the unscattered light in the system's Fourier plane forms an image of this annulus. SLIM proceeds in a similar manner to FPM in that the condenser annulus of the phase contrast microscope is imaged onto a phase modulator while the scattered light spreads around the image of the annulus. By collecting a sequence of four images with different phase modulation on the unscattered light, the phase of the sample can be reconstructed. A SLIM system is commercially available from Phi Optics (http://phioptics.com). iSLIM is the analogue of diffraction phase microscopy in a phase contrast microscope. A diffraction grating is used to split the output light from the microscope into two beams. One of the beams is allowed to pass through a mask identical to the condenser annulus (only passing unscattered light) while the entire other beam is passed. The beams interfere on the image sensor in a way analogous to HPM, and the Hilbert transform is used to reconstruct the phase.

4.4 Propagation-Based Methods

Propagation-based methods are distinct from interferometric methods in that they do not require light to be separated into two parts that are forced to interfere. Instead, these methods rely on the fact that propagation of light through an optical system is described by the interference of light with itself. This self-interference depends critically on the phase of the light leaving the sample, so that sample phase can be reconstructed from propagated intensity measurements by computationally inverting the propagation equations. These methods have the benefit of being extremely robust to vibrations in comparison to other techniques since they utilize the optics of a conventional bright-field microscope. They also are compatible with LED illumination and can be utilized in many instances with white-light illumination. They also generally require the least modification of a conventional bright-field microscope. Many require only accurate measurement of defocus distance. Several methods rely on or are substantially improved by controlling the direction from which the sample is illuminated which is often achieved by substituting a low-cost array of regularly spaced LEDs at the position of the condenser lens on a conventional bright-field microscope.

Propagation-based methods can be broadly classified by whether the equations of propagation are linear or nonlinear. While the relationship between phase and the propagation of intensity is fundamentally governed by nonlinear differential equations, the solution of such equations for phase given measurements of intensity is not straightforward and typically require a significant number of images acquired under different conditions (varying illumination, sample position, defocus, *etc.*). Algorithms to solve nonlinear equations often start from an accurate initial guess (often obtained from a linear approximation) and improve the phase estimate iteratively. However, the phase produced also tends to be sensitive to the initial guess and so a given set of measurements is not guaranteed to produce a unique phase reconstruction.

For this reason, approximations are also commonly made to linearize the propagation equations. Linear models have the advantages of often being directly solvable for phase in a single step and producing provably unique solutions. Phase can also be reconstructed with fewer images (sometimes with only a single image), making phase imaging based on linear models faster. However, the approximations made to linearize the models place significant limitations on the reconstructions, often restricting the class of samples that can be accurately reconstructed.

Linear models (TIE, CTF and WOTF). Quantitative phase imaging based on the transport of intensity equation (TIE) (Barty, Nugent, Paganin, & Roberts, 1998; Petruccelli, Tian, & Barbastathis, 2013; Streibl, 1984; Teague, 1983) relies on comparing images acquired under a slight (typically tens of microns) defocus. Under this approximation, the difference between an under- and over-focused image leads to a second order, elliptic, partial differential equation for phase which can be solved computationally. A single defocused image may be used for nearly-transparent specimens. TIE does not require significant temporal coherence, so broadband or white light sources may be used. It also allows spatial coherence requirements to be relaxed significantly in comparison with interferometric methods. Finally, since it relies only on defocus in an imaging configuration, it is easily implemented at low cost in conventional bright-field microscopes, although a precise measure of the defocus (typically on the scale of microns) is required. The drawbacks of TIE are that it has significant difficulty capturing slowly-varying phase features and as a result it tends to enhance low-frequency noise features leading to inaccurate phase measurements if noise is not carefully handled. Multiple-image schemes and computational techniques can alleviate these issues (Chakraborty & Petruccelli, 2017; Jingshan, Claus, Dauwels, Tian, & Waller, 2014; Kostenko, Batenburg, King, Offerman, & van Vliet, 2013; Tian, Petruccelli, & Barbastathis, 2012; Waller, Tian, & Barbastathis, 2010; Zuo, Chen, Yu, & Asundi, 2013).

A second approximation that can be made is that the object is weakly varying (or that the phase is sufficiently slowly varying). This leads to either the contrast transfer function (CTF) model (Guigay, 1971; Zabler, Cloetens, Guigay, Baruchel, & Schlenker, 2005) (by ignoring the objective's pupil function and considering coherent illumination) or the weak object transfer function (WOTF) model (by including both the objective's pupil and the condenser diaphragm) (Sheppard, 2004; Streibl, 1985). Because the pupil function plays a critical role in determining the resolution limit of optical imaging, the WOTF is likely to produce more accurate phase reconstructions than TIE, although for only the limited class of weakly varying objects. Finally, the WOTF allows improved phase measurements by varying the illumination from the standard bright-field configuration. Methods have been proposed to produce differential phase contrast (DPC) measurements from two images acquired with asymmetric illumination (Mehta & Sheppard, 2009; Tian & Waller, 2015). Additionally, annular illumination has been proposed to improve phase reconstruction resolution (Li et al., 2017a).

An advantage of propagation-based techniques over many interferometric techniques is that the former are compatible with spatially partially coherent illumination (Gureyev, Nesterets, Paganin, Pogany, & Wilkins, 2006; Li et al., 2017a; Petruccelli et al., 2013) which can in theory achieve twice the transverse resolution of spatially coherent illumination (Zuo et al., 2017).

Nonlinear propagation-based models (Ptychography and Fourier ptychography). As previously mentioned, producing accurate, quantitative phase from nonlinear models is significantly more difficult than for linear models and requires more data and typically a reasonably accurate initial phase estimate which typically comes from first applying a linear model. It also tends to require significant computational power and as such is only recently becoming widely studied as a tool for microscopy.

Ptychographic microscopy (PM) (Godden, Suman, Humphry, Rodenburg, & Maiden, 2014; Suman et al, 2016) and Fourier ptychographic (FP) microscopy (Zheng, Horstmeyer, & Yang, 2013; Tian, Li, Ramchandran, & Waller, 2014; Yeh, et al. 2015) attempt to reconstruct the phase by solving a nonlinear equation modeling light propagation through an optical system. Both methods produce high resolution intensity and phase reconstructions. The difference between the two is whether the image is formed in-focus or in the Fourier plane of the sample. In ptychography, the measurement is obtained in the sample's Fourier plane, typically while scanning the sample across the microscope's field of view, although a recent advancement allows scanning the wavelength of the illumination (Bai, Vettil, Pan, Liu, & Zhu, 2017). The sample must be moved in small enough increments so that there is overlap between the field of view in each acquired image. A reconstruction process is then applied to find the sample phase and intensity most consistent with the set of acquired images.

FPM on the other hand modulates the illumination, typically using an array of LEDs in place of the condenser optics. The sample is illuminated from different angles while acquiring in-focus images. The scattering angles of light accepted by the imaging objective correspond to different phase structures in the sample. By illuminating in a way that scattering angles are sampled with enough overlap, a reconstruction algorithm can be applied to find the phase compatible with all measured images. One advantage of FP is that its geometry is naturally compatible with DPC phase imaging and a system that can produce fast phase measurements of weakly scattering objects using DPC or higher resolution phase measurements using the slower FP method.

4.5 Spatially Resolved Refractive Index Measurements

Eq. (15) approximates the phase measured by QPI as the integrated refractive index through the cell in the direction of the illumination. This can provide accurate information about the cell if it is thinner than the depth of field of the imaging optics or if there are no overlapping sub-cellular structures that must be disentangled. However, with some modification, QPI methods can provide 3D refractive index measurements. Most 3D methods utilize both more rigorous models than Eq. (15), typically employing the Born or Rytov scattering approximations (Chen & Stamnes, 2009). The basic idea of most 3D techniques is to illuminate the sample from different angles and capture the scattered field. This can be achieved either by varying the angles of the illumination, which typically limits the range of angles that can be interrogated, or by fixing the illumination and rotating the sample. The 3D profile of the cell's refractive index can be reconstructed by using tomographic reconstruction techniques.

As originally proposed, optical diffraction tomography (ODT) utilized holography to retrieve the phase image for each illumination angle and combined those results to yield a 3D refractive index profile (Wolf, 1969). Despite some promising early results (Streibl, 1985; Noda, Kawata, & Minami, 1990; Devaney & Schatzberg, 1992), it has been recent advances in QPI methods and computational power that has made the technique practical for live cell imaging (Haeberlé, Belkebir, Giovaninni, & Sentenac, 2010; Sung et al., 2009). For example, computational regularization methods can be employed to reduce artifacts that are introduced by the limited illumination scan angles in many systems (Sung & Dasari, 2011) as well as to include more robust scattering models for improved refractive index reconstruction (Kamilov et al., 2015). Despite advances in computational algorithms, many images are required for high resolution, and so ODT is slower than conventional QPI methods.

Since phase tomography requires only that phase be retrieved for many illumination/sample rotation angles, tomographic reconstruction is not limited to holographic data acquisition. ODT has been extended to systems utilizing SLIM (Kim et al., 2014b) which allows rapid imaging using white-light illumination. It has also recently been investigated for Fourier ptychography in which the illumination direction is necessarily varied as a part of the QPI algorithm (Horstmeyer, Chung, Xiaoze, Zheng, & Yang, 2016), as well as DPC (Chen, Tian, & Waller, 2016) and TIE (Li et al., 2017b).

5. BIOLOGICAL QUESTIONS

5.1 Heterogeneity of Refractive Index

Refractive index is not uniform across the cell (Choi et al., 2007; Yamauchi, Iwai, Miwa, & Yamashita, 2008). It varies from 1.36 to 1.39 in different parts of the cytoplasm (which corresponds to approximately 0.15–0.3 g/mL protein) and may reach even higher local values (Liu et al., 2017). Isolated mitochondria have n ≈ 1.41 (Haseda et al., 2015), as one would expect from their high protein concentration (see below), and the estimates for lysosomes gave an astonishing 1.6 (Wilson, Cottrell, & Foster, 2007). The refractive indices of organelles seem to correlate with their viscosity (Kuimova, Yahioglu, Levitt, & Suhling, 2008; Liu et al., 2014).

The expected consequences of this heterogeneity can be examined. By rewriting Eq. (5), we find the condition relating molar concentrations of intracellular osmolytes:

$$[A_S] + z_L[A_L] + 2[Cl] \approx P \tag{18}$$

Because P should be the same in all parts of the cell, local variations in $[A_L]$ (reported by refractive index) must be accompanied by opposite changes in at least one of the other three quantities: $[A_S]$, $[Cl^-]$, or z_L (which is a function of pH).

Lysosomes. Lysosomal swelling can be a sign of toxicity (Di Sant'Agnese, Jensen, Levin, & Miller, 1983) or necrotic damage, such as observed in ischemia (Becker & Barron, 1961; Decker & Wildenthal, 1980). Their swelling can be mediated by a variety of mechanisms: through osmotic influx of water (Arai, Yasuda, Isohashi, Okamoto, & Ohkuma, 2002; Solheim & Seglen, 1983; Zahlten, Hochberg, Stratman, & Lardy, 1972), accumulation of organic material (Huang et al., 1997), or disrupted balance between fusion and fission (Durchfort et al., 2012).

Without attempting to literally interpret the value 1.6 for the lysosomal refractive index (which would give an impossibly high protein concentration), we can at least accept that lysosomes are very densely packed with protein. The high luminal concentrations of the main ions, including chloride (Ishida, Nayak, Mindell, & Grabe, 2013), leaves relatively little room for additional terms in Eq. (18); the concentration of amino acids in lysosomes has been measured at 1 mM (Lloyd, 1996), and we are not aware of quantitative studies of other lysosomal osmolytes. Thus, it is possible that

$[A_S]$ in lysosomes is small. The magnitude of the second term, $z_L[A_L]$ should also be small despite a high protein concentration. Of course, mass/volume concentrations do not have to correlate with molecular concentrations: the average molecular mass of proteins may simply be larger in the compartment with higher refractive index. We checked the sizes of known lysosomal proteins (Lübke, Lobel, & Sleat, 2009) (without considering their abundance) and arrived at the average molecular weight of ~70 kDa, which is indeed larger than the global average 40 kDa (Brocchieri & Karlin, 2005). The potentially more important factor is the low intraluminal pH (Stauber & Jentsch, 2013), which should drastically reduce z_L or even change its sign. Thus, alkalinization of lysosomes is expected to result in water accumulation and a concomitant decrease in their refractive index due to dilution of proteins. We are not aware of experimental data that would support or refute this conclusion. Although the previous authors (Ishida et al., 2013) have calculated that alkalinization should result not in swelling but in a slight shrinkage, the charged membrane-impermeant molecules were not included in their model. Experiments have shown that cell swelling produces alkalinization of lysosomes (Busch et al., 1996), but the inverse process has not apparently been studied, possibly due to the small sizes of lysosomes. Lysosomal storage disorders are expected to result in a higher than normal refractive index of lysosomes due to accumulation of undigested macromolecules (Cooper, 2000).

Although viscosity probes for lysosomes have been synthesized (Wang, Xiao, Tian, & Deng, 2013; Guo et al., 2018), their biological use has not been sufficiently developed as yet. An increase in lysosomal viscosity observed in response to dexamethasone (Wang et al., 2013) could possibly have reasons other than changes in protein concentration. As mentioned earlier, protein concentration is not the only factor affecting viscosity.

Mitochondria. Mitochondria are very labile structures capable of more than doubling of their volume under pathological conditions (Gao, Pu, Luo, & Chang, 2001; Safiulina, Veksler, Zharkovsky, & Kaasik, 2006), as well as fission, fusion, and shape changes (Arnoult, 2007). Proteins occupy >60% of the mitochondrial matrix (Verkman, 2002), but this value should obviously depend on the extent of swelling; the concentrations of sodium, potassium and chloride in mitochondria are not very different from those in the cytosol (Arrebola et al., 2006; Somlyo, Somlyo, & Shuman, 1979). Under physiological conditions, mitochondria control their volume mainly through ATP- and calcium-sensitive potassium channels (Kaasik, Safiulina, Zharkovsky, & Veksler, 2007). Phosphate serves as a counterion for potassium (Kowaltowski, Seetharaman, Paucek, & Garlid, 2001), and, indeed, mitochondria contain a relatively high concentration of inorganic phosphate (Akerboom, Bookelman, Zuurendonk, van der Meer, & Tager, 1978). Because of the export of protons by the respiratory chain, the internal pH of mitochondrial matrix is by ~1 unit higher than that of the cytosol.

The best-studied type of mitochondrial pathological swelling occurs as a result of formation of the mitochondrial permeability transition pore (MPTP), a channel-like structure that can assemble in the inner membrane and is now believed to be related to F-ATP synthase (Fontaine & Bernardi, 2016). MPTP causes disruption of oxidative phosphorylation, mitochondrial autophagy, and cell death through apoptosis or necrosis (Eisenhofer et al., 2010; Srinivasan, 2015). The opening of MPTP leads to free exchange of solutes under 1.5 kDa between the matrix space and the cytosol and ultimately to osmotic expansion of mitochondria. Aside from MPTP, the volume of mitochondria can be affected by hormones, metabolism, and the inner membrane potential (Halestrap, 1989; Kaasik et al., 2007). A detailed model of mitochondrial volume regulation has been constructed (Bazil, Buzzard, & Rundell, 2010).

Detecting changes in the mitochondrial refractive index could conceivably be employed for studying diverse intracellular processes, including, for example, differentiating between swelling and fusion. Most measurements of mitochondrial volume and water have been performed on isolated mitochondria (Halestrap, 1989). Fluorescence microscopy, sometimes aided by digital deconvolution (Kaasik et al., 2007), has been employed for in situ measurements; superresolution methods (Jakobs & Wurm, 2014) could prove even more valuable for measurements of structures as small as mitochondria or lysosomes. Mitochondria-specific viscosity sensors are available (Jiang et al., 2014; Liu et al., 2013), and an increase in viscosity during apoptosis (Jiang et al., 2014) or following addition of ionophores (Yang et al., 2013) has been reported. As with lysosomes, it is not entirely clear how to relate these facts to specific biochemical changes. It would be interesting to see if digital holography or other spatially resolved QPI methods for direct refractive index measurements can be sensitive enough for observation of biologically significant processes within small organelles.

Nuclei. Water spreads easily across the nuclear membrane (Guo et al., 2017), but its permeability to ions is controlled by channels (Leanza et al., 2013; Matzke, Weiger, & Matzke, 2010). Numerous data on the DNA content of different cell types have been compiled and analyzed by Gillooly, Hein, and Damiani (2015). The DNA/protein ratio in the nuclei of human cells is typically 0.2–0.4 or can be smaller in some cases (Cheung et al., 2013; Zeskind et al., 2007). The refractive increment of DNA is similar or slightly less than that of proteins (Fu et al., 2009), and the refractive index of nuclei is less than of the cytosol (Choi et al., 2007; Liu et al., 2016; Mudrak et al., 2018). It appears that the overall concentrations of macromolecules within the cytosol and nuclei are similar.

The question is whether this situation can change under biologically relevant conditions. The nuclear refractive index has been found to vary during the cell cycle in a way proportional to the amount of DNA (Bista et al., 2011). This observation could be explained by an increase in the DNA mass that is not matched by expansion of the nucleus. However, the cells used in that study were ethanol-fixed, and ethanol fixation causes dehydration and precipitation of proteins. Therefore, these results may also reflect different reactivity of DNA and protein with the fixative, and it remains to be shown that refractive index in these cases carries more information that conventional DNA staining. The same group reported that nuclear refractive index in processed (fixed, paraffin-embedded and stained) histological samples of breast biopsies is higher in malignant cells than in normal tissue (Wang, Bista, Bhargava, Brand, & Liu, 2010).

Membrane blebs. An example of the use of refractive index to understand a biological process comes from a study of membrane blebs in cells undergoing necrotic death (Rana et al., 2017). Blebs are organelle-free protrusions surrounded only by the plasma membrane. Unlike lysosomes and mitochondria, they are not isolated from the rest of the cell by a semipermeable membrane. By using TIE microscopy combined with volume measurements, a dramatic (up to 20-fold) loss of protein has been observed in necrotic blebs. That might create an osmotic imbalance by several mM, because small molecules are expected to be at the same concentration in the bleb and in the main cell body and therefore cannot compensate for the loss of protein. To explain this observation, a massive protein aggregation in the cell body was hypothesized and then verified by a biochemical assay. Aggregation should draw soluble protein out of the bleb and, at the same time, reduce the molar concentration of proteins in the main body, so that osmotic equilibrium can be maintained.

5.2 Long-term Volume Maintenance

Cell volume maintenance is one of the classical subjects of cell physiology (Burg, Ferraris, & Dmitrieva, 2007; Lang et al., 1998). The typical experimental design involves subjecting cells to a step change in extracellular osmolarity, forcing them to rapidly adjust their volume; subsequently, many cell types restore their initial volume, at least partially. Failure to expand after hypertonic shrinkage makes cells vulnerable to apoptosis (Friis et al., 2005; Maeno, Takahashi, & Okada, 2006; Yurinskaya et al., 2012).

Monitoring of cell volume in anisosmotic experiments is usually carried out for 10–30 min; prolonged exposure to anisosmotic media is usually avoided in these types of studies. However, culturing cells in hyperosmotic solutions is a standard practice in biotechnology because it has been found that media with osmolarities 400 mOsm/kg or even higher promote antibody production in hybridoma cell lines. Interestingly, cells kept under hyperosmotic conditions eventually become larger than cells growing in normal media (Ozturk & Palsson, 1991; Takagi, Hayashi, & Yoshida, 2000; Zhang, Garcia, Baldi, Hacker, & Wurm, 2010; Kiehl, Shen, Khattak, Jian Li, & Sharfstein, 2011). Results obtained with a crowding sensor suggest that this enlargement may be partly due to water accumulation (Gnutt et al., 2017). Slow adaptation seems to be an important aspect of cell volume regulation that has been given little attention by the mainstream cell volume research.

When studying long-term behavior of cells growing in suspension, flow-based analysis should be adequate. The problem arises when working with adherent cultures. Attached cells behave differently from suspended cells in many ways and should be analyzed in their native state, without prior detachment (Model, 2018). Because analysis of large cell populations under the microscope can be labor-intensive, a good alternative is to repeatedly observe the same cells as they evolve in time; however, their volumes during prolonged incubations would be affected not only by osmolarity but also by continuing growth and division. It is therefore more logical to conduct such studies using protein concentration as a readout parameter. Furthermore, as we have argued previously (Mudrak et al. 2018), protein density may act as a volume sensor during slow adaptation. Indeed, the main factors that are suspected to sense and provide feedback on the cell volume are mechanical tension, concentration of ions and macromolecular crowding (Hoffmann et al., 2009). But mechanical stretch or compression are not expected to persist over the course of many hours in growing and mobile cells. Likewise, ion concentrations should remain more or less stable during volume recovery and can hardly guide the cell toward the set point. On the other hand, macromolecular crowding (i.e., protein concentration), which affects numerous biochemical reactions (Ellis, 2001; Luby-Phelps, 2000; Minton, 2001), can could easily serve as a persistent signal indicating deviation from the normal structure. In this case, protein concentration becomes not only a convenient, but a biologically relevant parameter.

5.3 Apoptotic Shrinkage

Apoptotic cell death is indispensable for normal body functioning, embryonic development, numerous pathologies and medical interventions. When apoptosis was first recognized as a distinct type of cell death, it was nicknamed "shrinkage necrosis" (Kerr, 1971) because cells undergoing this type of death become smaller (Elmore, 2007; Lejeune & Linder, 2018; Saraste & Pulkki, 2000). This observation was remarkable because, when exposed to an unfavorable environment, cells are expected to accumulate water, increase their volume and eventually lose membrane integrity. Water accumulation is a natural cellular response to severe membrane damage and to cessation of the Na, K pump activity (Armstrong, 2003) and is a hallmark of unregulated necrotic death (Zong & Thompson, 2006). If the opposite happens during apoptosis, one is tempted to conclude that shrinkage is an essential and purposeful reaction of a cell to its imminent demise.

Apoptotic cell shrinkage has at least three aspects that are not always clearly differentiated in the literature: (1) loosening of substrate attachment, resulting in smaller-looking cells under a microscope; (2) loss of water with preservation of dry mass, and (3) loss of dry mass (i.e., splitting into fragments), which may or may not be accompanied by dehydration. The last two processes result in a volume decrease. A decrease in cell volume often amounts to 10%–15% of the initial volume but, occasionally, shrinkage by as much as 40%–85% has been reported (Beauvais, Michel, & Dubertret, 1995; Hessler et al., 2005; Bortner & Cidlowski, 2007; Bortner, Sifre, & Cidlowski, 2008; Ernest, Habela, & Sontheimer, 2008; l'Hoste et al., 2010; Khmaladze et al., 2012; Khmaladze, 2017; Sharikova et al., 2017).

The loss of water as a mechanism of apoptotic shrinkage has received most attention in the literature and has become almost synonymous with the term "apoptotic volume decrease" (AVD). Dehydration during apoptosis has been observed by measurements of buoyant density (e.g., Cohen et al., 1992; Grover et al., 2011; Patterson et al., 1995; Wyllie & Morris, 1982; Yamada & Ohyama, 1988; Yurinskaya et al., 2005) and by combined phase/volume measurements (Model & Schonbrun, 2013; Model et al., 2018).

AVD is believed to result from activation of K^+ and Cl^- channels, followed by osmotically driven water, in a manner similar to cell volume recovery in hypotonic solutions (Kunzelmann, 2016; Lang & Hoffmann, 2012). This hypothesis is supported by inhibition of the AVD by ion channel blockers and by activation of K^+ and Cl^- currents during apoptosis (Ernest et al., 2008; Nietsch, Roe, Fiekers, Moore, & Lidofsky, 2000; Okada et al., 2006; Porcelli et al., 2004). By using X-ray microanalysis, Fernández-Segura, Cañizares, Cubero, Warley, and Campos (1999) showed a loss of osmotically active ions in UV-irradiated U937 cells with preservation of the amount of phosphorus, which argues for the primary role of dehydration in apoptotic shrinkage.

At the same time, not all cell density measurements are in agreement with the dehydration nature of the AVD (Benson, Heer, Dive, & Watson, 1996; Yurinskaya et al., 2005), and in many cases dehydration has been claimed without sufficient experimental evidence. In the vast majority of experiments, only the total cell volume has been measured, for example, by electronic sizing or, much less directly, by light scattering. These methods do not make it possible to unequivocally attribute shrinkage either to dehydration or to fragmentation. As pointed out earlier, the cases of extreme volume decrease, such as by 40%–85%, can hardly be explained by the K^+/Cl^- mechanism alone, because many cell types simply do not contain enough chloride to provide for water loss exceeding 10%–15% (Bregestovski, Waseem, & Mukhtarov, 2009). The analysis of the few available data on actual $[Cl^-]$ changes during apoptosis reveals only a minimal loss of chloride, sufficient for no more than a 0–10% volume decrease (Model, 2014); in one case, cell shrinkage was even accompanied by Cl^- accumulation (Poulsen et al., 2010). These discrepancies have never been sufficiently addressed.

In this regard, we would like to mention the interesting results obtained with Cl^- ionophores ("anionophores"). These compounds, as well as the natural bacterial pigment prodigiosin, cause accumulation of intracellular Cl^-, apoptosis and shrinkage (Ko et al., 2014; Saha et al., 2016; Busschaert et al., 2017). Thus, their action is opposite to what is expected from the standard model of the AVD, in which shrinkage is mediated by an efflux, and not an influx, of Cl^- and K^+. There is currently not enough information to fully explain the biological effects of anionophores. One possibly relevant fact is that these chemicals also mediate Cl^-/H^+ symport (Busschaert et al., 2017; Wu et al., 2016), which is expected to partially neutralize the negative charges on macromolecules, favoring cell shrinkage. Moreover, anionophores are not strictly specific to Cl^-, and its accumulation may be balanced by undetected export of intracellular anions. This is all the more likely since the equilibrium potential for Cl^- in cells containing at least 10–15 mM Cl^- is less negative than the typical membrane potential, and the observed influx of Cl^- should be coupled to another, energetically more favorable, transport process.

Although the importance of K^+ for apoptosis has been abundantly documented, it may have effects other than osmotic (Cain, Langlais, Sun, Brown, & Cohen, 2001; Hughes, Bortner, Purdy, & Cidlowski, 1997; Thompson, Langlais, Cain, Conley, & Cohen, 2001), and it can be difficult to know whether inhibition of the AVD by channel blockers is a primary or secondary effect. One cannot deny, of course, that ion-mediated dehydration may indeed be a common apoptotic phenomenon, but its prevalence and extent cannot be established without direct experimentation, and alternative mechanisms of both volume decrease and water loss might be considered.

Fragmentation and separation of apoptotic bodies would be the most obvious candidate for dehydration-independent shrinkage mechanism. Fragmentation has been documented on numerous micrographs, as well as by light scatter (Fig. 2). It is frequently stated that formation of apoptotic bodies occurs only during the final stages of apoptosis, but this is not always the case: for example, the protein kinase inhibitor staurosporine, which is frequently used to study AVD, tends to cause rapid fragmentation in diverse cell types. The loss of cytoplasmic material, rather than just of ions and water, has been reported by Benson et al. (1996). Although small apoptotic bodies can often be identified on particle analyzers or on flow cytometric scattergrams, the cells *from which* those apoptotic bodies have detached are also expected to become smaller, and their reduced size can be easily misinterpreted as a result of dehydration. Fragmentation might explain the inefficiency of K^+ blockers in some cases (Nobel et al., 2000), the absence of intermediate cell sizes (Beauvais et al., 1995; McCarthy & Cotter, 1997) or the previously cited cases of extreme shrinkage. Particularly striking data in this regard come from QPI experiments (Khmaladze, 2017; Khmaladze et al., 2012; Mugnano, Calabuig, Grilli, Miccio, & Ferraro, 2015; Sharikova et al., 2017;

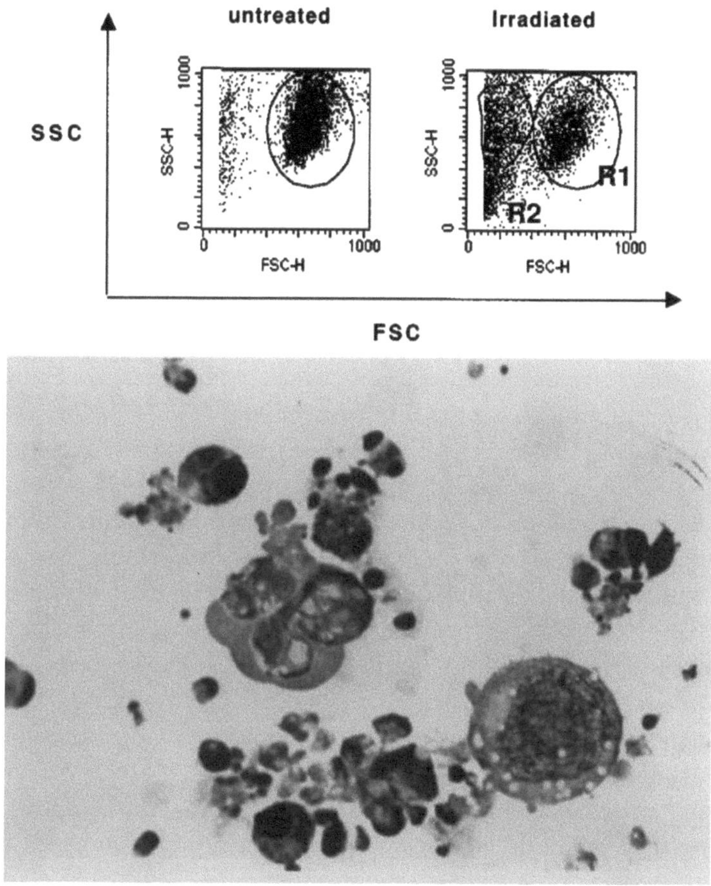

Figure 2 Flow cytometric scatter plots and a micrograph of apoptotic cells. UV irradiation of HL-60 cells produced a large increase in the R2 population. A microscopic image of the same sample shows extensive fragmentation. It is very likely that R2 represented cell fragments, rather than shrunken whole cells (as was the authors' interpretation). That would also explain the absence of intermediate cell sizes noted in the paper. *From McCarthy, J. V., & Cotter, T. G. (1997). Cell shrinkage and apoptosis: a role for potassium and sodium ion efflux.* Cell Death & Differentiation, 4, *756–770, with permission.*

Zhang et al., 2017): a decrease in the phase delay obtained by these methods can only be interpreted as the loss of dry mass. Our data (Model & Schonbrun, 2013; Model et al., 2018) show that both dehydration and fragmentation can occur during apoptosis. Thus, fragmentation appears to be a major, and underappreciated, factor in apoptotic shrinkage. Combined phase/volume or density/volume measurements would be the most direct way to assess the individual contributions of dehydration and fragmentation to the AVD.

Turning now specifically to apoptotic dehydration, the role of small organic molecules may be worth revisiting. Transport of organic osmolytes plays an active role in osmoregulation in microorganisms, plants and marine animals (Wehner, Olsen, Tinel, Kinne-Saffran, & Kinne, 2003). In mammalian cells, organic osmolytes (such as sorbitol, betaine, taurine, creatine or inositol) are important in the kidney medulla, where cells have to rapidly adjust to widely varying osmolarity of the environment. In many types of mammalian cells, the main organic osmolyte is taurine, a neutral compound chemically related to amino acids. Intracellular concentrations of taurine can be between 10 and 20 mM (Hoffmann & Hendil, 1976; Piez & Eagle, 1958), which is 2–3 orders of magnitude higher than its extracellular concentration. Thus, its passive efflux can have a measurable effect on cell volume. The idea that taurine might be involved in the AVD has been around for some time; however, the experimental evidence of any link between taurine and AVD is apparently limited to a single publication (Poulsen et al., 2010).

Control of cell volume by taurine is a process that operates in healthy cells, but apoptotic cells are anything but healthy. Activation of organic cation transport during initial phases of apoptosis has been demonstrated (Gibbons, Kharel, Robinson, Synowicki, & Model, 2016); it has the effect of causing temporary cell swelling. It is equally possible that apoptotic membranes become permeable to certain neutral molecules or to organic anions during their dehydration phase; apoptosis-stimulated efflux of glutathione (Franco & Cidlowski, 2006) and of other small molecules (Lee, Woo, & Lee, 2018) are known to occur. If the increase in permeability is not very selective and many organic molecules are allowed to pass, cell shrinkage may result. Most known transporters of organic anions are either of the exchange type or passive facilitators (Sekine, Cha, & Endou, 2000). An exchange of an organic for inorganic ion would not alter the osmolarity; however, one may hypothesize that, since a passive efflux of organic anions is favored both by concentration gradient and by electric potential, it may result in cell depolarization and stimulate, in turn, an efflux of potassium and, ultimately, of water.

Another mechanism of reducing intracellular concentration of organic osmolytes could operate through molecular aggregation, in a way similar to what has been described for necrotic blebs (Rana et al., 2017). It is known that free radicals are actively generated during apoptosis (Halliwell &

Gutteridge, 2007) and may cause crosslinking between amino acids, peptides or nucleotides by a variety of mechanisms (Stadtman & Levine, 2003; Dizdaroglu, Jaruga, Birincioglu, & Rodriguez, 2002; Davis, 2016). Formation of glutathione dimers from monomers is one example (Nobel, Aronson, van den Dobbelsteen, & Slater, 2000). Crosslinking between small molecules or between small molecules and proteins would reduce the osmolarity of the cytosol and may have an effect similar to that of ion loss. High-resolution mass-spectrometry may help resolve this question.

6. CONCLUSION

Macromolecules are the reason why cells have a volume. Notwithstanding this obvious fact, mechanisms on the macromolecular level have not been much discussed in the cell volume community. We believe this should to be reconsidered: some reasons have been listed, others may come to light after more research. But any experimental study depends on the right tools, and we presented an extended (though not complete) review of methods potentially useful for the task, with particular emphasis on quantitative phase imaging techniques. Some of them require common but expensive technology while others (such as TIE) can be easily realized in any reasonably equipped laboratory. QPI techniques all have a computational component, and we expect recent advances in computational power to continue to catalyze new research in this area. The other theme that has been illustrated with several examples in this review is the benefit of analyzing the data in the framework of a quantitative physiological model.

SYMBOLS

$\mathbf{A_L}$	large organic molecules (or their molar amount)
$\mathbf{A_s}$	small organic molecules (or their molar amount)
$\mathbf{A_T}$	all organic matter (or its molar amount)
$\mathbf{A_T}$	mass of A_T
$\mathbf{c}$	mass/volume concentration of any solute
$\mathbf{d_A}$	density of organic matter
$\mathbf{h}$	cell thickness
$\mathbf{n_{cell}}$	refractive index of a cell
$\mathbf{n_0}$	refractive index of a protein-free medium
$\mathbf{P}$	external osmolarity
$\mathbf{V_W}$	volume occupied by intracellular water
$\mathbf{V_{cell}}$	total cell volume
$\mathbf{W}$	ratio of V_W/V_{cell}

w_m	ratio of the mass of intracellular water to the total mass of a cell
z_{av}	average (negative) charge of A_T
z_L	average (negative) charge of A_L
γ	refractive index increment
λ	wavelength of light
τ	fluorescence lifetime
ϕ	phase of light

ABBREVIATIONS

AVD	apoptotic volume decrease
CT	computed tomography
CTF	contrast transfer function
DHM	digital holographic microscopy
DPC	differential phase contrast
DPM	diffraction phase microscopy
FCS	fluorescence correlation spectroscopy
FLIM	fluorescence lifetime imaging microscopy
FP	Fourier ptychographic microscopy
FPM	Fourier phase microscopy
FRAP	fluorescence recovery after photobleaching
FRET	fluorescence resonance energy transfer
HPM	Hilbert phase microscopy
iSLIM	instantaneous spatial light interference microscopy
MPTP	mitochondrial permeability transition pore
ODT	optical diffraction tomography
PM	ptychographic microscopy
QPI	quantitative phase imaging
SLIM	spatial light interference microscopy
SPR	surface plasmon resonance
TIE	transport of intensity equation
TIR	total internal reflection
WOTF	weak object transfer function

ACKNOWLEDGMENTS

We gratefully acknowledge advice from Daniel Axelrod (University of Michigan), Alexey Vereninov (Institute of Cytology, St. Petersburg, Russia), Cees Otto (University of Twente), David Sleat (Rutgers University), Philip Gale (University of Sydney), and Natalie Busschaert (Tulane University). The work was partially supported by Kent State University Research Council (MM).

REFERENCES

Adar, F., Lee, F., Mamedov, S., & Whitley, A. (2010). Experimental evaluation of the depth resolution of a Raman microscope. *Microscopy and Microanalysis, 16*(S2), 360–361.

Akerboom, T. P., Bookelman, H., Zuurendonk, P. F., van der Meer, R., & Tager, J. M. (1978). Intramitochondrial and extramitochondrial concentrations of adenine nucleotides and inorganic phosphate in isolated hepatocytes from fasted rats. *European Journal of Biochemistry, 84*, 413–420.

Alfieri, R. R., Cavazzoni, A., Petronini, P. G., Bonelli, M. A., Caccamo, A. E., Borghetti, A. F., et al. (2002). Compatible osmolytes modulate the response of porcine endothelial cells to hypertonicity and protect them from apoptosis. *The Journal of Physiology, 540*, 499–508.

Arai, K., Yasuda, N., Isohashi, F., Okamoto, K., & Ohkuma, S. (2002). Inhibition of weak-base amine-induced lysis of lysosomes by cytosol. *The Journal of Biochemistry, 132*, 529–534.

Armstrong, C. M. (2003). The Na/K pump, Cl ion, and osmotic stabilization of cells. *Proceedings of the National Academy of Sciences of the United States of America, 100*, 6257–6262.

Arnoult, D. (2007). Mitochondrial fragmentation in apoptosis. *Trends in Cell Biology, 17*, 6–12.

Arrebola, F., Fernández-Segura, E., Campos, A., Crespo, P. V., Skepper, J. N., & Warley, A. (2006). Changes in intracellular electrolyte concentrations during apoptosis induced by UV irradiation of human myeloblastic cells. *American Journal of Physiology. Cell Physiology, 290*, C638–C649.

Axelrod, J. J., & Axelrod, D. (2018). Scattering of evanescent illumination by sample inhomogeneities in TIRF microscopy: A theoretical study. *Biophysical Journal, 114*, 533a.

Bai, Y., Vettil, S. P., Pan, X., Liu, C., & Zhu, J. (2017). Ptychographic microscopy via wavelength scanning. *APL Photonics, 2*, 056101.

Baraona, E., Leo, M. A., Borowsky, S. A., & Lieber, C. S. (1975). Alcoholic hepatomegaly: Accumulation of protein in the liver. *Science, 190*, 794–795.

Barer, R., & Tkaczyk, S. (1954). Refractive index of concentrated protein solutions. *Nature, 173*, 21–822.

Barty, A., Nugent, K. A., Paganin, D., & Roberts, A. (1998). Quantitative optical phase microscopy. *Optics Letters, 23*, 817–819.

Bauer, N. J., Wicksted, J. P., Jongsma, F. H., March, W. F., Hendrikse, F., & Motamedi, M. (1998). Noninvasive assessment of the hydration gradient across the cornea using confocal Raman spectroscopy. *Investigative Ophthalmology and Visual Science Vis Sci, 39*, 831–835.

Baumgart, F., Rossi, A., & Verkman, A. S. (2012). Light inactivation of water transport and protein-protein interactions of aquaporin-Killer Red chimeras. *The Journal of General Physiology, 139*, 83–91.

Baust, J. G., Gao, D., & Baust, J. M. (2009). Cryopreservation: An emerging paradigm change. *Organogenesis, 5*, 90–96.

Bazil, J. N., Buzzard, G. T., & Rundell, A. E. (2010). Modeling mitochondrial bioenergetics with integrated volume dynamics. *PLoS Computational Biology, 6*, e1000632.

Beauvais, F., Michel, L., & Dubertret, L. (1995). Human eosinophils in culture undergo a striking and rapid shrinkage during apoptosis. Role of K^+ channels. *Journal of Leukocyte Biology, 57*, 851–855.

Becker, N. H., & Barron, K. D. (1961). The cytochemistry of anoxic and anoxic-ischemic encephalopathy in rats. I. Alterations in neuronal lysosomes identified by acid phosphatase activity. *American Journal of Pathology, 38*, 161–175.

Bennett, B. D., Kimball, E. H., Gao, M., Osterhout, R., Van Dien, S. J., & Rabinowitz, J. D. (2009). Absolute metabolite concentrations and implied enzyme active site occupancy in *Escherichia coli*. *Nature Chemical Biology, 5*, 593–599.

Benson, R. S., Heer, S., Dive, C., & Watson, A. J. (1996). Characterization of cell volume loss in CEM-C7A cells during dexamethasone-induced apoptosis. *American Journal of Physiology, 270*, C1190–C1203.

Bereiter-Hahn, J., Fox, C. H., & Thorell, B. (1979). Quantitative reflection contrast microscopy of living cells. *Journal of Cell Biology, 82*, 767–779.

Bhaduri, B., Pham, H., Mir, M., & Popescu, G. (2012). Diffraction phase microscopy with white light. *Optics Letters, 37*, 1094–1096.

Bhaduri, B., Tangella, K., & Popescu, G. (2013). Fourier phase microscopy with white light. *Optics Express, 4*, 1434–1441.

Bista, R. K., Uttam, S., Wang, P., Staton, K., Choi, S., Bakkenist, C. J., et al. (2011). Quantification of nanoscale nuclear refractive index changes during the cell cycle. *Journal of Biomedical Optics, 16*, 070503.

Boersma, A. J., Zuhorn, I. S., & Poolman, B. (2015). A sensor for quantification of macromolecular crowding in living cells. *Nature Methods, 12*, 227–229.

Bohannon, K. P., Holz, R. W., & Axelrod, D. (2017). Refractive index imaging of cells with variable-angle near-total internal reflection (TIR) microscopy. *Microscopy and Microanalysis, 23*, 978–988.

Bortner, C. D., & Cidlowski, J. A. (2002). Apoptotic volume decrease and the incredible shrinking cell. *Cell Death & Differentiation, 9*, 1307–1310.

Bortner, C. D., & Cidlowski, J. A. (2007). Cell shrinkage and monovalent cation fluxes: Role in apoptosis. *Archives of Biochemistry and Biophysics, 462*, 176–188.

Bortner, C. D., Sifre, M. I., & Cidlowski, J. A. (2008). Cationic gradient reversal and cytoskeleton-independent volume regulatory pathways define an early stage of apoptosis. *The Journal of Biological Chemistry, 283*, 7219–7229.

Bratbak, G., & Dundas, I. (1984). Bacterial dry matter content and biomass estimations. *Applied and Environmental Microbiology, 48*, 755–757.

Bregestovski, P., Waseem, T., & Mukhtarov, M. (2009). Genetically encoded optical sensors for monitoring of intracellular chloride and chloride-selective channel activity. *Frontiers in Molecular Neuroscience, 2*, 15.

Brocchieri, L., & Karlin, S. (2005). Protein length in eukaryotic and prokaryotic proteomes. *Nucleic Acids Research, 33*, 3390–3400.

Bryan, A. K., Hecht, V. C., Shen, W., Payer, K., Grover, W. H., & Manalis, S. R. (2014). Measuring single cell mass, volume, and density with dual suspended microchannel resonators. *Lab on a Chip, 14*, 569–576.

Burg, M. B., Ferraris, J. D., & Dmitrieva, N. I. (2007). Cellular response to hyperosmotic stress. *Physiological Reviews, 87*, 1441–1474.

Burton, R. F. (1983). The composition of animal cells: Solutes contributing to osmotic pressure and charge balance. *Comparative Biochemistry and Physiology B, 76*, 663–671.

Busch, G. L., Wiesinger, H., Gulbins, E., Wagner, H. J., Hamprecht, B., & Lang, F. (1996). Effect of astroglial cell swelling on pH of acidic intracellular compartments. *Biochimica et Biophysica Acta, 1285*, 212–218.

Busschaert, N., Park, S. H., Baek, K. H., Choi, Y. P., Park, J., Howe, E. N. W., Hiscock, J. R., Karagiannidis, L. E., Marques, I., Félix, V., Namkung, W., Sessler, J. L., Gale, P. A., & Shin, I. (2017). A synthetic ion transporter that disrupts autophagy and induces apoptosis by perturbing cellularchloride concentrations. *Nat Chem, 9*, 667–675.

Cain, K., Langlais, C., Sun, X. M., Brown, D. G., & Cohen, G. M. (2001). Physiological concentrations of K^+ inhibit cytochrome c-dependent formation of the apoptosome. *The Journal of Biological Chemistry, 276*, 41985–41990.

Canepa, A., Filho, J. C., Gutierrez, A., Carrea, A., Forsberg, A. M., Nilsson, E., et al. (2002). Free amino acids in plasma, red blood cells, polymorphonuclear leukocytes, and muscle in normal and uraemic children. *Nephrology Dialysis Transplantation, 17*, 413–421.

Caspers, P. J., Lucassen, G. W., Carter, E. A., Bruining, H. A., & Puppels, G. J. (2001). In vivo confocal Raman microspectroscopy of the skin: Noninvasive determination of molecular concentration profiles. *Journal of Investigative Dermatology, 116*, 434–442.

Chakraborty, T., & Petruccelli, J. C. (2017). Source diversity for transport of intensity phase imaging. *Optics Express, 25*, 9122–9137.

Charrière, F., Marian, A., Montfort, F., Kuehn, J., Colomb, T., Cuche, E., et al. (2016). Cell refractive index tomography by digital holographic microscopy. *Optics Letters, 31*, 178–180.

Chen, B., & Stamnes, J. J. (2009). Validity of diffraction tomography based on the first Born and the first Rytov approximations. *Applied Optics, 37*, 2996–3006.

Chen, M., Tian, L., & Waller, L. (2016). 3D differential phase contrast microscopy. *Biomedical Optics Express, 7*, 3940–3950.

Cheung, M. C., Evans, J. G., McKenna, B., & Ehrlich, D. J. (2011). Deep ultraviolet mapping of intracellular protein and nucleic acid in femtograms per pixel. *Cytometry, Part A, 79*, 920–932.

Cheung, M. C., LaCroix, R., McKenna, B. K., Liu, L., Winkelman, J., & Ehrlich, D. J. (2013). Intracellular protein and nucleic acid measured in eight cell types using deep-ultraviolet mass mapping. *Cytometry, Part A, 83*, 540–551.

Choi, W., Fang-Yen, C., Badizadegan, K., Oh, S., Lue, N., Dasari, R. R., et al. (2007). Tomographic phase microscopy. *Nature Methods, 4*, 717–719.

Cocuron, J. C., Tsogtbaatar, E., & Alonso, A. P. (2017). High-throughput quantification of the levels and labeling abundance of free amino acids by liquid chromatography tandem mass spectrometry. *Journal of Chromatography A, 1490*, 148–155.

Cohen, G. M., Sun, X. M., Snowden, R. T., Dinsdale, D., & Skilleter, D. N. (1992). Key morphological features of apoptosis may occur in the absence of internucleosomal DNA fragmentation. *Biochemical Journal, 286*, 331–334.

Conway, E. J. (1957). Nature and significance of concentration relations of potassium and sodium ions in skeletal muscle. *Physiological Reviews, 37*, 84–132.

Cooper, G. M. (2000). *The cell: A molecular approach* (2nd ed.). Sunderland: Sinauer Associates.

Darling, T. N., Burrows, C. M., & Blum, J. J. (1990). Rapid shape change and release of ninhydrin-positive substances by Leishmania major promastigotes in response to hypoosmotic stress. *The Journal of Protozoology, 37*, 493–499.

Davidson, A. F., & Higgins, A. Z. (2013). Detection of volume changes in calcein-stained cells using confocal microscopy. *Journal of Fluorescence, 23*, 393–398.

Davis, M. J. (2016). Protein oxidation and peroxidation. *Biochemical Journal, 473*, 805–825.

Davis, H. G., Wilkins, M. H. F., Chayen, J., & La Cour, L. F. (1954). The use of the interference microscope to determine dry mass in living cells and as a quantitative cytochemical method. *Quarterly Journal of Microscopical Science, 95*, 271–304.

de Paoli, F. V., Overgaard, K., Pedersen, T. H., & Nielsen, O. B. (2007). Additive protective effects of the addition of lactic acid and adrenaline on excitability and force in isolated rat skeletal muscle depressed by elevated extracellular K^+. *The Journal of Physiology, 581*, 829–839.

Decker, R. S., & Wildenthal, K. (1980). Lysosomal alterations in hypoxic and reoxygenated hearts. I. Ultrastructural and cytochemical changes. *American Journal of Pathology, 98*, 425–444.

Devaney, A. J., & Schatzberg, A. (1992). Coherent optical tomographic microscope. *Proceedings of SPIE, 1767*, 62–71.

Di Sant'Agnese, P. A., Jensen, K. D., Levin, A., & Miller, R. K. (1983). Placental toxicity of cadmium in the rat: An ultrastructural study. *Placenta, 4*, 149–163.

Ding, H., & Popescu, G. (2010). Instantaneous spatial light interference microscopy. *Optics Express, 18*, 1569–1575.

Dizdaroglu, M., Jaruga, P., Birincioglu, M., & Rodriguez, H. (2002). Free radical-induced damage to DNA: Mechanisms and measurement. *Free Radical Biology and Medicine, 32*, 1102–1115.

Durchfort, N., Verhoef, S., Vaughn, M. B., Shrestha, R., Adam, D., Kaplan, J., et al. (2012). The enlarged lysosomes in *beige*$_j$ cells result from decreased lysosome fission and not increased lysosome fusion. *Traffic, 13*, 108–119.

Egawa, M., Hirao, T., & Takahashi, M. (2007). In vivo estimation of stratum corneum thickness from water concentration profiles obtained with Raman spectroscopy. *Acta Dermato-Venereologica, 87*, 4–8.

Eisenhofer, S., Toókos, F., Hense, B. A., Schulz, S., Filbir, F., & Zischka, H. (2010). A mathematical model of mitochondrial swelling. *BMC Research Notes, 3*, 67.

Ellis, R. J. (2001). Macromolecular crowding: Obvious but underappreciated. *Trends in Biochemical Sciences, 26*, 597–604.

Elmore, S. (2007). Apoptosis: A review of programmed cell death. *Toxicologic Pathology, 35*, 495–516.

Ernest, N. J., Habela, C. W., & Sontheimer, H. (2008). Cytoplasmic condensation is both necessary and sufficient to induce apoptotic cell death. *Journal of Cell Science, 121*, 290–297.

Esteva-Font, C., Phuan, P. W., Anderson, M. O., & Verkman, A. S. (2013). A small molecule screen identifies selective inhibitors of urea transporter UT-A. *Chemical Biology, 20*, 1235–1244.

Feijó Delgado, F., Cermak, N., Hecht, V. C., Son, S., Li, Y., Knudsen, S. M., et al. (2013). Intracellular water exchange for measuring the dry mass, water mass and changes in chemical composition of living cells. *PLoS One, 8*, e67590.

Fernández-Segura, E., Cañizares, F. J., Cubero, M. A., Warley, A., & Campos, A. (1999). Changes in elemental content during apoptotic cell death studied by electron probe X-ray microanalysis. *Experimental Cell Research, 253*, 454–462.

Fontaine, E., & Bernardi, P. (2016). Lethal and nonlethal functions of the permeability transition pore. In D. Hockenbery (Ed.), *Mitochondria and cell death. Cell death in biology and diseases* (pp. 1–15). New York: Humana Press.

Franco, R., & Cidlowski, J. A. (2006). SLCO/OATP-like transport of glutathione in FasL-induced apoptosis: Glutathione efflux is coupled to an organic anion exchange and is necessary for the progression of the execution phase of apoptosis. *The Journal of Biological Chemistry, 281*, 29542–29557.

Fraser, J. A., & Huang, C. L. (2004). A quantitative analysis of cell volume and resting potential determination and regulation in excitable cells. *The Journal of Physiology, 559*, 459–478.

Fraser, J. A., & Huang, C. L. (2007). Quantitative techniques for steady-state calculation and dynamic integrated modelling of membrane potential and intracellular ion concentrations. *Progress in Biophysics and Molecular Biology, 94*, 336–372.

Fraser, J. A., Middlebrook, C. E., Usher-Smith, J. A., Schwiening, C. J., & Huang, C. L. (2005). The effect of intracellular acidification on the relationship between cell volume and membrane potential in amphibian skeletal muscle. *The Journal of Physiology, 563*, 745–764.

Friedman, M. (2004). Applications of the ninhydrin reaction for analysis of amino acids, peptides, and proteins to agricultural and biomedical sciences. *Journal of Agricultural and Food Chemistry, 52*, 385–406.

Friis, M. B., Friborg, C. R., Schneider, L., Nielsen, M. B., Lambert, I. H., Christensen, S. T., et al. (2005). Cell shrinkage as a signal to apoptosis in NIH 3T3 fibroblasts. *The Journal of Physiology, 567*, 427–443.

Fu, D., Choi, W., Sung, Y., Oh, S., Yaqoob, Z., Park, Y., et al. (2009). Ultraviolet refractometry using field-based light scattering spectroscopy. *Optics Express, 17*, 18878–18886.

Fumarola, C., Zerbini, A., & Guidotti, G. G. (2001). Glutamine deprivation-mediated cell shrinkage induces ligand-independent CD95 receptor signaling and apoptosis. *Cell Death & Differentiation, 8*, 1004–1013.

Gao, W., Pu, Y., Luo, K. Q., & Chang, D. C. (2001). Temporal relationship between cytochrome c release and mitochondrial swelling during UV-induced apoptosis in living HeLa cells. *Journal of Cell Science, 114*, 2855–2862.

Gass, J., Dakoff, A., & Kim, M. K. (2003). Phase imaging without the 2π ambiguity by multiwavelength digital holography. *Optics Letters, 28*, 1141–1143.
Germond, A., Fujita, H., Ichimura, T., & Watanabe, T. M. (2016). Design and development of genetically encoded fluorescent sensors to monitor intracellular chemical and physical parameters. *Biophysical Reviews, 8*, 121–138.
Gibbons, B. A., Kharel, P., Robinson, L. C., Synowicki, R. A., & Model, M. A. (2016). Volume measurements and fluorescent staining indicate an increase in permeability for organic cation transporter substrates during apoptosis. *Experimental Cell Research, 344*, 112–119.
Gillooly, J. F., Hein, A., & Damiani, R. (2015). Nuclear DNA content varies with cell size across human cell types. *Cold Spring Harbor Perspectives in Biology, 7*, a019091.
Gnutt, D., Brylski, O., Edengeiser, E., Havenith, M., & Ebbinghaus, S. (2017). Imperfect crowding adaptation of mammalian cells towards osmotic stress and its modulation by osmolytes. *Molecular Biosystems, 13*, 2218–2221.
Gnutt, D., Gao, M., Brylski, O., Heyden, M., & Ebbinghaus, S. (2015). Excluded-volume effects in living cells. *Angewandte Chemie, 54*, 2548–2551.
Godden, T. M., Suman, R., Humphry, M. J., Rodenburg, J. M., & Maiden, A. M. (2014). Ptychographic microscope for three-dimensional imaging. *Optics Express, 22*, 12513–12523.
Gregg, J. L., McGuire, K. M., Focht, D. C., & Model, M. A. (2010). Measurement of the thickness and volume of adherent cells using transmission-through-dye microscopy. *Pflügers Archiv - European Journal of Physiology, 460*, 1097–1104.
Grinstein, S., & Foskett, J. K. (1990). Ionic mechanisms of cell volume regulation in leukocytes. *Annual Review of Physiology, 52*, 399–414.
Grover, W. H., Bryan, A. K., Diez-Silva, M., Suresh, S., Higgins, J. M., & Manalis, S. R. (2011). Measuring single-cell density. *Proceedings of the National Academy of Sciences of the United States of America, 108*, 10992–10996.
Guigas, G., Kalla, C., & Weiss, M. (2007). Probing the nanoscale viscoelasticity of intracellular fluids in living cells. *Biophysical Journal, 93*, 316–323.
Guigay, J. P. (1971). On Fresnel diffraction by one-dimensional periodic objects, with application to structure determination of phase objects. *Optica Acta, 18*, 677–682.
Guo, B., Jing, J., Nie, L., Xin, F., Gao, C., Yang, W., et al. (2018). A lysosome targetable versatile fluorescent probe for imaging viscosity and peroxynitrite with different fluorescence signals in living cells. *Journal of Materials Chemistry B, 6*, 580–585.
Guo, M., Pegoraro, A. F., Mao, A., Zhou, E. H., Arany, P. R., Han, Y., et al. (2017). Cell volume change through water efflux impacts cell stiffness and stem cell fate. *Proceedings of the National Academy of Sciences of the United States of America, 114*, E8618–E8627.
Gureyev, T. E., Nesterets, Y. I., Paganin, D. M., Pogany, A., & Wilkins, S. W. (2006). Linear algorithms for phase retrieval in the Fresnel region. 2. Partially Coherent Illumination. *Optics Communications, 259*, 569–580.
Haeberlé, O., Belkebir, K., Giovaninni, H., & Sentenac, A. (2010). Tomographic diffractive microscopy: Basics, techniques and perspectives. *Journal of Modern Optics, 57*, 686–699.
Hagedorn, M., Kleinhans, F. W., Freitas, R., Liu, J., Hsu, E. W., Wildt, D. E., et al. (1997). Water distribution and permeability of zebrafish embryos, *Brachydanio rerio*. *The Journal of Experimental Zoology, 278*, 356–371.
Halestrap, A. P. (1989). The regulation of the matrix volume of mammalian mitochondria in vivo and in vitro and its role in the control of mitochondrial metabolism. *Biochimica et Biophysica Acta, 973*, 355–382.
Halliwell, B., & Gutteridge, J. M. C. (2007). *Free radicals in biology and medicine* (4th ed.). New York: Oxford University Press.

Haseda, K., Kanematsu, K., Noguchi, K., Saito, H., Umeda, N., & Ohta, Y. (2015). Significant correlation between refractive index and activity of mitochondria: Single mitochondrion study. *Biomedical Optics Express, 6*, 859–869.
Hendil, K. B., & Hoffmann, E. K. (1974). Cell volume regulation in Ehrlich ascites tumor cells. *Journal of Cellular Physiology, 84*, 115–125.
Hessler, J. A., Budor, A., Putchakayala, K., Mecke, A., Rieger, D., Banaszak Holl, M. M., et al. (2005). Atomic force microscopy study of early morphological changes during apoptosis. *Langmuir: the ACS Journal of Surfaces and Colloids, 21*, 9280–9286.
Hoffman, D., Kumar, A. M., Spitzer, A., & Gupta, R. K. (1986). NMR measurement of intracellular water volume in rat kidney proximal tubules. *Biochimica et Biophysica Acta, 889*, 355–360.
Hoffmann, E. K., & Hendil, K. B. (1976). The role of amino acids and taurine in isosmotic intracellular regulation in Erlich ascites mouse tumour cells. *Journal of Comparative Physiology, 108*, 279–286.
Hoffmann, E. K., Lambert, I. H., & Pedersen, S. F. (2009). Physiology of cell volume regulation in vertebrates. *Physiological Reviews, 89*, 193–277.
Horstmeyer, R., Chung, J., Xiaoze, O., Zheng, G., & Yang, C. (2016). Diffraction tomography with Fourier ptychography. *Optica, 8*, 827–835.
Hosokawa, N., Hara, Y., & Mizushima, N. (2007). Generation of cell lines with tetracycline-regulated autophagy and a role for autophagy in controlling cell size. *FEBS Letters, 581*, 2623–2629.
Huang, J. Q., Trasler, J. M., Igdoura, S., Michaud, J., Hanal, N., & Gravel, R. A. (1997). Apoptotic cell death in mouse models of GM2 gangliosidosis and observations on human Tay-Sachs and Sandhoff diseases. *Human Molecular Genetics, 6*, 1879–1885.
Hughes, F. M., Jr., Bortner, C. D., Purdy, G. D., & Cidlowski, J. A. (1997). Intracellular K^+ suppresses the activation of apoptosis in lymphocytes. *The Journal of Biological Chemistry, 272*, 30567–30576.
Huizinga, A., Bot, A. C., de Mul, F. F., Vrensen, G. F., & Greve, J. (1989). Local variation in absolute water content of human and rabbit eye lenses measured by Raman microspectroscopy. *Experimental Eye Research, 48*, 487–496.
l'Hoste, S., Chargui, A., Belfodil, R., Corcelle, E., Duranton, C., Rubera, I., et al. (2010). CFTR mediates apoptotic volume decrease and cell death by controlling glutathione efflux and ros production in cultured mice proximal tubules. *American Journal of Physiology. Renal Physiology, 298*, F435–F453.
Ikeda, T., Popescu, G., Dasari, R. R., & Feld, M. S. (2005). Hilbert phase microscopy for investigating fast dynamics in transparent systems. *Optics Letters, 30*, 1165–1167.
Ishida, Y., Nayak, S., Mindell, J. A., & Grabe, M. (2013). A model of lysosomal pH regulation. *The Journal of General Physiology, 141*, 705–720.
Jakobsson, E. (1980). Interactions of cell volume, membrane potential, and membrane transport parameters. *American Journal of Physiology. Cell Physiology, 238*, C196–C206.
Jakobs, S., & Wurm, C. A. (2014). Super-resolution microscopy of mitochondria. *Current Opinion in Chemical Biology, 20*, 9–15.
Jiang, N., Fan, J., Zhang, S., Wu, T., Wang, J., Gao, P., et al. (2014). Dual mode monitoring probe for mitochondrial viscosity in single cell. *Sensors and Actuators B, 190*, 685–693.
Jingshan, Z., Claus, R. A., Dauwels, J., Tian, L., & Waller, L. (2014). Transport of intensity phase imaging by intensity spectrum fitting of exponentially spaced defocus planes. *Optics Express, 22*, 18125–18137.
Jones, C., & Suhling, K. (2006). Mapping the refractive index sensing range of the GFP fluorescence decay with FLIM. *Proceedings of SPIE, 6098*, 1–11.
Kaasik, A., Safiulina, D., Zharkovsky, A., & Veksler, V. (2007). Regulation of mitochondrial matrix volume. *American Journal of Physiology. Cell Physiology, 292*, C157–C163.

Kamilov, U. S., Papadopoulos, I. N., Shoreh, M. H., Goy, A., Vonesch, C., Unser, M., et al. (2015). Learning approach to optical tomography. *Optica, 2*, 517–522.

Kao, H. P., Abney, J. R., & Verkman, A. S. (1993). Determinants of the translational mobility of a small solute in cell cytoplasm. *The Journal of Cell Biology, 120*, 175–184.

Kay, A. R. (2017). How cells can control their size by pumping ions. *Frontiers in Cell and Developmental Biology, 5*, 41.

Kemp, G. (2007). Muscle cell volume and pH changes due to glycolytic ATP synthesis. *The Journal of Physiology, 582*, 461–465.

Kemper, B., Carl, D., Schnekenburger, J., Bredebusch, I., Schafer, M., Domschke, W., et al. (2006). Investigation of living pancreas tumor cells by digital holographic microscopy. *Journal of Biomedical Optics, 11*, 033001.

Kerr, J. F. (1971). Shrinkage necrosis: A distinct mode of cellular death. *The Journal of Pathology, 105*, 13–20.

Khmaladze, A. (2017). Examining live cell culture cultures during apoptosis by digital holographic phase imaging and Raman spectroscopy. *Journal of Physics: Conference Series, 909*, 012001.

Khmaladze, A., Kim, M., & Lo, C. M. (2008). Phase imaging of cells by simultaneous dual-wavelength reflection digital holography. *Optics Express, 16*, 10900–10911.

Khmaladze, A., Matz, R. L., Epstein, T., Jasensky, J., Banaszak Holl, M. M., & Chen, Z. (2012). Cell volume changes during apoptosis monitored in real time using digital holographic microscopy. *Journal of Structural Biology, 178*, 270–278.

Kiehl, T. R., Shen, D., Khattak, S. F., Jian Li, Z., & Sharfstein, S. T. (2011). Observations of cell size dynamics under osmotic stress. *Cytometry, Part A, 79*, 560–569.

Kim, M. K. (2010). Principles and techniques of digital holographic microscopy. *SPIE Reviews, 1*, 018005-1–018005-50.

Kim, M. K. (2011). *Digital holographic microscopy: Principles, techniques, and applications*. New York: Springer.

Kimelberg, H. K. (2005). Astrocytic swelling in cerebral ischemia as a possible cause of injury and target for therapy. *Glia, 50*, 389–397.

Kim, Y., Shim, H., Kim, K., Park, H., Heo, J. H., Yoon, J., et al. (2014a). Common-path diffraction optical tomography for investigation of three-dimensional structures and dynamics of biological cells. *Optics Express, 22*, 10398–10407.

Kim, T., Zhou, R., Mir, M., Babacan, S. D., Carney, P. S., Goddard, L. L., et al. (2014b). White-light diffraction tomography of unlabelled live cells. *Nature Photonics, 8*, 256–263.

Ko, S. K., Kim, S. K., Share, A., Lynch, V. M., Park, J., Namkung, W., Van Rossom, W., Busschaert, N., Gale, P. A., Sessler, J. L., & Shin, I. (2014). Synthetic ion transporters can induce apoptosis by facilitating chloride anion transport into cells. *Nat Chem, 6*, 885–892.

Kosaihira, A., & Ona, T. (2008). Rapid and quantitative method for evaluating the personal therapeutic potential of cancer drugs. *Analytical and Bioanalytical Chemistry, 391*, 1889–1897.

Kostenko, A., Batenburg, K. J., King, A., Offerman, S. E., & van Vliet, L. J. (2013). Total variation minimization approach in in-line x-ray phase-contrast tomography. *Optics Express, 21*, 12185–12196.

Kowaltowski, A. J., Seetharaman, S., Paucek, P., & Garlid, K. D. (2001). Bioenergetic consequences of opening the ATP-sensitive K^+ channel of heart mitochondria. *American Journal of Physiology. Heart and Circulatory Physiology, 280*, H649–H657.

Kramer, P. J. (1988). Measurement of plant water status: Historical perspectives and current concerns. *Irrigation Science, 9*, 275–287.

Kreis, R., Ernst, T., & Ross, B. D. (1993). Development of the human brain: In vivo quantification of metabolite and water content with proton magnetic resonance spectroscopy. *Magnetic Resonance in Medicine, 30*, 424–437.

Kubitschek, H. E., & Friske, J. A. (1986). Determination of bacterial cell volume with the Coulter Counter. *Journal of Bacteriology, 168*, 1466–1467.

Kuimova, M. K., Botchway, S. W., Parker, A. W., Balaz, M., Collins, H. A., Anderson, H. L., et al. (2009). Imaging intracellular viscosity of a single cell during photoinduced cell death. *Nature Chemistry, 1*, 69–73.

Kuimova, M. K., Yahioglu, G., Levitt, J. A., & Suhling, K. (2008). Molecular rotor measures viscosity of live cells via fluorescence lifetime imaging. *Journal of the American Chemical Society, 130*, 6672–6673.

Kunzelmann, K. (2016). Ion channels in regulated cell death. *Cellular and Molecular Life Sciences: CMLS, 73*, 2387–2403.

Lang, F. (2007). Mechanisms and significance of cell volume regulation. *Journal of the American College of Nutrition, 26*, 613S–623S.

Lang, F., Busch, G. L., Ritter, M., Völkl, H., Waldegger, S., Gulbins, E., et al. (1998). Functional significance of cell volume regulatory mechanisms. *Physiological Reviews, 78*, 247–306.

Lang, F., & Hoffmann, E. K. (2012). Role of ion transport in control of apoptotic cell death. *Comprehensive Physiology, 2*, 2037–2061.

Law, R. O. (1991). Amino acids as volume-regulatory osmolytes in mammalian cells. *Comparative Biochemistry and Physiology. A, Comparative Physiology, 99*, 263–277.

Law, R. O., & Turner, D. P. (1987). Are ninhydrin-positive substances volume-regulatory osmolytes in rat renal papillary cells? *The Journal of Physiology, 386*, 45–61.

Leanza, L., Biasutto, L., Managò, A., Gulbins, E., Zoratti, M., & Szabò, I. (2013). Intracellular ion channels and cancer. *Frontiers in Physiology, 4*, 227.

Lee, H., Woo, E. R., & Lee, D. G. (2018). Apigenin induces cell shrinkage in *Candida albicans* by membrane perturbation. *FEMS Yeast Research, 18*, 1–9.

Lejeune, E., & Linder, C. (2018). Understanding the relationship between cell death and tissue shrinkage via a stochastic agent-based model. *Journal of Biomechanics*. https://doi.org/10.1016/j.jbiomech.2018.03.019. pii: S0021–9290(18)30197-0.

Lew, V. L., & Bookchin, R. M. (1986). Volume, pH, and ion-content regulation in human red cells: Analysis of transient behavior with an integrated model. *Journal of Membrane Biology, 92*, 57–74.

Lew, V. L., Freeman, C. J., Ortiz, O. E., & Bookchin, R. M. (1991). A mathematical model of the volume, pH, and ion content regulation in reticulocytes. Application to the pathophysiology of sickle cell dehydration. *Journal of Clinical Investigation, 87*, 100–112.

Li, J., Chen, Q., Zhang, J., Zhang, Y., Lu, L., & Zuo, C. (2017a). Efficient quantitative phase microscopy using programmable annular LED illumination. *Biomedical Optics Express, 8*, 4687.

Li, J., Chen, Q., Zhang, J., Zhang, Z., Zhang, Y., & Zuo, C. (2017b). Optical diffraction tomography microscopy with transport of intensity equation using a light-emitting diode array. *Optics and Lasers in Engineering, 95*, 26–34.

Limozin, L., & Sengupta, K. (2009). Quantitative reflection interference contrast microscopy (RICM) in soft matter and cell adhesion. *Chemphyschem: a European Journal of Chemical Physics and Physical Chemistry, 10*, 2752–2768.

Liu, P. Y., Chin, L. K., Ser, W., Chen, H. F., Hsieh, C. M., Lee, C. H., et al. (2016). Cell refractive index for cell biology and disease diagnosis: Past, present and future. *Lab on a Chip, 16*, 634–644.

Liu, P. Y., Hsieh, C. M., Chin, L. K., Leprince-Wang, Y., Ser, W., Bourouina, T., et al. (2017). *3D refractive index mapping of single cells*. San Jose, USA: CLEO: Applications and Technology.

Liu, T., Liu, X., Spring, D. R., Qian, X., Cui, J., & Xu, Z. (2014). Quantitatively mapping cellular viscosity with detailed organelle information via a designed PET fluorescent probe. *Scientific Reports, 4*, 5418.

Liu, F., Wu, T., Cao, J., Cui, S., Yang, Z., Qiang, X., et al. (2013). Ratiometric detection of viscosity using a two-photon fluorescent sensor. *Chemistry: a European Journal, 19*, 1548–1553.

Lloyd, J. B. (1996). The taxonomy of lysosomes and related structures. In J. B. Lloyd, & R. W. Mason (Eds.), *Subcellular biochemistry: Vol. 27. Biology of the lysosome* (pp. 1–13). New York: Springer.

Lübke, T., Lobel, P., & Sleat, D. E. (2009). Proteomics of the lysosome. *Biochimica et Biophysica Acta, 1793*, 625–635.

Luby-Phelps, K. (2000). Cytoarchitecture and physical properties of cytoplasm: Volume, viscosity, diffusion, intracellular surface area. *International Review of Cytology, 192*, 189–221.

Lue, N., Choi, W., Popescu, G., Ikeda, T., Dasari, R. R., Badizadegan, K., et al. (2007). Quantitative phase imaging of live cells using fast Fourier phase microscopy. *Applied Optics, 46*, 1836–1842.

Lue, N., Popescu, G., Ikeda, T., Dasari, R. R., Badizadegan, K., & Feld, M. S. (2006). Live cell refractometry using microfluidic devices. *Optics Letters, 31*, 2759–2761.

Machiyama, H., Morikawa, T. J., Okamoto, K., Watanabe, T. M., & Fujita, H. (2017). The use of a genetically encoded molecular crowding sensor in various biological phenomena. *Biophysics and Physicobiology, 14*, 119–125.

Maeno, E., Takahashi, N., & Okada, Y. (2006). Dysfunction of regulatory volume increase is a key component of apoptosis. *FEBS Letters, 580*, 6513–6517.

Mann, C. J., Yu, L. F., & Kim, M. K. (2006). Movies of cellular and sub-cellular motion by digital holographic microscopy. *BioMedical Engineering Online, 5*, 2110.

Mann, C. J., Yu, L. F., Lo, C. M., & Kim, M. K. (2005). High-resolution quantitative phase-contrast microscopy by digital holography. *Optics Express, 13*, 8693–8698.

Marguet, P., Depeursinge, C., & Magistretti, P. J. (2014). Review of quantitative phase-digital holographic microscopy: Promising novel imaging technique to resolve neuronal network activity and identify cellular biomarkers of psychiatric disorders. *Neurophotonics, 1*, 020901.

Marquet, P., Rappaz, B., Magistretti, P. J., Cuche, E., Emery, Y., Colomb, T., et al. (2005). Digital holographic microscopy: A noninvasive contrast imaging technique allowing quantitative visualization of living cells with subwavelength axial accuracy. *Optics Letters, 30*, 468–470.

Matzke, A. J., Weiger, T. M., & Matzke, M. (2010). Ion channels at the nucleus: Electrophysiology meets the genome. *Molecular Plant, 3*, 642–652.

Maughan, D. W., & Godt, R. E. (2001). Protein osmotic pressure and the state of water in frog myoplasm. *Biophysical Journal, 80*, 435–442.

McCarthy, J. V., & Cotter, T. G. (1997). Cell shrinkage and apoptosis: A role for potassium and sodium ion efflux. *Cell Death & Differentiation, 4*, 756–770.

Mehta, S. B., & Sheppard, C. J. R. (2009). Quantitative phase-gradient imaging at high resolution with asymmetric illumination-based differential phase contrast. *Optics Letters, 34*, 1924–1926.

Miermont, A., Waharte, F., Hu, S., McClean, M. N., Bottani, S., Léon, S., et al. (2013). Severe osmotic compression triggers a slowdown of intracellular signaling, which can be explained by molecular crowding. *Proceedings of the National Academy of Sciences of the United States of America, 110*, 5725–5730.

Miettinen, T. P., & Björklund, M. (2015). Mevalonate pathway regulates cell size homeostasis and proteostasis through autophagy. *Cell Reports, 13*, 2610–2620.

Minton, A. P. (2001). The influence of macromolecular crowding and macromolecular confinement on biochemical reactions in physiological media. *The Journal of Biological Chemistry, 276*, 10577–10580.

Minton, A. P. (2008). Effective hard particle model for the osmotic pressure of highly concentrated binary protein solutions. *Biophysical Journal, 94*, L57–L59.

Model, M. A. (2014). Possible causes of apoptotic volume decrease: An attempt at quantitative review. *American Journal of Physiology. Cell Physiology, 306*, C417–C424.

Model, M. A. (2018). Methods for cell volume measurement. *Cytometry, 93A*, 281–296.

Model, M. A., & Schonbrun, E. (2013). Optical determination of intracellular water in apoptotic cells. *The Journal of Physiology, 591*, 5843–5849.

Model, M. A., Mudrak, N. J., Rana, P. S., & Clements, R. J. (2018). Staurosporine-induced apoptotic water loss is cell- and attachment-specific. *Apoptosis*. Accepted.

Moh, K. J., Yuan, X. C., Bu, J., Zhu, S. W., & Gao, B. Z. (2008). Surface plasmon resonance imaging of cell-substrate contacts with radially polarized beams. *Optics Express, 16*, 20734–20741.

Morikawa, T. J., Fujita, H., Kitamura, A., Horio, T., Yamamoto, J., Kinjo, M., et al. (2016). Dependence of fluorescent protein brightness on protein concentration in solution and enhancement of it. *Scientific Reports, 6*, 22342.

Mourant, J. R., Canpolat, M., Brocker, C., Esponda-Ramos, O., Johnson, T. M., Matanock, A., et al. (2000). Light scattering from cells: The contribution of the nucleus and the effects of proliferative status. *Journal of Biomedical Optics, 5*, 131–137.

Mudrak, N. J., Rana, P. S., & Model, M. A. (2018). Calibrated brightfield-based imaging for measuring intracellular protein concentration. *Cytometry, 93A*, 297–304.

Mugnano, M., Calabuig, A., Grilli, S., Miccio, L., & Ferraro, P. (2015). Monitoring cell morphology during necrosis and apoptosis by quantitative phase imaging. *Proceedings of SPIE, 9529*, 952909–952911.

Nietsch, H. H., Roe, M. W., Fiekers, J. F., Moore, A. L., & Lidofsky, S. D. (2000). Activation of potassium and chloride channels by tumor necrosis factor alpha. Role in liver cell death. *The Journal of Biological Chemistry, 275*, 20556–20561.

Nobel, C. S., Aronson, J. K., van den Dobbelsteen, D. J., & Slater, A. F. (2000). Inhibition of Na^+/K^+-ATPase may be one mechanism contributing to potassium efflux and cell shrinkage in CD95-induced apoptosis. *Apoptosis: an International Journal on Programmed Cell Death, 5*, 153–163.

Noda, T., Kawata, S., & Minami, S. (1990). Three-dimensional phase contrast imaging by an annular illumination microscope. *Applied Optics, 29*, 3810–3815.

Ocon, P., Acerete, C., & Reboiras, M. D. (1987). Activity coefficients of KCl in highly concentrated protein solutions. *European Biophysical Journal, 14*, 477–484.

Okada, Y., Shimizu, T., Maeno, E., Tanabe, S., Wang, X., & Takahashi, N. (2006). Volume-sensitive chloride channels involved in apoptotic volume decrease and cell death. *Journal of Membrane Biology, 209*, 21–29.

Orlov, S. N., Platonova, A. A., Hamet, P., & Grygorczyk, R. (2013). Cell volume and monovalent ion transporters: Their role in cell death machinery triggering and progression. *American Journal of Physiology. Cell Physiology, 305*, C361–C372.

Ovádi, J., & Saks, V. (2004). On the origin of intracellular compartmentation and organized metabolic systems. *Molecular and Cellular Biochemistry, 256-257*, 5–12.

Ozturk, S. S., & Palsson, B. O. (1991). Effect of medium osmolarity on hybridoma growth, metabolism, and antibody production. *Biotechnology and Bioengineering, 37*, 989–993.

Park, Y., Popescu, G., Badizadegan, K., Dasari, R. R., & Feld, M. S. (2006). Diffraction phase and fluorescence microscopy. *Optics Express, 14*, 8263–8268.

Patterson, S. D., Grossman, J. S., D'Andrea, P., & Latter, G. I. (1995). Reduced numatrin/B23/nucleophosmin labeling in apoptotic Jurkat T-lymphoblasts. *The Journal of Biological Chemistry, 270*, 9429–9436.

Petruccelli, J. C., Tian, L., & Barbastathis, G. (2013). The transport of intensity equation for optical path length recovery using partially coherent illumination. *Optics Express, 21*, 14430–14441.

Pfister, H., & Pauly, H. (1972). Chemical potential of KCl and its ion constituents in concentrated protein salt solutions. *Journal of Polymer Science, 39*, 179–189.

Piez, K. A., & Eagle, H. (1958). The free amino acid pool of cultured human cells. *The Journal of Biological Chemistry, 231*, 533–545.

Popescu, G. (2011). *Quantitative phase imaging of cells and tissues*. New York: McGraw Hill.

Popescu, G., Deflores, L. P., Vaughan, J. C., Badizadegan, K., Iwai, H., Dasari, R. R., et al. (2004). Fourier phase microscopy for investigation of biological structures and dynamics. *Optics Letters, 29*, 2503–2505.

Popescu, G., Ikeda, T., Best, C., Badizadegan, K., Dasari, R. R., & Feld, M. S. (2006a). Erythrocyte structure and dynamics quantified by Hilbert phase microscopy. *Journal of Biomedical Optics, 10*, 060503.

Popescu, G., Ikeda, T., Dasari, R. R., & Feld, M. S. (2006b). Diffraction phase microscopy for quantifying cell structure and dynamics. *Optics Letters, 31*, 775–777.

Popescu, G., Park, Y., Choi, W., Dasari, R. R., Feld, M. S., & Badizadegan, K. (2008). Imaging red blood cell dynamics by quantitative phase microscopy. *Blood Cells, Molecules & Diseases, 41*, 10–16.

Porcelli, A. M., Ghelli, A., Zanna, C., Valente, P., Ferroni, S., & Rugolo, M. (2004). Apoptosis induced by staurosporine in ECV304 cells requires cell shrinkage and upregulation of Cl^- conductance. *Cell Death & Differentiation, 11*, 655–662.

Poulsen, K. A., Andersen, E. C., Hansen, C. F., Klausen, T. K., Hougaard, C., Lambert, I. H., et al. (2010). Deregulation of apoptotic volume decrease and ionic movements in multidrug-resistant tumor cells: Role of chloride channels. *American Journal of Physiology. Cell Physiology, 298*, C14–C25.

Puchkov, E. O. (2013). Intracellular viscosity: Methods of measurement and role in metabolism. *Biochemistry (Moscow) Supplement Series A, 7*, 270–279.

Rana, P. S., Mudrak, N. J., Lopez, R., Lynn, M., Kershner, L., & Model, M. A. (2017). Phase separation in necrotic cells. *Biochemical and Biophysical Research Communications, 492*, 300–303.

Rappaz, B., Barbul, A., Hoffmann, A., Boss, D., Korenstein, R., Depeursinge, C., et al. (2009). Spatial analysis of erythrocyte membrane fluctuations by digital holographic microscopy. *Blood Cells, Molecules, and Diseases, 42*, 228–232.

Rappaz, B., Marquet, P., Cuche, E., Emery, Y., Depeursinge, C., & Magistretti, P. J. (2005). Measurement of the integral refractive index and dynamic cell morphometry of living cells with digital holographic microscopy. *Optics Express, 13*, 9361–9373.

Riqing, Z., & Nanming, Z. (1997). Intracellular water volume measurement using NMR. *Tsinghua Science and Technology, 2*, 516–517.

Safiulina, D., Veksler, V., Zharkovsky, A., & Kaasik, A. (2006). Loss of mitochondrial membrane potential is associated with increase in mitochondrial volume: Physiological role in neurones. *Journal of Cellular Physiology, 206*, 347–353.

Saha, T., Hossain, M. S., Saha, D., Lahiri, M., & Talukdar, P. J. (2016). Chloride-Mediated Apoptosis-Inducing Activity of Bis(sulfonamide) Anionophores. *Am Chem Soc, 138*, 7558–7567.

Saraste, A., & Pulkki, K. (2000). Morphologic and biochemical hallmarks of apoptosis. *Cardiovascular Research, 45*, 528–537.

Savitz, D., Sidel, V. W., & Solomon, A. K. (1964). Osmotic properties of human red cells. *The Journal of General Physiology, 48*, 79–94.

Schmid, G., & Blanch, H. W. (1992). Extra- and intracellular metabolite concentrations for murine hybridoma cells. *Applied Microbiology and Biotechnology, 36*, 621–625.

Schneider, K., Krömer, J. O., Wittmann, C., Alves-Rodrigues, I., Meyerhans, A., Diez, J., et al. (2009). Metabolite profiling studies in *Saccharomyces cerevisiae*: An assisting tool to prioritize host targets for antiviral drug screening. *Microbial Cell Factories, 8*, 12.

Sekine, T., Cha, S. H., & Endou, H. (2000). The multispecific organic anion transporter (OAT) family. *Pflügers Archiv - European Journal of Physiology, 440*, 337–350.

Sharikova, A., Saide, G., Sfakis, L., Park, J. Y., Desta, H., Maloney, M. C., et al. (2017). Monitoring of live cell cultures during apoptosis by phase imaging and Raman spectroscopy. *Proceedings of SPIE, 10074*, 100740V–100741V.

Sharpe, P. T. (1988). Methods of cell separation. In *Laboratory techniques in biochemistry and molecular biology* (Vol. 18). Amsterdam: Elsevier.

Sheppard, C. J. R. (2004). Defocused transfer function for a partially coherent microscope and application to phase retrieval. *Journal of the Optical Society of America. A, Optics, Image Science, and Vision, 21*, 828–831.

Shinar, H., & Navon, G. (1985). The determination of intracellular water space by NMR. *FEBS Letters, 193*, 75–78.

Shumway, S. E., & Gabbott, P. A. (1977). The effect of fluctuating salinity on the concentrations of free amino acids and ninhydrin-positive substances in the adductor muscles of eight species of bivalve mollusks. *Journal of Experimental Marine Biology and Ecology, 29*, 131–150.

Soleimaninejad, H., Chen, M. Z., Lou, X., Smith, T. A., & Hong, Y. (2017). Measuring macromolecular crowding in cells through fluorescence anisotropy imaging with an AIE fluorogen. *Chemical Communications, 53*, 2874–2877.

Solenov, E., Watanabe, H., Manley, G. T., & Verkman, A. S. (2004). Sevenfold-reduced osmotic water permeability in primary astrocyte cultures from AQP-4-deficient mice, measured by a fluorescence quenching method. *American Journal of Physiology. Cell Physiology, 286*, C426–C432.

Solheim, A. E., & Seglen, P. O. (1983). Structural and physical changes in lysosomes from isolated rat hepatocytes treated with methylamine. *Biochimica et Biophysica Acta, 763*, 284–291.

Somlyo, A. P., Somlyo, A. V., & Shuman, H. (1979). Electron probe analysis of vascular smooth muscle. Composition of mitochondria, nuclei, and cytoplasm. *The Journal of Cell Biology, 81*, 316–335.

Srinivasan, B. (2015). Mitochondrial permeability transition pore: An enigmatic gatekeeper. *New Horizons in Science & Technology (NHS&T), 1*, 47–51.

Stadtman, E. R., & Levine, R. L. (2003). Free radical-mediated oxidation of free amino acids and amino acid residues in proteins. *Amino Acids, 25*, 207–218.

Stauber, T., & Jentsch, T. J. (2013). Chloride in vesicular trafficking and function. *Annual Review of Physiology, 75*, 453–477.

Streibl, N. (1984). Phase imaging by the transport equation of intensity. *Optics Communications, 49*, 6–10.

Streibl, N. (1985). Three-dimensional imaging by a microscope. *Journal of the Optical Society of America A, 2*, 121–127.

Suhling, K., Davis, D. M., & Phillips, D. (2002). The influence of solvent viscosity on the fluorescence decay and time-resolved anisotropy of green fluorescent protein. *Journal of Fluorescence, 12*, 91–95.

Suhling, K., Hirvonen, L. M., Levitt, J. A., Chung, P.-H., Tregidgo, C., Le Marois, A., et al. (2015). Fluorescence lifetime imaging (FLIM): Basic concept and some recent developments. *Medical Photonics, 27*, 3–40.

Suman, R., Smith, G., Hazel, K. E. A., Kasprowicz, R., Coles, M., O'Toole, P., et al. (2016). Label-free imaging to study phenotypic behavioural traits of cells in complex co-cultures. *Scientific Reports, 6*, 22032.

Sung, Y., Choi, W., Fang-Yen, C., Badizadegan, K., Dasari, R. R., & Feld, M. S. (2009). Optical diffraction tomography for high resolution live cell imaging. *Optics Express, 17*, 266–277.

Sung, Y., & Dasari, R. R. (2011). Deterministic regularization of three-dimensional optical diffraction tomography. *JOSA A, 28*, 1554–1561.

Swaminathan, R., Hoang, C. P., & Verkman, A. S. (1997). Photobleaching recovery and anisotropy decay of green fluorescent protein GFP-S65T in solution and cells: Cytoplasmic viscosity probed by green fluorescent protein translational and rotational diffusion. *Biophysical Journal, 72*, 1900–1907.

Takagi, M., Hayashi, H., & Yoshida, T. (2000). The effect of osmolarity on metabolism and morphology in adhesion and suspension Chinese hamster ovary cells producing tissue plasminogen activator. *Cytotechnology, 32*, 171–179.

Teague, M. R. (1983). Deterministic phase retrieval: A green's function solution. *Journal of the Optical Society of America, 73*, 1434–1441.

Terashima, K., Takeuchi, A., Sarai, N., Matsuoka, S., Shim, E. B., Leem, C. H., et al. (2006). Modelling Cl^- homeostasis and volume regulation of the cardiac cell. *Philosophical Transactions of the Royal Society A: Mathematical, Physical & Engineering Sciences, 364*, 1245–1265.

Terryn, C., Michel, J., Kilian, L., Bonhomme, P., & Balossier, G. (2000). Comparison of intracellular water content measurements by dark-field imaging and EELS in medium voltage TEM. *The European Physical Journal - Applied Physics, 11*, 215–226.

Theillet, F. X., Binolfi, A., Frembgen-Kesner, T., Hingorani, K., Sarkar, M., Kyne, C., et al. (2014). Physicochemical properties of cells and their effects on intrinsically disordered proteins (IDPs). *Chemical Reviews, 114*, 6661–6714.

Theisen, A., Johann, C., Deacon, M. P., & Harding, S. E. (2000). *Refractive increment data-book for polymer and biomolecular scientists.* Nottingham: Nottingham University Press.

Thompson, G. J., Langlais, C., Cain, K., Conley, E. C., & Cohen, G. M. (2001). Elevated extracellular $[K^+]$ inhibits death-receptor- and chemical-mediated apoptosis prior to caspase activation and cytochrome c release. *Biochemical Journal, 357*, 137–145.

Tian, L., Li, X., Ramchandran, K., & Waller, L. (2014). Multiplexed coded illumination for Fourier ptychography with an LED array microscope. *Biomedical Optics Express, 5*, 2376–2389.

Tian, L., Petruccelli, J. C., & Barbastathis, G. (2012). Nonlinear diffusion regularization for transport of intensity phase imaging. *Optics Letters, 37*, 4131.

Tian, L., & Waller, L. (2015). Quantitative differential phase contrast imaging in an LED array microscope. *Optics Express, 23*, 11394–11403.

Tolansky, S. (1968). *Interference microscopy for the biologist.* Springfield, IL: Thomas.

Traut, T. W. (1994). Physiological concentrations of purines and pyrimidines. *Molecular and Cellular Biochemistry, 140*, 1–22.

Trusiak, M., Mico, V., Garcia, J., & Patorski, K. (2016). Quantitative phase imaging by single-shot Hilbert–Huang phase microscopy. *Optics Letters, 41*, 4344–4347.

Usher-Smith, J. A., Fraser, J. A., Bailey, P. S., Griffin, J. L., & Huang, C. L. (2006). The influence of intracellular lactate and H^+ on cell volume in amphibian skeletal muscle. *The Journal of Physiology, 573*, 799–818.

Uzunbajakava, N., Lenferink, A., Kraan, Y., Willekens, B., Vrensen, G., Greve, J., et al. (2003). Nonresonant Raman imaging of protein distribution in single human cells. *Biopolymers, 72*, 1–9.

van Manen, H. J., Verkuijlen, P., Wittendorp, P., Subramaniam, V., van den Berg, T. K., Roos, D., et al. (2008). Refractive index sensing of green fluorescent proteins in living cells using fluorescence lifetime imaging microscopy. *Biophysical Journal, 94*, L67–L69.

Vasquez, B., Ishibashi, F., & Howard, B. V. (1982). Measurement of intracellular volume in monolayers of cultured cells. *In Vitro, 18*, 643–649.

Vereninov, A. A., Goryachaya, T. S., Moshkov, A. V., Vassilieva, I. O., Yurinskaya, V. E., Lang, F., et al. (2007). Analysis of the monovalent ion fluxes in U937 cells under the balanced ion distribution: Recognition of ion transporters responsible for changes in cell ion and water balance during apoptosis. *Cell Biology International, 31*, 382–393.

Verkman, A. S. (2002). Solute and macromolecule diffusion in cellular aqueous compartments. *Trends in Biochemical Sciences, 27*, 27–33.

Vilker, V. L., Colton, C. K., & Smith, K. A. (1981). The osmotic pressure of concentrated protein solutions: Effect of concentration and ph in saline solutions of bovine serum albumin. *Journal of Colloid and Interface Science, 79*, 548–566.

Waller, L., Tian, L., & Barbastathis, G. (2010). Transport of intensity phase-amplitude imaging with higher order intensity derivatives. *Optics Express, 18*, 12552–12561.

Wang, L., Xiao, Y., Tian, W., & Deng, L. (2013). Activatable rotor for quantifying lysosomal viscosity in living cells. *Journal of the American Chemical Society, 135*, 2903–2906.

Wang, P., Bista, R., Bhargava, R., Brand, R. E., & Liu, Y. (2010). Spatial-domain low-coherence quantitative phase microscopy for cancer diagnosis. *Optics Letters, 35*, 2840–2842.

Wang, Z., Millet, L., Mir, M., Ding, H., Unarunotai, S., Rogers, J., et al. (2011). Spatial light interference microscopy (SLIM). *Optics Express, 19*, 1016–1026.

Wehner, F., Olsen, H., Tinel, H., Kinne-Saffran, E., & Kinne, R. K. H. (2003). Cell volume regulation: Osmolytes, osmolyte transport, and signal transduction. *Reviews of Physiology, Biochemistry & Pharmacology, 148*, 1–80.

Wilson, J. D., Cottrell, W. J., & Foster, T. H. (2007). Index-of-refraction-dependent subcellular light scattering observed with organelle-specific dyes. *Journal of Biomedical Optics, 12*, 014010.

Wolf, E. (1969). Three-dimensional structure determination of semi-transparent objects from holographic data. *Optics Communications, 1*, 153–156.

Wu, X., Judd, L. W., Howe, E. N. W., Withecombe, A. M., Soto-Cerrato, V., Li, H., Busschaert, N., Valkenier, H., Pérez-Tomás, R., Sheppard, D. N., Jiang, Y.-B., Davis, A. P., & Gale, P. A. (2016). Nonprotonophoric electrogenic Cl– transport mediated by valinomycin-like carriers. *Chem, 1*, 127–146.

Wyllie, A. H., & Morris, R. G. (1982). Hormone-induced cell death. Purification ad properties of thymocytes undergoing apoptosis after glucocorticoid treatment. *American Journal of Pathology, 109*, 78–87.

Xue, L., Lai, J., Wang, S., & Li, Z. (2011). Single-shot slightly-off-axis interferometry based Hilbert phase microscopy of red blood cells. *Biomedical Optics Express, 2*, 987–995.

Yadav, S., Shire, S. J., & Kalonia, D. S. (2011). Viscosity analysis of high concentration bovine serum albumin aqueous solutions. *Pharmaceutical Research, 28*, 1973–1983.

Yamada, T., & Ohyama, H. (1988). Radiation-induced interphase death of rat thymocytes is internally programmed (apoptosis). *International Journal of Radiation Biology & Related Studies in Physics, Chemistry & Medicine, 53*, 65–75.

Yamauchi, T., Iwai, H., Miwa, M., & Yamashita, Y. (2008). Low-coherent quantitative phase microscope for nanometer-scale measurement of living cells morphology. *Optics Express, 16*, 12227–12238.

Yanase, Y., Hiragun, T., Kaneko, S., Gould, H. J., Greaves, M. W., & Hide, M. (2010). Detection of refractive index changes in individual living cells by means of surface plasmon resonance imaging. *Biosensors & Bioelectronics, 26*, 674–681.

Yanase, Y., Suzuki, H., Tsutsui, T., Hiragun, T., Kameyoshi, Y., & Hide, M. (2007). The SPR signal in living cells reflects changes other than the area of adhesion and the formation of cell constructions. *Biosensors & Bioelectronics, 22*, 1081–1086.

Yang, Z., He, Y., Lee, J. H., Park, N., Suh, M., Chae, W. S., et al. (2013). A self-calibrating bipartite viscosity sensor for mitochondria. *Journal of the American Chemical Society, 135*, 9181–9185.

Yeh, L.-H., Dong, J., Zhong, J., Tian, L., Chen, M., Tang, G., et al. (2015). Experimental robustness of Fourier ptychography phase retrieval algorithms. *Optics Express, 23*, 33214–33240.

Yurinskaya, V., Goryachaya, T., Guzhova, I., Moshkov, A., Rozanov, Y., Sakuta, G., et al. (2005). Potassium and sodium balance in U937 cells during apoptosis with and without cell shrinkage. *Cellular Physiology and Biochemistry, 16*, 155–162.

Yurinskaya, V. E., Moshkov, A. V., Wibberley, A. V., Lang, F., Model, M. A., & Vereninov, A. A. (2012). Dual response of human leukemia U937 cells to hypertonic shrinkage: Initial regulatory volume increase (RVI) and delayed apoptotic volume decrease (AVD). *Cellular Physiology and Biochemistry, 30*, 964–973.

Zabler, S., Cloetens, P., Guigay, J. P., Baruchel, J., & Schlenker, M. (2005). Optimization of phase contrast imaging using hard x rays. *Review of Scientific Instruments, 76*, 703–705.

Zahlten, R. N., Hochberg, A. A., Stratman, F. W., & Lardy, H. A. (1972). Glucagon-stimulated phosphorylation of mitochondrial and lysosomal membranes of rat liver in vivo. *Proceedings of the National Academy of Sciences of the United States of America, 69*, 800–804.

Zangle, T. A., & Teitell, M. A. (2014). Live-cell mass profiling: An emerging approach in quantitative biophysics. *Nature Methods, 11*, 1221–1228.

Zeskind, B. J., Jordan, C. D., Timp, W., Trapani, L., Waller, G., Horodincu, V., et al. (2007). Nucleic acid and protein mass mapping by live-cell deep-ultraviolet microscopy. *Nature Methods, 4*, 567–569.

Zhang, X., Garcia, I. F., Baldi, L., Hacker, D. L., & Wurm, F. M. (2010). Hyperosmolarity enhances transient recombinant protein yield in Chinese hamster ovary cells. *Biotechnology Letters, 32*, 1587–1592.

Zhang, Q., Zhong, L., Tang, P., Yuan, Y., Liu, S., Tian, J., & Lu, X. (2017). Quantitative refractive index distribution of single cell by combining phase-shifting interferometry and AFM imaging. *Scientific Reports, 7*, 2532.

Zhao, H., Brown, P. H., & Schuck, P. (2011). On the distribution of protein refractive index increments. *Biophysical Journal, 100*, 2309–2317.

Zheng, G., Horstmeyer, R., & Yang, C. (2013). Wide-field, high-resolution Fourier ptychographic microscopy. *Nature Photonics, 7*, 739–745.

Zong, W. X., & Thompson, C. B. (2006). Necrotic death as a cell fate. *Genes & Development, 20*, 1–15.

Zuo, C., Chen, Q., Yu, Y., & Asundi, A. (2013). Transport-of-intensity phase imaging using Savitzky-Golay differentiation filter–theory and applications. *Optics Express, 21*, 5346–5362.

Zuo, C., Sun, J., Li, J., Zhang, J., Asundi, A., & Chen, Q. (2017). High-resolution transport-of-intensity quantitative phase microscopy with annular illumination. *Scientific Reports, 7*, 7654.

SECTION 3

Cell Volume Regulation in the Airways

CHAPTER NINE

Slippery When Wet: Airway Surface Liquid Homeostasis and Mucus Hydration

Megan J. Webster* and Robert Tarran*,§,1

*Marsico Lung Institute, University of North Carolina at Chapel Hill, Chapel Hill, NC, United States
§Department of Cell Biology & Physiology, University of North Carolina at Chapel Hill, Chapel Hill, NC, United States
[1]Corresponding author: E-mail: robert_tarran@med.unc.edu

Contents

Current Topics in Membranes, Volume 81
ISSN 1063-5823
https://doi.org/10.1016/bs.ctm.2018.08.004

Abstract

The ability to regulate cell volume is crucial for normal physiology; equally the regulation of extracellular fluid homeostasis is of great importance. Alteration of normal extracellular fluid homeostasis contributes to the development of several diseases including cystic fibrosis. With regard to the airway surface liquid (ASL), which lies apically on top of airway epithelia, ion content, pH, mucin and protein abundance must be tightly regulated. Furthermore, airway epithelia must be able to switch from an absorptive to a secretory state as required. A heterogeneous population of airway epithelial cells regulate ASL solute and solvent composition, and directly secrete large mucin molecules, antimicrobials, proteases and soluble mediators into the airway lumen. This review focuses on how epithelial ion transport influences ASL hydration and ASL pH, with a specific focus on the roles of anion and cation channels and exchangers. The role of ions and pH in mucin expansion is also addressed. With regard to fluid volume regulation, we discuss the roles of nucleotides, adenosine and the short palate lung and nasal epithelial clone 1 (SPLUNC1) as soluble ASL mediators. Together, these mechanisms directly influence ciliary beating and in turn mucociliary clearance to maintain sterility and to detoxify the airways. Whilst all of these components are regulated in normal airways, defective ion transport and/or mucin secretion proves detrimental to lung homeostasis as such we address how defective ion and fluid transport, and a loss of homeostatic mechanisms, contributes to the development of pathophysiologies associated with cystic fibrosis.

1. INTRODUCTION

Ion and water flow across cell membranes contributes to the size and shape of mammalian cells (Jiang & Sun, 2013). As described throughout this book, the regulation of cell volume is complex and influences numerous physiological processes. Equally, the ability of cells to regulate the volume of their extracellular environment is also of great importance, as evident from the pathophysiology of diseases associated with altered fluid retention. Examples include cystic fibrosis (CF), chronic obstructive pulmonary disease (COPD), edema, hyponatremia, Liddle's syndrome and secretory diarrhea (Ghosh, Boucher, & Tarran, 2015; Haq, Gray, Garnett, Ward, & Brodlie, 2016; Matalon, Bartoszewski, & Collawn, 2015; Rotin & Schild, 2008; Thiagarajah & Verkman, 2013). Here, we discuss the importance of regulating extracellular fluid volume and composition within the airways and describe how epithelial cells lining the respiratory tract contribute toward regulation of extracellular solute and solvent composition. Additionally, we will describe how defects in epithelial ion and water transport and a failure of extracellular volume homeostasis contribute to the development of chronic lung disease, with a specific focus on the genetic disease CF.

1.1 Airway Anatomy

The pulmonary system facilitates the uptake of O_2 and the excretion of CO_2. This gas exchange process is vital to ensure that sufficient O_2 is available to meet the body's metabolic needs and that waste CO_2 is removed to prevent respiratory acidosis. The respiratory tract is divided anatomically into the upper (i.e., the nasopharyngeal) and lower airways (i.e., the trachea, bronchi and bronchioles) and the alveolar region. Functionally, the pulmonary system can be divided into the conducting and respiratory zones (Patwa & Shah, 2015), with the former serving to sterilize, detoxify and humidify the air before gas exchange takes place in the more delicate respiratory zone. The airways undergo a series of dichotomous branchings, giving rise to approximately 21–23 airway generations. The conducting airways include the cartilaginous trachea and bronchi, which contain submucosal glands, as well as the terminal bronchioles. They lie proximal to the 17th branching of the respiratory tree and do not take part in gas exchange (Saint-Criq & Gray, 2017; Wittekindt, 2017). Next come the respiratory bronchioles, alveolar ducts and alveoli (Hyde, Hamid, & Irvin, 2009; Patwa & Shah, 2015). Alveoli, localized most distally, are surrounded by a dense mesh of capillaries, with every alveolus having its own blood supply, and are the site of gas exchange (Saint-Criq & Gray, 2017; Wittekindt, 2017). The large, expandable, surface area of the alveoli, and their proximity to pulmonary endothelia, is optimal for efficient gas exchange.

1.2 Epithelial Heterogeneity

Like other organs, the airway lumen is lined by a continuous, heterogeneous layer of epithelial cells which, together with their secretions, are in contact with the outside world and provide a frontline barrier to defend against inhaled toxins and pathogens. Airway epithelia have a relatively slow turnover time of at least 100 days under normal conditions (Rawlins & Hogan, 2006). The large airways contain pseudostratified surface epithelia and possess submucosal glands, while the more distal bronchioles are simple cuboidal epithelia (Boucher, 1994; Hyde et al., 2009). The epithelium of the trachea and bronchi is primarily composed of ciliated cells, goblet cells and basal cells. Ciliated cells express cilia, hair-like microtubule based cell organelles which extend from the apical cell membrane into the airway surface liquid (ASL) (Yaghi & Dolovich, 2016). Cilia beat in a coordinated pattern of active and recovery strokes to propel ASL/mucus along the mucociliary escalator, up the respiratory tract and toward the mouth for expulsion (Yaghi & Dolovich, 2016). Goblet cells were named for their histological appearance after fixation and are mucin-containing cells that are interspersed between ciliated cells. Under normal conditions, the large airways are ~80% ciliated cells and ~20% goblet cells (Tarran et al., 2001). However, during chronic inflammation, goblet cell metaplasia can occur and this ratio can be reversed (Davis & Dickey, 2008). Submucosal glands enter onto the surface of large airways (discussed in detail in Section 3.5) (Hyde et al., 2009; Widdicombe & Wine, 2015). Both ciliated cells and the serous cells of submucosal glands contribute to ion and fluid secretion (Widdicombe, 2002; Widdicombe & Wine, 2015). With distal progression, the airway epithelium changes from a pseudostratified/columnar, to a simple columnar phenotype (Hyde et al., 2009). For example, in the conducting bronchi and bronchioles, goblet cells, which secrete mucins, are the predominant form of non-ciliated epithelia. As conducting bronchioles transition into respiratory bronchioles, club cells (formerly Clara cells) appear (Lumsden, McLean, & Lamb, 1984). Club cells are non-mucus, secretory cells which produce a surfactant like secretion in the respiratory bronchioles, and form part of the simple cuboidal epithelia of this region of the airways (Harkema, Nikula, & Haschek, 2013, pp. 1935—2003; Widdicombe, 2002). A thorough review of epithelial cell types can be found elsewhere (Boucher, 1994).

Only two cell types are present within the alveoli: simple squamous alveolar epithelia Type 1 cells, and cuboidal alveolar epithelia Type 2 cells. Type 1 cells lie in close proximity to the surrounding capillary mesh network and are extremely thin in order to facilitate diffusion and exchange of O_2/CO_2. Type 2 cells are responsible for the secretion of pulmonary surfactant, which is required to prevent alveolar collapse and allow expansion of alveoli during respiration (Barkauskas et al., 2013). In addition, Type 2 cells maintain stem-cell like characteristics and are able to de-differentiate and proliferate to replace damaged Type 1 cells following lung injury (Barkauskas et al., 2013). Such stem cell characteristics are vital to allow rapid repair of damaged alveolar surfaces to maintain a barrier between the interstitial and air filled spaces (Wittekindt, 2017).

1.3 Transcellular and Paracellular Transport in the Airways

Airway epithelia serve as a barrier between the interstitial fluid filled compartment and the air filled respiratory tract (Wittekindt, 2017). However, ions and water must transverse this barrier to regulate the composition and volume of the ASL (Verkman, 2007a). Ions (solutes) and water (solvent) can transverse the airway epithelial barrier via two routes: the transcellular and paracellular pathways. The transcellular pathway involves the movement of ions and water across the basolateral and apical cell membranes via a series of ATP-consuming pumps as well as secondary active transporters and passive channels (Hollenhorst, Richter, & Fronius, 2011). Conversely, the paracellular movement of ions occurs via intercellular junctions between neighboring cells (Brune, Frank, Schwingshackl, Finigan, & Sidhaye, 2015). Together, the transcellular and paracellular pathways of the pulmonary epithelia regulate the volume and composition of the ASL (Boucher, 2007a; Wittekindt, 2017).

Intercellular junctions are subdivided into four types; tight junctions, adherens junctions, gap junctions and desmosomes. Together, the tight and adherens junctions maintain epithelial polarity and divide the apical and basolateral compartments whilst gap junctions allow for intercellular communication and desmosomes help hold cells together (Brune et al., 2015). Tight junctions are localized apically, and are comprised of multimeric protein complexes including occludin and various claudin proteins (Van Itallie & Anderson, 2004). Several claudin isoforms have been identified in the airways, including claudin1, claudin2, claudin3, claudin4, claudin5, claudin7 & claudin8 (Wittekindt, 2017). Specific expression of different claudin proteins determines the relative permeability of airway tight junctions to different ions. For example, claudin2 is cation-selective, claudin7 is anion selective and can also increase the barrier function, and claudin1 is barrier forming (Gunzel & Yu, 2013). Airway epithelia and have tight junctions that are electrically "leaky" and are highly permeable to water, anions and cations. This phenotypic trait is vital to maintain an isotonic ASL and to enable both absorption and secretion (as discussed later). Indeed, freshly-isolated human bronchial epithelia have a transepithelial electrical resistance (TEER) of approximately 100 $\Omega.cm^2$ (Boucher, 1994), while primary human bronchial epithelial cultures typically have a TEER of $\sim$300 $\Omega.cm^2$ (Tarran et al., 2001). In contrast, extensively passaged airway cultures often exhibit unphysiologically high TEER values in excess of 1000 $\Omega.cm^2$ (Bishop et al., 2018). Alteration, and loss, of epithelial barrier integrity is known to confer respiratory distress syndrome and acute lung injury (Koval, 2013), contributing to mortality and morbidity. Similarly, a loss of intercellular barrier selectively has been observed during pneumococcal pneumonia which leads to a disruption of junctional proteins and an uncontrolled movement of ions and water from the serosal to mucosal compartments (Peter et al., 2017; Wittekindt, 2017).

1.4 Submucosal Glands

Submucosal glands are predominantly localized within the cartilaginous/large airways that secrete ions, water, mucus and other proteins in order to contribute toward ASL volume homeostasis/composition (Verkman, Song, & Thiagarajah, 2003). Submucosal glands have four distinct regions, each with a different physiological function (Meyrick, Sturgess, & Reid, 1969). Most distally, submucosal glands contain serous acini and tubules that secrete ions, fluid, antimicrobials and other proteins (Meyrick et al., 1969; Verkman et al., 2003). As secreted fluids move toward the airway lumen, mucous tubules add high molecular weight mucin molecules via exocytosis of densely packed granules (Meyrick et al., 1969; Verkman et al., 2003). Fluid and mucous are collected in tubules lined by tall, columnar, non-ciliated epithelia proposed to alter the composition of submucosal gland secretions (Verkman et al., 2003). The fourth, and most proximal region of the submucosal gland is an extension of the airway epithelium and contains ciliated epithelia (Meyrick et al., 1969).

Submucosal glands are innervated by the parasympathetic nervous system (Nadel & Davis, 1980). Serous and mucous cells have previously been shown to be stimulated selectively and in combination (Nadel & Davis, 1980; Ueki, German, & Nadel, 1980). Glandular fluid ion composition and viscosity is unchanged by muscarinic stimulation, suggesting innervation and stimulation of both serous and mucous cells (Ueki et al., 1980; Verkman et al., 2003). Alternatively, α-adrenergic and β -adrenergic stimulation is reported to be selective for serous and mucous cells respectively (Verkman et al., 2003), thus altering secreted fluid composition to produce fluid with a low and high viscosity and protein content respectively (Verkman et al., 2003). It is therefore believed that both ciliated and goblet cells of the epithelium lining, and submucosal glands contribute to ASL homeostasis within the airways and can rapidly provide additional ASL volume in the large airways that can increase mucus clearance rates.

2. AIRWAY SURFACE LIQUID (ASL) COMPOSITION

With every inhalation, the lung is exposed to and samples the external environment. However, the lung must prevent infection by inhaled bacteria and viruses and must also clear inhaled pollutants such as tobacco smoke and diesel exhaust. As such, mechanisms exist within the airways to promote both sterility and detoxification, many of which are contained in the ASL. In addition to salt and water, proteomic studies have revealed that the ASL contains over 1000 proteins including antimicrobial peptides/proteins, anti-oxidants, proteases/protease inhibitors and mucins, which work in concert with the underlying epithelia in order to achieve optimal clearance and sterility. Resident antimicrobials within the ASL include cationic defensins, lysozyme (Parker & Prince, 2011), β-defensins (Brune et al., 2015), lactoferin, cathelicidins (Stoltz, Meyerholz, & Welsh, 2015) and the short palate lung and nasal epithelial clone 1 (SPLUNC1) (Ahmad et al., 2016). Increased recruitment and activity of leukocytes during infection can help eradicate pathogens by phagocytosis, NETosis, increased proteolysis and oxidative burst (Parker & Prince, 2011). The ASL protein composition can also change during chronic infection and/or inflammation. However, whether this is advantageous or leads to defects in ASL function remains to be determined. Thorough reviews of ASL antimicrobial and leukocyte activity can be found elsewhere (Parker & Prince, 2011; Pettersen & Adler, 2002; Zawrotniak & Rapala-Kozik, 2013); here we will focus on epithelial cell regulation of ASL ion content and fluid volume.

The ASL lies atop the mucosal surface of the airway epithelium. The ASL is divided into two distinct phases: the periciliarly layer (PCL) which, as its name suggests, bathes the cilia and acts as a lubricant for efficient ciliary beating and the overlying mucus layer, which traps inhaled particles and pathogens which are then removed by mucociliary clearance. The mucus-PCL model has undergone a revision since immunohistochemistry showed the presence of the cell surface tethered mucins MUC1 and MUC4 (Kesimer et al., 2013). These cell tethered mucins project into the airway lumen, and form a polyanionic mesh on which the overlying mucus layer rests, holding the free moving mucus at an optimal distance from the cell surface, giving a PCL height of approximately 7–8 μm. Thus, the current model suggests that the fixed negative charges on the ciliary/tethered mucins repel the negatively charged secreted mucins to maintain the distinct PCL vs. mucus layers (Button et al., 2012). As such the mucociliary escalator removes mucus and inhaled particles, pathogens, allergens and pollutants from the lungs. As will be discussed later, dehydration of the ASL causes the PCL to collapse and the mucus, which invades this layer prevents normal ciliary function leading to a failure of associated mucus clearance that is contributory to airway infection. Thus, ASL volume regulation is critical for lung health.

When the homeostatic mechanisms regulating fluid volume/hydration of both the mucus layer and the PCL are defective, for example in the genetic disease CF (discussed in section 5.0) or in the acquired disease COPD (De Rose, Molloy, Gohy, Pilette, & Greene, 2018; Ghosh et al., 2015), mucocilliary clearance mechanisms fail, and dehydrated mucus accumulates within the airways. It has been proposed that ASL is initially absorbed from the mucus layer followed by absorption of PCL (Tarran et al., 2001). As water is isotonically absorbed from the mucus layer, the % solids content increases, which together with continued mucin secretion from goblet cells, leads to the formation of a thick, highly viscous mucus. This increase in viscosity, in combination with reduced PCL volume, causes adhesion of secreted mucins such as MUC5AC and MUC5B to cell tethered mucins such as MUC1 and MUC4, causing ciliary stasis and a loss of mucocilliary clearance (Button, Anderson, & Boucher, 2016; Ghosh et al., 2015). Of note, increases in mucus % solids content have been shown to inversely correlate with FEV_1 (Anderson et al., 2015).

3. AIRWAY SURFACE LIQUID VOLUME REGULATION

Both the transcellular and paracellular movement of ions and water are required for optimal hydration of the ASL, as illustrated in Figs. 1 and 2. The airway epithelia are leaky and highly water permeable (Matsui, Davis, Tarran, & Boucher, 2000; Verkman, 2007a). ASL is isotonic to plasma and ASL volume is set by the mass of solutes, not their concentrations which typically change little. Thus, during Cl^- secretion, Cl^- moves into the ASL, with a counter ion (e.g. Na^+) and water moving paracellularly, so the mass of ions and volume increase, but their concentration is unchanged. The

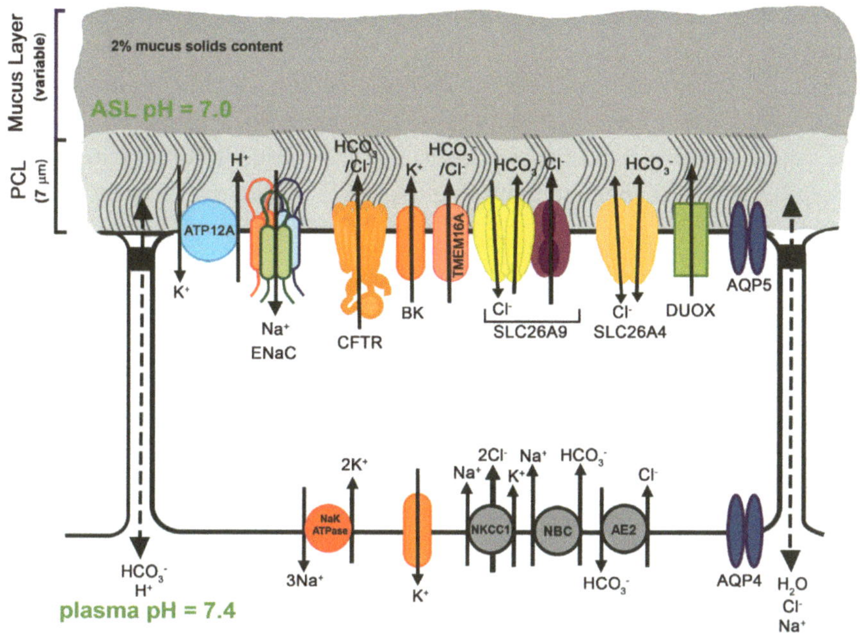

Figure 1 ***Schematic showing expression of ion channels, exchangers and pumps in normal human airway epithelia.*** Basolaterally, electrochemical gradients are established by the Na^+/K^+ ATPase, with K^+ being recycled via a basolateral K^+ channel which also helps establish the membrane potential. For Na^+ absorption, Na^+ enters the cell via ENaC and is pumped out basolaterally via the Na^+/K^+ ATPase, with Cl^- and water following paracellularly. For Cl^- secretion, 2 Cl^-s enter the cell basolaterally via NKCC, with K^+ and Na^+ being recycled basolaterally. Cl^- can then be secreted apically via CFTR or TMEM16A. SLC26A4 and SLC26A9 are also present apically and may also play a role in transepithelial Cl^- secretion, with the counter ion and water either moving paracellularly/and or K+ being secreted via BK channels as the counter ion. AQP4 and AQP5 are present, but their roles in water movement remain to be determined. Basolateral NBC and AE2 may play a role in regulating cytosolic pH.

primary ions involved in ASL hydration are Na^+ and Cl^-, which are both present at ~100–130 mM (Knowles et al., 1997; Matsui et al., 1998). However, K^+ (~20 mM) and HCO_3^- (~10 mM) are also present and physiologically relevant to ASL volume homeostasis, although due to their lower ASL concentrations, they most likely play a regulatory role, rather than exert a major osmotic influence. In contrast, ASL glucose is present at very low concentrations (3–20 fold lower than in plasma; i.e. ~0.5 mM) due to uptake by SGLT and GLUT-type transporters and likely has little effect on ASL volume (Garnett, Baker, & Baines, 2012).

Airway epithelial ion transport is determined by electrochemical gradients: the basolateral Na^+/K^+ ATPase is the ATP-consuming pump that effluxes 3 Na^+ ions in exchange for 2 K^+ ions, and is the primary mechanism by which ion gradients are established. For Na^+ absorption, Na^+ crosses the apical membrane via the epithelia Na^+ channel (ENaC), due to the favorable gradient generated by the Na^+/K^+ ATPase and then Na^+ is pumped out basolaterally with K^+ being recycled via a basolateral K^+ channel. For Cl^- secretion, Cl^- is taken into the cell via the Na^+, K^+, 2 Cl^- co-transporter (NKCC1) with Na^+ exiting by the Na^+/K^+ ATPase and K^+ being recycled via the basolateral K^+ channel (Figs. 1 and 2).

Apically, Cl^- ions are secreted via CFTR or the Ca^{2+} activated Cl^- channel (CaCC) as discussed in more detail in the next section. Due to the high concentrations of NaCl in the ASL, the roles of ENaC, CFTR

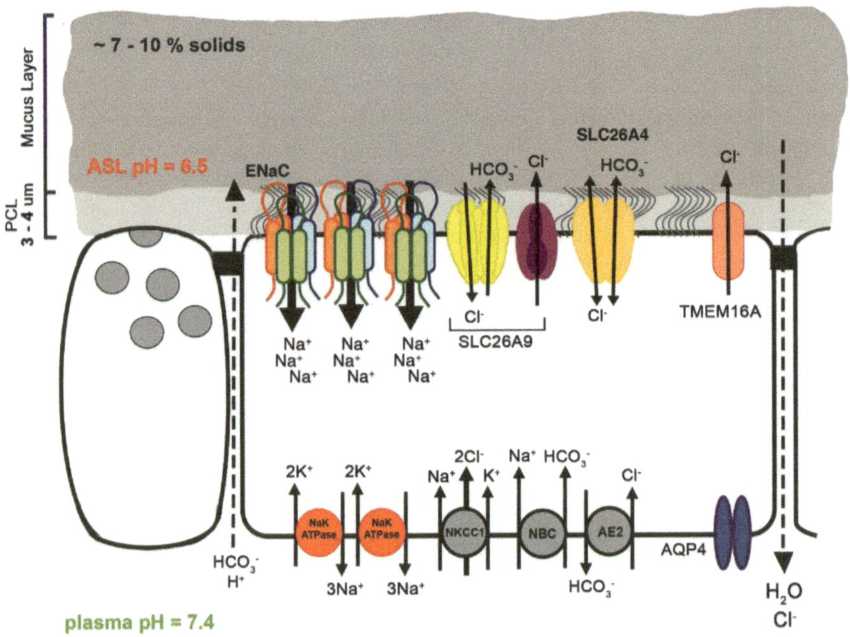

Figure 2 ***Schematic showing various ion channels, pumps and exchangers in CF airway epithelia.*** CF-affected airway epithelia lack functional CFTR and Cl^-/HCO_3^- secretion is reduced. TMEM16A and SLC family members are unable to generate sufficient Cl^- secretion to cope with the absence of CFTR, leading to reduced ASL volume and moderate ASL acidification. Loss of CFTR mediated inhibition of ENaC allows for upregulation of ENaC expression and activity resulting in increased Na^+ absorption which further drives dehydration.

and CaCC in ASL volume regulation are arguably the best defined. However, apical big conductance (BK) K^+ channels and a series of pumps and exchangers including the non-gastric H^+/K^+-ATPase (ATP12a), SLC26A9 and SLC26A4 are expressed and may also play a role in ASL volume homeostasis, although arguably, their roles are less well understood.

3.1 K^+ Channels

K^+ channels are expressed both basolaterally and apically in airway epithelia where they play roles in both K^+ recycling and K^+ secretion respectively (Cotton, 2000). Over 40 different types of K^+ channels have been detected at the expression level in the airways (Bardou, Trinh, & Brochiero, 2009). Indeed, basolateral K^+ channels are involved in the generation of cellular electrochemical gradients and are required for continued operation of both the Na^+/K^+-ATPase and NKCC1 (Cotton, 2000). ATP has been shown to induce K^+ secretion across airway epithelia, which may affect ASL volume (Clarke, Chinet, & Boucher, 1997). More recently, the big conductance (BK) K^+ channels have been found to be apically expressed in ciliated cells (Manzanares et al., 2015). BK channels are activated by changes in membrane voltage and/or intracellular $[Ca^{2+}]$(53) and likely underlie the changes in K^+ secretion observed by Clarke et al. (Clarke et al., 1997). More information on the structure and regulation of BK channels can be found elsewhere (Kis, Krick, Baumlin, & Salathe, 2016). Encoded by the *KCNMA1* gene, the α-subunit of the BK channel assembles as a homotetramer to form a central channel pore. The α-subunits form the cytosolic and extracellular domains that are augmented by voltage-sensing and a pore-gate domains that regulate BK channel activity. BK channels possess further 4 β- and 4 γ-domains which influence channel kinetics and gating, and modulate the voltage and Ca^{2+} dependence (Kis et al., 2016; Latorre & Brauchi, 2006).

Since they are Ca^{2+}-sensitive, BK channels are also activated by extracellular ATP and UTP via the G-protein G_q-coupled P_2Y_2 receptor (Lazarowski & Boucher, 2009). The efflux of K^+ out of the cell then generates an electrochemical gradient for Cl^- secretion via TMEM16A (discussed in Section 3.4) and potentially via CFTR (Kis et al., 2016). Since there is no chemical gradient for Cl^- secretion into the ASL, hyperpolarization must occur in order to create a favorable electrical driving force for Cl^- secretion. Whilst inhibition of ENaC, either experimentally using amiloride, or physiologically using SPLUNC1 can achieve the required apical membrane hyperpolarization, activation of K^+ channels and K^+

secretion can have a similar effect. K^+ is not present in sufficient quantities in the ASL to have a direct osmotic effect on ASL volume (<20 mM). Thus, activation of K^+ channels constitutes an additional way of driving Cl^- secretion that can in turn affect ASL volume. Indeed, prolonged inhibition of BK channels, or knockdown of *KCNMA1* has previously been shown to cause ASL dehydration (Manzanares et al., 2011). Furthermore, TGF-β has previously been shown to reduce BK channel activity and cause ASL dehydration due to downregulation of the leucine-rich repeat containing protein 26 γ-subunit (LRRC26), which shifts the activation potential of BK channels towards the membrane potential in non-excitatory cells (Kis et al., 2016; Manzanares et al., 2015). Inhibition of TGF-β by pirfenidone, reversed this BK induced ASL dehydration (Manzanares et al., 2015). Similarly, inhibition of the two pore K^+ channels with bupivacaine and quinidine inhibited both Cl^- secretion and Na^+ absorption, suggesting a crucial role of this class of channel in maintaining membrane potentials and in regulating ASL volume (Zhao, Xiong, Wilber, Cohen, & Kreindler, 2012).

3.2 The Epithelial Na^+ Channel (ENaC)

ENaC is the rate-limiting step for transepithelial Na^+ absorption in the airways. An amiloride-sensitive Na^+ channel was first cloned from rat distal colon in 1993 (Canessa, Horisberger, & Rossier, 1993) and was subsequently cloned from human lung (Voilley et al., 1994). The proposed topology of ENaC followed shortly after (Snyder, McDonald, Stokes, & Welsh, 1994). Hydrophobic analysis, identification of N-glycosylation sites and mapping of the amino- and carboxy-terminals of ENaC suggested that each subunit had two membrane spanning domains, intracellular N- and C- termini and a large extracellular domain, as shown in Fig. 4 (Snyder et al., 1994). Whilst initially thought to be a tetramer, homology studies based on the crystal structure of ASIC1 have suggested that ENaC is a heterotrimer, composed of α-, β-, and γ- ENaC subunits at a 1:1:1 ratio (Stockand, Staruschenko, Pochynyuk, Booth, & Silverthorn, 2008). Each ENaC subunit is encoded by a different gene; α- SCNN1A, β- SCNN1B and γ- SCNN1G, which form a channel with a central pore that is highly selective for Na^+ and Li^+ over K^+, and sensitive to the small molecule inhibitor amiloride (Sheng, McNulty, Harvey, & Kleyman, 2001; Stockand et al., 2008). ENaC activity depends on both the number (N) of channels present at the apical membrane and the open probability (P_o) of those channels.

The lack of functional CFTR in CF patients results in upregulation of amiloride-sensitive potential differences *in vivo* in the airways of both humans and mice, and the upregulation of amiloride-sensitive currents in freshly isolated excised airway tissues (Boucher, Stutts, Knowles, Cantley, & Gatzy, 1986; Grubb, Vick, & Boucher, 1994; Knowles, Gatzy, & Boucher, 1981). Several studies have investigated the effects of CFTR expression and activity on ENaC. Co-transfection studies in *Xenopus laevis* oocytes, using both mouse and human ENaC and CFTR, showed reduced surface and whole cell expression levels of ENaC compared with ENaC transfection alone (Yan, Samaha, Ramkumar, Kleyman, & Rubenstein, 2004). Furthermore, increased Na^+ absorption and amiloride sensitive currents have been shown in cultured CF human bronchial epithelial cells relative to cultures derived from non-CF patients (Rubenstein et al., 2011). Although the mechanism by which CFTR regulates ENaC activity remains to be established, studies have shown an interaction between the carboxyl terminal of α- and β- ENaC, and the R domain of CFTR (Ji et al., 2000). Alternatively, others have suggested the presence of two Cl^- inhibitory sites, located extracellularly between the interface of the α-/γ- and β-/γ-subunits (Collier & Snyder, 2011). It could be that multiple mechanisms work in unison to reduce Na^+ absorption when CFTR is expressed, in order to facilitate secretion and an increase in ASL volume.

In addition to CFTR mediated regulation of ENaC, ENaC has been shown to self-regulate (Chraibi & Horisberger, 2002), be regulated by pH (Collier & Snyder, 2009) and be cleaved, and activated by proteases (Caldwell, Boucher, & Stutts, 2005; Rossier & Stutts, 2009). Indeed, two pools of ENaC are expressed in airway epithelia, basally active channels and inactive, near-silent, channels (Caldwell, Boucher, & Stutts, 2004). Whilst basally active channels are cleaved intracellularly by furin-type covertases during channel maturation (Hughey et al., 2004; Vallet, Chraibi, Gaeggeler, Horisberger, & Rossier, 1997), near-silent channels bypass this cleavage and are susceptible to extracellular cleavage by serine proteases (Caldwell et al., 2004, 2005). Extracellular channel activating proteases (CAPs) cleave α and γ ENaC within their extracellular domains to increase ENaC open probability (P_o) (Caldwell et al., 2005). In addition to trypsin, proteases expressed in airway epithelia including prostasin, TMPRSS4, matriptase, as well as proteases in the ASL including cathepsin B and neutrophil elastase have been shown to cleave and activate ENaC (Bruns et al., 2007; Caldwell et al., 2005; Donaldson et al., 2002; Tan et al., 2014; Tarran et al., 2006; Vallet et al., 1997). This can become problematic during chronic inflammation when extracellular protease levels, including neutrophil elastase and cathepsins are increased in the airway lumen. Indeed, the increased presence of free elastase in the airways has been reported in CF children as young as 4 years (Birrer et al., 1994; Carroll, Greene, Taggart, McElvaney, & O'Neill, 2005). Increased levels and activity of elastase and cathepsins may cleave ENaC extracellularly, activating the channel and promoting Na^+ hyperabsorption. Na^+ hyperabsorption can subsequently cause ASL dehydration, as discussed in Section 5.2.

3.3 The Cystic Fibrosis Transmembrane Regulator (CFTR)

The CFTR anion channel, named following the discovery of its involvement in CF, is encoded by the *CFTR* gene (Quinton, 1983; Riordan, Rommens, & Kerem, 1989). Early studies found increased Cl^- content in the sweat of CF patients, that was attributed to a Cl^- channel defect (Quinton, 1983) and altered Cl^- permeability was also identified in across CF airway epithelia (Knowles et al., 1983; Knowles et al., 1983). CFTR was then cloned from epithelial cells using unbiased chromosome walking (Riordan et al., 1989). CFTR conducts anions with the preference $NO_3^- > Br^- \geq Cl^- > I^- > F^-$ and is also permeable to larger anions such lactobionate (Gray et al., 1990; Sheppard & Welsh, 1999). CFTR also conducts HCO_3^-, with a permeability ratio $P_{HCO_3^-}/P_{Cl^-} \sim 0.2$ (Gray et al., 1990; Poulsen, Fischer, Illek, & Machen, 1994). CFTR-mediated HCO_3^- secretion is importance with for ASL pH homeostasis, and is discussed later in this chapter (Fig 3).

Like other ATP binding-cassette (ABC) transporters CFTR has 2 membrane spanning domains (MSDs) which each traverse the membrane 6 times to form the channel pore (Sheppard & Welsh, 1999) (Fig. 4), and two intracellular nucleotide binding domains (NBDs). However unlike other family members, CFTR has a unique regulatory 'R' domain (Sheppard & Welsh, 1999). CFTR has a low single-channel conductance ranging between 4–10 pS (Berger et al., 1991; Gray, Greenwell, & Argent, 1988). Gating of CFTR is dependent of ATP binding at the NBDs (Hwang & Sheppard, 2009). It has been proposed that ATP binding initiates dimerization of NBD1 and NBD2 to open the channel, whilst subsequent hydrolysis of ATP at NBD2 accelerates CFTR channel closing (Hwang & Sheppard, 2009). In addition to channel gating, CFTR activity is regulated

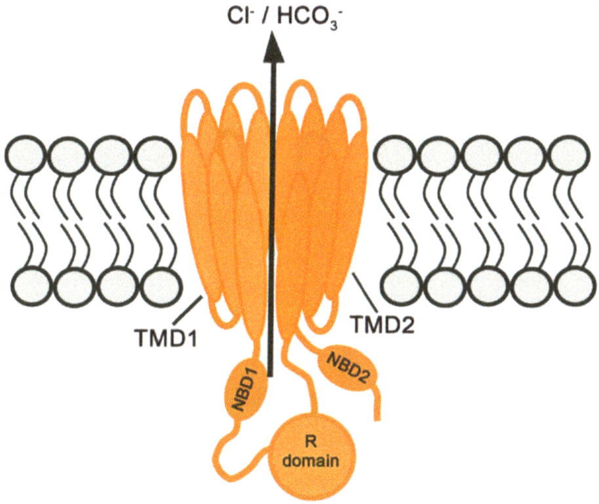

Figure 3 ***Schematic of the CFTR ion channel.*** CFTR is composed of two transmembrane domains (TMD1 and TMD2), each containing six membrane spanning α-helices. Two intracellular nucleotide binding domains (NBD) are located intracellularly and are the site of ATP hydrolysis. A regulatory 'R' domain connects TMD1 and TMD2, and is the site of phosphorylation and in turn, regulation of channel activity.

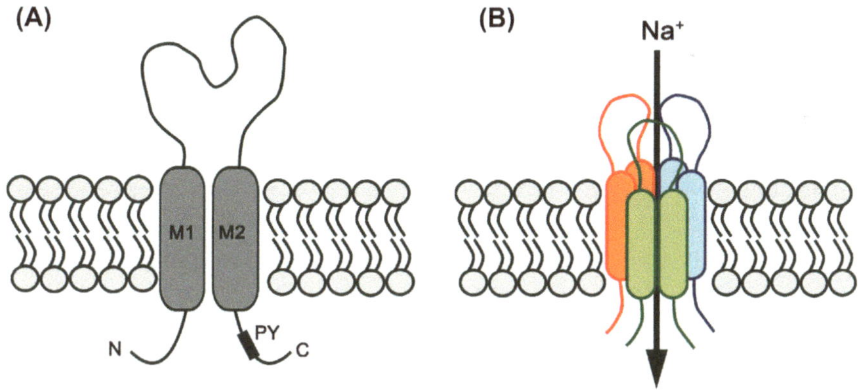

Figure 4 ***Schematic showing structure of a single ENaC subunit, and the heterotrimeric ENaC channel.*** (A) Schematic showing a single ENaC subunit, with two membrane spanning domains, intracellular C- and N- termini, NEDD4.2-binding PY motif and a large extracellular domain protruding from the apical membrane. (B) Schematic showing arrangement of three ENaC subunits, α- β- and γ-to form a central pore through which Na^+ permeates.

by phosphorylation of the regulatory 'R' domain. This phosphorylation occurs in a PKA dependent manner at consensus phosphorylation sequences (Cheng et al., 1991). Of the ten dibasic PKA consensus sequences present within the R domain, site-directed mutagenesis showed *in vivo* PKA dependent phosphorylation of five serine residues (Cheng et al., 1991; Picciotto, Cohn, Bertuzzi, Greengard, & Nairn, 1992). PKA dependent phosphorylation has previously been demonstrated to significantly increase channel gating, and conductance (Winter & Welsh, 1997) whilst other groups have reported phosphorylation is a pre-requisite for NBD dimerization (Li et al., 1996). CFTR is highly expressed in serous cells of the submucosal glands (Engelhardt et al., 1992), a trait also apparent in the commonly used Calu-3 serous lung cell line (Shen, Finkbeiner, Wine, Mrsny, & Widdicombe, 1994). However, CFTR is also present and active in ciliated cells throughout the airways but was found to be poorly expressed in non-cilated cells of the distal airways (Kreda et al., 2005).

3.4 Ca^{2+}-Activated Cl^- Channels/TMEM16A

In addition to CFTR, airway epithelia also have an additional Ca^{2+}-activated Cl^- channel (CaCC). CaCC is characterized by activation by cytosolic Ca^{2+}, outward rectification and inactivation at depolarizing potentials (Ferrera, Zegarra-Moran, & Galietta, 2011). CaCC is permeable to Cl^-, but also to HCO_3^- (Jung et al., 2013). TMEM16A, also called Anoctamin1 (Ano1) was independently identified as CaCC by 3 separate laboratories (Caputo et al., 2008; Schroeder, Cheng, Jan, & Jan, 2008; Yang et al., 2008). TMEM16A is a dimer that has 8 membrane spanning domains and intracellular N- and C- termini (Ferrera et al., 2011; Sheridan et al., 2011). Whilst TMEM16A$^{-/-}$ mice show reduced CaCC activity in their airways, whether or not TMEM16A is the only CaCC is controversial. Galietta et al. found that TMEM16A expression is greatly upregulated by cytokines such as IL-4 (Caputo et al., 2008). Uninflamed airway epithelia possess a significant CaCC current. However, TMEM16 may be poorly expressed in the absence of pro-inflammatory cytokines. Therefore, other groups have proposed that TMEM16A is not the only CaCC and that other, as yet unidentified proteins, serve as CaCC under basal conditions (Namkung, Phuan, & Verkman, 2011).

CaCC likely contributes toward Cl^- and ASL secretion in airway epithelia. CaCC-mediated secretion is often studied after stimulation of $P2Y_2$ receptors by UTP or ATP. In addition to the well characterized effects on CaCC, stimulation of $P2Y_2$ receptors also causes an increase in ASL secretion across primary normal and CF airway cultures. Interestingly, whilst

CFTR maintains activity over time, so long as cAMP is present, CaCC/TMEM16A activity tends to be more transient and the effect on fluid secretion is shorter lasting (Tarran et al., 2001, 2002). Interestingly, this increase in fluid secretion is also capable of increasing mucus transport in CF airway cultures (Tarran et al., 2001).

CFTR has been shown to negatively regulate CaCCs and a moderate upregulation of CaCC has been detected in CF airway epithelia (Tarran et al., 2002). Using FRET, CFTR and TMEM16A were shown to associate (Sheridan et al., 2011). Ousingsawat et al. showed an interaction between TMEM16A and CFTR by co-immunoprecipitation (Ousingsawat, Kongsuphol, Schreiber, & Kunzelmann, 2011). Using both cell lines (16HBE human airway epithelia) and primary HBECs it was shown that activation of endogenously expressed CFTR or overexpressing of CFTR in cells lacking endogenous expression significantly inhibited TMEM16A Cl^- secretion (Ousingsawat et al., 2011). In agreement, others have reported increased CFTR mediated inhibition of TMEM16A activity. Furthermore, increased secretory responses to UTP and histamine have been shown in excised CF tissues and HBECs cultured from CF patients (Clarke, Paradiso, & Boucher, 1992; Mall et al., 2003). Thus, under basal conditions Cl^- secretion mediated by TMEM16A may be minimal in comparison with CFTR and it may be upregulated following inflammation or if CFTR activity is modified.

3.5 SLC26 Exchangers

Several members of the solute carrier (SLC) family function as anion exchangers, and have been shown to transport a wide variety of ions including Cl^- and HCO_3^-. Two SLC26 family members, SLC26A9 and SLC26A4 are expressed at the apical membrane of airway epithelia (Saint-Criq & Gray, 2017) (Figs. 1 and 2). SLC26A9 was initially reported to be a Cl^-/HCO_3^- exchanger. However, other researchers found that SLC26A9 had minimal permeability to HCO_3^- and suggested that SLC26A9 instead functions as a Cl^- channel (Dorwart, Shcheynikov, Wang, Stippec, & Muallem, 2007). SLC26A4 (pendrin) is also expressed in airway epithelia. However, unlike SLC26A9, SLC26A4 is poorly expressed under basal conditions and is upregulated following increases in infection and inflammation, for example after exposure to IL-4, IL-13 and IL-17A (Adams et al., 2014; Nofziger, Dossena, Suzuki, Izuhara, & Paulmichl, 2011). Thus, while SLC26A9 may play a role in basal Cl^- secretion/ASL maintenance, SLC26A4 may facilitate additional ASL hydration during inflammation. Furthermore, it has been speculated that SLC26A4 upregulation facilities mucin secretion, although how this would occur is not fully understood (Nakao et al., 2008).

SLC26 family members have intracellular N- and C- termini with up to 14 membrane spanning regions (Alper & Sharma, 2013). SLC proteins also express an intracellular 'sulfate transporter and anti-sigma factor antagonist' (STAS) domain which regulates not only cell surface expression of SLC26 proteins but may also allow them to interact with other proteins (Alper & Sharma, 2013). For example, SLC26A9 activity may be positively regulated by CFTR, likely via interactions between its STAS domain and CFTR's R-domain (Ko et al., 2004) (Avella, Loriol, Boulukos, Borgese, & Ehrenfeld, 2011; Chang et al., 2009). Constitutive Cl^- secretion, attributable to SLC26A9, has been reported in normal HBECs that was lost in HBECs derived from CF patients (Bertrand, Zhang, Pilewski, & Frizzell, 2009). Interestingly, the interaction between SLC26A9 and CFTR appears to be reciprocal, with SLC26A9 increasing the P_o of CFTR up to 6-fold (Ko et al., 2004). Although further work will be required to fully determine how SLC26A9 influences ASL ion and fluid content, it is likely that under basal conditions, in the presence of CFTR, SCL26A9 secretes Cl^- ions, either via exchange for HCO_3^- or via channel mediated secretion, to hydrate the ASL.

In addition to being expressed on the apical membrane, Cl^-/HCO_3^- and Na^+/H^+ exchangers have also been described on the basolateral membrane of airway epithelia (Paradiso, 1997). However, their role in ASL volume regulation has not been tested. They may play a role in regulating cytosolic pH rather than in regulating ASL volume.

3.6 Aquaporins

Airway epithelia are highly water permeable and express a number of aquaporins on both their apical and basolateral membranes. Indeed, along with the kidney's proximal tubule, it is estimated that airway epithelia are amongst the most water permeable of epithelia throughout the body. A major function of the conducting airways is to humidify inhaled air to 100% humidity before it reaches the delicate alveolar regions. Thus, whilst little humidification is needed during the summer in North Carolina, in dry environments, water can evaporate from the ASL to humidify the air. Subsequently, additional water can then move across the epithelia via aquaporins without the need for active ion transport processes. This water movement, called insensible water loss, can be responsible for up to 500 mls of water being lost from the lungs per day.

The mammalian AQP family has some 12 members, of which 4 have been detected in the respiratory tract (Verkman, Matthay, & Song, 2000). In the large airways, AQP3 and AQP5 expression has been detected at the basolateral and apical membranes of ciliated cells of the surface epithelia and gland acinar cells (Nielsen, King, Christensen, & Agre, 1997; Verkman, 2007b). AQP4 is expressed at the basal membrane of ciliated ducts (Frigeri, Gropper, Turck, & Verkman, 1995; Nielsen et al., 1997). AQP1 and AQP5 are expressed in microvascular endothelia and the apical membrane of alveolar type 1 cells respectively (Nielsen et al., 1993, 1997; Schnitzer & Oh, 1996; Verkman, 2002). Despite expression of AQPs in the airways their role in fluid transport remains elusive. Knockout mouse models lacking aquaporins showed up to a 10-fold reduction in water permeability (Verkman et al., 2000). Despite this, aquaporin knockout has little effect on airway physiology. In contrast, aquaporin knockout in the kidney has led to more severe phenotypes, including 10-fold greater fluid consumption and nephrogenic diabetes after AQP3 knockdown (Ma et al., 2000). For example, knockout of AQP1 and 4, either alone or in combination did not affect ASL salt concentration or ASL volume (Jayaraman, Song, Vetrivel, Shankar, & Verkman, 2001; Song, Yang, Matthay, Ma, & Verkman, 2000). A lack of change following AQP deletion may i) suggest a minimal role of these transporters in airway fluid regulation or ii) indicate that paracellular water transport in the airways compensates for the loss of expressed AQPs.

3.7 ASL Volume Regulators

Soluble mediators have previously been shown to regulate ASL volume (Tarran et al., 2006) and mucociliary clearance (Davis & Lazarowski, 2008). For example, nucleotides and adenosine are present at physiologically relevant concentrations within the ASL (Lazarowski et al., 2004). The mechanics of breathing, with repetitive expansion and contraction of the airways, imparts mechanical stress. This includes stretching of the airways, compressive shear and air-flow induced shear stress due to tidal oscillations (Tarran et al., 2006). We previously demonstrated that physiologic levels of shear stress increased apical ATP release (Tarran et al., 2005, 2006). Indeed, ATP levels were elevated to values capable of activating the G_q/phospholipase C-coupled $P2Y_2$ receptor and were capable of stimulating Cl^-/ASL secretion (Tarran et al., 2005). As will be discussed in Section 4.1, mucin molecules are stored in granules within goblet cells and are released onto the apical epithelial surface by exocytosis. In addition to phasic shear stress mediated release of ATP, it has been proposed that ATP is simultaneously released with mucin molecules. Kreda et al. showed that activation of the $P2Y_2$ receptor increased mucin secretion and luminal ATP release (Kreda et al., 2007).

ATP release from airway epithelia is rapid (~370 fmol.min.cm^2) (Davis & Lazarowski, 2008). However, ASL ATP is rapidly degraded by a series of ectoenyzmes that are either active on the extracellular side of the apical membrane of airway epithelia, are secreted into the ASL or from glandular secretions. Indeed, the metabolism of ATP has previously been shown to approximate the rate of ATP release (Lazarowski et al., 2004). Whilst this may prevent chronic activation of $P2Y_2$ receptors, sharp increases in ATP due to shear stress elevates ATP to levels capable of activating $P2Y_2$ receptors (Davis & Lazarowski, 2008). Activation of $P2Y_2$ receptors promotes the mobilization of Ca^{2+} from intracellular stores, which in turn activates CaCC/TMEM16A and BK channels to increase KCl secretion with concomitant water movement to increase ASL volume (Lazarowski & Boucher, 2009). ATP is degraded to ADP, AMP and then to adenosine, which is present at sufficient concentrations (100–200 nM) to activate adenosine 2B (A2B) receptors (Lazarowski et al., 2004). These A2B receptors are expressed on the apical surface of airway epithelial cells and since they are G_s-linked (Lazarowski & Boucher, 2009), their activation increases intracellular cAMP levels, which via PKA phosphorylation activates CFTR mediated Cl^-/HCO_3^- section (Lazarowski & Boucher, 2009).

ASL fluid volumes may directly influence the concentration of nucleotides and nucleosides in the ASL. By way of example, under secretory conditions in which ASL volumes increase, ASL ATP and adenosine concentrations would tend to be reduced leading to a reduction in CaCC/TMEM16A and CFTR mediated Cl^- secretion. However, when ASL volume diminishes, i.e. when airway epithelia exist in an absorptive state, ATP and ADO concentrations will increase. Thus, the changes in ATP/adenosine may serve as a feedback mechanism to limit changes in ASL volume induced by either secretion or absorption. Note, in addition to activating secretion, stimulation of $P2Y_2$ and A2B receptors will also tend to inactivate ENaC and prolong the secretory event (Boucher et al., 1986; Devor & Pilewski, 1999). In primary HBECs, the combined addition of apyryase (to degrade ATP) and 8-SPT (to inhibit A2B receptors) leads to a complete inhibition of endogenous ASL/Cl^- secretion (Tarran et al., 2005). However, additional, second messengers that can regulate ASL/Cl^-

secretion may also exist *in vivo*. For example, vasointestinal peptide (VIP) regulates CFTR trafficking and cAMP-mediated ASL/Cl^- secretion (Alcolado et al., 2014; Lee et al., 2013).

It is important to note that ATP and adenosine are physiologically active under thin film conditions where native ASL is intact. In comparison, when cultures are placed in Ussing chambers, where the native ASL is removed and replaced with up to 10 mL of Ringer solution, these soluble mediators are removed. Here, due to the unphysiologically large volumes, even though the epithelia continue to release ATP, it does not reach physiologically relevant concentrations. Such considerations also hold true for other experimental setups where ASL is washed away. Thus, in some circumstances, it is useful to remove ASL and replace it with a defined medium. However, it is also important to consider how this relates to ASL volume regulation under thin film conditions where native ASL is present. For example, using the Ussing chamber analogy, when ATP is added at for example, 10 μM in 10 mls, the effective concentration over the epithelia is clamped since the degradative capacity of the epithelia is overwhelmed and high ATP likely leads to prolonged receptor desensitization. In contrast, under thin film conditions, the ecto-enzymes that are on the apical membrane and/ or secreted can rapidly degrade ATP. Thus, 10 μM ATP in 10 μLs would likely have a half-life of 30 s, therefore within 3 min, only ~1% of the original ATP would be present (Tarran et al., 2001).

3.8 SPLUNC1 as a Soluble Mediator of ASL Volume

The endogenous, epithelial secreted, SPLUNC1 protein has multiple functions within the airways. SPLUNC1 has surfactant-like properties (Bartlett et al., 2011), binds and internalizes the plasma membrane Ca^{2+} channel Orai1 (Wu et al., 2017), has antimicrobial activity against gram-negative bacteria (Ahmad et al., 2016) and binds to and internalizes ENaC (Hobbs et al., 2013; Kim et al., 2018). It is with regards to its latter function that we shall focus here. We previously observed that washing HBECs lead to an increase in ENaC activity that waned over 24 h (Tarran et al., 2006). Based on this information, we hypothesized that there was a soluble ENaC inhibitor that was present within the ASL. Using a proteomic approach, we identified SPLUNC1(144). Indeed, SPLUNC1 is a soluble mediator of ASL volume regulation and stable shRNA SPLUNC1 knockdown in HBECs leads to ASL volume depletion that can be restored by addition of recombinant SPLUNC1 (Garcia-Caballero et al., 2009). We have localized the

ENaC inhibitory domain to SPLUNC1's N-terminus, which has been termed the S18 region, because it is from SPLUNC1 and was initially thought to have 18 residues, although further work showed it most likely requires only 8 residues to inhibit ENaC (Hobbs et al., 2013; Scott et al., 2017). Unfortunately, due to species differences, murine SPLUNC1 does not have an S18 region (Garland et al., 2013) and likely does not regulate ENaC, so ASL volume studies with the SPLUNC1 knockout mice are not possible. However, SPLUNC1 is a CF gene modifier, and CF patients who have polymorphisms that reduce SPLUNC1 expression do worse than control CF patients with normal SPLUNC1 expression, suggesting that SPLUNC1 plays an important role in airway physiology (Saferali et al., 2015).

SPLUNC1 binds extracellularly to ENaC and leads to channel internalization. Interestingly, its mechanism of action is becoming clear. SPLUNC1 is an allosteric modulator of ENaC activity that leads to NEDD4.2-dependent binding and ubiquitination of the channel (Hobbs et al., 2013; Kim et al., 2018). Subunit internalization prevents Na^+ and ASL absorption. Interestingly, since it is a channel internalizer, it has a different mechanism of action to (1) amiloride, that blocks ENaC's pore, and (2) protease inhibitors such as aprotinin that prevent ENaC cleavage. Indeed, inhibition of ENaC using a SPLUNC1- derived peptide (SPX-101) has an extended duration of action because ENaC has been removed from the cell surface (Scott et al., 2017).

3.9 Switching From a Secretory to an Absorptive State

Surface epithelia of the conducting airways can switch between absorptive and secretory states. The surface epithelia also have the capacity to modify glandular secretions and modify mucins produced by goblet cells and submucosal glands. Physiologically, ASL absorption is necessary under basal conditions since ASL/mucus move up the airways and as two airways converge, there will be an excess of fluid that needs to be dealt with (i.e., absorbed) in order to maintain a constant ASL height. How switching from absorptive to secretory modes occurs *in vivo* is not fully understood. However, we postulate that whilst basal absorption predominates, Cl^-/ASL secretion occurs regionally in order to increase ASL height where needed. For example, during local irritation or inflammation, secretion-induced increases in ASL height would serve to increase mucus transport and help clear inhaled particles/pathogens etc. It is interesting that the inflammatory cytokine IL-1β leads to a large increase in CFTR-mediated

Cl^- secretion (Gray et al., 2004), while IL-4 enhances TMEM16A expression and leads to a functional increase in CacC activity (Caputo et al., 2008), suggesting that during times of inflammation in normal subjects, Cl^- secretion is enhanced. Similarly, inactive (silent) ENaC is also present apically and may represent a reserve pool of ENaC that can be activated as needed to increase the absorptive capacity and reduce ASL volume (Boucher, 2007a; Caldwell et al., 2005). Thus, there is likely a reserve of absorptive and secretory capacity that is available *in vivo*. For example, addition of ENaC antagonists or the protease inhibitor camostat lead to an increase in mucus clearance rates in sheep (Astrand et al., 2015; Coote et al., 2009), whilst addition of proteases, which activate ENaC reduces mucus clearance rates (Sabater, Lee, & Abraham, 2005), suggesting that ENaC is about 50% active under normal conditions.

It has previously been proposed that some epithelia can have "hot spots", i.e. areas that only absorb or only secrete. For example, in the collecting duct of the kidney, principal cells are the site of Na^+ absorption while intercalated cells mediate acid secretion and bicarbonate reabsorption (Edwards & Crambert, 2017). However, ENaC has been detected by immunohistochemistry in the apical membrane of ciliated cells of the surface epithelia from adult lung and was not present in basal cells or goblet cells (Gaillard et al., 2000). These authors also found ENaC subunit expression in serous gland cells. Similarly, CFTR is also highly expressed in the ciliated cells of the surface epithelia and also in gland serous cells (Kreda et al., 2005). Importantly, these data suggest that surface epithelia do not have specialized absorbing cells or secreting cells. Functionally, using the patch clamp technique, amiloride-sensitive and cAMP-stimulated currents have been detected in the same single freshly isolated ciliated cells (Tarran et al., 1998). The idea of regional differences in ion transport and distinct regions of absorbing and secreting cells has recently been mooted for porcine small airways (Flores-Delgado, Lytle, & Quinton, 2016; Shamsuddin & Quinton, 2012). However, to date, this has yet to be studied in humans.

Experimentally, the difference between ASL absorption versus ASL secretion must be considered when designing ASL volume experiments. For example, if HBEC's mucosal surfaces are extensively washed and a bolus of solution is added apically (e.g. 20–50 $\mu L/cm^2$), SPLUNC1 is washed away, ENaC-led Na^+ absorption is highly active and there is little CFTR- or CaCC-mediated Cl^- secretion during this phase (i.e. $<12-$ h after washing). This is ideal for studying Na^+ absorption, but since

Cl^- secretion is minimal during this time, misleading data could be obtained if one is studying CFTR for example. In contrast, if the cultures are allowed to go undisturbed for 24 h, soluble mediators such as ATP, adenosine and SPLUNC1 increase, leading to decreased ENaC activity, little Na^+ absorption and more Cl^- secretion which is better for studying CFTR activity. We have shown that after extensive mucosal washing, it takes >12 h for SPLUNC1 levels to return to normal μM levels in the ASL (Hobbs et al., 2013), which is consistent with the time frame for changes from absorption to secretion. Alternatively, after mucosal washing, ENaC can be rapidly inhibited with aprotinin or another non-amiloride antagonist (amiloride has too short a half-life in the ASL to be useful under thin-film conditions; half-life of amiloride in the ASL is ~9 min). Indeed, after washing the cultures, adenosine only induces CFTR-mediated ASL secretion in the presence of aprotinin when ENaC is inhibited (Tarran et al., 2006).

4. PH HOMEOSTASIS OF THE ASL

As discussed in Section 3.1, CFTR conducts multiple anions, with differing selectivity and preference including the key physiological pH buffering ion HCO_3^- (Gray et al., 1990; Sheppard & Welsh, 1999) (Gray et al., 1990; Poulsen et al., 1994). The importance of HCO_3^- secretion first became evident when examining CF pancreatic disease. The pancreas is the organ most severely affected by CF, with pancreatic destruction beginning *in utero* (Borowitz, 2015; Durie, 1989). Bicarbonate concentrations in pancreatic secretions are high, with up to 140 mM being reported (Borowitz, 2015). High HCO_3^- concentrations are necessary to neutralize secreted acidic digestive enzymes, and to prevent pancreatic destruction (Borowitz, 2015). The permeability of CFTR to HCO_3^- is significantly lower than the permeability of CFTR to Cl^-. It was therefore suggested that Cl^- secretion via CFTR in the pancreas generates a Cl^- gradient across the apical membrane. Establishment of this gradient allows SLC26A6 to exchange Cl^- for HCO_3^- to generate high luminal HCO_3^- content (Wilschanski & Novak, 2013). Correlation between CFTR mutation severity and pancreatic HCO_3^- has been established, indicating a role for defective HCO_3^- secretion in the establishment of CF disease (Choi et al., 2001).

The role of HCO_3^- in the airways has been less studied than in the pancreas and is more controversial. However, even though the ASL HCO_3^- concentration is less than that in plasma or pancreatic juices, it likely acts as a buffer of ASL pH and can influence ASL volume and possibly CF lung disease pathology (Figuiera, Webster, & Tarran, 2018). The paracellular pathway is permeable to HCO_3^-, so one would expect [HCO_3^-] and ASL pH to be similar to plasma (i.e. 25 mM and pH 7.4). However, normal ASL has a pH of 7.0. Thus, using the Henderson–Hasselbach equation and assuming a CO_2 partial pressure of 5%, the ASL HCO_3^- concentration is calculated at 10 mM, or approximately 15 mM less than the plasma HCO_3^- concentration. This estimate is broadly consistent with anion gap studies performed by Knowles and colleagues where they measured ASL Na^+, K^+ and Cl^- by flame photometry and found [$Na^+ + K^+$] to be ~ 10 mM greater than [Cl^-], with the resulting "anion gap" assumed to be due to HCO_3^-. Since the partial pressure of CO_2 in the airway lumen is much lower on inspiration (0.4%) than during expiration (up to 5%), it has been hypothesized that ASL CO_2 levels may rapidly change during each cycle of breathing. In contrast, Thornell et al. recently postulated that carbonic anhydrase, the enzyme responsible for catabolizing CO_2 and H_2O into H^+ and HCO_3^-, is minimally expressed in the airways, and so, despite constant fluctuations in luminal CO_2 levels, overall pH values are not so influenced by these parameters (Thornell et al., 2018). Interestingly, they also predicted that ASL pH was slightly more alkaline during respiratory-like fluctuations in ASL CO_2 levels, indicating that steady state CO_2 may be slightly less than 5%. In addition to HCO_3^- secretion, protons are also secreted, either by the dual oxidases (DUOX) and/or the H^+/K^+-ATPase (ATP12a) that are also expressed apically on airway epithelia (Coakley et al., 2003; Fischer, 2009). As such, HCO_3^- buffers ASL pH by sequestering H^+ ions to form H_2O and CO_2 (Fischer, 2009).

Calu-3 cells are a model bronchial, adenocarcinomata-derived cell line that form polarized monolayers and highly express CFTR (Daniel & Burnett, 1991; Rubenstein, 2018). Since they express CFTR, these cells have been used extensively to study airway bicarbonate transport. Based on data obtained with this cell line, Garnett et al. suggested that altered activity of the SLC26A4 exchanger, not CFTR, caused reduced ASL pH (Garnett et al., 2011). In contrast, Huang et al. found little involvement of SLC26A4 in this process. Possible differences in these findings could have arisen from these groups using different sub-clonal populations. IL-13 and

IL-17A can upregulate the expression of SLC26A4 (Adams et al., 2014; Gray, 2004), further suggesting that bicarbonate transport is highly variable, particularly under inflammatory conditions associated with airway disease. Using primary HBECs, Haggie et al. found a role for pendrin in ASL volume regulation of Cl^-/HCO_3^- exchange in IL-13-treated but not vehicle treated cultures (Haggie et al., 2016), suggesting that SLC26A4/pendrin expression is variable and plays a greater role in regulating ASL volume and pH under inflamed conditions. Indeed, SLC26A4/pendrin knockout mice are more susceptible to bacterial lung infections, providing further evidence of a role for pendrin, via pH and volume regulation in innate defense under conditions of inflammation (Scanlon et al., 2014). Interestingly, SLC26A4/pendrin knockout mice had reduced inflammation, suggesting a hitherto unrecognized role of SLC26A4/pendrin in modulating not only ASL volume/pH but inflammation itself. Whether inhibition of lung inflammation in SLC26A4/pendrin knockout mice were directly related to altered ASL volume/pH or whether SLC26A4/pendrin has other roles remains to be determined. For example, one could hypothesize that alkalinized ASL alters macrophage/neutrophil function. Indeed, an acid pH has been shown to impair neutrophil chemotaxis and respiratory activity (Lardner, 2001). However, more studies will be required in order to better understand these phenomena in the airways.

4.1 Role of pH in Mucin Expansion

Mucins are high molecular weight glycoporteins that are expressed as tethered mucins at the cell surface (i.e. MUC1, MUC4) or as secreted proteins (e.g. MUC5B, MUC5AC). Mucins exert large effects on mucus rheology and confer its non-Newtonian visco-elastic properties (Voynow & Rubin, 2009). Mucin molecules are cross-linked by divalent cations and stored in dense, highly packed granules which undergo exocytosis to release contents onto the apical surface (Quinton, 2008). Upon release, mucin molecules expand by up to 800 times (Verdugo, 1991). This expansion is dependent on the exchange or sequestration of divalent cations for monovalent cations such as Na^+, which abolishes the cross-linking. Thus, to allow tight packing of mucin molecules within the granules, their negative electrical charges are shielded by Ca^{2+} ions to limit electrostatic repulsion (Quinton, 2008). Such shielding is achieved by the presence of high Ca^{2+} concentrations within granules and Ca^{2+} concentrations greater than 200 mM have been reported (Verdugo, Deyrup-Olsen, Aitken, Villalon, & Johnson, 1987). In addition to the exchange of Ca^{2+} ions for monovalent cations including Na^+, K^+

and H^+ (Nguyen, Chin, & Verdugo, 1998), more recent work has proposed a role for HCO_3^- in sequestering of mucin bound Ca^{2+} (Borowitz, 2015). HCO_3^- present within ASL complexes with Ca^{2+}, and neutralizes mucin bound H^+ to form H_2CO_3, $CaHCO_3^+$ and $CaCO_3$, thus uncovering mucin negative charges to facilitate electrostatically-induced mucin expansion (Borowitz, 2015). Efficient expansion of mucin molecules and regulation of mucus viscosity is vital to ensure trapping of inhaled particles and pathogens whilst preventing mucus adherence and accumulation within the airways. However, these researchers have proposed that ASL pH levels greater than 7.4 are required for HCO_3^- to play a role in modifying Ca^{2+}-mucin interactions, and as discussed above, ASL pH is typically less than pH 7.4. Furthermore, recent studies with human airway mucus have found that while pH/[HCO_3^-] plays a detectable role in modifying mucus viscoelasticity, that these parameters are dominated by the mucin concentration, suggesting that ASL volume regulation plays a larger role regulating mucin rheology than does ASL pH (Hill et al., 2018).

5. CF LUNG DISEASE

CF was first identified as a pancreatic disease in the 1930's (Anderson et al., 1991) and was later associated with lung pathophysiology during a New York heatwave (Elborn, 2016). It is now known that CF is a multi-organ autosomal recessive disease which affects the pancreas, lungs, sweat glands, intestine, liver and both the male and female reproductive tracts (Elborn, 2016). During the last seven decades, the management and treatment of CF has developed significantly, increasing the median survival age from just a few months in the 1950's to over 40 years today (Dodge, 2015). Causes of CF related mortality have also changed as a result of increased understanding and development of CF treatments. Meconium ileus and malnutrition, secondary to pancreatic deficiency and GI dysfunction, were previously the main causes of infant mortality. However, lung disease is currently the most common cause of mortality and morbidity amongst CF patients (Elborn, 2016; Gibson, Burns, & Ramsey, 2003).

5.1 CF Genetics

CF is caused by mutations in the *CFTR* gene, with over 2000 mutations having been described (Elborn, 2016). CFTR mutations are sub-divided into six classes, dependent on the effect of the mutation on CFTR protein expression and function. Mutation classes are summarized in Table 1. In patients of northern European ancestry, deletion of the amino acid phenylalanine at position 508 (ΔF508), causes reduced CFTR expression, CFTR misfolding and retention in the endoplasmic reticulum, deceased channel open probability and poor membrane stability. Thus, ΔF508 CFTR is simultaneously a Class II, III and VI mutation. The ΔF508 mutation accounts for > 66% of CF cases worldwide (Bobadilla, Macek, Fine, & Farrell, 2002; Farrell, 2008). In contrast, another severe CFTR mutation, G551D traffics to the plasma membrane but does not open (low P_o). G551D and R117H are more common in Ireland. Class 1 mutations, so called nonsense mutations, that code for premature stop codons such as W1282X and result in no CFTR protein being translated (see Table 1) account for ~36% of the CF population in Israel (Estivill, Bancells, & Ramos, 1997). With regard to CF lung disease, two main alternative hypotheses have been presented in an attempt to explain the initiating events of CF lung disease (Boucher, 2007b), namely the ASL dehydration hypothesis and the bacterial defect hypothesis. Since both of these are influenced by ASL volume/pH regulation and airways ion transport, they will be discussed here. A third hypothesis, where CF-affected tissues are hyperinflammatory has also been proposed, but this has been largely discounted since CF HBECs and CF piglets do not appear to be

Table 1 Summary of CFTR Class Mutations and Their Effects on CFTR Ion Channel Activity

Mutation class	Effect on CFTR	Effect on CFTR Function	Mutation Examples
I	CFTR is not synthesised	No Activity	G542X, R553X, W1282X
II	CFTR trafficking defect	No Activity	ΔI507, ΔF508, N1303K
III	Defective CFTR Regulation	No Activity	V520F, S554R, G551D
IV	Reduced CFTR conductance	Reduced Chloride Transport	R117H, R334W, S1235R
V	Reduced CFTR production	Reduced Chloride Transport	A455E, 1689−996A>6, 2657+5G > A
VI	Accelerated Degradation of CFTR	Reduced Chloride Transport	ΔF508, Q1412X

hyperinflammatory, and if anything, CF may be a disease of immunosuppression and impaired immune response (Bartlett et al., 2016; Becker et al., 2004).

5.2 The CF Dehydration Hypothesis

The most widely accepted hypothesis proposes that Cl^- hyposecretion due to the lack of functional CFTR, along with Na^+ hyperabsorption, due to abnormally active ENaC, cause ASL dehydration, which in turn causes mucus stasis and mucus plugging that lead to a failure to clear inhaled bacteria from the airways (Boucher, 2007a). Cl^- hyposecretion is directly due to the lack of CFTR activity. However, increased Na^+ absorption occurs initially due to a loss of CFTR mediated ENaC inhibition in the airways (Boucher, 2007b). Preceding the discovery of defective Cl^- secretion by CFTR in CF, it was shown that voltage (V_t) across nasal epithelia in CF patients was more negative than that of controls, a finding attributed to increased Na^+ absorption (Knowles et al., 1981). Additionally, increased rates of Na^+ absorption across freshly-isolated CF epithelia, with an approximate 200–300% increase in Na^+ absorption were reported (Boucher et al., 1986; Stutts, Rossier, & Boucher, 1997) along with an increase in O_2 consumption in CF airways, attributed to the increased Na^+/K^+ ATPase activity required to extrude Na^+ basolaterally and maintain appropriate intracellular Na^+ levels (Stutts, Knowles, Gatzy, & Boucher, 1986). As shown in Figs. 1 and 2, Na^+ hyperabsorption promotes co-absorption of Cl^- and H_2O via paracellular junctions to maintain isotonicity. The subsequent ASL dehydration initially causes a decrease in CF airway surface liquid height/volume (Matsui et al., 1998). However, as mucin secretion continues in the absence of anion secretion/mucus clearance, one sees the formation of a viscous mucus layer, leading to an increase in ASL/mucus height, but not PLC height. Crucially, this increase in mucus layer height is accompanied with an increased mucus percentage solids. As height continues to increase and mucus accumulates within the airways, mucus plugs are formed and airway obstruction occurs. Importantly, since there is some Cl^-/ASL secretion in CF airways due to CaCC/TMEM16A, this process is heterogeneous. Further, since CaCC activation by nucleotides is vulnerable to viral infections (Tarran et al., 2005), we hypothesize that ASL dehydration/mucus accumulation will also be heterogeneous and will occur at different rates in different regions of the CF lung. Indeed, since pulmonary ventilation differs between different lobes, shear-stress-induced ATP release is also predicted to be different. For example, airflow is least in the upper respiratory zones of the lung, which have a poor ventilation/perfusion (V/Q) ratio and this is where CF lung disease is typically worse.

More recently, it has been shown that CF pigs fail to secrete ASL in response to bacterial infections (Luan et al., 2017). These observations suggest that local secretory response may occur following bacterial inhalation and that CF airways will fail to clear bacteria. Defective mucociliary clearance therefore allows for a failure of bacterial clearance, subsequent chronic bacterial colonization and concomitant chronic inflammation. Indeed, neutrophil-dominated infiltration has been reported just weeks following birth (Birrer et al., 1994). Neutrophil recruitment culminates in increased release of neutrophil elastase within the airways. Under normal physiological conditions anti-proteases such as α1-antitrypsin (α1AT) and secretory leukoprotease inhibitor (SLPI) present within the ASL inhibit neutrophil elastase activity. However, increased release of neutrophil elastase in CF lungs shifts the physiological protease/anti-protease balance such that free neutrophil elastase protein levels and activity are significantly increased (Armstrong et al., 1997; Birrer et al., 1994; Ranganathan, Hall, Sly, Stick, & Douglas, 2017). Other proteases are also upregulated in CF airways and are also likely contributory to lung damage, as reviewed elsewhere (Nichols & Chmiel, 2015; Voynow, Fischer, & Zheng, 2008). In addition to being contributory to the development of bronchiectasis, increases in concentration and activity of neutrophil elastase and other proteases also increases ENaC activity via i) increased extracellular cleavage and activation of ENaC (Caldwell et al., 2005) and ii) via degradation of the secreted ENaC inhibitory protein SPLUNC1 and thus contribute to changes in ASL volume (Hobbs et al., 2013; Jiang et al., 2013).

5.3 ASL Acidification as the Initial Event in CF Lung Disease

Defective ion transport, and a loss of homeostasis, resulting in acidification of ASL pH is proposed to reduce antimicrobial activity (Bahar & Ren, 2013; Bals, Weiner, & Wilson, 1999; Bechinger, 1996; Berkebile & McCray, 2014; Pezzulo et al., 2012; Zasloff, 1992), prevent the ENaC inhibitory action of SPLUNC1 (Garland et al., 2013; Hobbs et al., 2013) and alter mucin rheology (Dorschner et al., 2006; Quinton, 2017). Differences in ASL pH between normal and CF patients have been detected in human nasal epithelia (Garland et al., 2013) and in CF piglets that also develop CF lung disease (Pezzulo et al., 2012). However, whether or not CF ASL pH is different from normal has become controversial, with an Australian group recently reporting no difference in ASL pH (Schultz et al., 2017). A debate about ASL pH has recently been published in the Journal of Physiology (Figuiera et al., 2018; Stick & Schultz, 2018). Indeed, due to the high proton and bicarbonate permeability of the paracellular pathway, the pH difference is likely only ~0.4 pH units and thus is hard to measure experimentally. For example, in our experience, whilst we can detect a pH difference between normal vs. CF primary HBECs using fluorescent dyes, in order to detect this change, all solutions need to be pre-warmed and pre-gassed since changes in temperature or CO_2 can drastically affect the reading.

As we mentioned in Section 3.8, SPLUNC1 is an allosteric, negative regulator of ENaC (Kim et al., 2018). Interestingly, SPLUNC1 fails to function in CF ASL due to the moderate decrease in pH (146). Indeed, raising [HCO_3^-] was sufficient to restore SPLUNC1 activity/ASL regulation, and increasing ASL [HCO_3^-] after SPLUNC1 had been stably knocked down did not increase ASL height. We found that peptides of SPLUNC1's ENaC inhibitory "S18" domain were pH-independent and inhibited ENaC equally well at acidic, neutral and basic pH and in CF ASL (Garland et al., 2013; Hobbs et al., 2013). From a molecular perspective, we found that SPLUNC1 has pH-sensitive salt bridges that likely cause a conformational change in SPLUNC1, hindering the S18 region's ability to inhibit ENaC (Garland et al., 2013). More recently, we have found that mutation of SPLUNC1's two cysteine residues to alanines, or aldehyde binding to these residues also prevents ENaC/ASL volume regulation (Moore et al., 2018). Thus, SPLUNC1's S18 region appears to be particularly sensitive to conformational changes in the main body of SPLUNC1. Importantly, these data highlight how the ASL pH and ASL dehydration hypotheses are not mutually exclusive. That is, changes in ASL pH via SPLUNC1, can lead to ASL dehydration. Indeed, the amiloride-sensitive nasal P.D. is also pH-sensitive *in vivo* and a small reduction in pH of the perfusate lead to an increase in the amiloride-sensitive voltage, suggesting that small pH changes are relevant for ASL volume regulation in humans.

The initiating steps in CF lung disease continue to be debated. However, it is likely that ASL dehydration, reduced ASL pH, resultant impairment in antimicrobial activity and altered mucin rheology all contribute toward establishment of pathology in CF. CF lung disease demonstrates the importance of epithelial cells in maintaining ASL homeostasis, and illustrates how changes in extracellular volume influence protective mechanisms within the airways.

6. CONCLUSIONS

In the airways, ion transport is vital to maintain the ASL volume homeostasis needed for lung health and innate defense. Ion transport is influenced by channel/transporter expression levels and their activity, paracellular movement of ions and water, abundance of soluble mediators present within the ASL and the secretion and expansion of mucin molecules. Whilst all of these components can be efficiently regulated in normal airways, defective channel activity and/or mucin secretion and mucus clearance can prove detrimental to lung homeostasis. The genetic disease, CF, exemplifies the effects of defective ion transport, and illustrates how changes in Cl^- and Na^+ transport influence airway hydration, inflammation, infection and airway obstruction. Given that the loss of CFTR culminates in severe airway pathophysiology, which is currently the leading cause of mortality amongst CF patients, substantial effort is ongoing to treat both the underlying ion defects of CF, Cl^- and Na^+, and associated CF pathophysiology. Ultimately, concerted efforts by researchers and pharmaceutical companies aim to increase the mean survival age of all CF patients and improve their quality of life through therapeutic intervention. Whilst CF is a multi-organ disease, restoring ASL homeostasis will help achieve this goal and is predicted to increase patient survival.

REFERENCES

Abraham, V., Spielman, D., Kolls, J. K., Rubenstein, R. C., Conner, G. E., et al. (2014). IL-17A induces Pendrin expression and chloride-bicarbonate exchange in human bronchial epithelial cells. *PLoS One, 9*(8), e103263.

Ahmad, S., Tyrrell, J., Walton, W. G., Tripathy, A., Redinbo, M. R., & Tarran, R. (2016). Short palate, lung, and nasal epithelial clone 1 has antimicrobial and antibiofilm activities against the burkholderia cepacia complex. *Antimicrobial Agents and Chemotherapy, 60*(10), 6003–6012.

Alcolado, N. G., Conrad, D. J., Poroca, D., Li, M., Alshafie, W., Chappe, F. G., et al. (2014). Cystic fibrosis transmembrane conductance regulator dysfunction in VIP knockout mice. *American Journal of Physiology. Cell Physiology, 307*(2), C195–C207.

Alper, S. L., & Sharma, A. K. (2013). The SLC26 gene family of anion transporters and channels. *Molecular Aspects of Medicine, 34*(2–3), 494–515.

Anderson, M. P., Berger, H. A., Rich, D. P., Gregory, R. J., Smith, A. E., & Welsh, M. J. (1991). Nucleoside triphosphates are required to open the CFTR chloride channel. *Cell, 67*(4), 775–784.

Anderson, W. H., Coakley, R. D., Button, B., Henderson, A. G., Zeman, K. L., Alexis, N. E., et al. (2015). The relationship of mucus concentration (hydration) to mucus osmotic pressure and transport in chronic bronchitis. *American Journal of Respiratory and Critical Care Medicine, 192*(2), 182–190.

Armstrong, D. S., Grimwood, K., Carlin, J. B., Carzino, R., Gutierrez, J. P., Hull, J., et al. (1997). Lower airway inflammation in infants and young children with cystic fibrosis. *American Journal of Respiratory and Critical Care Medicine, 156*(4 Pt 1), 1197–1204.

Astrand, A. B., Hemmerling, M., Root, J., Wingren, C., Pesic, J., Johansson, E., et al. (2015). Linking increased airway hydration, ciliary beating, and mucociliary clearance through ENaC inhibition. *American Journal of physiology. Lung Cellular and Molecular Physiology, 308*(1), L22–L32.

Avella, M., Loriol, C., Boulukos, K., Borgese, F., & Ehrenfeld, J. (2011). SLC26A9 stimulates CFTR expression and function in human bronchial cell lines. *Journal of Cellular Physiology, 226*(1), 212–223.

Bahar, A. A., & Ren, D. (2013). Antimicrobial peptides. *Pharmaceuticals, 6*(12), 1543–1575.

Bals, R., Weiner, D. J., & Wilson, J. M. (1999). The innate immune system in cystic fibrosis lung disease. *Journal of Clinical Investigation, 103*(3), 303–307.

Bardou, O., Trinh, N. T., & Brochiero, E. (2009). Molecular diversity and function of K+ channels in airway and alveolar epithelial cells. *American Journal of physiology. Lung Cellular and Molecular Physiology, 296*(2), L145–L155.

Barkauskas, C. E., Cronce, M. J., Rackley, C. R., Bowie, E. J., Keene, D. R., Stripp, B. R., et al. (2013). Type 2 alveolar cells are stem cells in adult lung. *The Journal of Clinical Investigation, 123*(7), 3025–3036.

Bartlett, J., Gakhar, L., Penterman, J., Singh, P., Mallampalli, R. K., Porter, E., et al. (2011). PLUNC: A multifunctional surfactant of the airways. *Biochemical Society Transactions, 39*(4), 1012–1016.

Bartlett, J. A., Ramachandran, S., Wohlford-Lenane, C. L., Barker, C. K., Pezzulo, A. A., Zabner, J., et al. (2016). Newborn cystic fibrosis pigs have a blunted early response to an inflammatory stimulus. *American Journal of Respiratory and Critical Care Medicine, 194*(7), 845–854.

Bechinger, B. (1996). Towards membrane protein design: pH-sensitive topology of histidine-containing polypeptides. *Journal of Molecular Biology, 263*(5), 768–775.

Becker, M. N., Sauer, M. S., Muhlebach, M. S., Hirsh, A. J., Wu, Q., Verghese, M. W., et al. (2004). Cytokine secretion by cystic fibrosis airway epithelial cells. *American Journal of Respiratory and Critical Care Medicine, 169*(5), 645–653.

Berger, H. A., Anderson, M. P., Gregory, R. J., Thompson, S., Howard, P. W., Maurer, R. A., et al. (1991). Identification and regulation of the cystic fibrosis transmembrane conductance regulator-generated chloride channel. *Journal of Clinical Investigation, 88*(4), 1422–1431.

Berkebile, A. R., & McCray, P. B., Jr. (2014). Effects of airway surface liquid pH on host defense in cystic fibrosis. *Int J Biochem Cell Biol, 52*, 124–129.

Bertrand, C. A., Zhang, R., Pilewski, J. M., & Frizzell, R. A. (2009). SLC26A9 is a constitutively active, CFTR-regulated anion conductance in human bronchial epithelia. *The Journal of General Physiology, 133*(4), 421–438.

Birrer, P., McElvaney, N. G., Rudeberg, A., Sommer, C. W., Liechti-Gallati, S., Kraemer, R., et al. (1994). Protease-antiprotease imbalance in the lungs of children with cystic fibrosis. *American Journal of Respiratory and Critical Care Medicine, 150*(1), 207–213.

Bishop, E., Haswell, L., Adamson, J., Costigan, S., Thorne, D., & Gaca, M. (2018). An approach to testing undiluted e-cigarette aerosol in vitro using 3D reconstituted human airway epithelium. *Toxicology in Vitro*. S0887-2333(18)30011-0.

Bobadilla, J. L., Macek, M., Jr., Fine, J. P., & Farrell, P. M. (2002). Cystic fibrosis: A worldwide analysis of CFTR mutations–correlation with incidence data and application to screening. *Human Mutation, 19*(6), 575–606.

Borowitz, D. (2015). CFTR, bicarbonate, and the pathophysiology of cystic fibrosis. *Pediatric Pulmonology, 50*(Suppl 40), s24–s30.

Boucher, R. C. (1994). Human airway ion transport. Part one. *American Journal of Respiratory and Critical Care Medicine, 150*(1), 271–281.

Boucher, R. C. (2007a). Airway surface dehydration in cystic fibrosis: Pathogenesis and therapy. *Annual Review of Medicine, 58*, 157–170.

Boucher, R. C. (2007b). Evidence for airway surface dehydration as the initiating event in CF airway disease. *Journal of Internal Medicine, 261*(1), 5–16.

Boucher, R. C., Stutts, M. J., Knowles, M. R., Cantley, L., & Gatzy, J. T. (1986). Na+ transport in cystic fibrosis respiratory epithelia. Abnormal basal rate and response to adenylate cyclase activation. *Journal of Clinical Investigation, 78*(5), 1245–1252.

Brune, K., Frank, J., Schwingshackl, A., Finigan, J., & Sidhaye, V. K. (2015). Pulmonary epithelial barrier function: Some new players and mechanisms. *American Journal of physiology. Lung Cellular and Molecular Physiology, 308*(8), L731–L745.

Bruns, J. B., Carattino, M. D., Sheng, S., Maarouf, A. B., Weisz, O. A., Pilewski, J. M., et al. (2007). Epithelial Na+ channels are fully activated by furin- and prostasin-dependent release of an inhibitory peptide from the gamma-subunit. *Journal of Biological Chemistry, 282*(9), 6153–6160.

Button, B., Anderson, W. H., & Boucher, R. C. (2016). Mucus hyperconcentration as a unifying aspect of the chronic bronchitic phenotype. *Ann Am Thorac Soc, 13*(Suppl 2), S156–S162.

Button, B., Cai, L. H., Ehre, C., Kesimer, M., Hill, D. B., Sheehan, J. K., et al. (2012). A periciliary brush promotes the lung health by separating the mucus layer from airway epithelia. *Science, 337*(6097), 937–941.

Caldwell, R. A., Boucher, R. C., & Stutts, M. J. (2004). Serine protease activation of near-silent epithelial Na+ channels. *American Journal of Physiology. Cell Physiology, 286*(1), C190–C194.

Caldwell, R. A., Boucher, R. C., & Stutts, M. J. (2005). Neutrophil elastase activates near-silent epithelial Na+ channels and increases airway epithelial Na+ transport. *American Journal of Physiology - Lung Cellular and Molecular Physiology, 288*(5), L813–L819.

Canessa, C. M., Horisberger, J. D., & Rossier, B. C. (1993). Epithelial sodium channel related to proteins involved in neurodegeneration. *Nature, 361*(6411), 467–470.

Caputo, A., Caci, E., Ferrera, L., Pedemonte, N., Barsanti, C., Sondo, E., et al. (2008). TMEM16A, a membrane protein associated with calcium-dependent chloride channel activity. *Science, 322*(5901), 590–594.

Carroll, T. P., Greene, C. M., Taggart, C. C., McElvaney, N. G., & O'Neill, S. J. (2005). Interleukin-1, neutrophil elastase, and lipopolysaccharide: Key proinflammatory stimuli regulating inflammation in cystic fibrosis. *Current Respiratory Medicine Review, 1*.

Chang, M. H., Plata, C., Sindic, A., Ranatunga, W. K., Chen, A. P., Zandi-Nejad, K., et al. (2009). Slc26a9 is inhibited by the R-region of the cystic fibrosis transmembrane conductance regulator via the STAS domain. *Journal of Biological Chemistry, 284*(41), 28306–28318.

Cheng, S. H., Rich, D. P., Marshall, J., Gregory, R. J., Welsh, M. J., & Smith, A. E. (1991). Phosphorylation of the R domain by cAMP-dependent protein kinase regulates the CFTR chloride channel. *Cell, 66*(5), 1027–1036.

Choi, J. Y., Muallem, D., Kiselyov, K., Lee, M. G., Thomas, P. J., & Muallem, S. (2001). Aberrant CFTR-dependent HCO3- transport in mutations associated with cystic fibrosis. *Nature, 410*(6824), 94–97.

Chraibi, A., & Horisberger, J. D. (2002). Na self inhibition of human epithelial Na channel: Temperature dependence and effect of extracellular proteases. *The Journal of General Physiology, 120*(2), 133–145.

Clarke, L. L., Chinet, T., & Boucher, R. C. (1997). Extracellular ATP stimulates K+ secretion across cultured human airway epithelium. *American Journal of Physiology, 272*(6 Pt 1), L1084–L1091.

Clarke, L. L., Paradiso, A. M., & Boucher, R. C. (1992). Histamine-induced Cl- secretion in human nasal epithelium: Responses of apical and basolateral membranes. *American Journal of Physiology - cell Physiology, 263*(6), C1190–C1199.

Coakley, R. D., Grubb, B. R., Paradiso, A. M., Gatzy, J. T., Johnson, L. G., Kreda, S. M., et al. (2003). Abnormal surface liquid pH regulation by cultured cystic fibrosis bronchial epithelium. *Proc Natl Acad Sci U S A, 100*(26), 16083–16088.

Collier, D. M., & Snyder, P. M. (2009). Extracellular protons regulate human ENaC by modulating Na+ self-inhibition. *Journal of Biological Chemistry, 284*(2), 792–798.

Collier, D. M., & Snyder, P. M. (2011). Identification of epithelial Na(+) channel (ENaC) intersubunit Cl(−) inhibitory residues suggests a trimeric αγβ channel architecture. *The Journal of Biological Chemistry, 286*(8), 6027–6032.

Coote, K., Atherton-Watson, H. C., Sugar, R., Young, A., MacKenzie-Beevor, A., Gosling, M., et al. (2009). Camostat attenuates airway epithelial sodium channel function in vivo through the inhibition of a channel-activating protease. *The Journal of Pharmacology and Experimental Therapeutics, 329*(2), 764–774.

Cotton, C. U. (2000). Basolateral potassium channels and epithelial ion transport. *American Journal of Respiratory Cell and Molecular Biology, 23*(3), 270–272.

Daniel, M. R., & Burnett, H. E. (1991). Immunocytochemical investigation of the tissue of origin of two lung cancer cell lines. *International Journal of Experimental Pathology, 72*(4), 397–405.

Davis, C. W., & Dickey, B. F. (2008). Regulated airway goblet cell mucin secretion. *Annual Review of Physiology, 70*, 487–512.

Davis, C. W., & Lazarowski, E. (2008). Coupling of airway ciliary activity and mucin secretion to mechanical stresses by purinergic signaling. *Respiratory Physiology & Neurobiology, 163*(1–3), 208–213.

De Rose, V., Molloy, K., Gohy, S., Pilette, C., & Greene, C. M. (2018). Airway epithelium dysfunction in cystic fibrosis and COPD. *Mediators of Inflammation, 2018*, 1309746.

Devor, D. C., & Pilewski, J. M. (1999). UTP inhibits Na+ absorption in wild-type and DeltaF508 CFTR-expressing human bronchial epithelia. *American Journal of Physiology, 276*(4 Pt 1), C827–C837.

Dodge, J. A. (2015). A millennial view of cystic fibrosis. *Dev Period Med, 19*(1), 9–13.

Donaldson, S. H., Hirsh, A., Li, D. C., Holloway, G., Chao, J., Boucher, R. C., et al. (2002). Regulation of the epithelial sodium channel by serine proteases in human airways. *Journal of Biological Chemistry, 277*(10), 8338–8345.

Dorschner, R. A., Lopez-Garcia, B., Peschel, A., Kraus, D., Morikawa, K., Nizet, V., et al. (2006). The mammalian ionic environment dictates microbial susceptibility to antimicrobial defense peptides. *FASEB Journal: Official Publication of the Federation of American Societies for Experimental Biology, 20*(1), 35–42.

Dorwart, M. R., Shcheynikov, N., Wang, Y., Stippec, S., & Muallem, S. (2007). SLC26A9 is a Cl(-) channel regulated by the WNK kinases. *J Physiol, 584*(Pt 1), 333–345.

Durie, P. R. (1989). The pathophysiology of the pancreatic defect in cystic fibrosis. *Acta Psychiatrica Scandinavica Supplementum, 363*, 41–44.

Edwards, A., & Crambert, G. (2017). Versatility of NaCl transport mechanisms in the cortical collecting duct. *American Journal of Physiology - Renal Physiology, 313*(6), F1254–F1263.

Elborn, J. S. (2016). Cystic fibrosis. *The Lancet, 388*(10059), 2519–2531.

Engelhardt, J. F., Yankaskas, J. R., Ernst, S. A., Yang, Y., Marino, C. R., Boucher, R. C., et al. (1992). Submucosal glands are the predominant site of CFTR expression in the human bronchus. *Nature Genetics, 2*(3), 240–248.

Estivill, X., Bancells, C., & Ramos, C. (1997). Geographic distribution and regional origin of 272 cystic fibrosis mutations in European populations. The Biomed CF Mutation Analysis Consortium. *Human Mutation, 10*(2), 135–154.

Farrell, P. M. (2008). The prevalence of cystic fibrosis in the European Union. *Journal of Cystic Fibrosis, 7*(5), 450–453.

Ferrera, L., Zegarra-Moran, O., & Galietta, L. J. (2011). Ca2+-activated Cl- channels. *Comparative Physiology, 1*(4), 2155–2174.

Figuiera, M. F., Webster, M. J., & Tarran, R. (2018). Mucosal acidification drives early progressive lung disease in cystic fibrosis. *Journal of Physiology, Paris, 596*(16), 3433–3437.

Fischer, H. (2009). Mechanisms and function of DUOX in epithelia of the lung. *Antioxidants and Redox Signaling, 11*(10), 2453–2465.

Flores-Delgado, G., Lytle, C., & Quinton, P. M. (2016). Site of fluid secretion in small airways. *American Journal of Respiratory Cell and Molecular Biology, 54*(3), 312–318.

Frigeri, A., Gropper, M. A., Turck, C. W., & Verkman, A. S. (1995). Immunolocalization of the mercurial-insensitive water channel and glycerol intrinsic protein in epithelial cell plasma membranes. *Proc Natl Acad Sci U S A, 92*(10), 4328–4331.

Gaillard, D., Hinnrasky, J., Coscoy, S., Hofman, P., Matthay, M. A., Puchelle, E., et al. (2000). Early expression of beta- and gamma-subunits of epithelial sodium channel during human airway development. *American Journal of physiology. Lung Cellular and Molecular Physiology, 278*(1), L177–L184.

Garcia-Caballero, A., Rasmussen, J. E., Gaillard, E., Watson, M. J., Olsen, J. C., Donaldson, S. H., et al. (2009). SPLUNC1 regulates airway surface liquid volume by protecting ENaC from proteolytic cleavage. *Proceedings of the National Academy of Sciences, 106*(27), 11412–11417.

Garland, A. L., Walton, W. G., Coakley, R. D., Tan, C. D., Gilmore, R. C., Hobbs, C. A., et al. (2013). Molecular basis for pH-dependent mucosal dehydration in cystic fibrosis airways. *Proceedings of the National Academy of Sciences, 110*(40), 15973–15978.

Garnett, J. P., Baker, E. H., & Baines, D. L. (2012). Sweet talk: Insights into the nature and importance of glucose transport in lung epithelium. *The European Respiratory Journal, 40*(5), 1269–1276.

Garnett, J. P., Hickman, E., Burrows, R., Hegyi, P., Tiszlavicz, L., Cuthbert, A. W., et al. (2011). Novel role for pendrin in orchestrating bicarbonate secretion in cystic fibrosis transmembrane conductance regulator (CFTR)-expressing airway serous cells. *Journal of Biological Chemistry, 286*(47), 41069–41082.

Ghosh, A., Boucher, R. C., & Tarran, R. (2015). Airway hydration and COPD. *Cellular and Molecular Life Sciences: CMLS, 72*(19), 3637–3652.

Gibson, R. L., Burns, J. L., & Ramsey, B. W. (2003). Pathophysiology and management of pulmonary infections in cystic fibrosis. *American Journal of Respiratory and Critical Care Medicine, 168*(8), 918–951.

Gray, M. A. (2004). Bicarbonate secretion: It takes two to tango. *Nature Cell Biology, 6*(4), 292–294.

Gray, T., Coakley, R., Hirsh, A., Thornton, D., Kirkham, S., Koo, J. S., et al. (2004). Regulation of MUC5AC mucin secretion and airway surface liquid metabolism by IL-1beta in human bronchial epithelia. *American Journal of physiology. Lung Cellular and Molecular Physiology, 286*(2), L320–L330.

Gray, M. A., Greenwell, J. R., & Argent, B. E. (1988). Secretin-regulated chloride channel on the apical plasma membrane of pancreatic duct cells. *The Journal of Membrane Biology, 105*(2), 131–142.

Gray, M. A., Pollard, C. E., Harris, A., Coleman, L., Greenwell, J. R., & Argent, B. E. (1990). Anion selectivity and block of the small-conductance chloride channel on pancreatic duct cells. *American Journal of Physiology, 259*(5 Pt 1), C752–C761.

Grubb, B. R., Vick, R. N., & Boucher, R. C. (1994). Hyperabsorption of Na+ and raised Ca(2+)-mediated Cl- secretion in nasal epithelia of CF mice. *American Journal of Physiology, 266*(5 Pt 1), C1478–C1483.

Gunzel, D., & Yu, A. S. (2013). Claudins and the modulation of tight junction permeability. *Physiological Reviews, 93*(2), 525–569.

Haggie, P. M., Phuan, P. W., Tan, J. A., Zlock, L., Finkbeiner, W. E., & Verkman, A. S. (2016). Inhibitors of pendrin anion exchange identified in a small molecule screen increase airway surface liquid volume in cystic fibrosis. *FASEB Journal: Official Publication of the Federation of American Societies for Experimental Biology, 30*(6), 2187–2197.

Haq, I. J., Gray, M. A., Garnett, J. P., Ward, C., & Brodlie, M. (2016). Airway surface liquid homeostasis in cystic fibrosis: Pathophysiology and therapeutic targets. *Thorax, 71*(3), 284–287.

Harkema, J. R., Nikula, K. J., & Haschek, W. M. (2013). *Haschek and Rousseaux's handbook of toxicologic pathology* (3rd ed.). Boston: Academic Press.

Hill, D. B., Long, M. R., Markovetz, M. R., Kissner, W. J., Garabine, I. C., & Button, B. (2018). *What's new in cystic fibrosis, bpd and other congenital pediatric lung diseases?* (p. A2316) San Diego, USA: American Thoracic Society.

Hobbs, C. A., Blanchard, M. G., Alijevic, O., Tan, C. D., Kellenberger, S., Bencharit, S., et al. (2013). Identification of the SPLUNC1 ENaC-inhibitory domain yields novel strategies to treat sodium hyperabsorption in cystic fibrosis airway epithelial cultures. *American Journal of physiology. Lung Cellular and Molecular Physiology, 305*(12), L990–L1001.

Hollenhorst, M. I., Richter, K., & Fronius, M. (2011). Ion transport by pulmonary epithelia. *Journal of Biomedicine and Biotechnology, 2011*, 174306.

Hughey, R. P., Bruns, J. B., Kinlough, C. L., Harkleroad, K. L., Tong, Q., Carattino, M. D., et al. (2004). Epithelial sodium channels are activated by furin-dependent proteolysis. *Journal of Biological Chemistry, 279*(18), 18111–18114.

Hwang, T.-C., & Sheppard, D. N. (2009). Gating of the CFTR Cl(−) channel by ATP-driven nucleotide-binding domain dimerisation. *The Journal of Physiology, 587*(Pt 10), 2151–2161.

Hyde, D. M., Hamid, Q., & Irvin, C. G. (2009). Anatomy, pathology, and physiology of the tracheobronchial tree: Emphasis on the distal airways. *The Journal of Allergy and Clinical Immunology, 124*(6), S72–S77.

Jayaraman, S., Song, Y., Vetrivel, L., Shankar, L., & Verkman, A. S. (2001). Noninvasive in vivo fluorescence measurement of airway-surface liquid depth, salt concentration, and pH. *Journal of Clinical Investigation, 107*(3), 317–324.

Jiang, H., & Sun, S. X. (2013). Cellular pressure and volume regulation and implications for cell mechanics. *Biophysical Journal, 105*(3), 609–619.

Jiang, D., Wenzel, S. E., Wu, Q., Bowler, R. P., Schnell, C., & Chu, H. W. (2013). Human neutrophil elastase degrades SPLUNC1 and impairs airway epithelial defense against bacteria. *PLoS One, 8*(5), e64689.

Ji, H. L., Chalfant, M. L., Jovov, B., Lockhart, J. P., Parker, S. B., Fuller, C. M., et al. (2000). The cytosolic termini of the beta- and gamma-ENaC subunits are involved in the functional interactions between cystic fibrosis transmembrane conductance regulator and epithelial sodium channel. *Journal of Biological Chemistry, 275*(36), 27947–27956.

Jung, J., Nam, J. H., Park, H. W., Oh, U., Yoon, J. H., & Lee, M. G. (2013). Dynamic modulation of ANO1/TMEM16A HCO3(-) permeability by Ca2+/calmodulin. *Proc Natl Acad Sci U S A, 110*(1), 360–365.

Kesimer, M., Ehre, C., Burns, K. A., Davis, C. W., Sheehan, J. K., & Pickles, R. J. (2013). Molecular organization of the mucins and glycocalyx underlying mucus transport over mucosal surfaces of the airways. *Mucosal Immunology, 6*(2), 379–392.

Kim, C. S., Ahmad, S., Wu, T., Walton, W. G., Redinbo, M. R., & Tarran, R. (2018). SPLUNC1 is an allosteric modulator of the epithelial sodium channel. *FASEB Journal: Official Publication of the Federation of American Societies for Experimental Biology, 32*(5), 2478–2491.

Kis, A., Krick, S., Baumlin, N., & Salathe, M. (2016). Airway hydration, apical K(+) secretion, and the large-conductance, Ca(2+)-activated and voltage-dependent potassium (BK) channel. *Annals of the American Thoracic Society, 13*(Suppl 2), S163–S168.

Knowles, M., Gatzy, J., & Boucher, R. (1981). Increased bioelectric potential difference across respiratory epithelia in cystic fibrosis. *The New England Journal of Medicine, 305*(25), 1489–1495.

Knowles, M., Gatzy, J., & Boucher, R. (1983). Relative ion permeability of normal and cystic fibrosis nasal epithelium. *Journal of Clinical Investigation, 71*(5), 1410–1417.

Knowles, M. R., Robinson, J. M., Wood, R. E., Pue, C. A., Mentz, W. M., Wager, G. C., et al. (1997). Ion composition of airway surface liquid of patients with cystic fibrosis as compared with normal and disease-control subjects. *Journal of Clinical Investigation, 100*(10), 2588–2595.

Knowles, M. R., Stutts, M. J., Spock, A., Fischer, N., Gatzy, J. T., & Boucher, R. C. (1983). Abnormal ion permeation through cystic fibrosis respiratory epithelium. *Science, 221*(4615), 1067–1070.

Koval, M. (2013). Claudin heterogeneity and control of lung tight junctions. *Annual Review of Physiology, 75*(1), 551–567.

Ko, S. B., Zeng, W., Dorwart, M. R., Luo, X., Kim, K. H., Millen, L., et al. (2004). Gating of CFTR by the STAS domain of SLC26 transporters. *Nature Cell Biology, 6*(4), 343–350.

Kreda, S. M., Mall, M., Mengos, A., Rochelle, L., Yankaskas, J., Riordan, J. R., et al. (2005). Characterization of wild-type and deltaF508 cystic fibrosis transmembrane regulator in human respiratory epithelia. *Molecular Biology of the Cell, 16*(5), 2154–2167.

Kreda, S. M., Okada, S. F., van Heusden, C. A., O'Neal, W., Gabriel, S., Abdullah, L., et al. (2007). Coordinated release of nucleotides and mucin from human airway epithelial Calu-3 cells. *J Physiol, 584*(Pt 1), 245–259.

Lardner, A. (2001). The effects of extracellular pH on immune function. *Journal of Leukocyte Biology, 69*(4), 522–530.

Latorre, R., & Brauchi, S. (2006). Large conductance Ca2+-activated K+ (BK) channel: Activation by Ca2+ and voltage. *Biological Research, 39*(3), 385–401.

Lazarowski, E. R., & Boucher, R. C. (2009). Purinergic receptors in airway epithelia. *Current Opinion in Pharmacology, 9*(3), 262–267.

Lazarowski, E. R., Tarran, R., Grubb, B. R., van Heusden, C. A., Okada, S., & Boucher, R. C. (2004). Nucleotide release provides a mechanism for airway surface liquid homeostasis. *The Journal of Biological Chemistry, 279*(35), 36855–36864.

Lee, R. J., Chen, B., Doghramji, L., Adappa, N. D., Palmer, J. N., Kennedy, D. W., et al. (2013). Vasoactive intestinal peptide regulates sinonasal mucociliary clearance and synergizes with histamine in stimulating sinonasal fluid secretion. *The FASEB Journal, 27*(12), 5094–5103.

Li, C., Ramjeesingh, M., Wang, W., Garami, E., Hewryk, M., Lee, D., et al. (1996). ATPase activity of the cystic fibrosis transmembrane conductance regulator. *Journal of Biological Chemistry, 271*(45), 28463–28468.

Luan, X., Belev, G., Tam, J. S., Jagadeeshan, S., Hassan, N., Gioino, P., et al. (2017). Cystic fibrosis swine fail to secrete airway surface liquid in response to inhalation of pathogens. *Nature Communications, 8*(1), 786.

Lumsden, A. B., McLean, A., & Lamb, D. (1984). Goblet and Clara cells of human distal airways: Evidence for smoking induced changes in their numbers. *Thorax, 39*(11), 844–849.

Mall, M., Gonska, T., Thomas, J., Schreiber, R., Seydewitz, H. H., Kuehr, J., et al. (2003). Modulation of Ca2+-activated Cl- secretion by basolateral K+ channels in human normal and cystic fibrosis airway epithelia. *Pediatric Research, 53*(4), 608–618.

Manzanares, D., Gonzalez, C., Ivonnet, P., Chen, R.-S., Valencia-Gattas, M., Conner, G. E., et al. (2011). Functional apical large conductance, Ca(2+)-activated, and voltage-dependent K(+) channels are required for maintenance of airway surface liquid volume. *The Journal of Biological Chemistry, 286*(22), 19830–19839.

Manzanares, D., Krick, S., Baumlin, N., Dennis, J. S., Tyrrell, J., Tarran, R., et al. (2015). Airway surface dehydration by transforming growth factor beta (TGF-beta) in cystic fibrosis is due to decreased function of a voltage-dependent potassium channel and can Be rescued by the drug pirfenidone. *Journal of Biological Chemistry, 290*(42), 25710–25716.

Ma, T., Song, Y., Yang, B., Gillespie, A., Carlson, E. J., Epstein, C. J., et al. (2000). Nephrogenic diabetes insipidus in mice lacking aquaporin-3 water channels. *Proc Natl Acad Sci U S A, 97*(8), 4386–4391.

Matalon, S., Bartoszewski, R., & Collawn, J. F. (2015). Role of epithelial sodium channels in the regulation of lung fluid homeostasis. *American Journal of physiology. Lung Cellular and Molecular Physiology, 309*(11), L1229–L1238.

Matsui, H., Davis, C. W., Tarran, R., & Boucher, R. C. (2000). Osmotic water permeabilities of cultured, well-differentiated normal and cystic fibrosis airway epithelia. *Journal of Clinical Investigation, 105*(10), 1419–1427.

Matsui, H., Grubb, B. R., Tarran, R., Randell, S. H., Gatzy, J. T., Davis, C. W., et al. (1998). Evidence for periciliary liquid layer depletion, not abnormal ion composition, in the pathogenesis of cystic fibrosis airways disease. *Cell, 95*(7), 1005–1015.

Meyrick, B., Sturgess, J. M., & Reid, L. (1969). A reconstruction of the duct system and secretory tubules of the human bronchial submucosal gland. *Thorax, 24*(6), 729–736.

Moore, P. J., Reidel, B., Ghosh, A., Sesma, J., Kesimer, M., & Tarran, R. (2018). Cigarette smoke modifies and inactivates SPLUNC1, leading to airway dehydration. *The FASEB Journal*. fj201800345R.

Nadel, J. A., & Davis, B. (1980). Parasympathetic and sympathetic regulation of secretion from submucosal glands in airways. *Federation Proceedings, 39*(13), 3075–3079.

Nakao, I., Kanaji, S., Ohta, S., Matsushita, H., Arima, K., Yuyama, N., et al. (2008). Identification of pendrin as a common mediator for mucus production in bronchial asthma and chronic obstructive pulmonary disease. *The Journal of Immunology: Official Journal of the American Association of Immunologists, 180*(9), 6262–6269.

Namkung, W., Phuan, P. W., & Verkman, A. S. (2011). TMEM16A inhibitors reveal TMEM16A as a minor component of calcium-activated chloride channel conductance in airway and intestinal epithelial cells. *Journal of Biological Chemistry, 286*(3), 2365–2374.

Nguyen, T., Chin, W. C., & Verdugo, P. (1998). Role of Ca2+/K+ ion exchange in intracellular storage and release of Ca2+. *Nature, 395*(6705), 908–912.

Nichols, D. P., & Chmiel, J. F. (2015). Inflammation and its genesis in cystic fibrosis. *Pediatric Pulmonology, 50*(Suppl 40), S39–S56.

Nielsen, S., King, L. S., Christensen, B. M., & Agre, P. (1997). Aquaporins in complex tissues. II. Subcellular distribution in respiratory and glandular tissues of rat. *American Journal of Physiology, 273*(5 Pt 1), C1549–C1561.

Nielsen, S., Smith, B. L., Christensen, E. I., & Agre, P. (1993). Distribution of the aquaporin CHIP in secretory and resorptive epithelia and capillary endothelia. *Proceedings of the National Academy of Sciences of the United States of America, 90*(15), 7275–7279.

Nofziger, C., Dossena, S., Suzuki, S., Izuhara, K., & Paulmichl, M. (2011). Pendrin function in airway epithelia. *Cellular Physiology and Biochemistry: International Journal of Experimental Cellular Physiology, Biochemistry, and Pharmacology, 28*(3), 571–578.

Ousingsawat, J., Kongsuphol, P., Schreiber, R., & Kunzelmann, K. (2011). CFTR and TMEM16A are separate but functionally related Cl- channels. *Cellular Physiology and Biochemistry, 28*(4), 715–724.

Paradiso, A. M. (1997). ATP-activated basolateral Na+/H+ exchange in human normal and cystic fibrosis airway epithelium. *American Journal of Physiology, 273*(1 Pt 1), L148–L158.

Parker, D., & Prince, A. (2011). Innate immunity in the respiratory epithelium. *American Journal of Respiratory Cell and Molecular Biology, 45*(2), 189–201.

Patwa, A., & Shah, A. (2015). Anatomy and physiology of respiratory system relevant to anaesthesia. *Indian Journal of Anaesthesia, 59*(9), 533–541.

Peter, A., Fatykhova, D., Kershaw, O., Gruber, A. D., Rueckert, J., Neudecker, J., et al. (2017). Localization and pneumococcal alteration of junction proteins in the human alveolar-capillary compartment. *Histochemistry and Cell Biology, 147*(6), 707–719.

Pettersen, C. A., & Adler, K. B. (2002). Airways inflammation and COPD: Epithelial-neutrophil interactions. *Chest, 121*(5 Suppl), 142S–150S.

Pezzulo, A. A., Tang, X. X., Hoegger, M. J., Abou Alaiwa, M. H., Ramachandran, S., Moninger, T. O., et al. (2012). Reduced airway surface pH impairs bacterial killing in the porcine cystic fibrosis lung. *Nature, 487*(7405), 109–113.

Picciotto, M. R., Cohn, J. A., Bertuzzi, G., Greengard, P., & Nairn, A. C. (1992). Phosphorylation of the cystic fibrosis transmembrane conductance regulator. *Journal of Biological Chemistry, 267*(18), 12742–12752.

Poulsen, J. H., Fischer, H., Illek, B., & Machen, T. E. (1994). Bicarbonate conductance and pH regulatory capability of cystic fibrosis transmembrane conductance regulator. *Proc Natl Acad Sci U S A, 91*(12), 5340–5344.

Quinton, P. M. (1983). Chloride impermeability in cystic fibrosis. *Nature, 301*(5899), 421–422.

Quinton, P. M. (2008). Cystic fibrosis: Impaired bicarbonate secretion and mucoviscidosis. *The Lancet, 372*(9636), 415–417.

Quinton, P. M. (2017). Both ways at once: Keeping small airways clean. *Physiology, 32*(5), 380–390.

Ranganathan, S. C., Hall, G. L., Sly, P. D., Stick, S. M., & Douglas, T. A. (2017). Early lung disease in infants and preschool children with cystic fibrosis. What have we learned and what should we do about it? *American Journal of Respiratory and Critical Care Medicine, 195*(12), 1567–1575.

Rawlins, E. L., & Hogan, B. L. (2006). Epithelial stem cells of the lung: Privileged few or opportunities for many? *Development, 133*(13), 2455–2465.

Riordan, J. R., Rommens, J. M., & Kerem, B. (1989). Identification of the cystic fibrosis gene: Cloning and characterization of complementary DNA. *Science, 245*(4925), 1437.

Rossier, B. C., & Stutts, M. J. (2009). Activation of the epithelial sodium channel (ENaC) by serine proteases. *Annual Review of Physiology, 71*(1), 361–379.

Rotin, D., & Schild, L. (2008). ENaC and its regulatory proteins as drug targets for blood pressure control. *Current Drug Targets, 9*(8), 709–716.

Rubenstein, R. C. (2018). Editorial focus: CFTR-dependent bicarbonate secretion by Calu-3 cells. *Physiol Rep, 6*(10), e13691.

Rubenstein, R. C., Lockwood, S. R., Lide, E., Bauer, R., Suaud, L., & Grumbach, Y. (2011). Regulation of endogenous ENaC functional expression by CFTR and DeltaF508-CFTR in airway epithelial cells. *American Journal of physiology. Lung Cellular and Molecular Physiology, 300*(1), L88–L101.

Sabater, J. R., Lee, T. A., & Abraham, W. M. (2005). Comparative effects of salmeterol, albuterol, and ipratropium on normal and impaired mucociliary function in sheep. *Chest, 128*(5), 3743–3749.

Saferali, A., Obeidat, M., Berube, J. C., Lamontagne, M., Bosse, Y., Laviolette, M., et al. (2015). Polymorphisms associated with expression of BPIFA1/BPIFB1 and lung disease severity in cystic fibrosis. *American Journal of Respiratory Cell and Molecular Biology, 53*(5), 607–614.

Saint-Criq, V., & Gray, M. A. (2017). Role of CFTR in epithelial physiology. *Cellular and Molecular Life Sciences: CMLS, 74*(1), 93–115.

Scanlon, K. M., Gau, Y., Zhu, J., Skerry, C., Wall, S. M., Soleimani, M., et al. (2014). Epithelial anion transporter pendrin contributes to inflammatory lung pathology in mouse models of Bordetella pertussis infection. *Infection and Immunity, 82*(10), 4212–4221.

Schnitzer, J. E., & Oh, P. (1996). Aquaporin-1 in plasma membrane and caveolae provides mercury-sensitive water channels across lung endothelium. *American Journal of Physiology - heart and Circulatory Physiology, 270*(1), H416–H422.

Schroeder, B. C., Cheng, T., Jan, Y. N., & Jan, L. Y. (2008). Expression cloning of TMEM16A as a calcium-activated chloride channel subunit. *Cell, 134*(6), 1019–1029.

Schultz, A., Puvvadi, R., Borisov, S. M., Shaw, N. C., Klimant, I., Berry, L. J., et al. (2017). Airway surface liquid pH is not acidic in children with cystic fibrosis. *Nature Communications, 8*(1), 1409.

Scott, D. W., Walker, M. P., Sesma, J., Wu, B., Stuhlmiller, T. J., Sabater, J. R., et al. (2017). SPX-101 is a novel epithelial sodium channel-targeted therapeutic for cystic fibrosis that restores mucus transport. *American Journal of Respiratory and Critical Care Medicine, 196*(6), 734–744.

Shamsuddin, A. K., & Quinton, P. M. (2012). Surface fluid absorption and secretion in small airways. *J Physiol, 590*(15), 3561–3574.

Shen, B. Q., Finkbeiner, W. E., Wine, J. J., Mrsny, R. J., & Widdicombe, J. H. (1994). Calu-3: A human airway epithelial cell line that shows cAMP-dependent Cl-secretion. *American Journal of Physiology, 266*(5 Pt 1), L493–L501.

Sheng, S., McNulty, K. A., Harvey, J. M., & Kleyman, T. R. (2001). Second transmembrane domains of ENaC subunits contribute to ion permeation and selectivity. *Journal of Biological Chemistry, 276*(47), 44091–44098.

Sheppard, D. N., & Welsh, M. J. (1999). Structure and function of the CFTR chloride channel. *Physiological Reviews, 79*(1), S23–S45.

Sheridan, J. T., Worthington, E. N., Yu, K., Gabriel, S. E., Hartzell, H. C., & Tarran, R. (2011). Characterization of the oligomeric structure of the Ca(2+)-activated Cl- channel Ano1/TMEM16A. *Journal of Biological Chemistry, 286*(2), 1381–1388.

Snyder, P. M., McDonald, F. J., Stokes, J. B., & Welsh, M. J. (1994). Membrane topology of the amiloride-sensitive epithelial sodium channel. *Journal of Biological Chemistry, 269*(39), 24379–24383.

Song, Y., Yang, B., Matthay, M. A., Ma, T., & Verkman, A. S. (2000). Role of aquaporin water channels in pleural fluid dynamics. *American Journal of Physiology. Cell Physiology, 279*(6), C1744–C1750.

Stick, S. M., & Schultz, A. (2018). CrossTalk opposing view: mucosal acidification does not drive early progressive lung disease in cystic fibrosis. *Journal of Physiology, 596*(16), 3439–3441. https://doi.org/10.1113/JP275426.

Stockand, J. D., Staruschenko, A., Pochynyuk, O., Booth, R. E., & Silverthorn, D. U. (2008). Insight toward epithelial Na+ channel mechanism revealed by the acid-sensing ion channel 1 structure. *IUBMB Life, 60*(9), 620–628.

Stoltz, D. A., Meyerholz, D. K., & Welsh, M. J. (2015). Origins of cystic fibrosis lung disease. *New England Journal of Medicine, 372*(4), 351–362.

Stutts, M. J., Knowles, M. R., Gatzy, J. T., & Boucher, R. C. (1986). Oxygen consumption and ouabain binding sites in cystic fibrosis nasal epithelium. *Pediatric Research, 20*(12), 1316–1320.

Stutts, M. J., Rossier, B. C., & Boucher, R. C. (1997). Cystic fibrosis transmembrane conductance regulator inverts protein kinase a-mediated regulation of epithelial sodium channel single channel kinetics. *Journal of Biological Chemistry, 272*(22), 14037–14040.

Tan, C. D., Hobbs, C., Sameni, M., Sloane, B. F., Stutts, M. J., & Tarran, R. (2014). Cathepsin B contributes to Na+ hyperabsorption in cystic fibrosis airway epithelial cultures. *J Physiol, 592*(23), 5251–5268.

Tarran, R., Button, B., & Boucher, R. C. (2006). Regulation of normal and cystic fibrosis airway surface liquid volume by phasic shear stress. *Annual Review of Physiology, 68*, 543–561.

Tarran, R., Button, B., Picher, M., Paradiso, A. M., Ribeiro, C. M., Lazarowski, E. R., et al. (2005). Normal and cystic fibrosis airway surface liquid homeostasis. The effects of phasic shear stress and viral infections. *Journal of Biological Chemistry, 280*(42), 35751–35759.

Tarran, R., Gray, M. A., Evans, M. J., Colledge, W. H., Ratcliff, R., & Argent, B. E. (1998). Basal chloride currents in murine airway epithelial cells: Modulation by CFTR. *American Journal of Physiology, 274*(4 Pt 1), C904–C913.

Tarran, R., Grubb, B. R., Gatzy, J. T., Davis, C. W., & Boucher, R. C. (2001). The relative roles of passive surface forces and active ion transport in the modulation of airway surface liquid volume and composition. *The Journal of General Physiology, 118*(2), 223–236.

Tarran, R., Grubb, B. R., Parsons, D., Picher, M., Hirsh, A. J., Davis, C. W., et al. (2001). The CF salt controversy: In vivo observations and therapeutic approaches. *Molecular Cell, 8*(1), 149–158.

Tarran, R., Loewen, M. E., Paradiso, A. M., Olsen, J. C., Gray, M. A., Argent, B. E., et al. (2002). Regulation of murine airway surface liquid volume by CFTR and Ca2+-activated Cl- conductances. *The Journal of General Physiology, 120*(3), 407–418.

Tarran, R., Trout, L., Donaldson, S. H., & Boucher, R. C. (2006). Soluble mediators, not cilia, determine airway surface liquid volume in normal and cystic fibrosis superficial airway epithelia. *The Journal of General Physiology, 127*(5), 591–604.

Thiagarajah, J. R., & Verkman, A. S. (2013). Chloride channel-targeted therapy for secretory diarrheas. *Current Opinion in Pharmacology, 13*(6), 888–894.

Thornell, I. M., Li, X., Tang, X. X., Brommel, C. M., Karp, P. H., Welsh, M. J., et al. (2018). Nominal carbonic anhydrase activity minimizes airway-surface liquid pH changes during breathing. *Physiol Rep, 6*(2).

Ueki, I., German, V. F., & Nadel, J. A. (1980). Micropipette measurement of airway submucosal gland secretion. Autonomic effects. *American Review of Respiratory Disease, 121*(2), 351–357.

Vallet, V., Chraibi, A., Gaeggeler, H. P., Horisberger, J. D., & Rossier, B. C. (1997). An epithelial serine protease activates the amiloride-sensitive sodium channel. *Nature, 389*(6651), 607–610.

Van Itallie, C. M., & Anderson, J. M. (2004). The role of claudins in determining paracellular charge selectivity. *Proceedings of the American Thoracic Society, 1*(1), 38–41.

Verdugo, P. (1991). Mucin exocytosis. *American Review of Respiratory Disease, 144*(3 Pt 2), S33–S37.

Verdugo, P., Deyrup-Olsen, I., Aitken, M., Villalon, M., & Johnson, D. (1987). Molecular mechanism of mucin secretion: I. The role of intragranular charge shielding. *Journal of Dental Research, 66*(2), 506–508.

Verkman, A. S. (2002). Aquaporin water channels and endothelial cell function. *Journal of Anatomy, 200*(6), 617–627.

Verkman, A. S. (2007a). Role of aquaporins in lung liquid physiology. *Respiratory Physiology & Neurobiology, 159*(3), 324–330.

Verkman, A. S. (2007b). Role of aquaporins in lung liquid physiology. *Respiratory Physiology & Neurobiology, 159*(3), 324–330.

Verkman, A. S., Matthay, M. A., & Song, Y. (2000). Aquaporin water channels and lung physiology. *American Journal of physiology. Lung Cellular and Molecular Physiology, 278*(5), L867–L879.

Verkman, A. S., Song, Y., & Thiagarajah, J. R. (2003). Role of airway surface liquid and submucosal glands in cystic fibrosis lung disease. *American Journal of Physiology. Cell Physiology, 284*(1), C2–C15.

Voilley, N., Lingueglia, E., Champigny, G., Mattéi, M. G., Waldmann, R., Lazdunski, M., et al. (1994). The lung amiloride-sensitive Na+ channel: Biophysical properties, pharmacology, ontogenesis, and molecular cloning. *Proceedings of the National Academy of Sciences of the United States of America, 91*(1), 247–251.

Voynow, J. A., Fischer, B. M., & Zheng, S. (2008). Proteases and cystic fibrosis. *Int J Biochem Cell Biol, 40*(6–7), 1238–1245.

Voynow, J. A., & Rubin, B. K. (2009). Mucins, mucus, and sputum. *Chest, 135*(2), 505–512.

Widdicombe, J. H. (2002). Regulation of the depth and composition of airway surface liquid. *Journal of Anatomy, 201*(4), 313–318.

Widdicombe, J. H., & Wine, J. J. (2015). Airway gland structure and function. *Physiological Reviews, 95*(4), 1241–1319.

Wilschanski, M., & Novak, I. (2013). The cystic fibrosis of exocrine pancreas. *Cold Spring Harb Perspect Med, 3*(5), a009746.

Winter, M. C., & Welsh, M. J. (1997). Stimulation of CFTR activity by its phosphorylated R domain. *Nature, 389*, 294.

Wittekindt, O. H. (2017). Tight junctions in pulmonary epithelia during lung inflammation. *Pflügers Archiv, 469*(1), 135–147.

Wu, T., Huang, J., Moore, P. J., Little, M. S., Walton, W. G., Fellner, R. C., et al. (2017). Identification of BPIFA1/SPLUNC1 as an epithelium-derived smooth muscle relaxing factor. *Nature Communications, 8*, 14118.

Yaghi, A., & Dolovich, M. B. (2016). Airway epithelial cell cilia and obstructive lung disease. *Cells, 5*(4), 40.

Yang, Y. D., Cho, H., Koo, J. Y., Tak, M. H., Cho, Y., Shim, W. S., et al. (2008). TMEM16A confers receptor-activated calcium-dependent chloride conductance. *Nature, 455*(7217), 1210–1215.

Yan, W., Samaha, F. F., Ramkumar, M., Kleyman, T. R., & Rubenstein, R. C. (2004). Cystic fibrosis transmembrane conductance regulator differentially regulates human and mouse epithelial sodium channels in Xenopus oocytes. *Journal of Biological Chemistry, 279*(22), 23183–23192.

Zasloff, M. (1992). Antibiotic peptides as mediators of innate immunity. *Current Opinion in Immunology, 4*(1), 3–7.

Zawrotniak, M., & Rapala-Kozik, M. (2013). Neutrophil extracellular traps (NETs) - formation and implications. *Acta Biochimica Polonica, 60*(3), 277–284.

Zhao, K. Q., Xiong, G., Wilber, M., Cohen, N. A., & Kreindler, J. L. (2012). A role for two-pore K(+) channels in modulating Na(+) absorption and Cl(-) secretion in normal human bronchial epithelial cells. *American Journal of physiology. Lung Cellular and Molecular Physiology, 302*(1), L4–L12.

CHAPTER TEN

Physiology of the Gut: Experimental Models for Investigating Intestinal Fluid and Electrolyte Transport

Isha Dey and Neil A. Bradbury[1]
Department of Physiology and Biophysics, Chicago Medical School, North Chicago, IL, United States
[1]Corresponding author: E-mail: neil.bradbury@rosalindfranklin.edu

Contents

Current Topics in Membranes, Volume 81
ISSN 1063-5823
https://doi.org/10.1016/bs.ctm.2018.08.003

Abstract

Once thought to be exclusively an absorptive tissue, the intestine is now recognized as an important secretory tissue, playing a key role in body ion and fluid homeostasis. Given the intestine's role in fluid homeostasis, it is not surprising that important clinical pathologies arise from imbalances in fluid absorption and secretion. Perhaps the most important examples of this can be seen in enterotoxigenic secretory diarrheas with extreme fluid secretion, and Cystic Fibrosis with little or no fluid secretion. A mechanistic understanding of the cellular pathways regulating ion and fluid transport has been obtained from a variety of approaches and model systems. These have ranged from the intact intestine to a single intestinal epithelial cell type. Although for many years a reductionist approach has held sway for investigating intestinal transport, the growing realization that physiologic processes should really be examined within a physiological context has seen a marked increase in studies using models that are essentially mini-intestines in a dish. The aim of this chapter is to provide a historical context for our understanding of intestinal ion and fluid transport, and to highlight the model systems that have been used to acquire this knowledge.

Abbreviations

ASL	Airway surface liquid
CFTR	Cystic fibrosis transmembrane conductance regulator
ENaC	Epithelial sodium channel
ISCs	Intestinal stem cells
Lgr5	Leucine-rich repeat-containing G-protein coupled receptor 5
LMTK2	Lemur tyrosine kinase 2
NBF	Nucleotide binding fold
ORT	Oral rehydration therapy
PHA1	Pseudohypoaldosteronism type 1
PKA	Cyclic AMP dependent protein kinase
PKG	Cyclic GMP dependent protein kinase
STa	*Escherichia coli* heat-stable enterotoxin
VX-809	Lumacaftor (vertex pharmaceuticals)

1. INTRODUCTION

1.1 Transport Problems

Epithelial cells perform two critical functions in the body, firstly as a barrier between the body and the outside world, and secondly as a selective vectorial transporter of nutrients, fluid and electrolytes. Epithelial tissues are remarkable in that they can move substances across themselves either in an absorptive or a secretory fashion, depending upon the specific asymmetric distribution of transport proteins, giving rise to the particular physiological function of that tissue. For example, pancreatic acini, salivary acini, and sweat glands all secrete a primary isotonic fluid into a lumen using vectorial transport of ions, whereas tissues like the small intestine function to work in an absorptive direction, absorbing nutrients from the gut lumen into the blood supply in a unidirectional manner. However, even within a specific epithelial tissue, secretory and absorptive components can be seen within the same tissue. For example, the sweat gland contains a secretory coil responsible for the production of a primary isotonic fluid, contiguous with a duct epithelium that is absorptive in nature. As fluid travels up the sweat duct, Na and Cl ions are absorbed in excess of water, resulting in a hypotonic fluid appearing on the surface of the skin. Remarkably, the sweat coils of the average person are capable of transporting 0.8 to 1.4 L of fluid from the blood to the skin surface every day, and a maximum sweat rate of 10–14 L per day. One of the more impressive examples of large scale vectorial absorption is derived from the observation that although just before birth the lungs are fluid filled, that fluid must be removed across the lung epithelium to generate an air-filled space in order to facilitate proper gas exchange. This absorption of fluid is governed by a specific asymmetric distribution of ion transporters between the basal and apical membranes, driving ion and fluid transport in a unidirectional absorptive manner. Genetic modification of epithelial transport proteins in the lung epithelium of mice, such as the epithelial sodium channel ENaC, can prevent removal of prenatal lung fluid, and pups essentially die from drowning within one or two days of birth (Hummler et al., 1996).

Several important clinical conditions are associated with abnormalities in ion and fluid transport across epithelia, where patients have either too much or too little fluid absorption or secretion. Sitting on top of the upper airways is a thin layer of liquid called the Airway Surface Liquid (ASL) that coats the airway epithelial cells. The height of this ASL is maintained at roughly 9 microns, a height that allows for the normal beating of airway cilia (Fig. 1 left panel) (Donaldson & Boucher, 2003; Matsui et al., 1998). Above the ASL is a mucus layer that traps bacteria, dust and viral particles; mucus that can be moved up the airways along the mucocilliary escalator as the cilia beat in waves in the ASL. It is now clear that a normal ASL height is maintained by a careful balance of fluid secretion and absorption. Many respiratory diseases, such as the flu (Kunzelmann et al., 2000) or infections with respiratory syncytial virus (Kunzelmann, Sun, Meanger, King, & Cook, 2007), can lead to a reduction in fluid reabsorption in the upper airways, leading to the characteristic "runny nose" as the excess fluid not absorbed

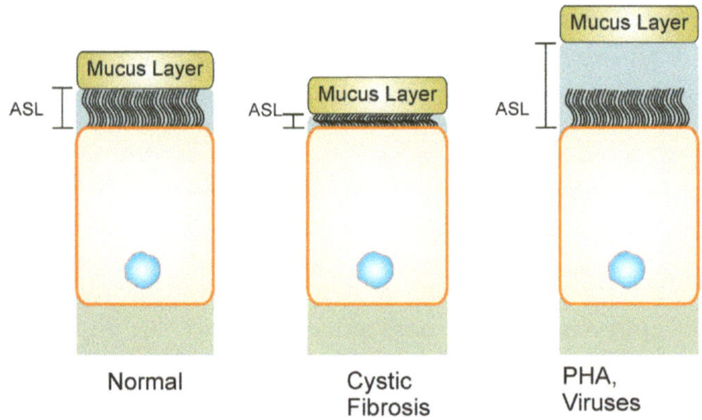

Figure 1 ***Fluid transport in the airways.*** Under normal conditions, a layer of fluid (airway surface liquid or ASL) is maintained on the airway surface through a balance of absorption and secretion (Normal). Cilia beat in the ASL to propel a mucus layer (containing inhaled viruses, bacteria, dust particles etc.) upwards on a mucocilliary escalator to the back of the throat where it is swallowed. In Cystic Fibrosis (CF) an imbalance between fluid secretion and absorption results in a loss of ASL height and a collapse of the cilia and preventing efficient airway clearance (Cystic Fibrosis). When secretion of fluid exceeds absorption, ASL height expands. This can be a minor annoyance or can lead to pulmonary distress.

is transported out of the airways by mucocilliary clearance. The inability of the airways to absorb efficiently fluid from the airways can also be seen in the rare genetic disease pseudohypoaldosteronism type I (PHA1), where an imbalance between fluid secretion and absorption leads to a net movement of water in a secretory direction, increasing ASL height (Fig. 1 right panel) (Amin et al., 2013; Belot et al., 2008; Riepe, 2009). As the imbalance between secretion and absorption of fluid continues, and ASL height increases, respiratory distress ensues, and can result in serious health care issues, especially in premature infants (Barker, Gowen, Lawson, & Knowles, 1997). In contrast to instances where fluid absorption is in deficit relative to fluid secretion, there are also clinical conditions where fluid absorption across the epithelium is excessive and pathological. For example, cystic fibrosis is a disease associated with impaired mucocilliary clearance because fluid is absorbed in excess of secretion, causing not only an impairment of airway clearance, but also the production of dehydrated inspissated mucus in the airways with resultant chronic infections and inflammation (Boucher, 2007; Bradbury, 2016; Rosenstein, 1998) (Fig. 1 middle panel).

Not just in the airways does an imbalance between fluid secretion and absorption have pathological consequences, the same is also true for fluid transport across the epithelia of the gastrointestinal tract. In the intestine, infection by colonizing pathogenic microorganisms such as *Escherichia coli*, results in sustained secretion of fluid, leading to secretory diarrhea a condition with huge financial and personal impact. The seriousness of toxigenic secretory diarrhea is most easily seen in *Vibrio cholera* which causes massive unrelenting intestinal fluid secretion that is deadly if the lost fluid is not replaced (Satitsri, Pongkorpsakol, Srimanote, Chatsudthipong, & Muanprasat, 2016). In contrast to clinical pathologies arising from an excess of fluid secretion, a lack of appropriate fluid transport from salivary glands, and also tear glands, as seen in Sjogren's syndrome, where lack of fluid transport can cause dehydration of the oral cavity and predisposition to oral infections. Similarly, loss of fluid production by the tear glands leads to dry eye and eye infections (Nair & Singh, 2017; Zeng et al., 2017). In the exocrine pancreas, loss of fluid secretion from the acinar cells results in an inability to flush out digestive enzymes from the pancreas, leading to autodigestion of the pancreas and an absence of enzymes from the intestine, resulting in malnutrition, as seen in cystic fibrosis (Ozen Alahdab & Duman, 2016; Welsh & Ramsey, 2001). Failure to secrete fluid into the intestine can be especially problematic in newborns with cystic fibrosis, where the first bowel movement can be prevented by intestinal dehydration and blockage, as seen in meconium ileus, a condition the requires urgent surgical intervention to prevent necrosis of intestinal tissues (Sathe & Houwen, 2017). Given the above short description of pathologies associated with imbalances in fluid absorption and secretion, it is clearly important that such processes are subject to tight regulation. This review will focus on model systems that have been utilized to study intestinal ion and fluid transport, highlighting the new approach of enteroids as a model for investigating fluid transport across epithelia.

1.2 The Small Intestine

The adult intestinal epithelium can be broadly divided into two structurally and functionally distinct regions, the crypt and the villus. At a cellular level, the intestine is comprised of six cell types segmentally distributed along the crypt-villus axis (Fig. 2). At the base of the crypt are Paneth cells, that not only secrete antimicrobial peptides and play a key role in host–microbiome interactions, but also help provide much of the environment for maintaining a stem cell niche (Clevers & Bevins, 2013). Interspersed between the Paneth cells are intestinal stem cells (ISCs) from which all the other intestinal cells arise (Barker et al., 2007). Thus, the crypt represents

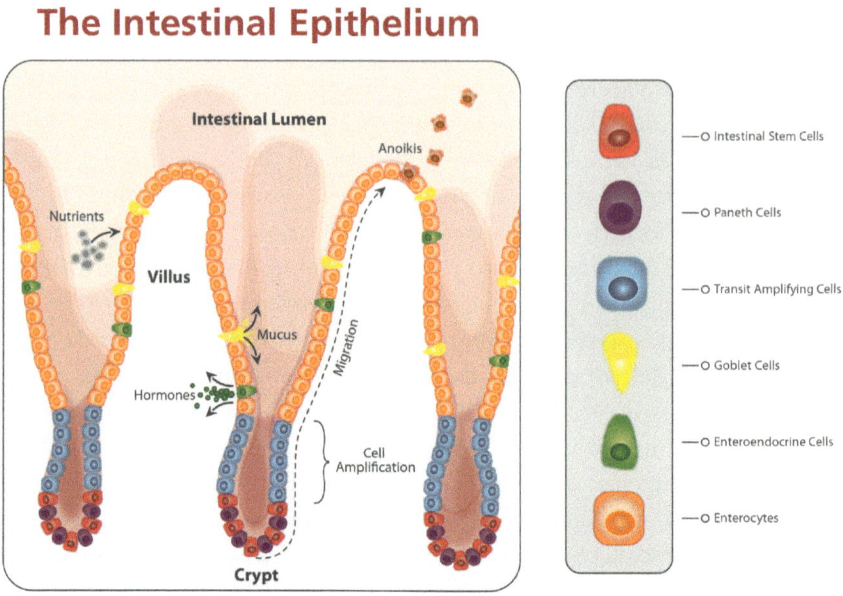

Figure 2 ***Diagram of the small intestinal epithelium highlighting the identity and spatial arrangement of key epithelial cell types.*** The LGR5+ stem cells retain their capacity to self-renew and regenerate the intestinal epithelium primarily due to their position within a stem cell niche. The intestinal niche has been well-characterized, and shown to consist of spatial gradients of high WNT and Epithelial Growth Factor (EGF), while Bone Morphogenetic Protein (BMP) signals are inhibited. Understanding of these niche signals heavily informed development of the culture conditions for intestinal organoids. Figure courtesy of StemCell Technologies, Canada.

the site of major cell division from which the entire crypt-villus axis is populated. As the stem cells divide, they move upward and out of the crypt, moving along signaling gradients triggering differentiation to the mature cell types populating the intestine. Mature cells include the enterocytes that make up the bulk of the cell type of the villus epithelium, and are responsible for nutrient absorption. Goblet cells secrete mucus into the intestinal lumen to help protect the epithelial lining of the intestine, and aid in the movement of luminal contents through the intestine. Enteroendocrine cells remarkably constitute the largest endocrine system in the body, and secrete multiple regulatory hormones to regulate metabolism and post-prandial secretion and motility (Moran, Leslie, Levison, Worthington, & McLaughlin, 2008). As new cells arise in the crypt, all cells continue to migrate towards the tip of the villus, where the eventually undergo anoikis and are sloughed off into the intestinal lumen. Superimposed on the architecture of the crypt-villus structure are two additional strategies employed to amplify surface area. On the microscopic end, each enterocyte has a highly invaginated apical membrane containing passive and ion-coupled transport proteins and digestive enzymes, referred to as the brush border. On the macroscopic side, the intestine displays a large undulating organization, resulting from structures called the Folds of Kerckring. Were the intestine to be a simple tube lined with epithelial cells, the intestine would have a surface area of approximately 24 square yards, yet augmented by large folds of Kerckring, villi and microvilli, the final intestinal surface area is some 250 square yards in an adult, roughly the size of a tennis court. It is across this epithelium that not only are nutrients transported, but also ions and fluid.

2. EPITHELIAL ION AND FLUID SECRETION

The driving forces for transepithelial water movement are differential ion concentrations across the epithelium, leading to the generation of an osmotic gradient. Depending on the tissue and the asymmetric distribution of ion transporters between the apical and basolateral membranes, the osmotic gradient can be set up to favour either secretion or absorption of fluid. A model describing the production of primary fluid by acinar cells, sweat coil, airways surface and intestinal cells has been proposed (Frizzell, Field, & Schultz, 1979; Frizzell, Rechkemmer, & Shoemaker, 1986; Frizzell & Schultz, 1979; Petersen, 1986; Welsh, 1986; Welsh & Liedtke, 1986) (Fig. 3). The driving force for transepithelial Na and Cl secretion is ultimately from the Na/K-ATPase in the basolateral membrane, which uses the energy derived from ATP hydrolysis to extrude Na ions from the cytoplasm to the basal interstitial fluid, generating a large inwardly directed Na ion gradient. Na can re-enter the cell down this electrochemical gradient on a basolateral cotransporter which couples Na, K and Cl transport in a ratio of 1:1:2, resulting in the accumulation of Cl above its equilibrium potential. Upon appropriate stimulation, apical chloride channels are opened, allowing electrogenic secretion of chloride down a favourable electrochemical gradient. In secretory tissues, sodium follows chloride through a paracellular route in response to the electronegativity generated in the epithelial lumen. The luminal accumulation of sodium and chloride ions establishes an osmotic gradient across the epithelium, leading to osmotic water flow from a serosal to mucosal milieu. Opening of apical chloride channels would not, on its own, result in prolonged chloride secretion since

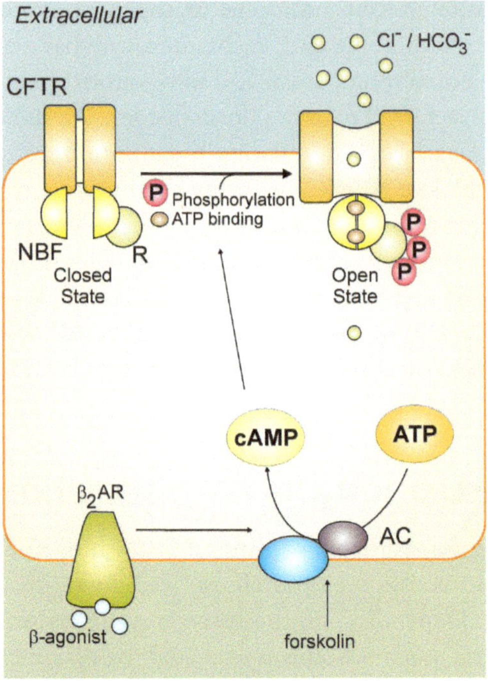

Figure 3 ***Model for chloride/CFTR mediated ion and fluid transport.*** The basolateral sodium pump generates an inwardly directed Na gradient to bring in Cl across the basolateral membrane on the Na/K/2Cl cotransporter. Na and K are recycled across the basolateral membrane leading to an intracellular accumulation of Cl that can exit the apical membrane through CFTR Cl channels. Increased luminal negative charge causes Na to move increasing osmolytes in the luminal domain with a resultant flow of water from the serosal to luminal milieu.

exit of chloride, along with obligatory potassium entry on the basolateral membrane would depolarize the membrane and decrease the driving force for chloride exit (Devor, Singh, Frizzell, & Bridges, 1996; Petersen, 1986; Singh et al., 2000; Welsh, 1986). Thus increased potassium channel activity is also required to maintain apical chloride (and hence fluid) transport.

Evidence for this model was obtained primarily from animal tissues showing an ionic (i.e., osmolyte) dependence of fluid secretion (Case, Hunter, Novak, & Young, 1984; Frizzell, Dugas, & Schultz, 1975; Petersen, 1986; Schultz & Frizzell, 1972; Seow, Lingard, & Young, 1986; Welsh, 1986), dependence on the sodium pump, as shown by ouabain sensitivity (Case, et al., 1984; Quinton, 1981; Welsh, 1986), inhibition by the loop diuretics furosemide and bumetenide, which block the Na/K/2Cl co-transporter (Martinez & Cassity, 1985; Sato, 1977; Seow, et al., 1986;

Welsh, 1986), as well as an apical chloride conductance (Petersen, 1986; Welsh & Liedtke, 1986), now known to be the Cystic Fibrosis anion channel, CFTR (Drumm et al., 1990; Gabriel, Brigman, Koller, Boucher, & Stutts, 1994).

2.1 CFTR Anion Channels and Cystic Fibrosis

The **C**ystic **F**ibrosis **T**ransmembrane conductance **R**egulator (CFTR) protein is an anion channel essential for fluid and electrolyte homeostasis in many organs, including the gastrointestinal tract and the airways (Bradbury, 2016; Harris & Kirk, 2016; Welsh & Ramsey, 2001), where it facilitates the vectorial transport of Cl and HCO_3 ions. CFTR is predominantly found on the apical membrane of epithelial cells, though the sweat duct is unique in having CFTR present in both apical and basolateral membranes. CFTR is a member of the ABC superfamily of transporters, comprised of two six-membered transmembrane segments, two nucleotide binding folds (NBFs) and an "R" or regulatory domain that is a substrate for PKA and PKC dependent phosphorylation (Liu, Zhang, Csanady, Gadsby, & Chen, 2017; Zhang & Chen, 2016). Phosphorylation of the R domain by cAMP-dependent protein kinase (PKA) is the preferential mode of activation in vivo, a process that is mimicked experimentally by using forskolin. Other kinases have also been shown to phosphorylate and regulate CFTR, including protein kinase C (PKC) and cyclic guanosine monophosphate (cGMP)-dependent protein kinase (PKG) II (Ahsan et al., 2017; French et al., 1995) (Fig. 4). CFTR is associated with two devastating intestinal diseases, Cystic Fibrosis (CF) and toxin-induced secretory diarrhea; the former resulting from too little CFTR activity, and the latter from too much CFTR activity.

Since the cloning of CFTR in 1989, over 2000 mutations in the gene have been identified (Riordan et al., 1989) (http://genet.sickkids.on.ca). Patients bearing mutations on both CFTR alleles display severe pulmonary and gastrointestinal dysfunction, and have a life expectancy of around 40 years. Although over 2000 mutations have been described in CFTR, they nonetheless broadly fall into two categories, (1) *no or limited apical CFTR protein expression* or (2) *CFTR protein present in the apical membrane, but a protein with no or limited protein function* (Veit et al., 2016). To complicate things, the most common mutation, ΔF508 as limited channel function (Ameen, Silvis, & Bradbury, 2007; Bradbury, 2016). Other prominent mutations include G551D CFTR, which reaches the cell surface but has drastically reduced gating activity (Yang et al., 1993), and G542X, which results in a premature

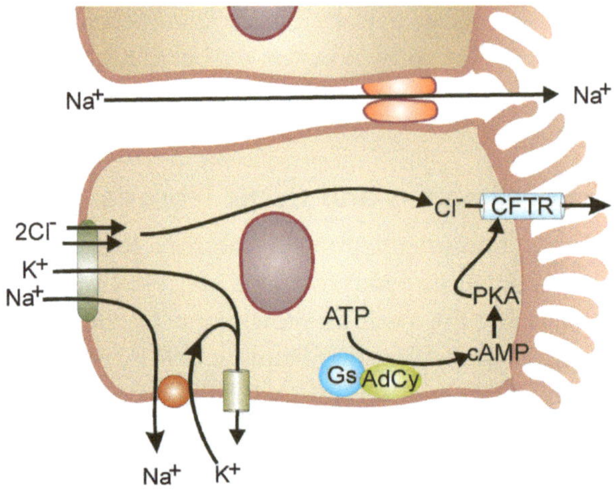

Figure 4 ***CFTR dependent Cl transport.*** CFTR is composed of two nucleotide binding folds (NBFs), one regulatory domain (R domain) and two membrane spanning domains. Phosphorylation of the R domain triggers dimerization of the NBFs in an ATP-dependent manner, leading to channel opening and transport of Cl. Dephosphorylation of the R domain and ATP hydrolysis causes separation of the NBFs and channel closing. Binding of the R domain and ATP hydrolysis causes separa$_2$-adrenergic receptor leads to activation of adenylate cyclase and increased cAMP-dependent protein kinase activity. Bacterial toxins such as cholera, or plant alkaloids such as forskolin can directly activate adenylate cyclase. Other bacterial toxins can increase intracellular cGMP levels leading to increased cGMP-dependent protein kinase activity, which can also activate CFTR.

stop codon (Xue et al., 2017) and no functional protein being produced. Much of the pathology associated with CF arises from the lack of fluid secretion that is required to rehydrate mucins secreted by goblet cells in the intestine and airways, and a lack of fluid needed to flush out digestive enzymes from the pancreas. However, some studies have suggested that a lack of chloride in the ASL also leads to reduced defensin activity in the airways, contributing to chronic bacterial colonization (Abou Alaiwa et al., 2014; McCray, Zabner, Jia, Welsh, & Thorne, 1999), although there is still debate as to whether an ionic gradient can exist across the airways, which are quite permeable. What is agreed, is that loss of fluid in the airways results in the accumulation of thick sticky mucus that harbours bacteria, and chronically activated immune cells, leading to poor ventilation, bacterial infection and replacement of lung tissue with fibrous tissue. Although the specific details are different in the intestine, the general principle of reduced fluid secretion still applies. Lack of fluid secretion from pancreatic acini, for example, preventing the normal "flushing" of digestive enzymes from the acini into the small intestine, with a lack of pancreatic enzymes resulting in malabsorption of nutrients and the remaining in situ pancreatic enzymes causing autodigestion of the pancreas itself.

Since CF is a lethal disease, the question has often arisen as to why there is such a high incidence and prevalence of mutations in the CFTR gene. The carrier frequency for CFTR mutations is roughly 1:25 for individuals of Northern European descent, making it the most common lethal genetic disease of Caucasians. Theories for this high prevalence have focused on the relationship between CFTR and cholera induced secretory diarrhea. The theory proposes that since cholera causes massive unregulated intestinal fluid secretion, and such secretion is dependent upon CFTR, then heterozygotes in a population (who would only have half the normal level of CFTR function comparted to homozygous wild-type individuals), would therefore necessarily have a selective advantage in resisting death as a result of cholera exposure (Rodman & Zamudio, 1991). Indeed, mouse CFTR knock-out studies would support this notion (Gabriel, et al., 1994), where loss of CFTR is associated with increased survival following cholera exposure. Thus heterozygous individuals could potentially live through a cholera epidemic and pass on their genetics to their offspring. While scientifically this seems a reasonable theory, it is contradicted by history. Cholera epidemics did not strike Northern Europe until the 19th century, long after the appearance of CFTR mutations. Although cholera can be ruled out as a selective agent, other intestinal pathogens, such as *E. coli* have affected Europeans for many centuries, and may have exerted selective pressure (Pier et al., 1998; Quinton, 1999). However, studies by Cuthbert and colleagues found no evidence for a heterozygous advantage of mice bearing CFTR mutations in acute responses to toxin induced secretory diarrhea (Cuthbert, Halstead, Ratcliff, Colledge, & Evans, 1995). Another intriguing theory postulates that indeed CFTR mutations afford a selective advantage when individuals succumb to secretory diarrhea, but it is argued that the diarrhea arises not from cholera exposure but rather exposure to lactose (Modiano, Ciminelli, & Pignatti, 2007). Northern Europeans share a high prevalence of CFTR mutations and a geographical and historical demographic associated with a high incidence of post-weaning milk consumption (Swallow, 2003). Lactose intolerance due to post-weaning loss of intestinal lactase activity can lead to secretory diarrhea (Swallow & Hollox, 2000), and restricted CFTR activity would potentially lessen this effect. Since farming of cattle in Northern Europe goes back over several thousand years (Klein, 2000), the time lines for lactase persistence and the genesis of CFTR mutations overlap.

3. INTESTINAL ION AND FLUID TRANSPORT

The gastrointestinal tract is composed of a variety of organs whose coordinated function is necessary for life. The intestine and organs that drain into it (gall bladder, pancreas, and stomach) all work in concert to facilitate the dual function of digesting and absorbing nutrients, while at the same time preventing the uptake of toxins and pathogens. Key to the function and physiology of the intestine is the biology of columnar epithelial cells that line the length of the intestine. The intestine represents a barrier between the body and the outside world, and nearly all molecules that enter the body (oxygen molecules obviously passing through lung alveoli) pass across the intestinal epithelium. To facilitate this process, the architecture of the intestine is designed to maximize surface area for transport processes. Although we tend to think of the kidney as the primary system for water and electrolyte homeostasis in the body, the intestine is actually the second most important organ system in such homeostasis. However, whereas the kidney can compensate for large fluctuations in body fluid content, the capacity of the gastrointestinal tract to handle such changes and maintain volume homeostasis and plasma osmolarity is limited, and can often be overwhelmed, especially in pathophysiologic processes, including viral and bacterial diarrheal diseases, inflammatory diseases, and physical damage caused by ischaemia or radiation.

Excessive uncontrolled secretion of fluid into the intestine, resulting in diarrhea, claims several million lives annually. Painfully, most of these deaths could have been prevented with adequate fluid replacement. Less common, but no less important, are conditions arising from inadequate fluid secretion in the GI tract, that can be seen as exocrine pancreatic duct blockage (with resultant malnutrition) as seen in cystic fibrosis. An even more immediate problem in neonates with cystic fibrosis is meconium ileus resulting from a lack of hydration in the intestine. The first stool released by neonates is the meconium (composed of material ingested in utero including intestinal epithelial cells, amniotic fluid, bile, mucus and lanugo). Without appropriate intestinal secretions, the meconium can become dehydrated and compacted resulting in intestinal blockage (Borowitz & Gelfond, 2013; Waldhausen &

Richards, 2018), with potential intestinal necrosis. All segments of the intestine from duodenum through the distal colon have physiological mechanisms for both the secretion and absorption of electrolytes and fluid. Indeed, all transport processes, whether for fluid or nutrients are ultimately dependent on ion fluxes across the cell membrane.

3.1 Absorption

The connection between ion and fluid transport was first proposed in the 1940s, and refined by studies showing the uptake of ions along with small molecules such as sugars and amino acids across the brush border of the small intestine. Specifically, uptake of glucose and galactose, and amino acids in the small intestine is coupled to the movement of sodium ions (Read, Levin, & Holdsworth, 1975; Schultz, Fuisz, & Curran, 1966). The movement of sodium into the cell downhill along its electrochemical gradient can be used to transport organic solutes in an uphill fashion from an area of low concentration to one of high concentration. Once inside the cell, the organic molecules exit across the basolateral membrane down their chemical gradient. In contrast to individual amino acids, small oligopeptides can be brought into the cell using an inwardly directed proton gradient, rather than sodium (Ganapathy & Leibach, 1985; Thwaites et al., 2002). Once inside the cell the oligopeptides are hydrolyzed to individual amino acids and exit passively across the basolateral membrane. Even though oligopeptides utilize a proton gradient, that proton gradient is still dependent upon sodium, as protons are extruded across the apical membrane coupled to inward sodium transport on a Na/H antiporter. The sodium gradient is therefore the driving force for sugar, amino acid and polypeptide absorption. The sodium gradient is in turn, of course, generated by the basolateral Na/K-ATPase. It can be appreciated therefore that as organic nutrients are absorbed, so is salt (NaCl) since chloride is obligatorily absorbed to maintain electroneutrality. The transepithelial movement of salt from the lumen to the basal and lateral intercellular space results in water following osmotically. The observation that water absorption was coupled to the absorption of Na and organic solutes is the theoretical basis for oral rehydration therapy (ORT) in severe diarrhea.

3.2 Secretion

Remarkably even up to the late 1960s, the gut was generally accepted to be solely an absorptive tissue, with the occasional report of active fluid secretion across the intestinal epithelium generally being dismissed as artefact

(Gregory, 1965; Tidball, 1961). Indeed, the 1965 *Annual Review of Physiology* published a paper on "Secretory Mechanisms of the Digestive Tract," which managed to discuss the topic without mentioning intestinal fluid secretion at all (Gregory, 1965). Indeed, intestinal fluid accumulation arising from cholera was proposed to arise entirely from an inability of the intestine to reabsorb sodium, and hence a lack of osmotic fluid reabsorption (Huber & Phillips, 1960; Phillips, 1964), and not as a consequence on increased fluid secretion. Seminal work on intestinal fluid secretion did indeed arise from studies on cholera induced secretory diarrhea. Reports from Calcutta (now Kolkata) (De, 1959) and Bombay (now Mumbai) (Dutta, Panse, & Kulkarni, 1959) revealed that cell-free extracts of *V. cholera* cultures placed on the intestines of rabbits would elicit fluid secretion, demonstrating that a substance released from the bacteria (we now know as cholera toxin) stimulated fluid accumulation in the intestine. Small intestine biopsies from acute and convalescing cholera patients revealed that the intestinal epithelium was quite intact, showing that the toxin was eliciting fluid accumulation through a normal cellular response rather than tissue degradation (Gangarosa, Beisel, Benyajati, Sprinz, & Piyaratn, 1960). However, cholera researchers were as biased as other intestinal physiologists and were not inclined to the view that cholera toxin could in any way actively stimulate the secretion of water and electrolytes. Alternative explanations were argued, including the ability of cholera toxin to prevent intestinal fluid absorption of fluid (gastric and pancreatic secretions) already present in the intestinal lumen. By the late 1960s and early 1970s, physiologists started applying a technique first employed two decades earlier to study why frogs (which essentially live in distilled water) don't constantly absorb fluid osmotically and swell to the point of exploding! Such studies envisioned and implemented by Hans Ussing led to the concept of active ion transport processes (Ussing & Zerahn, 1951). Ussing is also known for having invented the Ussing Chamber to study ion transport processes. Utilizing Ussing's chamber approach, rabbit ileum was stripped of its underlying muscle to yield an isolated mucosa that was placed in Ussing chambers, allowing isolated access to both serosal and luminal compartments. Agents that increased intracellular cAMP had two effects on this preparation, firstly the stimulation of active Cl secretion and secondly, the inhibition of active electroneutral NaCl absorption (Field, 1971; Field, Plotkin, & Silen, 1968). Sodium dependent absorption of sugars and amino acids was unaffected by these manipulations. It is now known that cholera toxin directly and chronically activates adenylate cyclase through a covalent modification, to continuously increase cellular cAMP and cause active Cl secretion (Field, Fromm, al-Awqati, & Greenough, 1972; Kimberg, Field, Johnson, Henderson, & Gershon, 1971). Through subsequent studies, as described above (*Epithelial Ion and Fluid secretion*) the molecular mechanisms and transporters underlying cAMP dependent Cl secretion were identified.

3.3 Coupling of Physiologic Fluid Absorption With Pathophysiologic Fluid Secretion

Although by no means universally accepted, there are models that best seem to fit the available data regarding intestinal fluid transport (Kockerling & Fromm, 1993; O'Loughlin et al., 1996). In such models, in the small intestine, most (but possibly not all) secretion arises from the crypts, whereas all absorption occurs in the villi. Almost all of the secretory chloride channel, CFTR, molecules are localized to the crypts (Crawford et al., 1991; Trezise & Buchwald, 1991), although intriguingly, a subset of villus cells expresses CFTR at very high levels (Ameen, Ardito, Kashgarian, & Marino, 1995; Jakab, Collaco, & Ameen, 2013). Cholera and enterotoxins such as *E. coli* toxin stimulate chloride secretion without damaging the intestinal mucosa, which means the rest of the transport properties, such as sodium dependent nutrient absorption, remain intact. Even in viral enteritis where mucosal injury does occur, considerable absorptive capacity remains.

Although cholera patients had been given a simple oral salt solution to replenish diarrheal fluid loss, it did little to increase blood volume, and really only added to diarrheal volume. This is because cAMP inhibits nutrient-independent (Na/H and Cl/HCO3 exchanges) salt absorption. By the 1960s, the concept of sodium-dependent nutrient uptake in the small intestine was emerging, such that researchers in Dacca and Calcutta found that oral administration of an isosmolar solution of salt and sugar (equimolar amounts of sodium and glucose) allowed patients to absorb fluid (Hirschhorn et al., 1968; Pierce et al., 1968). The effects of this oral rehydration therapy (ORT) were so great that the British medical journal *The Lancet* opined "The discovery that sodium transport and glucose transport are coupled in the small intestine so that glucose accelerates absorption of solute and water is potentially the most important medical advance this century" (Water with sugar and salt, 1978).

4. MODEL SYSTEMS FOR STUDYING INTESTINAL ION TRANSPORT

Researchers have sought out numerous approaches to investigate fluid and electrolyte transport in the gut. One of the earliest models, and perhaps arguably the most physiologically relevant, was the intestinal loop, housed within the abdominal cavity of a live animal. As more reductionist approaches arose, the intestine was isolated and removed from the animal and placed in Ussing chambers as an ex vivo tissue, where the bathing solutions could be carefully controlled. The ultimate reductionist approach is the use of a single intestinal cell type in the form of an immortalized cell line that can also be mounted in Ussing chambers. More recently as knowledge of intestinal stem cells (ISCs), and markers for such cells, has been obtained, there has been a move back towards a more physiologically relevant (though still experimentally amenable) system, known as the intestinal organoid. All these models are used in transport research, and have provided invaluable information concerning the development, physiology and pathophysiology of the intestine.

4.1 Intestinal Loops

Perhaps the simplest of all experimental models for studying intestinal fluid transport is the intact in vivo intestine (Richardson & Kuhn, 1986), referred to as the sealed adult mouse (or rat is so desired) model. In this experimental approach, mice are given reagents to modulate intestinal ion and fluid transport (drugs, toxins etc) by feeding through an oro-gastric tube. After a determined period of time, the mice are sacrificed and the entire intestine from pylorus to caecum carefully removed. Fluid accumulation can be determined either by fluid collection or by difference in tissue wet and dry weight. Such approaches have been used to investigate the sensitivity of intestines to enterotoxins (Richardson & Kuhn, 1986), and to confirm the role of the CFTR chloride channel in cholera induced intestinal fluid accumulation (Gabriel, et al., 1994; Ma et al., 2002). In the latter case, instilled cholera toxin was able to stimulate fluid secretion in wild-type (wt) mice, but cholera toxin induced fluid secretion was significantly impaired in *cftr*$^{-/-}$ mice; *cftr*$^{+/-}$ mice had responses midway between wt and *cftr*$^{-/-}$ *mice*. Although cystic fibrosis represents the extreme in reduced intestinal fluid secretion, it is by no means the only important pathology. Constipation is a common clinical complaint in both children and adults, negatively impacting quality of life. Roughly $7 billion in annual health care costs are associated with constipation, with around $800 million spent each year on laxatives (Mugie, Di Lorenzo, & Benninga, 2011). Although the FDA approved drug linaclotide is available for patients (linaclotide is a peptide agonist of the guanylate cyclase C receptor, leading to increased intestinal secretion (Busby et al., 2010)), there is still a need for safe effective treatments for constipation. Using a loperamide-induced mouse model of constipation, Verkman and colleagues injected a CFTR activator ($CFTR_{act}$-J027) into closed intestinal loops. Fluid accumulation at 90 min of treatment was 140% greater in $CFTR_{act}$-J027 treated loops compared to vehicle injection arguing for a therapeutic proof-of-concept of such an approach for treating constipation (Cil et al., 2016).

In addition to identifying pharmacological therapies to modulate intestinal fluid transport, intestinal loops can also be used to investigate the underlying molecular mechanisms underlying transport processes. For example, using rat intestinal loops, we have shown that CFTR is virtually absent from the apical membrane of resting intestine (Golin-Bisello, Bradbury, & Ameen, 2005). Such an observation is consistent with low levels of fluid secretion seen under these conditions. Since CFTR is needed in the apical membrane to facilitate ion and fluid transport, CFTR must be trafficked to the cell surface in the presence of secretagogues. Indeed this is the case, where *E. coli* heat-stable enterotoxin (*STa*) stimulation of marked fluid secretion is accompanied by a concomitant trafficking of CFTR from intracellular compartments to the apical surface (Ahsan et al., 2017; Golin-Bisello, et al., 2005).

A slightly more complicated experimental model for studying ion and fluid transport in the gut than the sealed adult model, is the intestinal loop. Under anaesthesia the abdomen is opened with a small incision, and the intestine teased out, carefully preserving the blood supply to the tissue. A segment of intestine is then ligated with suture thread to form an enclosed sac. Material (STa, cholera toxin, forskolin etc) can then be injected into the loop to stimulate transport. The loop can then either be left outside under moist warmth, or replaced in the abdomen and the abdomen sutured before allowing the animal to recover from anaesthesia. At the end of the experiment, the animal is sacrificed and the loop excised. Fluid accumulation, for example, can be determined by measuring tissue wet and dry weight and comparing to an adjacent untreated loop. Studies in the 1960s demonstrated, using this technique, cholera dependent fluid accumulation in the intestine of the rabbit (Leitch, Burrows, & Stolle, 1967; Love, 1969). Importantly, detailed mechanistic information regarding ion and fluid transport

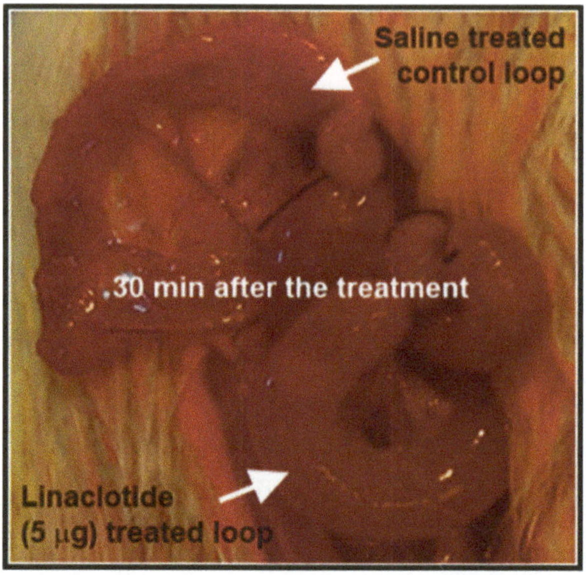

Figure 5 ***Intestinal loops and chloride secretion.*** In rat intestinal loops, linaclotide induces apical trafficking and recruitment of CFTR with increased ion and fluid transport. Different segments of intestinal loops were treated with saline of linaclotide for 30 mins, followed by dilation of the treated intestinal segment as a result of ion and fluid transport (Ahsan et al., 2017).

can be readily achieved using this approach (Collaco et al., 2013; Golin-Bisello, Bradbury & Ameen., 2005; Li et al., 2005; Zhang, Zhang, Zhang, & Naren, 2017) (Fig. 5).

4.2 Everted Sacs

An extension of the intestinal loop model that allows for more careful control over luminal and serosal solutions was developed by Wiseman and Wilson in 1954, as the *everted sac* model (Hamilton, 2014; Wilson & Wiseman, 1954). Under anaesthesia, the segments of the small intestine can be removed and divided into duodenum, jejunum and ileum. The intestinal segments are then gently everted so that the luminal aspect is facing outwards and the serosal aspect interior. One end is ligated and the interior filled with fluid and solute before ligated the other end. Placed in oxygenated media, transport across the tissue can be assessed by sampling the inside fluid at the start and end of the experiment. Such a system has proven to be remarkably useful in pharmacokinetic determination of drug absorption and metabolism (Barthe, Woodley, Kenworthy, & Houin, 1998; Ugolev,

Bagiian, & Ekkert, 1980). The advantages of this model are a relatively large surface area for transport, however tissue viability still remains an issue, with limitations of around 2 h for physiological viability (Barthe, et al., 1998). Another potential complicating factor is the presence of the underlying muscularis layer, which might yield underestimates of solute transport due to a tendency of compounds to bind with, or be absorbed by, muscle cells. Nonetheless, the everted gut model is still a sensitive and specific method that is still useful for investigating mechanisms of drug transport (Gruden, 1977; Kitagawa, Sato, Sato, & Sasaki, 1996), as well as the role of formulations in drug transport (Mainardes, Chaud, Gremiao, & Evangelista, 2006; Wang, Jiang, Li, & Ren, 2007), and as an important laboratory exercise in demonstrating intestinal carbohydrate absorption (Hamilton & Butt, 2013).

4.3 Ussing Chambers

Although sealed intestines and intestinal loops can provide a quantitative measure of fluid transport, such a system is not easily amenable for analysis of the ionic components underlying the transport properties of the tissue; although transport of large anions such as bicarbonate (duo to its effect on pH) have been measured successfully (Singh et al., 2009). Intestinal segments can be excised from animals, stripped of the underlying muscle layers, and the mucosa mounted in Ussing chambers to allow isolated individual access to both apical and basolateral domains, and allowing ionic analysis of transport systems (Fig. 6). Some tissues are easier to strip of the underlying muscle layer, but nonetheless when mounted provide the ability to characterize both transcellular and paracellular pathways. Freshly isolated crypts or muscle-stripped intestinal sheets have been utilized to elucidate many important transport processes, including cell volume changes (Mignen, Le Gall, Harvey, & Thomas, 1999; Valverde et al., 1993), extracellular pH regulation (Chu & Montrose, 1995), as well as absorption (Geibel, Rajendran, & Binder, 2001; Guan et al., 2006) and secretion (Greger et al., 1997). Typical data seen for stripped rat intestine is seen in Fig. 7A. Under resting conditions very little ion transport (and therefore little short circuit current; Isc) is observed. Stimulation of transport by forskolin causes a marked rapid increase in Isc, indicative of a substantial increase in ion transport (Fuller, Bridges, & Benos, 1994). However, it should be noted that it is not formally possible just from an increase in Isc to determine if the increase is due to increased anion secretion, or increased cation absorption. The lack of effect on Isc following addition of amiloride (a selective blocker of the epithelial sodium channel ENaC) argues against ENaC dependent sodium transport,

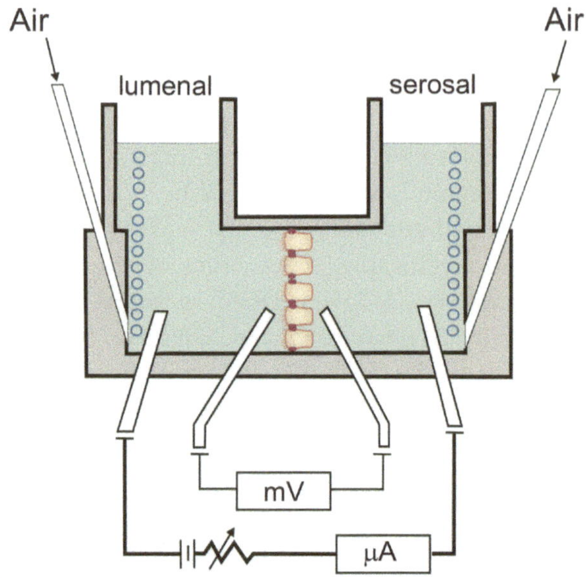

Figure 6 ***Typical Ussing Chamber set-up.*** Tissue is mounted in an Ussing chamber with access to the apical and basolateral membranes. Tissues are maintained at 37°C and aerated. Voltage electrodes are placed close to the tissue to monitor changes in trans-epithelial voltage. Current electrodes are also placed to facilitate short-circuiting of the tissue.

and suggests that chloride secretion is responsible for the increased Isc. Addition of the loop diuretic furosemide (an inhibitor of the Na/K/2Cl co-transporter), further argues for chloride secretion as the dominant component of forskolin induced transport. That CFTR was the chloride conductive pathway in intestinal tissues was shown using stripped intestines from *cftr*$^{+/+}$ and *cftr*$^{-/-}$ mice, where genetic loss of CFTR results in no forskolin stimulated transport or Isc (Fig. 7B) (Robert et al., 2010). Removal of the underlying serosa of gastrointestinal tissues is not always needed, and can in some cases damage the mucosal layer. For example, ion fluxes in response to prostaglandin stimulation across gallbladder tissues mounted in Ussing chambers have been identified (Fig. 7C) (Moser et al., 2007).

Ussing chambers can also be used to monitor bidirectional ion movement, by applying components to either the apical or basolateral bathing solutions. Using this approach, calcium transport across rat colonic epithelium in a mucosal to serosal direction was observed to be saturable (Bridges, Rummel, & Schreiner, 1987), indicating a transcellular transporter

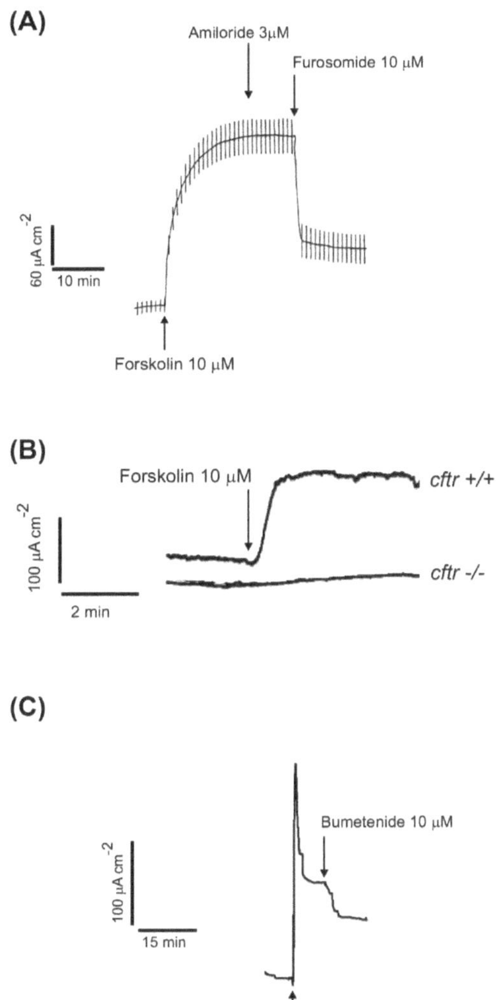

Figure 7 ***Typical Ussing chamber current traces.*** (A) Short-circuit current measurements in stripped rat colon after exposure of tissue to 10 hoforskolin. Segments of distal rat colon, stripped of the underlying muscle layers, were mounted in Ussing chambers. Tissue was maintained at 37°C. Forskolin induced current was insensitive to amiloride but was sensitive to furosemide, indicating the major portion of stimulated current was carried by chloride from (Fuller, et al., 1994). (B). Stripped ileum from wt and ΔF508 Ussing chambers and stimulated with 10 μ forskolin. A substantial increase in short-circuit current was seen in tissues from wt CFTR mice, but not ΔF508 (Robert, et al., 2010). (C) Unstripped prairie dog gall bladders were mounted in Ussing chambers, and displayed a forskolin-induced increase in short-circuit current that was sensitive to the loop diuretic bumetanide. *Adapted from Moser, A. J., Gangopadhyay, A., Bradbury, N. A., Peters, K. W., Frizzell, R. A., & Bridges, R.J. (2007). Electrogenic bicarbonate secretion by prairie dog gallbladder. American* Journal of Physiology-Gastrointestinal and Liver Physiology, 292*(6), G1683—G1694.*

dependent movement, whereas serosal to mucosal calcium transport is dependent only on the size of the gradient, indicating a passive paracellular pathway (Karbach, Bridges, & Rummel, 1986).

4.4 Monolayers

When stripped intestines are placed in Ussing chambers, the tissue maintains its gross morphology of the crypt and villus axis. Multiple cell types are present, often functioning in an absorptive and secretory manner simultaneously. This can sometimes make it hard to dissect out what is actually happening with regard to individual transport processes, since it is net transport that is measured. It is possible to tease out individual transport pathways using selective inhibitors of certain transporters, but very few inhibitors are entirely specific. To get around this issue, investigators have utilized models comprised of a single cell type. Historically, researchers have utilized transformed cancer-derived intestinal cell lines. These grow on permeable supports as polarized epithelial cells, and are capable of carrying out vectorial transport for sodium absorption and chloride and HCO_3^- secretion. The most widely used of such models are T84, Caco-2 and HT20 cells (Ahsan et al., 2017; Barrett, 1993; Broughman, Sun, Umar, Sellin, & Morris, 2006). Indeed, the study of intestinal transport pathways was greatly aided by a report from Dharmsathaphorn and colleagues (Dharmsathaphorn, McRoberts, Mandel, Tisdale, & Masui, 1984) describing the human intestinal cell line T84. Although these cells are indeed intestinal in origin, the cells were initially isolated from a colonic tumour that had metastasized to the lungs (Shah, McCormack, & Bradbury, 2014). T84 cells can be grown continuously in cell culture, and display differentiated properties of crypt epithelial cells, and perform regulated vectorial ion transport (Dharmsathaphorn & Barrett, 1990). While such single cell systems have greatly aided intestinal transport studies, they nonetheless have several drawbacks that limit their utility. Firstly, through passaging and subcloning, cell lines used in one laboratory may not reflect the proteome or transport phenotype of the same cell line in another laboratory (Oltra-Noguera et al., 2015). Secondly, even within an individual laboratory, physiological aspects of cultured cells, such as growth rate, expression of transport proteins, transepithelial resistance, can change with passaging. Finally, the limited number of cell lines available fails to fully recapitulate the diversity of the cell populations found in the intact intestine, and do not reflect the contributions of the various segmental portions present in the human intestine. For these reasons, although monolayers have proven useful for many aspects of transport physiology, an alternative model for evaluating intestinal physiology was desperately needed.

4.5 3-D Cultures

Cell morphology and physiology are intimately associated with each other, where not only does the cells intrinsic protein complement play a role, but also a wide range of factors from the environment. Although many studies can be performed on cells and tissues in situ, some experiments require the isolation of specific cell types from their host organism. In the mid-19th century, English physiologist Sydney Ringer created his eponymous solution containing the chloride salts of sodium, potassium and calcium in defined ratios, which allowed him to keep an isolated perfused frog's heart beating (Miller, 2004). In 1885, German zoologist Wilhelm Roux used Ringer's solutions to cultivate tissue from a chick embryo, but the first real success arose from Harrison's studies on the growth of frog neurons using a medium derived from clotted lymph. Following up on Harrison's achievements, Alexis Carrel and his assistant Montrose Burrows published a series of papers in which they coined the phrase "*tissue culture*". For the next century, scientists continued to improve on culture techniques, establishing the need for various growth factors and other biological agents in the culture media. For the most part, cells were cultured in two dimensions, as sheets of cells grown on plastic or permeable supports. Such cultures have proven to be remarkably useful, and have been a mainstay of experimental research. Two dimensional sheets of cells, such as T84, certainly have their advantages (ease of maintenance and reasonably inexpensive), but they do have their drawbacks. Their simplicity arises from the system being composed of a single cell type grown as a flat monolayer. While this has many advantages for deducing the physiology of individual cell types, it does not fully recapitulate the native epithelium in the same way that can be achieved with intestinal loops or stripped mucosae in Ussing chambers. In contrast, intact loops and stripped mucosae, while physiologically relevant, are harder to control for, and access to specific compartments may be limited. What is needed is a system that has the benefits of a physiologically relevant 3-D architecture, with the simplicity of a few identified cell types. In recent years, there has been a move towards a three-dimensional culture that more faithfully replicates the environment that the cells are exposed to in situ. Moreover, it is possible to co-culture different cell types within a 3-D culture taking cell culture back to a more realistic milieu.

The concept of 3-D culture, in terms of its present incarnation, is actually a revamping of an experimental approach rather than a novel technology, since 3-D culture of cells and organ fragments to monitor physiologic functions is fairly old (McPherson, Dormer, Bradbury, Dodge, & Goodchild, 1986). However, only recently has a deeper understanding of extracellular matrix biology, tissue development, and stem cell markers blended to allow for the self-organizing properties of stem cells to regenerate realistic in vitro models of human tissues. Indeed, the generation of tissue organoids or mini organs from a variety of human tissues was selected as Nature magazine's Method of the Year in 2017 (Method of the Year 2017: Organoids, 2018). Although organoids have now been generated from a remarkable cadre of human tissues, including brain, liver, kidney, retina, breast, pancreas and airway, this review will focus primarily on organoids from the gastrointestinal tract (Baker, Tiriac, Clevers, & Tuveson, 2016; Barker, et al., 1997; Broutier et al., 2016; de-Lau et al., 2011; Jaks et al., 2008; Mansour et al., 2018; Takasato et al., 2015).

The intestinal epithelium is the most rapidly proliferating tissue in mammals, with the entire tissue replaced every 3—5 days. Such turnover is critical to the normal function of the intestine. Within the crypt-villus axis (Fig. 3) the proliferation zone is confined to the lower third of the structure, and has long been of interest to developmental biologists. Early studies by Cheng and Leblond (Cheng & Leblond, 1974) defined the "Unitarian hypothesis" regarding undifferentiated cells (crypt-base columnar or CBC) located just above the Paneth cells in the crypts, that served as multipotent stem cells from which all the intestinal differentiated lineages arose. Study of this CBC cell population, now termed intestinal stem cells (ISCs) was hampered by the lack of any defining cellular marker until the identification of Lgr5 in 2007 (Barker, et al., 2007). Leucine-rich repeat-containing G-protein-coupled receptor (Lgr5) is a G-protein coupled receptor found in ISCs, and lineage tracing experiments have demonstrated that all of the main cell types in the intestine originate from this stem cell population.

A critical step in the development of 3-D intestinal culture arose from Hans Clevers laboratory which showed that isolated Lgr5-EGFP cells could be clonogenically grown in vitro as 3-D structures when placed in a Matrigel matrix and exposed to an appropriate cocktail of growth factors (Sato et al., 2009). Although Lgr5+ stem cells were used initially to "regrow" intestines, it is now known that initiation of cultures with isolated crypts rather than stem cells is significantly more efficient (Fuller et al., 2012) than starting with stem cells alone. When isolated crypts are placed in Matrigel, along with appropriate growth factors, the crypts seal at the open end and initiate a process whereby new crypts are formed by a process of budding (Cummins et al., 2008; Cummins, Jones, & Thompson, 2006). As cells migrate from the crypt structures, they move to a central epithelial-lined cavity (villus-equivalent domain), where they eventually shed off into the lumen in an apoptotic process (Fig. 2). The 3-D structures produce the main cell lineages (Paneth, goblet, enteroendocrine and enterocyte) in roughly the same proportions as in vivo, with less than 2% variance in gene expression between freshly isolated crypts and passaged cells (Sato, et al., 2009). Remarkably, these structures can be passaged and frozen efficiently (Foulke-Abel et al., 2014; Mahe, Sundaram, Watson, Shroyer, & Helmrath, 2015).

4.6 Human Mini-Intestines

There are actually two types of mini-intestine models, depending upon the exact nature of the starting material. **Organoids** are initiated by iPS cells, whereas **enteroids** refer to those models derived from intestinal Lgr5+ cells (Stelzner et al., 2012). Organoids are derived from induced pluripotent stem cells (iPSC) or embryonic stem cells (ESCs) (McCracken, Howell, Wells, & Spence, 2011; Spence et al., 2011). Although initially displaying a low clonogenic efficiency (~1%) modifications to the original growth protocol have markedly enhanced growth of organoids from ISCs (Wang et al., 2013; Yin et al., 2014). Early protocols required recombinant growth factors for stem cell differentiation, though the use of serum and conditioned media can replace isolated growth factors, and significantly lower costs (Miyoshi & Stappenbeck, 2013). Enteroids are derived from small-intestinal stem cells (enteroids) or from colonic stem cells (colonoids). Rather than isolating individual iSPCs, Lgr5+ containing crypts are isolated, and form the basis for tissue regeneration (Fig. 8A). Enteroids have the advantage that culture efficiency (yield of enteroids versus number of plated units) is around 10–100 times greater when starting with intestinal crypts compared to isolated iSPCs, and crypts isolated from mice can typically reach culture efficiencies of 60%–80% (Fuller, et al., 2012). The pure epithelial 3-D cultures recapitulate the in vivo tissue architecture with respect to cell polarity, self-renewal kinetics, cell type composition and organization. Organoids are organized such that a central lumen is lined by the apical membranes of the epithelial cells, and when mature, contain multiple crypt-like domains that harbour the ISCs (Fig. 8B) (Supplemental Video 1). Organoids can be frozen down for biobanking, or subject to a gamut of proteomic, genomic and functional studies. Organoids can be easily passaged through simple

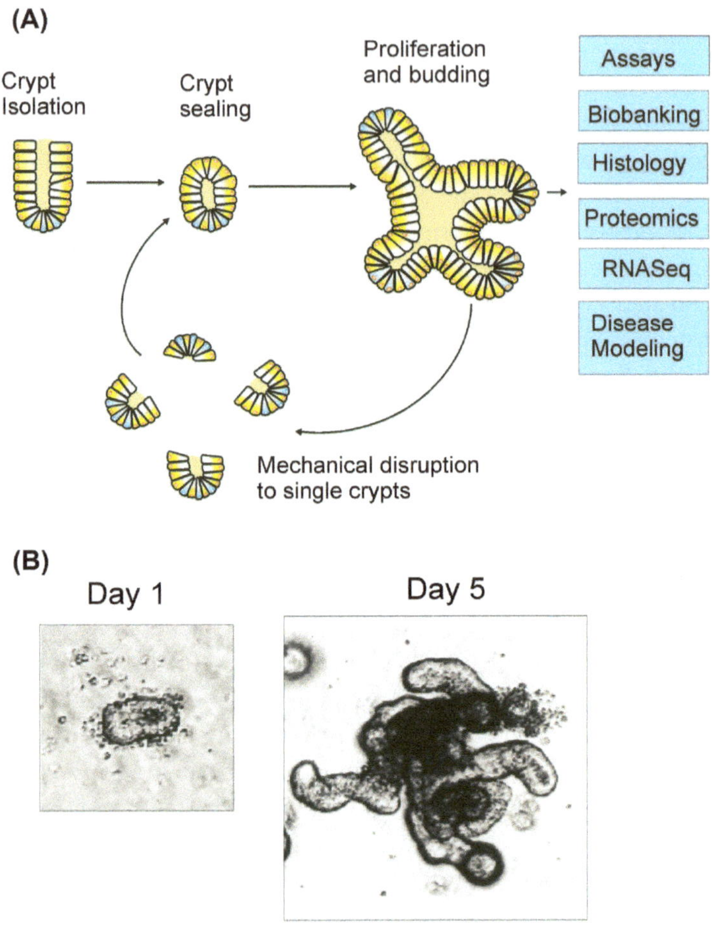

Figure 8 ***Schematic overview of enteroid replication.*** (A) Crypts isolated from intestinal segments self-organize into closed 3-D structures consisting of a single epithelial layer that recapitulates the essential features of the in vivo tissue architecture. Enteroids can be passaged by breaking back down to single crypts or subject to a variety of experimental manipulations. (B) Representative microscopic images from murine small intestinal organoids on days 1 and 5 after passaging. Scale bar 30 μ. *(A) Adapted from Dekkers, J. F., van der Ent, C. K., & Beekman, J. M. (2013). Novel opportunities for CFTR-targeting drug development using organoids.* Rare Diseases, 1, *e27112.*

mechanical disruption yielding isolated crypts than can reform into new enteroids (Sato & Clevers, 2013) Studies with enteroids to examine gastrointestinal pathologies have typically taken one of two approaches. Firstly, enteroids from healthy control subjects have been subjected to pathogenic factors that cause disease, such as bacterial toxins. Secondly, enteroids from patients with gastrointestinal disease have been used as models to mimic certain aspects of the disease to understand the molecular mechanisms underlying disease pathology. As an example of the first, enteroids from normal subjects have been exposed to pathogenic viral and bacterial agents (Noel et al., 2017; Zou et al., 2017). The second approach is currently best exemplified by enteroids from patients with cystic fibrosis, although investigations of other diseases such as microvillus inclusion disease or celiac disease are also likely to take advantage of intestinal enteroids. Microvillus inclusion disease (MVID) is commonly associated with mutations in myosin 5B (MyoVb) that leads to abnormalities in apical protein trafficking of structural proteins and apical transporters. Clinically, MVID is associated with life threatening watery diarrhea (with associated dehydration) and an inability of the intestine to absorb nutrients. CFTR dependent fluid transport is attenuated in MVID organoids, and as will be shown next, a decrease in luminal swelling of the organoids (Schneeberger et al., 2015; Weis et al., 2016). For those wishing to learn more about the technical aspects of organoid culture StemCell provides an excellent resource (www.stemcell.com).

5. A "SWELL" ING ASSAY

Owing to the physical nature of the experimental set up in Ussing chambers, transport of ions across intestinal cancer cells or stripped tissues is easily achieved, however such is not the case for any evaluation of fluid transport. Although vectorial fluid transport does occur, the fluid is essentially moved into an infinite volume of fluid already there (i.e., the bathing solution on each side of the Ussing chamber). Any energy driven fluid accumulation on the luminal side of the chamber will be observed only as a very small increase in total chamber volume of fluid. In contrast, the tied loop assay allows for ion and fluid transport into a fixed, finite, volume (i.e., the intestinal lumen). Since luminal volume increases, and because fluids cannot be compressed, luminal expansion, or swelling, results from stimulated ion and fluid transport. The organoid swelling assay is essentially a hybrid between the two approaches, using cells grown in culture, but also measuring fluid transport as changes in volume.

5.1 Enteroid Swelling

CFTR is arguably the dominant ion channel in crypt epithelial cells where it functions in the apical membrane of crypt cells, where its activation drives vectorial Cl and fluid transport. As discussed, vectorial ion transport is a necessary prerequisite for osmotic fluid transport into the gut lumen. The architecture of intestinal organoids is spherical rather than tubular, as seen in native intestines. This means that transport of ions and fluid in a secretory direction results in fluid accumulation in an enclosed system. Since the organoids are not rigid, this results in an increase or swelling in the luminal volume of the organoid (Dekkers, van der Ent, & Beekman, 2013; Dekkers, Wiegerinck, et al., 2013). Rather than using receptor mediated activation to stimulate ion and fluid transport (which can incur problems of receptor downregulation), the agonist forskolin is usually employed. Forskolin directly and irreversibly activates adenylate cyclase, leading to increased cAMP, PKA activation, phosphorylation of CFTR and ion and fluid transport (Bridges, Rummel, & Simon, 1983). Using phase-contrast microscopy, a lumen can be identified in early stage enteroids (Fig. 9). Upon addition of forskolin, ions and fluid are transported into the lumen causing it to swell and increasing the lumen diameter (which is an indirect assessment of luminal volume) (Supplemental Video 2). Rapid live cell 3-D confocal microscopy of enteroids should facilitate a direct determination of changes in enteroid volume, a strategy being developed in the author's laboratory. That swelling due to fluid transport has an ionic basis is seen by the effects of CFTR-inh172, a CFTR selective inhibitor (Thiagarajah, Song, Haggie, & Verkman, 2004). Preincubation of enteroids with CFTR-inh172 prior to forskolin treatment essentially abolishes the swelling response, preventing any increase in luminal volume (Fig. 9), strongly supporting CFTR and an ionic basis for fluid transport dependent swelling. Similar inhibitory effects can also be observed with the glycine hydrazide inhibitor GlyH-101 (Muanprasat et al., 2004), suggesting that it is a CFTR mediated phenomenon, and not a non-specific CFTR-inh172 artefact. Of course, as with all inhibitors, the saying that "all inhibitors are born selective and become less selective as they age" is as true for CFTR inhibitors as it is for other targets. While both CFTR-inh172 and GlyH-101 certainly do inhibit CFTR, they also interfere with other channels, including the calcium-dependent chloride conductance (CaCC) and the volume-sensitive outwardly rectifying chloride conductance (VSORC) (Melis et al., 2014). In addition, CFTR-inh172 has also been shown to inhibit mitochondrial function, independent of its CFTR inhibition (Kelly et al., 2010). The fact that CFTR-inh172 acts at an intracellular site on CFTR means that the inhibitor must permeate the cell membrane in order to reach its target (Kopeikin, Sohma, Li, & Hwang, 2010). Observations from studies on

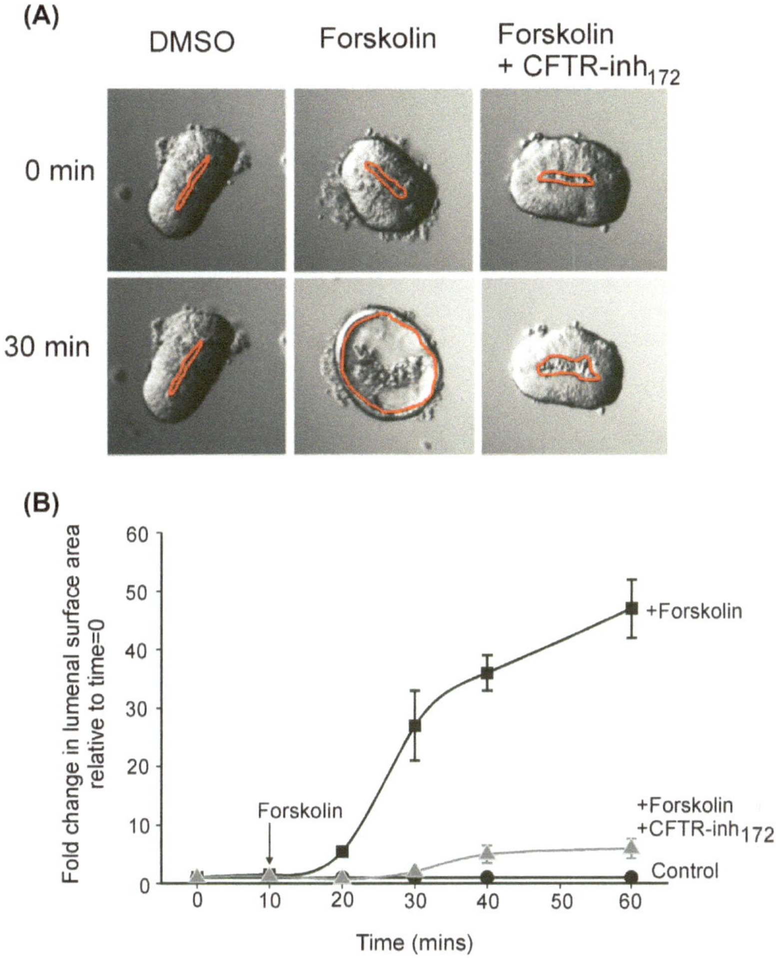

Figure 9 ***Ion dependent enteroid swelling.*** (A) Phase contrast microscopy was used to delineate the middle plane of the enteroid, and a line drawn covering the lumen. Following application of forskolin, ion and fluid transport increased the luminal diameter (luminal volume) and a new area drawn. In the presence of CFTR-inh172, little increase in luminal diameter was observed. (B) Data from multiple enteroids was accumulated to provide a mean rate of luminal increase with time, as a function of CFTR activity (*author's unpublished data*). Where appropriate, forskolin was added at the indicated time, in the presence of absence of CFTR-inh172. For control enteroids, no forskolin was applied.

intestinal tissues and T84 cells looking at CFTR mediated chloride secretion suggest that cell penetration of CFTR-inh172 in these cells can be fairly slow, which necessitates a pre-incubation with inhibitor, rather than an acute application.

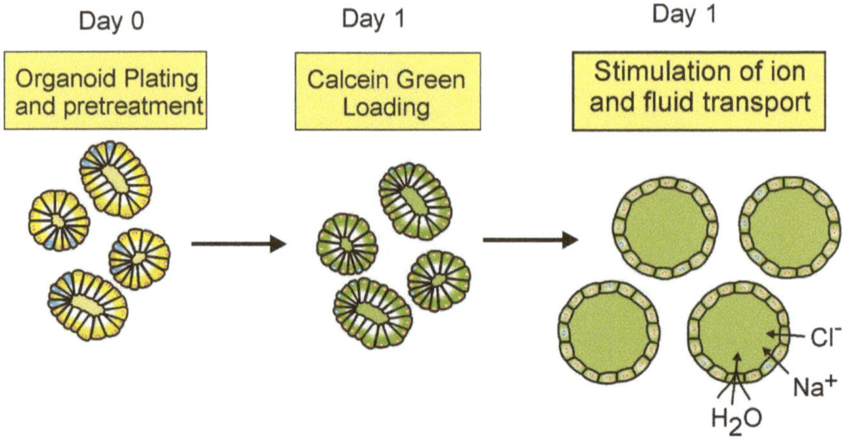

Figure 10 ***Calcein green monitoring of enteroid swelling.*** A representation of 2 day old enteroids used to measure forskolin induced swelling in the presence of calcium green. *Adapted from Dekkers, J. F., van der Ent, C. K., & Beekman, J.M. (2013). Novel opportunities for CFTR-targeting drug development using organoids.* Rare Diseases, 1, *e27112.*

Swelling can be quantified using ROI (region of interest analysis) (Supplemental Video 2) and manual measurements (Boj et al., 2017), or by automated fluorescent image analysis using fluorescent luminal markers (Fig. 10). Calcein green AM is a membrane permeant green fluorescent dye that can enter the cytosol, where esterases cleave the AM moiety leaving a non-membrane permeant fluorophore. This can be used to delineate the epithelium leading to easier determination of the boundary between epithelium and lumen (Dekkers, Wiegerinck, et al., 2013) (Fig. 10). This is oftentimes a viable approach, but the dye can also be extruded into the lumen, by an unknown process, so that the lumen too stains green, making delineation between lumen and epithelium less defined.

5.2 LMTK2 and Swelling

LMTK2 is a serine threonine kinase that resides in endosomal recycling compartments (Butler & Bradbury, 2015; Chibalina, Seaman, Miller, Kendrick-Jones, & Buss, 2007; Inoue et al., 2008; Nixon, Jia, White, & Bradbury, 2013; Wang & Brautigan, 2002), where it regulates a number of cellular processes including CFTR mediated Cl transport in airways (Luz et al., 2014), apoptosis (Conti et al., 2017), neurodegeneration (Bencze et al., 2018; Manser et al., 2012) and cancer development (Shah & Bradbury, 2015a, 2015b). Although LMTK2 is known as Lemur Tyrosine Kinase 2, the observation that LMTK2 is a serine/threonine kinase and not a tyrosine kinase, has led to the suggestion that LMTK2 be referred to as Lemur Tail Kinase 2, due to its long carboxyl tail reminiscent of the Madagascan primate (Bencze, et al., 2018). Since LMTK2 is a myosin VI binder (Tumbarello, Kendrick-Jones, & Buss, 2013), and myosin VI regulates CFTR traffic (Ameen & Apodaca, 2007; Swiatecka-Urban et al., 2004), it would not be unreasonable to invoke LMTK2 as a regulator of CFTR traffic and apical expression. Enteroids from *lmtk2*$^{-/-}$ mice display a markedly reduced forskolin induced chloride and fluid transport into the enteroid lumen (Fig. 11) compared to wild-type enteroids. A genetic loss of LMTK2 clearly reduces fluid transport into the gut lumen, and therefore is a novel target for therapeutic strategies aimed at treating and ameliorating secretory diarrheas.

5.3 CFTR, Cystic Fibrosis and Swelling

One of the main advantages of gastrointestinal-based enteroid systems is that cells (from CF and non-CF patients, or indeed other disease populations) are reasonably easily obtainable through endoscopic or suction biopsies. Such cells, as we have seen, provide an essentially limitless supply of experimental material, permitting biobanking and exchange of material with labs throughout the world (Sato et al., 2009, 2011). Once developed, enteroids are an attractive model system for testing pharmacological agents on a wide variety of diseases and genotypes, which can utilize close to native tissues (Boj, et al., 2017; Dekkers et al., 2016). Since CFTR chloride channels are so intimately associated with ion and fluid transport across epithelia (including airways and GI), it would seem reasonable that any mutations in CFTR that affect its function would also affect fluid transport and hence enteroid swelling. Indeed, swelling of intestinal organoids from patients with cystic fibrosis is markedly diminished comparted to non-CF controls (Fig. 12). The observation of reduced forskolin-induced swelling in intestinal organoids from CF patients opened up the possibility of using swelling as a screening assay for drugs that would be beneficial for patients (Dekkers, van der Ent, et al., 2013; Dekkers, Wiegerinck, et al., 2013). Enteroids appear to be sensitive to CFTR drugs, and yield responses that have a fairly large dynamic range, and as experience grows, the predictive performance of approved and investigational new drugs can be evaluated for personalized therapies. VX-809 (Lumacaftor, Vertex Pharmaceuticals) is an FDA approved drug that aids in the ER export and apical trafficking of the most common CFTR mutation, ΔF508. Overnight incubation of enteroids from patients with CF, when grown overnight in the presence of VX-809,

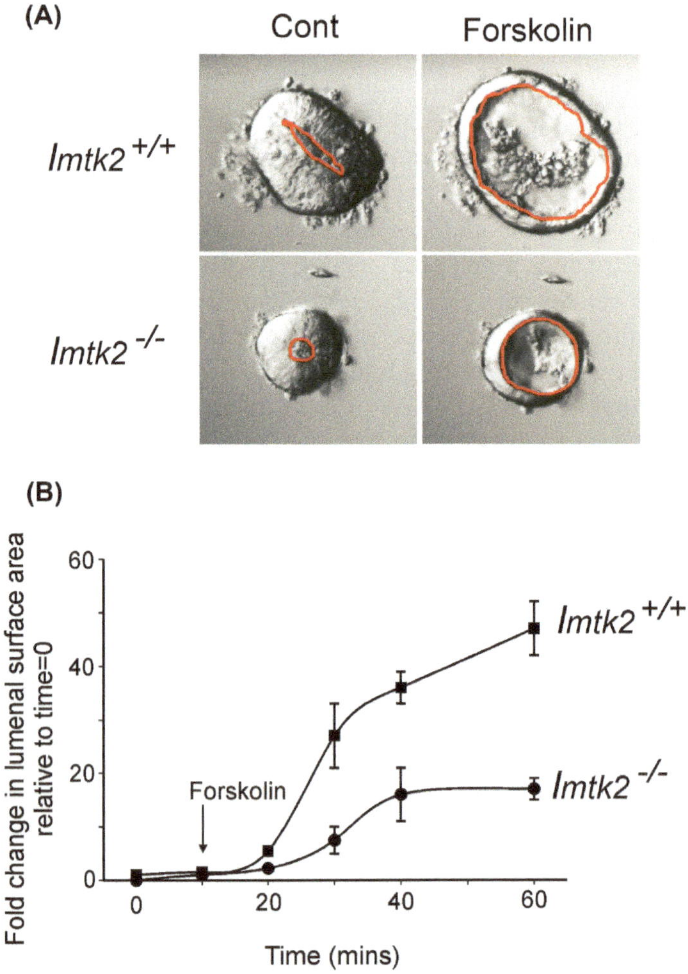

Figure 11 ***Loss of LMTK2 reduces enteroid swelling.*** Enteroids were derived from with wt (*lmtk2*$^{+/+}$) or LMTK2 knockout (*lmtk2*$^{-/-}$) mice. (A) Representative images from enteroids before and after forskolin stimulation. (B) Data from multiple enteroids was accumulated to provide a mean rate of luminal increase with time, as a function of CFTR activity (*author's unpublished data*).

show an increase in CFTR activity, as measured by increased fluid transport and enteroid swelling (Vijftigschild et al., 2016) (Fig. 12). The results obtained from swelling assays appear to recapitulate those from airway monolayers, Thus there is evidence that swelling assays may provide a viable adjunct to high throughput screening for novel CF therapeutics (Okiyoneda et al., 2013). With access to tissues (and resultant enteroids) from patients with rare mutations, it is feasible to now determine whether a particular

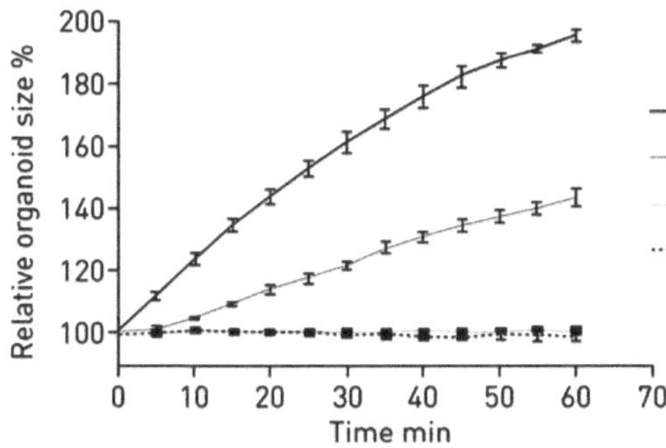

Figure 12 ***Forskolin dependent swelling in wt and F508del CFTR derived enteroids.*** CFTR wild-type (WT) and F508del (ΔF508) homozygous organoids were stimulated with forskolin or DMSO vehicle control and the relative increase in size was monitored over 60 min. CFTR Fdel508 homozygous organoids were pre-incubated with corrector VX-809 (Lumacaftor, Vertex Pharmaceuticals) for 24 h. *Adapted from Vijftigschild, L. A., Berkers, G., dekkers, J. F., Zomer-van Ommen, D. D., Matthers, E., Kruisselbrink, E., et al. (2016). β2-Adrenergic receptor agonists activate CFTR in intestinal organoids and subjects with cystic fibrosis. The European Respiratory Journal, 48, 768–779.*

drug therapy will likely have clinical benefit to the patient, based upon their enteroid swelling response to the test drug. The Human Individualized Therapy: HIT-CF program in Europe is designed to prescreen patients against approved CF drugs to determine likely efficacy in an individual patient (https://cordis.europa.eu/project/rcn/212684_en.html, https://www.ecfs.eu/sites/default/files/general-content-images/news/HIT_CF_final.pdf).

In a similar way that screening can be used to find drugs that increase CFTR activity for patients with CF, an identical approach can be used to identify drugs to inhibit CFTR activity for patients with secretory diarrhea. Wild-type CFTR expressing enteroids display swelling with stimuli such as cholera toxin, a response that can be mitigated with CFTR inhibitors; again showing proof-of-concept for enteroids as a screening system for both stimulating and inhibiting CFTR mediated ion and fluid transport (Clancy et al., 2018).

Although enteroids have a clear potential for drug screening and theratyping (the process of matching medications with genetic mutations), there are still some hurdles to overcome before its widespread application. Current assays are not easily transferable between labs, and assays require a considerable investment in supplies, equipment and personnel training. Culture reagents are commercially available, but they are still somewhat expensive, and universally accepted culture conditions remain to be determined.

Although enteroid swelling following forskolin exposure is a reflection of CFTR activity, it is still an indirect assessment of CFTR function. Converting enteroids to monolayers, as discussed next, may be a way to overcome this limitation and directly assess CFTR enteroid function using electrophysiological assays.

5.4 3-D Goes 2-D

Despite the obvious benefits of 3-D cell culture and organoid culture, there are some drawbacks. 3-D cultures require a scaffold on which to grow the culture, and the cultures require complex growth media containing somewhat expensive reagents, making 3-D cultures significantly more expense than 2-D cultures grown on plastic flasks. A second aspect of organoids, especially relevant to the study of toxigenic secretion, is the obligate architecture that puts the apical membrane in an enclosed space that has, experimentally, very limited access. The ability to study microbiome–epithelial interactions also requires access to the lumen. For this reason, several investigators are pulling open the enteroids and placing them down on a filter for use in Ussing chambers. This gives unlimited access to both the basolateral and apical environments. Although this is similar to traditional monolayer cultures of T84 cells, for example, this new approach has the benefit that it retains the cell type and proportions that were originally present in the enteroids, and therefore hopefully is a more realistic model of an intact epithelium. However, even with the benefits of this system, it still does not fully recapitulate the native environment, as there is a lack of neural innervation, immune cells, microbiome and vasculature. Whether future experimental approaches can add these components remains to be seen.

6. SUMMARY

The study of intestinal ion and fluid transport physiology has a long history. What is remarkable is the wide variety of experimental approaches that have been brought to bear upon the topic. Initial studies using intact intestines in vivo, for a brief period of time, gave way to the reductionist approach of single cell cultures as monolayers. As we grow in our appreciation of the complex interplay between the various cell types represented in our bodies, and the interaction with the microbiome, a more integrated approach is required to fully understand intestinal fluid transport and identify therapeutic approaches to correct such transport when it goes wrong.

Clearly no one approach is able to provide all the necessary information, and different experimental models are required depending upon the questions asked. Not only are intestinal organoids and enteroids providing insight into fundamental mechanisms regulating intestinal physiology, but they are also becoming immensely important in understanding intestinal pathophysiology (Zou et al., 2017). Moreover, newer strategies employing enteroids in high throughput imaging screens (HTS) will aid the evaluation of novel pharmacological agents in a relevant model that faithfully recapitulates the native environment. The American Physiological Society defines physiology as the study of "... how cells, tissues and organisms function." This history of research into the functions of the intestine surely is one of the most clear examples of the continued need for the discipline of physiology.

SUPPLEMENTARY DATA

Supplementary data related to this article can be found at https://doi.org/10.1016/bs.ctm.2018.08.003.

REFERENCES

Abou Alaiwa, M. H., Reznikov, L. R., Gansemer, N. D., Sheets, K. A., Horswill, A. R., Stoltz, D. A., et al. (2014). pH modulates the activity and synergism of the airway surface liquid antimicrobials beta-defensin-3 and LL-37. *Proceedings of the National Academy of Sciences of the United States of America, 111*(52), 18703–18708.

Ahsan, M. K., Tchernychev, B., Kessler, M. M., Solinga, R. M., Arthur, D., Linde, C. I., et al. (2017). Linaclotide activates guanylate cyclase-C/cGMP/protein kinase-II-dependent trafficking of CFTR in the intestine. *Physiol Rep, 5*(11).

Ameen, N., & Apodaca, G. (2007). Defective CFTR apical endocytosis and enterocyte brush border in myosin VI-deficient mice. *Traffic, 8*(8), 998–1006.

Ameen, N. A., Ardito, T., Kashgarian, M., & Marino, C. R. (1995). A unique subset of rat and human intestinal villus cells express the cystic fibrosis transmembrane conductance regulator. *Gastroenterol, 108*(4), 1016–1023.

Ameen, N., Silvis, M., & Bradbury, N. A. (2007). Endocytic trafficking of CFTR in health and disease. *Journal of Cystic Fibrosis, 6*(1), 1–14.

Amin, N., Alvi, N. S., Barth, J. H., Field, H. P., Finlay, E., Tyerman, K., et al. (2013). Pseudohypoaldosteronism type 1: Clinical features and management in infancy. *Endocrinol Diabetes Metab Case Rep, 2013*, 130010.

Baker, L. A., Tiriac, H., Clevers, H., & Tuveson, D. A. (2016). Modeling pancreatic cancer with organoids. *Trends Cancer, 2*(4), 176–190.

Barker, P. M., Gowen, C. W., Lawson, E. E., & Knowles, M. R. (1997). Decreased sodium ion absorption across nasal epithelium of very premature infants with respiratory distress syndrome. *The Journal of Pediatrics, 130*(3), 373–377.

Barker, N., van Es, J. H., Kuipers, J., Kujala, P., van den Born, M., Cozijnsen, M., et al. (2007). Identification of stem cells in small intestine and colon by marker gene Lgr5. *Nature, 449*(7165), 1003–1007.

Barrett, K. E. (1993). Positive and negative regulation of chloride secretion in T84 cells. *American Journal of Physiology. Cell Physiology, 265*, C859–C868.

Barthe, L., Woodley, J. F., Kenworthy, S., & Houin, G. (1998). An improved everted gut sac as a simple and accurate technique to measure paracellular transport across the small intestine. *European Journal of Drug Metabolism and Pharmacokinetics, 23*(2), 313–323.

Belot, A., Ranchin, B., Fichtner, C., Pujo, L., Rossier, B. C., Liutkus, A., et al. (2008). Pseudohypoaldosteronisms, report on a 10-patient series. *Nephrology Dialysis Transplantation, 23*(5), 1636–1641.

Bencze, J., Morotz, G. M., Seo, W., Bencs, V., Kalman, J., Miller, C. C. J., et al. (2018). Biological function of Lemur tyrosine kinase 2 (LMTK2): Implication in neurodegeneration. *Molecular Brain, 11*(1).

Boj, S. F., Vonk, A. M., Statia, M., Su, J., Vries, R. R., Beekman, J. M., et al. (2017). Forskolin-induced swelling in intestinal organoids: An in vitro assay for assessing drug response in cystic fibrosis patients. *Journal of Visualized Experiments, 120*. https://doi.org/10.3791/55159. e55159.

Borowitz, D., & Gelfond, D. (2013). Intestinal complications of cystic fibrosis. *Current Opinion in Pulmonary Medicine, 19*(6), 676–680.

Boucher, R. C. (2007). Airway surface dehydration in cystic fibrosis: Pathogenesis and therapy. *Annual Review of Medicine, 58*, 157–170.

Bradbury, N. A. (2016). Cystic fibrosis: A need for personalized medicine. In K. L. Hamilton, & D. C. Devor (Eds.), *Ion channels and transporters of epithelia in health and disease* (pp. 773–802). New York: Springer.

Bridges, R. J., Rummel, W., & Schreiner, J. (1987). In vitro effects of dexamethasone on sodium transport across rat colon. *Journal of Physiology, 383*, 69–77.

Bridges, R. J., Rummel, W., & Simon, B. (1983). Forskolin induced chloride secretion across the isolated mucosa of rat colon descendens. *Naunyn-Schmiedeberg's Archives of Pharmacology, 323*(4), 355–360.

Broughman, J. R., Sun, L., Umar, S., Sellin, J. H., & Morris, A. P. (2006). Chronic PKC-beta2 activation in HT-29 Cl.19a colonocytes prevents cAMP-mediated ion secretion by inhibiting apical membrane CFTR targeting. *American Journal of Physiology-Gastrointestinal and Liver Physiology, 291*(2), G331–G344.

Broutier, L., Andersson-Rolf, A., Hindley, C. J., Boj, S. F., Clevers, H., Koo, B. K., et al. (2016). Culture and establishment of self-renewing human and mouse adult liver and pancreas 3D organoids and their genetic manipulation. *Nature Protocols, 11*(9), 1724–1743.

Busby, R. W., Bryant, A. P., Bartolini, W. P., Cordero, E. A., Hannig, G., Kessler, M. M., et al. (2010). Linaclotide, through activation of guanylate cyclase C, acts locally in the gastrointestinal tract to elicit enhanced intestinal secretion and transit. *European Journal of Pharmacology, 649*(1–3), 328–335.

Butler, E. C., & Bradbury, N. A. (2015). Signal dependent ER export of lemur tyrosine kinase 2. *BMC Cell Biology, 16*, 26.

Case, R. M., Hunter, M., Novak, I., & Young, J. A. (1984). The anionic basis of fluid secretion by the rabbit mandibular salivary gland. *Journal of Physiology, 349*, 619–630.

Cheng, H., & Leblond, C. P. (1974). Origin, differentiation and renewal of the four main epithelial cell types in the mouse small intestin V. Unitarian theory of the origin of four epithelial cell types. *American Journal of Anatomy, 141*, 537–562.

Chibalina, M. V., Seaman, M. N., Miller, C. C., Kendrick-Jones, J., & Buss, F. (2007). Myosin VI and its interacting protein LMTK2 regulate tubule formation and transport to the endocytic recycling compartment. *Journal of Cell Science, 120*(Pt 24), 4278–4288.

Chu, S., & Montrose, M. H. (1995). Extracellular pH regulation in microdomains of colonic crypts: Effects of short-chain fatty acids. *Proceedings of the National Academy of Sciences of the United States of America, 92*(8), 3303–3307.

Cil, O., Phuan, P. W., Lee, S., Tan, J., Haggie, P. M., Levin, M. H., et al. (2016). CFTR activator increases intestinal fluid secretion and normalizes stool output in a mouse model of constipation. *Cell Mol Gastroenterol Hepatol, 2*(3), 317–327.

Clancy, J. P., Cotton, C. U., Donaldson, S. H., Solomon, G. M., VanDevanter, D. R., Boyle, M. P., et al. (2018). CFTR modulator theratyping: Current status, gaps and future directions. *Journal of Cystic Fibrosis*. doi: S1569-1993.

Clevers, H. C., & Bevins, C. L. (2013). Paneth cells: Maestros of the small intestinal crypts. *Annual Review of Physiology, 75*, 289–311.

Collaco, A. M., Jakab, R. L., Hoekstra, N. E., Mitchell, K. A., Brooks, A., & Ameen, N. A. (2013). Regulated traffic of anion transporters in mammalian Brunner's glands: A role for water and fluid transport. *Am. J.Physiol. Gastroenterol, 305*(3), G258–G275.

Conti, A., Majorini, M. T., Fontanella, E., Bardelli, A., Giacca, M., Delia, D., et al. (2017). Lemur tyrosine kinase 2 (LMTK2) is a determinant of cell sensitivity to apoptosis by regulating the levels of the BCL2 family members. *Cancer Letters, 389*, 59–69.

Crawford, I., Maloney, P. C., Zeitlin, P. L., Guggino, W. B., Hyde, S. C., Turley, H., et al. (1991). Immunocytochemical localization of the cystic fibrosis gene product CFTR. *Proceedings of the National Academy of Sciences of the United States of America, 88*(20), 9262–9266.

Cummins, A. G., Catto-Smith, A. G., Cameron, D. J., Couper, R. T., Davidson, G. P., Day, A. S., et al. (2008). Crypt fission peaks early during infancy and crypt hyperplasia broadly peaks during infancy and childhood in the small intestine of humans. *Journal of Pediatric Gastroenterology and Nutrition, 47*(2), 153–157.

Cummins, A. G., Jones, B. J., & Thompson, F. M. (2006). Postnatal epithelial growth of the small intestine in the rat occurs by both crypt fission and crypt hyperplasia. *Digestive Diseases and Sciences, 51*(4), 718–723.

Cuthbert, A. W., Halstead, J., Ratcliff, R., Colledge, W. H., & Evans, M. J. (1995). The genetic advantage hypothesis in cystic fibrosis heterozygotes: A murine study. *Journal of Physiology, 482*(Pt 2), 449–454.

De, S. N. (1959). Enterotoxicity of bacteria-free culture-filtrate of *Vibrio cholerae*. *Nature, 183*(4674), 1533–1534.

de Lau, W., Barker, N., Low, T. Y., Koo, B. K., Li, V. S., Teunissen, H., et al. (2011). Lgr5 homologues associate with Wnt receptors and mediate R-spondin signalling. *Nature, 476*(7360), 293–297.

Dekkers, J. F., Gogorza Gondra, R. A., Kruisselbrink, E., Vonk, A. M., Janssens, H. M., de Winter-de Groot, K. M., et al. (2016). Optimal correction of distinct CFTR folding mutants in rectal cystic fibrosis organoids. *The European Respiratory Journal, 48*(2), 451–458.

Dekkers, J. F., van der Ent, C. K., & Beekman, J. M. (2013). Novel opportunities for CFTR-targeting drug development using organoids. *Rare Diseases, 1*. e27112.

Dekkers, J. F., Wiegerinck, C. L., de Jonge, H. R., Bronsveld, I., Janssens, H. M., de Winter-de Groot, K. M., et al. (2013). A functional CFTR assay using primary cystic fibrosis intestinal organoids. *Nature Medicine, 19*(7), 939–945.

Devor, D. C., Singh, A. K., Frizzell, R. A., & Bridges, R. J. (1996). Modulation of Cl- secretion by benzimidazolones. I. Direct activation of a Ca(2+)-dependent K^+ channel. *American Journal of Physiology, 271*(5 Pt 1), L775–L784.

Dharmsathaphorn, K., & Barrett, K. E. (1990). Central regulation of duodenal bicarbonate secretion. *Gastroenterol, 99*(1), 278.

Dharmsathaphorn, K., McRoberts, J. A., Mandel, K. G., Tisdale, L. D., & Masui, H. (1984). A human colonic tumor cell line that maintains vectorial electrolyte transport. *American Journal of Physiology-Gastrointestinal and Liver Physiology, 246*(2 Pt 1), G204–G208.

Donaldson, S. H., & Boucher, R. C. (2003). Update on pathogenesis of cystic fibrosis lung disease. *Current Opinion in Pulmonary Medicine, 9*(6), 486–491.

Drumm, M. L., Pope, H. A., Cliff, W. H., Rommens, J. M., Marvin, S. A., Tsui, L. C., et al. (1990). Correction of the cystic fibrosis defect in vitro by retrovirus-mediated gene transfer. *Cell, 62*(6), 1227–1233.

Dutta, N. K., Panse, M. V., & Kulkarni, D. R. (1959). Role of cholera a toxin in experimental cholera. *Journal of Bacteriology, 78*, 594–595.

Field, M. (1971). Ion transport in rabbit ileal mucosa. II. Effects of cyclic 3′, 5′-AMP. *American Journal of Physiology, 221*(4), 992–997.

Field, M., Fromm, D., al-Awqati, Q., & Greenough, W. B., 3rd (1972). Effect of cholera enterotoxin on ion transport across isolated ileal mucosa. *Journal of Clinical Investigation, 51*(4), 796–804.

Field, M., Plotkin, G. R., & Silen, W. (1968). Effects of vasopressin, theophylline and cyclic adenosine monophosphate on short-circuit current across isolated rabbit ileal mucosa. *Nature, 217*(5127), 469–471.

Foulke-Abel, J., Kovbasnjuk, O., Zachos, N. C., Ettayebi, K., Blutt, S. E., de Jonge, H. R., et al. (2014). Human enteroids as an ex-vivo model of host-pathogen interactions in the gastrointestinal tract. *Exp Biol Med (Maywood), 239*(9), 1124–1134.

French, P. J., Bijman, J., Edixhoven, M., Vaandrager, A. B., Scholte, B. J., Lohmann, S. M., et al. (1995). Isotype-specific activation of cystic fibrosis transmembrane conductance regulator-chloride channels by cGMP-dependent protein kinase II. *Journal of Biological Chemistry, 270*(44), 26626–26631.

Frizzell, R. A., Dugas, M. C., & Schultz, S. G. (1975). Sodium chloride transport by rabbit gallbladder. Direct evidence for a coupled NaCl influx process. *The Journal of General Physiology, 65*(6), 769–795.

Frizzell, R. A., Field, M., & Schultz, S. G. (1979). Sodium-coupled chloride transport by epithelial tissues. *American Journal of Physiology, 236*(1).

Frizzell, R. A., Rechkemmer, G., & Shoemaker, R. L. (1986). Altered regulation of airway epithelial cell chloride channels in cystic fibrosis. *Science, 233*(4763), 558–560.

Frizzell, R. A., & Schultz, S. G. (1979). Models of electrolyte absorption and secretion by gastrointestinal epithelia. *International Review of Physiology, 19*, 205–225.

Fuller, C. M., Bridges, R. J., & Benos, D. J. (1994). Forskolin- but not ionomycin-evoked Cl^- secretion in colonic epithelia depends on intact microtubules. *American Journal of Physiology, 266*(3 Pt 1), C661–C668.

Fuller, M. K., Faulk, D. M., Sundaram, N., Shroyer, N. F., Henning, S. J., & Helmrath, M. A. (2012). Intestinal crypts reproducibly expand in culture. *The Journal of Surgical Research, 178*(1), 48–54.

Gabriel, S. E., Brigman, K. N., Koller, B. H., Boucher, R. C., & Stutts, M. J. (1994). Cystic fibrosis heterozygote resistance to cholera toxin in the cystic fibrosis mouse model. *Science, 266*(5182), 107–109.

Ganapathy, & Leibach, F. H. (1985). Is intestinal peptide transport energized by a proton gradient? *American Journal of Physiology, 249*(2 Pt 1), G153–G160.

Gangarosa, E. F., Beisel, W. R., Benyajati, C., Sprinz, H., & Piyaratn, P. (1960). The nature of the gastrointestinal lesion in asiatic cholera and its relation to pathogenesis: A biopsy study. *The American Journal of Tropical Medicine and Hygiene, 9*, 125–135.

Geibel, J. P., Rajendran, V. M., & Binder, H. J. (2001). Na(+)-dependent fluid absorption in intact perfused rat colonic crypts. *Gastroenterol, 120*(1), 144–150.

Golin-Bisello, F., Bradbury, N., & Ameen, N. (2005). STa and cGMP stimulate CFTR translocation to the surface of villus enterocytes in rat jejunum and is regulated by protein kinase G. *American Journal of Physiology. Cell Physiology, 289*(3), C708–C716.

Greger, R., Bleich, M., Leipzinger, J., Ecke, D., Mall, M., & Kunzelmann, K. (1997). Regulation of ion transport in colonic crypts. *News in Physiological Sciences, 12*, 62–66.

Gregory, R. A. (1965). Secretory mechanisms of the digestive tract. *Annual Review of Physiology, 27*, 395–414.

Gruden, N. (1977). Suppression of transduodenal manganese transport by milk diet supplemented with iron. *Nutrition and Metabolism, 21*(5), 305–309.
Guan, Y., Dong, J., Tackett, L., Meyer, J. W., Shull, G. E., & Montrose, M. H. (2006). NHE2 is the main apical NHE in mouse colonic crypts but an alternative Na^{+}-dependent acid extrusion mechanism is upregulated in NHE2-null mice. *Am. J. Physiol, Gastroenterol., 291*(4), G689–G699.
Hamilton, K. L. (2014). Even an old technique is suitable in the molecular world of science: The everted sac preparation turns 60 years old. *American Journal of Physiology. Cell Physiology, 306*(8), C715–C720.
Hamilton, K. L., & Butt, A. G. (2013). Glucose transport into everted sacs of the small intestine of mice. *Advances in Physiology Education, 37*(4), 415–426.
Harris, W. T., & Kirk, K. L. (2016). CFTR and cystic fibrosis. In K. Hamilton, & D. C. Devor (Eds.), *Ion channels and transporters of epithelia in health and disease* (Vol. 1, pp. 519–552). New York: Springer. this should be K. L. Hamilton & D.C. Devor.... Harris et al. 2016!.
Hirschhorn, N., Kinzie, J. L., Sachar, D. B., Northrup, R. S., Taylor, J. O., Ahmad, S. Z., et al. (1968). Decrease in net stool output in cholera during intestinal perfusion with glucose-containing solutions. *The New England Journal of Medicine, 279*(4), 176–181.
Huber, G. S., & Phillips, R. A. (1960). Cholera and the sodium pump. In *Paper presented at the SEATO Cholera Symposium*. Washington, D.C: N.I.H.
Hummler, E., Barker, P., Gatzy, J., Beermann, F., Verdumo, C., Schmidt, A., et al. (1996). Early death due to defective neonatal lung liquid clearance in alpha-ENaC-deficient mice. *Nature Genetics, 12*(3), 325–328.
Inoue, T., Kon, T., Ohkura, R., Yamakawa, H., Ohara, O., Yokota, J., et al. (2008). BREK/LMTK2 is a myosin VI-binding protein involved in endosomal membrane trafficking. *Genes to Cells, 13*(5), 483–495.
Jakab, R. L., Collaco, A. M., & Ameen, N. A. (2013). Characterization of CFTR high expresser cells in the intestine. *American Journal of Physiology, Gastroenterology, 305*(6), G453–G465.
Jaks, V., Barker, N., Kasper, M., van Es, J. H., Snippert, H. J., Clevers, H., et al. (2008). Lgr5 marks cycling, yet long-lived, hair follicle stem cells. *Nature Genetics, 40*(11), 1291–1299.
Karbach, U., Bridges, R. J., & Rummel, W. (1986). The role of the paracellular pathway in the net transport of calcium across the colonic mucosa. *Naunyn-Schmiedeberg's Archives of Pharmacology, 334*(4), 525–530.
Kelly, M., Trudel, S., Brouillard, F., Bouillaud, F., Colas, J., Nguyen-Khoa, T., et al. (2010). Cystic fibrosis transmembrane regulator inhibitors CFTR(inh)-172 and GlyH-101 target mitochondrial functions, independently of chloride channel inhibition. *The Journal of Pharmacology and Experimental Therapeutics, 333*(1), 60–69.
Kimberg, D. V., Field, M., Johnson, J., Henderson, A., & Gershon, E. (1971). Stimulation of intestinal mucosal adenyl cyclase by cholera enterotoxin and prostaglandins. *Journal of Clinical Investigation, 50*(6), 1218–1230.
Kitagawa, S., Sato, K., Sato, S., & Sasaki, M. (1996). Absorption of methochlorpromazine in rat small intestinal everted sac. *Biological & Pharmaceutical Bulletin, 19*(7), 998–1000.
Klein, R. G. (2000). Archeology and the evolution of human behavior. *Evolutionary Anthropology, 9*, 17–36.
Kockerling, A., & Fromm, M. (1993). Origin of cAMP-dependent Cl- secretion from both crypts and surface epithelia of rat intestine. *American Journal of Physiology, 264*(5 Pt 1), C1294–C1301.
Kopeikin, Z., Sohma, Y., Li, M., & Hwang, T. C. (2010). On the mechanism of CFTR inhibition by a thiazolidinone derivative. *The Journal of General Physiology, 136*(6), 659–671.

Kunzelmann, K., Beesley, A. H., King, N. J., Karupiah, G., Young, J. A., & Cook, D. I. (2000). Influenza virus inhibits amiloride-sensitive Na+ channels in respiratory epithelia. *Proceedings of the National Academy of Sciences of the United States of America, 97*(18), 10282–10287.

Kunzelmann, K., Sun, J., Meanger, J., King, N. J., & Cook, D. I. (2007). Inhibition of airway Na+ transport by respiratory syncytial virus. *Journal of Virology, 81*(8), 3714–3720.

Leitch, G. J., Burrows, W., & Stolle, L. C. (1967). Experimental cholera in the rabbit intestinal loop: Fluid accumulation and sodium pump inhibition. *The Journal of Infectious Diseases, 117*(3), 197–202.

Li, C., Dandridge, K. S., Di, A., Marrs, K. L., Harris, E. L., Roy, K., & Naren, A. P. (2005). Lysophosphatidic acid inhibits cholera toxin-induced secretory diarrhea through CFTR-dependent protein interactions. *The Journal of Experimental Medicine, 202*(7), 975–986.

Liu, F., Zhang, Z., Csanady, L., Gadsby, D. C., & Chen, J. (2017). Molecular structure of the human CFTR ion channel. *Cell, 169*(1), 85–95.

Love, A. H. (1969). Water and sodium absorption by the intestine in cholera. *Gut, 10*(1), 63–67.

Luz, S., Cihil, K. M., Brautigan, D. L., Amaral, M. D., Farinha, C. M., & Swiatecka-Urban, A. (2014). LMTK2-mediated phosphorylation regulates CFTR endocytosis in human airway epithelial cells. *Journal of Biological Chemistry, 289*(21), 15080–15093.

Mahe, M. M., Sundaram, N., Watson, C. L., Shroyer, N. F., & Helmrath, M. A. (2015). Establishment of human epithelial enteroids and colonoids from whole tissue and biopsy. *Journal of Visualized Experiments, 97*, 52483. https://doi.org/10.3791/52483.

Mainardes, R. M., Chaud, M. V., Gremiao, M. P., & Evangelista, R. C. (2006). Development of praziquantel-loaded PLGA nanoparticles and evaluation of intestinal permeation by the everted gut sac model. *Journal of Nanoscience and Nanotechnology, 6*(9–10), 3057–3061.

Manser, C., Guillot, F., Vagnoni, A., Davies, J., Lau, K. F., McLoughlin, D. M., et al. (2012). Lemur tyrosine kinase-2 signalling regulates kinesin-1 light chain-2 phosphorylation and binding of Smad2 cargo. *Oncogene, 31*(22), 2773–2782.

Mansour, A. A., Goncalves, J. T., Bloyd, C. W., Li, H., Fernandes, S., Quang, D., et al. (2018). An in vivo model of functional and vascularized human brain organoids. *Nature Biotechnology, 36*(5), 432–441.

Martinez, J. R., & Cassity, N. (1985). The chronically reserpinized rat as a model for cystic fibrosis: Abnormal Cl^- transport as the basis for reduced salivary fluid secretion. *Pediatric Research, 19*(7), 711–716.

Ma, T., Thiagarajah, J. R., Yang, H., Sonawane, N. D., Folllo, C., Galietta, L. J., et al. (2002). Thiazolidinone CFTR inhibitor identified by high-throughput screening blocks cholera toxin–induced intestinal fluid secretion. *Journal of Clinical Investigation, 110*(11), 1651–1658.

Matsui, H., Grubb, B. R., Tarran, R., Randell, S. H., Gatzy, J. T., Davis, C. W., et al. (1998). Evidence for periciliary liquid layer depletion, not abnormal ion composition, in the pathogenesis of cystic fibrosis airways disease. *Cell, 95*(7), 1005–1015.

McCracken, K. W., Howell, J. C., Wells, J. M., & Spence, J. R. (2011). Generating human intestinal tissue from pluripotent stem cells in vitro. *Nature Protocols, 6*(12), 1920–1928.

McCray, P. B., Jr., Zabner, J., Jia, H. P., Welsh, M. J., & Thorne, P. S. (1999). Efficient killing of inhaled bacteria in DeltaF508 mice: Role of airway surface liquid composition. *American Journal of Physiology, 277*(1 Pt 1), L183–L190.

McPherson, M. A., Dormer, R. L., Bradbury, N. A., Dodge, J. A., & Goodchild, M. C. (1986). Defective beta-adrenergic secretory responses in submandibular acinar cells from cystic fibrosis patients. *Lancet, 2*(8514), 1007–1008.

Melis, N., Tauc, M., Cougnon, M., Bendahhou, S., Giuliano, S., Rubera, I., et al. (2014). Revisiting CFTR inhibition: A comparative study of CFTRinh -172 and GlyH-101 inhibitors. *British Journal of Pharmacology, 171*(15), 3716–3727.

Method of the Year 2017: Organoids [Editorial]. *Nature Methods, 15*, (2018), 1.

Mignen, O., Le Gall, C., Harvey, B. J., & Thomas, S. (1999). Volume regulation following hypotonic shock in isolated crypts of mouse distal colon. *Journal of Physiology, 515*(Pt 2), 501–510.

Miller, D. J. (2004). Sydney Ringer; physiological saline, calcium and the contraction of the heart. *Journal of Physiology, 555*(Pt 3), 585–587.

Miyoshi, H., & Stappenbeck, T. S. (2013). In vitro expansion and genetic modification of gastrointestinal stem cells in spheroid culture. *Nature Protocols, 8*(12), 2471–2482.

Modiano, G., Ciminelli, B. M., & Pignatti, P. F. (2007). Cystic fibrosis and lactase persistence: A possible correlation. *European Journal of Human Genetics : EJHG, 15*(3), 255–259.

Moran, G. W., Leslie, F. C., Levison, S. E., Worthington, J., & McLaughlin, J. T. (2008). Enteroendocrine cells: Neglected players in gastrointestinal disorders? *Therapeutic Advances in Gastroenterology, 1*(1), 51–60.

Moser, A. J., Gangopadhyay, A., Bradbury, N. A., Peters, K. W., Frizzell, R. A., & Bridges, R. J. (2007). Electrogenic bicarbonate secretion by prairie dog gallbladder. *American Journal of Physiology-Gastrointestinal and Liver Physiology, 292*(6), G1683–G1694.

Muanprasat, C., Sonawane, N. D., Salinas, D., Taddei, A., Galietta, L. J., & Verkman, A. S. (2004). Discovery of glycine hydrazide pore-occluding CFTR inhibitors: Mechanism, structure-activity analysis, and in vivo efficacy. *The Journal of General Physiology, 124*(2), 125–137.

Mugie, S. M., Di Lorenzo, C., & Benninga, M. A. (2011). Constipation in childhood. *Nature Reviews. Gastroenterology & Hepatology, 8*(9), 502–511.

Nair, J. J., & Singh, T. P. (2017). Sjogren's syndrome: Review of the aetiology, Pathophysiology & Potential therapeutic interventions. *Journal of Clinical and Experimental Dermatology, 9*(4), e584–e589.

Nixon, A., Jia, Y., White, C., & Bradbury, N. A. (2013). Determination of the membrane topology of Lemur Tyrosines Kinase 2 (LMTK2) by fluorescence protease protection. *American Journal of Physiology. Cell Physiology, 304*(2), C164–C169.

Noel, G., Baetz, N. W., Staab, J. F., Donowitz, M., Kovbasnjuk, O., Pasetti, M. F., et al. (2017). A primary human macrophage-enteroid co-culture model to investigate mucosal gut physiology and host-pathogen interactions. *Scientific Reports, 7*, 45270.

O'Loughlin, E. V., Hunt, D. M., Bostrom, T. E., Hunter, D., Gaskin, K. J., Gyory, A., et al. (1996). X-ray microanalysis of cell elements in normal and cystic fibrosis jejunum: Evidence for chloride secretion in villi. *Gastroenterol, 110*(2), 411–418.

Okiyoneda, T., Velt, G., Dekkers, J. F., Bagdany, M., Soya, N., Xu, H., et al. (2013). Mechanism-based corrector combination restores ΔF508-CFTR folding and function. *Nature Chemical Biology, 9*(7), 444–454.

Oltra-Noguera, D., Mangas-Sanjuan, V., Centelles-Sanguesa, A., Gonzalez-Garcia, I., Sanchez-Castano, G., Gonzalez-Alvarez, M., et al. (2015). Variability of permeability estimation from different protocols of subculture and transport experiments in cell monolayers. *Journal of Pharmacological and Toxicological Methods, 71*, 21–32.

Ozen Alahdab, Y., & Duman, D. G. (2016). Pancreatic involvement in cystic fibrosis. *Minerva Medica, 107*(6), 427–436.

Petersen, O. H. (1986). Calcium-activated potassium channels and fluid secretion by exocrine glands. *American Journal of Physiology, 251*(1 Pt 1), G1–G13.

Phillips, R. A. (1964). Water and electrolyte losses in cholera. *Federation Proceedings, 23*, 705–712.

Pierce, N. F., Banwell, J. G., Rupak, D. M., Mitra, R. C., Caranasos, G. J., Keimowitz, R. I., et al. (1968). Effect of intragastric glucose-electrolyte infusion upon water and electrolyte balance in Asiatic cholera. *Gastroenterology, 55*(3), 333–343.

Pier, G. B., Grout, M., Zaidi, T., Meluleni, G., Mueschenborn, S. S., Banting, G., et al. (1998). *Salmonella typhi* uses CFTR to enter intestinal epithelial cells. *Nature, 393*(6680), 79–82.

Quinton, P. M. (1981). Effects of some ion transport inhibitors on secretion and reabsorption in intact and perfused single human sweat glands. *Pflügers Archiv, 391*(4), 309–313.

Quinton, P. M. (1999). Physiological basis of cystic fibrosis: A historical perspective. *Physiological Reviews, 79*(1 Suppl), S3–S22.

Read, N. W., Levin, R. J., & Holdsworth, C. D. (1975). Letter: Electrogenic absorption of glucose. *Lancet, 1*(7907), 631.

Richardson, S. H., & Kuhn, R. E. (1986). Studies on the genetic and cellular control of sensitivity to enterotoxins in the sealed adult mouse model. *Infection and Immunity, 54*(2), 522–528.

Riepe, F. G. (2009). Clinical and molecular features of type 1 pseudohypoaldosteronism. *Hormone Research, 72*(1), 1–9.

Riordan, J. R., Rommens, J. M., Kerem, B., Alon, N., Rozmahel, R., Grzelczak, Z., et al. (1989). Identification of the cystic fibrosis gene: Cloning and characterization of complementary DNA. *Science, 245*(4922), 1066–1073.

Robert, R., Carlile, G. W., Liao, J., Balghi, H., Lesimple, P., Liu, N., et al. (2010). Correction of the Delta phe508 cystic fibrosis transmembrane conductance regulator trafficking defect by the bioavailable compound glafenine. *Molecular Pharmacology,* 77(6), 922–930.

Rodman, D. M., & Zamudio, S. (1991). The cystic fibrosis heterozygote–advantage in surviving cholera? *Medical Hypotheses, 36*(3), 253–258.

Rosenstein, B. J. (1998). What is a cystic fibrosis diagnosis? *Clinics in Chest Medicine, 19*(3), 423–441.

Sathe, M., & Houwen, R. (2017). Meconium ileus in cystic fibrosis. *Journal of Cystic Fibrosis, 16*(Suppl 2), S32–S39.

Satitsri, S., Pongkorpsakol, P., Srimanote, P., Chatsudthipong, V., & Muanprasat, C. (2016). Pathophysiological mechanisms of diarrhea caused by the *Vibrio cholerae* O1 El tor variant: An in vivo study in mice. *Virulence,* 7(7), 789–805.

Sato, K. (1977). The physiology, pharmacology, and biochemistry of the eccrine sweat gland. *Reviews of Physiology, Biochemistry and Pharmacology, 79*, 51–131.

Sato, T., & Clevers, H. (2013). Growing self-organizing miniguts from a single intestinal stem cell: Mechanism and applications. *Science, 340*, 1190–1194.

Sato, T., Stange, D. E., Ferrante, M., Vries, R. G., Van Es, J. H., Van den Brink, S., et al. (2011). Long-term expansion of epithelial organoids from human colon, adenoma, adenocarcinoma, and Barrett's epithelium. *Gastroenterology, 141*(5), 1762–1772.

Sato, T., Vries, R. G., Snippert, H. J., van de Wetering, M., Barker, N., Stange, D. E., et al. (2009). Single Lgr5 stem cells build crypt-villus structures in vitro without a mesenchymal niche. *Nature, 459*(7244), 262–265.

Schneeberger, K., Vogel, G. F., Teunissen, H., van Ommen, D. D., Begthel, H., El Bouazzaoui, L., et al. (2015). An inducible mouse model for microvillus inclusion disease reveals a role for myosin Vb in apical and basolateral trafficking. *Proceedings of the National Academy of Sciences of the United States of America, 112*(40), 12408–12413.

Schultz, S. G., & Frizzell, R. A. (1972). An overview of intestinal absorptive and secretory processes. *Gastroenterology, 63*(1), 161–170.

Schultz, S. G., Fuisz, R. E., & Curran, P. F. (1966). Amino acid and sugar transport in rabbit ileum. *The Journal of General Physiology, 49*(5), 849–866.

Seow, K. T., Lingard, J. M., & Young, J. A. (1986). Anionic basis of fluid secretion by rat pancreatic acini in vitro. *American Journal of Physiology, 250*(2 Pt 1), G140–G148.

Shah, K., & Bradbury, N. A. (2015a). Kinase modulation of androgen receptor signaling: Implications for prostate cancer. *Cancer Cell Microenviron, 2*(4), e123. https://doi.org/10.14800/ccm.1023.

Shah, K., & Bradbury, N. A. (2015b). Lemur Tyrosine Kinase 2, a novel target in prostate cancer therapy. *Oncotarget, 6*(16), 14233–14246.

Shah, K., McCormack, C. E., & Bradbury, N. A. (2014). Do you know the sex of your cells? *American Journal of Physiology. Cell Physiology, 306*(1), C3–C18.

Singh, A. K., Devor, D. C., Gerlach, A. C., Gondor, M., Pilewski, J. M., & Bridges, R. J. (2000). Stimulation of Cl(-) secretion by chlorzoxazone. *The Journal of Pharmacology and Experimental Therapeutics, 292*(2), 778–787.

Singh, A. K., Riederer, B., Krabbenhoft, A., Rausch, B., Bonhagen, J., Lehmann, U., et al. (2009). Differential roles of NHERF1, NHERF2, and PDZK1 in regulating CFTR-mediated intestinal anion secretion in mice. *Journal of Clinical Investigation, 119*(3), 540–550.

Spence, J. R., Mayhew, C. N., Rankin, S. A., Kuhar, M. F., Vallance, J. E., Tolle, K., et al. (2011). Directed differentiation of human pluripotent stem cells into intestinal tissue in vitro. *Nature, 470*(7332), 105–109.

Stelzner, M., Helmarth, M., Dunn, J. C. Y., Henning, S. J., Houchen, C. W., Kuo, C., et al. (2012). A nomenclature for intestinal in vitro cultures. *American Journal of Physiology-Gastrointestinal and Liver Physiology, 302*(12), G1359–G1363.

Swallow, D. M. (2003). Genetics of lactase persistence and lactose intolerance. *Annual Review of Genetics, 37*, 197–219.

Swallow, D., & Hollox, E. J. (2000). The genetic polumorphism of intestinal lactase activity in adult humans. In C. R. Scriviver, A. L. Beadudet, D. Sly, & D. Valle (Eds.), *Metabolic basis of inherited disease* (pp. 1651–1663). New York: McGraw Hill.

Swiatecka-Urban, A., Boyd, C., Coutermarsh, B., Karlson, K. H., Barnaby, R., Aschenbrenner, L., et al. (2004). Myosin VI regulates endocytosis of the cystic fibrosis transmembrane conductance regulator. *Journal of Biological Chemistry, 279*(36), 38025–38031.

Takasato, M., Er, P. X., Chiu, H. S., Maier, B., Baillie, G. J., Ferguson, C., et al. (2015). Kidney organoids from human iPS cells contain multiple lineages and model human nephrogenesis. *Nature, 526*(7574), 564–568.

Thiagarajah, J. R., Song, Y., Haggie, P. M., & Verkman, A. S. (2004). A small molecule CFTR inhibitor produces cystic fibrosis-like submucosal gland fluid secretions in normal airways. *The FASEB Journal, 18*(7), 875–877.

Thwaites, D. T., Kennedy, D. J., Raldua, D., Anderson, C. M., Mendoza, M. E., Bladen, C. L., et al. (2002). H/dipeptide absorption across the human intestinal epithelium is controlled indirectly via a functional Na/H exchanger. *Gastroenterol, 122*(5), 1322–1333.

Tidball, C. S. (1961). Active chloride transport during intestinal secretion. *American Journal of Physiology, 200*, 309–312.

Trezise, A. E., & Buchwald, M. (1991). In vivo cell-specific expression of the cystic fibrosis transmembrane conductance regulator. *Nature, 353*(6343), 434–437.

Tumbarello, D. A., Kendrick-Jones, J., & Buss, F. (2013). Myosin VI and its cargo adaptors - linking endocytosis and autophagy. *Journal of Cell Science, 126*(Pt 12), 2561–2570.

Ugolev, A. M., Bagiian, A. A., & Ekkert, L. G. (1980). [New method of studying membrane hydrolysis and transport as well as metabolic processes in the small intestine in vitro (everted small intestinal sac with bilateral oxygenation)]. *Fiziologicheskll Zhurnal SSSR Imenl I M Sechenova, 66*(11), 1674–1677.

Ussing, H. H., & Zerahn, K. (1951). Active transport of sodium as the source of the electrical current in the short-circuited isolated frog skin. *Acta Physiologica Scandinavica, 23*, 110–127.

Valverde, M. A., O'Brien, J. A., Sepulveda, F. V., Ratcliff, R., Evans, M. J., & Colledge, W. H. (1993). Inactivation of the murine cftr gene abolishes cAMP-mediated but not Ca(2+)-mediated secretagogue-induced volume decrease in small-intestinal crypts. *Pflügers Archiv, 425*(5−6), 434−438.

Veit, G., Avramescu, R. G., Chiang, A. N., Houchk, S. A., Cai, Z., Peters, K. W., et al. (2016). From CFTR biology toward combinatorial pharmacotherapy: Expanded classification of cystic fibrosis mutations. *Molecular Biology of the Cell, 27*(3), 424−433.

Vijftigschild, L. A., Berkers, G., dekkers, J. F., Zomer-van Ommen, D. D., Matthers, E., Kruisselbrink, E., et al. (2016). β2-Adrenergic receptor agonists activate CFTR in intestinal organoids and subjects with cystic fibrosis. *The European Respiratory Journal, 48*, 768−779.

Waldhausen, J. H. T., & Richards, M. (2018). Meconium ileus. *Clinics in Colon and Rectal Surgery, 31*(2), 121−126.

Wang, H., & Brautigan, D. L. (2002). A novel transmembrane Ser/Thr kinase complexes with protein phosphatase-1 and inhibitor-2. *Journal of Biological Chemistry, 277*(51), 49605−49612.

Wang, L., Jiang, X., Li, C., & Ren, J. (2007). Investigation of the improved effects of 2-hydroxypropyl-beta-cyclodextrin on solubility, dissolution rate, and intestinal absorptive profile of tanshinone IIA in rats. *Archives of Pharmacal Research, 30*(8), 1020−1026.

Wang, F., Scoville, D., He, X. E., Mahe, M. M., Box, A., Perry, J. M., et al. (2013). Isolation and characterization of intestinal stem cells based on surface marker combinations and colony-forming assay. *Gastroenterol, 145*, 383−396.

Water with sugar and salt *Lancet, 2*(8084) (1978, 300−301.

Weis, G. V., Knowles, B. C., Choi, E., Goldstein, A. E., Williams, J. A., Manning, E. H., et al. (2016). Loss of MYO5B in mice recapitulates Microvillus Inclusion Disease and reveals an apical trafficking pathway distinct to neonatal duodenum. *Cell Mol Gastroenterol Hepatol, 2*(2), 131−157.

Welsh, M. J. (1986). The respiratory epithelium. In T. E. Andreoli, J. F. Hoffman, D. D. Fanestil, & S. G. Schultz (Eds.), *Physiology of membrane disorders* (pp. 751−766). New York: Plenum Press.

Welsh, M. J., & Liedtke, C. M. (1986). Chloride and potassium channels in cystic fibrosis airway epithelia. *Nature, 322*(6078), 467−470.

Welsh, M. J., & Ramsey, B. W. (2001). Cystic fibrosis. In C. R. Scriver, A. L. Beaudet, & D. Valle (Eds.), *The metabolic and molecular basis of inherited disease* (Vol. 3, pp. 5121−5188). New York: McGraw-Hill.

Wilson, T. H., & Wiseman, G. (1954). The use of sacs of everted small intestine for the study of the transference of substances from the mucosal to the serosal surface. *Journal of Physiology, 123*(1), 116−125.

Xue, X., Mutyam, V., Thakerar, A., Mobley, J., Bridges, R. J., Rowe, S. M., et al. (2017). Identification of the amino acids inserted during suppression of CFTR nonsense mutations and determination of their functional consequences. *Human Molecular Genetics, 26*(16), 3116−3129.

Yang, Y., Devor, D. C., Engelhardt, J. F., Ernst, S. A., Strong, T. V., Collins, F. S., et al. (1993). Molecular basis of defective anion transport in L cells expressing recombinant forms of CFTR. *Human Molecular Genetics, 2*(8), 1253−1261.

Yin, X., Farin, H. F., van Es, J. H., Clevers, H., Langer, R., & Karp, J. M. (2014). Niche-independent high-purity cultures of Lgr5+ intestinal stem cells and their progeny. *Nature Methods, 11*(1), 106−112.

Zeng, M., Szymczak, M., Ahuja, M., Zheng, C., Yin, H., Swaim, W., et al. (2017). Restoration of CFTR activity in ducts rescues acinar cell function and reduces inflammation in pancreatic and salivary glands of mice. *Gastroenterology, 153*(4), 1148−1159.

Zhang, Z., & Chen, J. (2016). Atomic structure of the cystic fibrosis transmembrane conductance regulator. *Cell, 167*(6), 1586–1597. e1589.
Zhang, W., Zhang, Z., Zhang, Y., & Naren, A. P. (2017). CFTR-NHERF2-LPA(2) complex in the airway and gut epithelia. *International Journal of Molecular Sciences, 18*(9), E1896. https://doi.org/10.3390/ijms 18091896.
Zou, W. Y., Blutt, S. E., Crawford, S. E., Ettayebi, K., Zeng, X. L., Saxena, K., et al. (2017). Human intestinal enteroids: New models to study gastrointestinal virus infections. *Methods in Molecular Biology*. https://doi.org/10.1007/7651_2017_1.

SECTION 4

Cell Volume Regulation in the Brain

CHAPTER ELEVEN

Cell Volume Control in Healthy Brain and Neuropathologies

Corinne S. Wilson* and Alexander A. Mongin*,§,1
*Department of Neuroscience and Experimental Therapeutics, Albany Medical College, Albany, NY, United States
§Department of Biophysics and Functional Diagnostics, Siberian State Medical University, Tomsk, Russian Federation
[1]Corresponding author: E-mail: MonginA@amc.edu

Contents

Current Topics in Membranes, Volume 81
ISSN 1063-5823
https://doi.org/10.1016/bs.ctm.2018.07.006

Abstract

Regulation of cellular volume is a critical homeostatic process that is intimately linked to ionic and osmotic balance in the brain tissue. Because the brain is encased in the rigid skull and has a very complex cellular architecture, even minute changes in the volume of extracellular and intracellular compartments have a very strong impact on tissue excitability and function. The failure of cell volume control is a major feature of several neuropathologies, such as hyponatremia, stroke, epilepsy, hyperammonemia, and others. There is strong evidence that such dysregulation, especially uncontrolled cell swelling, plays a major role in adverse pathological outcomes. To protect themselves, brain cells utilize a variety of mechanisms to maintain their optimal volume, primarily by releasing or taking in ions and small organic molecules through diverse volume-sensitive ion channels and transporters. In principle, the mechanisms of cell volume regulation are not unique to the brain and share many commonalities with other tissues. However, because ions and some organic osmolytes (e.g., major amino acid neurotransmitters) have a strong impact on neuronal excitability, cell volume regulation in the brain is a surprisingly treacherous process, which may cause more harm than good. This topical review covers the established and emerging information in this rapidly developing area of physiology.

Abbreviations

ADC	apparent diffusion coefficient [of water]
BBB	blood–brain barrier
CNS	central nervous system
CSF	cerebrospinal fluid
DRG	dorsal root ganglion
DWI	diffusion-weighted imaging
EGFP	enhanced green fluorescent protein
EM	electron microscopy
ENaC	epithelial Na^+ channels
GPCR	G protein-coupled receptor
HICCs	hypertonicity-induced cation channels
ICF	intracellular fluid
ISF	interstitial fluid
MRI	magnetic resonance imaging
PK	protein kinase
PKC	protein kinase C
RVD	regulatory volume decrease
RVI	regulatory volume increase
TRP	transient receptor potential [channels]
VRAC	volume-regulated anion channels

1. COMMON AND UNIQUE ASPECTS OF OSMOTIC BALANCE WITHIN THE CNS

Much like any other cell type in the human and animal body, brain cells are subjected to the actions of osmotic forces. The underlying mechanisms and principles are discussed in detail in the first chapter of this book (Delpire, 2018) and, therefore, are only briefly recapitulated here.

In the majority of animal cells, the surface lipid membranes have high passive permeability for water, although this permeability may strongly vary depending on the lipid composition (Fettiplace & Haydon, 1980). In contrast, the membrane lipid bilayers are, in practical terms, impermeant to ions and polar extra- and intracellular solutes. Owing to this semipermeability of the plasmalemma, water moves freely in and out of the cell as dictated by osmotic gradients set by membrane-impermeant molecules (Finkelstein, 1987). Any net accumulation or loss of solutes and metabolites is immediately followed by water flow and ensuing changes in hydrostatic pressure. Due to the absence of a rigid cell wall, animal cells are poorly equipped to withstand hydrostatic forces. Therefore, they swell or shrink, and cope with changes in cell volume and ionic homeostasis using several distinct strategies.

Moderate degrees of cell shrinkage produce relatively limited strain on cell architecture and, in the short-term, are accommodated by folding of the cell membrane. Moderate degrees of cell swelling are more problematic because the plasma membrane has a very limited stretch capacity and is tethered to extensive cytoskeletal networks. Nevertheless, upon encountering an acute osmotic challenge, many cell types can rapidly increase their volume up to several-fold using extensive membrane reserves and, in extreme cases, via exocytotic recruitment of intracellular membranes (Groulx, Boudreault, Orlov, & Grygorczyk, 2006; Morris & Homann, 2001). Movement of water and changes in cell volume dissipate osmotic gradients on the scale of seconds to several minutes, which is then followed by cell volume regulation.

Osmotic changes in cell volume are typically compensated via the act of active volume regulation. Cell volume control is a vital homeostatic property of living organisms, which emerged at early stages of evolution and is characteristic for organisms belonging to all branches of the phylogenetic tree (Chamberlin & Strange, 1989). The process of volume restoration is much slower than the initial osmotic swelling or shrinkage. Depending on conditions and cell type, it takes minutes to hours to complete. In principle, osmotic imbalances are counteracted via the accumulation or loss of small solute molecules. Osmolyte fluxes are mediated by diverse "volume-sensitive" membrane transporters and ion channels, the majority of which have been studied at length. Unfortunately, we know very little about how cells actually sense changes in their volume [the nature of cell volume sensor(s)] and have insufficient knowledge of how cell volume sensing is transduced to changes in solute transport [the nature of downstream intracellular signals]. For detailed discussion of these topics, the reader can be addressed to several comprehensive reviews (Hoffmann, Lambert, & Pedersen, 2009; Lang et al., 1998; Mongin & Orlov, 2001; Wehner, Olsen, Tinel, Kinne-Saffran, & Kinne, 2003).

1.1 Common Mechanisms of Regulatory Volume Decrease (RVD)

As schematically depicted in Fig. 1A, osmotic cell swelling is counteracted via the process of RVD. In general, volume recovery in swollen cells is mediated by the coupled efflux of cytosolic K^+ and the inorganic anions, Cl^- and bicarbonate (HCO_3^-). K^+ is the main intracellular osmolyte, but its movement must be accompanied by anions to preserve electroneutrality and sustain the electrochemical driving force for K^+. In most cell types, RVD is dominated by K^+ and Cl^- channels, which are activated independently but are coupled at the functional level via changes in the membrane potential. Among the diverse K^+ channels, many subtypes show sensitivity to cell swelling, including the two-pore domain TREKs (products of the *KCNK1* and *KCNK2* genes), TASKs (*KCNK3*, *5* and *9*), and TRAAK (*KCNK4*), the voltage-sensitive Kv1.1 and Kv1.3 (KCNA1 and KCNA3), Kv4.1 and Kv4.3 (*KCND1* and *KCND3*), Kv7.1, 4 and 5 (*KCNQ1*, *KCNQ4*, and *KCNQ5*), and the Ca^{2+}- and voltage-sensitive BK (*KCNMA1*) and IK1 (*KCNN4*) [reviewed in (Hoffmann et al., 2009; Lang et al., 1998; Wehner et al., 2003)]. It is important to note that due to the dominant nature of K^+ permeability in most mammalian cells, K^+ channels need not be cell-volume sensitive to contribute to cell volume control.

In contrast to K^+ conductance, the typical basal Cl^- permeability of the plasma membrane is low and must be augmented during RVD. This is accomplished via activation of a volume-sensitive chloride current ($I_{Cl,vol}$ or $I_{Cl,swell}$), which was initially characterized in the pioneer work done by the groups of Y. Okada, K. Strange, and B. Nilius. Because the biophysical properties of $I_{Cl,vol}$ are similar in different cell types, this led to the hypothesis

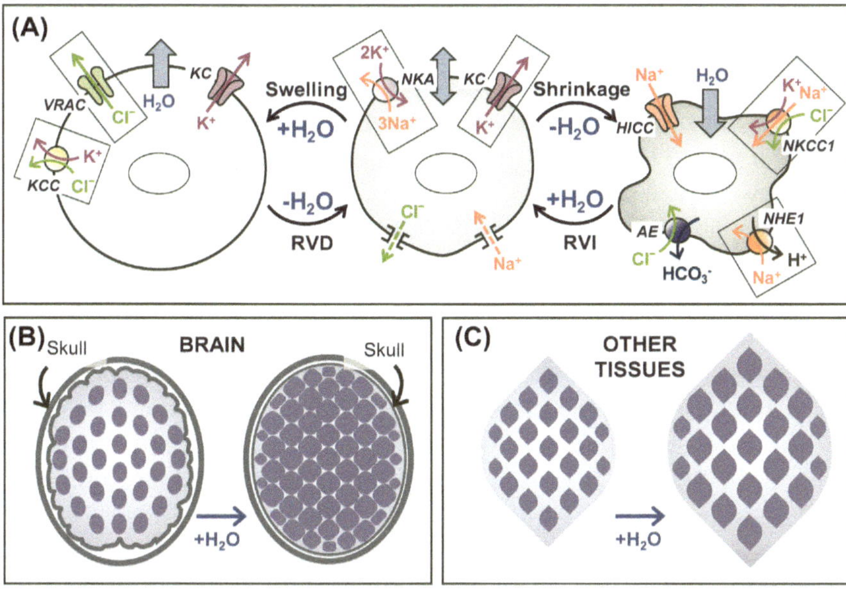

Figure 1 Basic principles of cell volume regulation in mammalian cells and major differences in the osmotic homeostasis within the brain and in other tissues. (A) Ion transport mechanisms responsible for cell volume regulation. *Center panel*: Under isosmotic conditions, the work of the Na^+/K^+-ATPase (NKA) and the dominant activity of K^+ channels (KC), set the transmembrane ionic gradients and negative membrane potential, as well as compensate for the persistent Na^+ uptake via a variety of mechanisms (Na^+ leak). Importantly, the negative membrane potential drives out intracellular Cl^-, offsetting the presence of the negatively charged impermeant macromolecules and metabolites in the cytosol. *Left panel*: Hypoosmotic cell swelling triggers activation of the volume-regulated Cl^-/anion channel (VRAC) and several K^+-Cl^- cotransporters (KCC). Cooperative activity of VRAC, KC, and KCC mediates the loss of cytosolic KCl and powers regulatory volume decrease (RVD). *Right panel*: Cell shrinkage upon exposure to hyperosmotic media stimulates the ubiquitous $Na^+-K^+-Cl^-$ cotransporter 1 (NKCC1) and/or the Na^+/H^+ exchanger 1 (NHE1). NHE works in cooperation with the volume-insensitive Cl^-/HCO_3^-anion exchangers (AEs). In some cell types, cell shrinkage also opens hypertonicity induced non-selective cation channels (HICCs). The combined activity of these transporters and channels leads to cytosolic accumulation of NaCl and KCl and mediates regulatory volume increase (RVI) (B and C) The major differences in cell volume control between the brain and other tissues are due to the fixed volume of (extracellular + intracellular) space in the CNS (B) In the brain, due to special limitations imposed by the rigid skull, cell swelling under hypoosmotic conditions or in pathologies occurs at the expense of the interstitial volume and may also compress blood vessels and cause ischemia. Also, due to the restricted ion transport across the blood–brain barrier, there is a "fixed" total pool of extracellular and intracellular ions. Therefore, during cell volume regulation, the electrochemical driving forces for ionic fluxes dissipate very quickly (C) In contrast to the brain, the majority of peripheral tissues are not restricted in terms of their osmotic expansion or shrinkage, and allow for the relatively rapid exchange of electrolytes between the interstitial space and the blood.

that they are mediated by the same ion channel or a group of similar channels. The putative $I_{Cl,vol}$ channel was christened volume-regulated anion channel (VRAC), volume-sensitive outwardly rectifying Cl^- channel (VSOR), or volume-sensitive organic osmolyte-anion channel (VSOAC) (Nilius et al., 1997; Okada, 1997; Strange, Emma, & Jackson, 1996). In this review we will use the acronym VRAC. Despite detailed studies of $I_{Cl,vol}$ currents in many cell types, the molecular identity of VRAC remained an enigma until a few years ago. In 2014, the laboratories of A. Patapoutian and T.J. Jentsch independently identified VRAC as a heteromeric complex of proteins from the family of leucine-rich repeat-containing 8 (LRRC8) (Qiu et al., 2014; Voss et al., 2014). The functional properties and physiological roles for the LRRC8-containing VRAC are discussed in several specialized reviews (Jentsch, 2016; Mongin, 2016; Pedersen, Okada, & Nilius, 2016; Stauber, 2015) and for the CNS tissue are additionally covered in Sections 2 and 8.

One of the properties of VRAC, which is reflected in their alternative name VSOAC, is the ability to conduct not only inorganic anions (Cl^-, HCO_3^-, and others), but also small, negatively charged and neutral organic molecules. The first evidence that Cl^- channels, which were later identified as VRAC, can constitute a route for transmembrane amino acid fluxes was collected in MDCK cells by U. Banderali and G. Roy, who measured anion currents mediated by aspartic acid, glutamic acid, and the amino sulfonic acid taurine in excised membrane patches (Banderali & Roy, 1992). Subsequent whole-cell electrophysiology studies measured amino acid currents in a variety of cell types subjected to hypoosmotic swelling (Jackson & Strange, 1993; Jackson, Morrison, & Strange, 1994; Manolopoulos, Voets, Declercq, Droogmans, & Nilius, 1997; Strange et al., 1996). Although the relative permeability of VRAC to organic osmolytes is a fraction of that for Cl^- (in the order of 0.05−0.3), the loss of organic molecules is strongly favored by their electrochemical gradients and may play a significant role in RVD. In some cell types, the loss of organic osmolytes accounts for about 50% of the total osmolyte loss during RVD, but such contribution is likely cell type specific [reviewed in (Wehner et al., 2003)]. What is essential in the context of the present discussion is that some of the VRAC-permeable osmolytes, such as glutamate, aspartate, taurine, and GABA, are major neurotransmitters in the CNS. Therefore, release of these molecules exerts a dramatic impact on neural functions, irrespective of their relative contribution to RVD [see (Mongin, 2016) and Sections 2 and 8].

Besides channel-mediated ion fluxes, in many cell types, RVD is supplemented by KCl loss via the electroneutral $K^+{-}Cl^-$ cotransporters (KCC1-4, encoded by *SLC12A4-6*). Discussion of the physiological roles of KCC and their contributions to cell volume control can be found in specialized reviews (Gamba, 2005; Kahle et al., 2008b, 2015; Lauf & Adragna, 2000), and in a brain-specific context in Section 2.

1.2 Generic Mechanisms of Regulatory Volume Increase (RVI)

Following osmotic or metabolic shrinkage, cells recover their volume through the process of RVI (see Fig. 1A). The two main ion transporters which drive RVI are the electroneutral $Na^+{-}K^+{-}2Cl^-$ cotransporter NKCC1 (gene product of *SLC12A2*) and the ubiquitous isoform of the Na^+/H^+-exchanger NHE1 (*SLC9A1*). Through their combined work, NKCC1 and NHE1 mediate the net accumulation of NaCl and KCl and drive the osmotically obligated movement of water inside the cell (Hoffmann et al., 2009; Lang et al., 1998; Mongin & Orlov, 2001; Wehner et al., 2003). Activity of the shrinkage-activated NHE1 (and other cell type-specific NHEs) causes alkalization of the cytosol and stimulates the pH-sensitive anion exchangers (AEs), which are not directly regulated by cell volume changes. Ubiquitous AE2 (*SLC4A2*) and the brain- and heart-enriched AE3 (*SLC4A3*) take Cl^- inside the cell in exchange for HCO_3^- out, and in such a way contribute to the accumulation of intracellular Cl^- and removal of alkaline equivalents (Alper, 1991; O'Neill, 1999). In contrast to NHEs, shrinkage-activated NKCC1 works "autonomously," but its efficacy in RVI depends on the preservation of the combined inwardly-directed gradient for [Na^+ plus K^+ plus Cl^-]. In certain cell types, hyperosmolar media strongly activate NKCC1, but this is not accompanied by RVI due to an insufficient electrochemical driving force for ionic movement. In contrast, the same cells can effectively regulate their volume in response to isosmotic cell shrinkage (e.g., caused by the receptor-mediated loss of intracellular electrolytes or the metabolic decrease of osmolyte content) or as a result of the so-called "RVI after RVD" protocol (when cells are returned to isosmotic medium after prolonged adaptation to hypoosmotic conditions) (O'Neill, 1999).

A relatively rare, but thermodynamically-effective, mechanism of RVI, which has been identified in only a few cell types, involves the opening of hypertonicity-induced non-selective cation channels (HICCs). HICCs mediate the rapid accumulation of monovalent cations, primarily Na^+, and very fast RVI (Bohmer et al., 2000; Koos et al., 2018; Wehner, Shimizu, Sabirov, & Okada, 2003; Wehner et al., 2006). The molecular composition of HICCs remains rather obscure, but depending on the cell type they may incorporate heteromeric assemblies of the transient receptor potential (TRP) channels from the melastatin family (e.g., TRPM2 and TRPM5), and/or one or several types of the "epithelial" Na^+ channels from the degenerin family (ENaCα through δ, *SCNN1A-D*) (Bondarava, Li, Endl, & Wehner, 2009; Koos et al., 2018).

The long-term adaptation to hyperosmotic conditions involves the intracellular accumulation of small, nonionic organic molecules, which have the collective name "idiogenic" or "compatible" osmolytes. The compatible osmolytes are beneficial for cell volume control because, unlike ions, they do not modify ionic strength and composition in the cytosol, and in such a way preserve the optimal conditions for intracellular enzymatic reactions. The relevant compounds include amino acids (e.g., glutamine and alanine), amino acid derivatives (e.g., the aminosulfonic acid taurine), methylamines (betaine and glycerophosphorylcholine), and polyalcohols (sorbitol and *myo*-inositol). Changes in the cytosolic levels of compatible osmolytes are in part due to the increase in expression of Na^+-dependent membrane transporters (such as SMIT for *myo*-inositol, BGT1 for betaine, and TauT for taurine), which lead to net osmolyte accumulation. Alternatively, there are increases in the expression of certain synthetic enzymes (aldose reductase for sorbitol, phospholipase for glycerophosphorylcholine, etc.) that lead to the net increase in osmolyte content inside of the cell. In all cases, the main driving mechanism for the long-term adaptation to hyperosmolarity are changes in gene expression, which are initiated by the osmosensitive transcription factor TonEBP, also known as OREBR and NFAT5 (Burg, Ferraris, & Dmitrieva, 2007; Hoffmann et al., 2009). As in the case of RVD, changes in transmembrane fluxes of certain organic osmolytes can have a dramatic impact on brain functions.

1.3 Donnan Cell Swelling

Besides the above-mentioned mechanisms responsible for "emergency" cell volume control (RVD and RVI), it is very important to recognize that under normotonic steady-state conditions all animal cells, including cells in the brain, are also constantly subjected to a persistent osmotic challenge. Due to the high content of impermeant organic macromolecules in the cytosol and intracellular organelles which create significant osmotic (or colloid) load, cells continuously accumulate water. The resulting osmotic behavior, which is often referred to as Donnan swelling or the Gibbs-Donnan effect, was theoretically predicted at the turn of the 20th century by J.W. Gibbs and experimentally tested by F.G. Donnan [reviewed in (Macknight & Leaf, 1977)]. Animal cells are capable of resisting Donnan swelling via mechanisms which were first recognized in the studies by E.B. Wilson (Wilson, 1954) and A. Leaf (Leaf, 1956). These investigators found that metabolically compromised cells and tissue slices steadily accumulate water, Na^+, and Cl^-, suggesting that cell volume maintenance involves an energy-dependent process. The subsequent work led to the development of the "double Donnan" or "pump-and-leak" model (Hoffmann et al., 2009; Lang et al., 1998; Macknight & Leaf, 1977). This model proposes that Donnan cell swelling is counteracted by the osmotic and electrical work of the Na^+/K^+-pump, which sets the negative membrane potential and drives the passive extrusion of intracellular Cl^- (Fig. 1A). The low intracellular $[Cl^-]$ compensates for the presence of the net-negatively charged macromolecules (e.g., proteins and polynucleotides) and numerous organic metabolites [see discussion in (Hoffmann et al., 2009; Lang et al., 1998) and the "neurocentric" view in (Wilson & Mongin, 2018)].

1.4 The Peculiarities of Cell Volume Regulation in the Brain

Several unique features of brain physiology strongly influence the execution and efficacy of cell volume control within the CNS. For the purposes of our discussion, it is helpful to briefly mention some of the relevant factors upfront:

The brain is contained within a rigid skull, which dramatically limits to what extent the CNS tissue and its functional components can expand or shrink (see Fig. 1B). In the context of fluid-electrolyte homeostasis, the brain can be divided into several distinct compartments [reviewed in (Kimelberg, 2004; Strange, 1992; Thrane, Rangroo Thrane, & Nedergaard, 2014)]. (1) An intracellular partition is filled with the intracellular fluid (**ICF**). Within the CNS, there are four major functionally distinct classes of cells: neurons, astrocytes, oligodendrocytes, and microglia; additionally, ciliated ependymal cells line the brain ventricles and spinal cord canal. (2) The interstitial extracellular fluid (**ECF**) bathes all brain cells and creates a milieu for their physiological activities. (3) The cerebrospinal fluid (**CSF**) is produced by the specialized epithelial cells of the choroid plexus in the brain ventricles; it serves as a precursor for ECF. CSF is moved through the ventricles into the subarachnoid space by positive hydrostatic pressure generated by the choroid plexus and with the assistance of ependymal cells. From there it is pushed via the periarterial spaces deep into the brain, where it is slowly equilibrated with ECF. The movement and equilibration of CSF and ECF is mediated by the glymphatic system, the idea of which has been largely shaped by M. Nedergaard and co-workers (Iliff et al., 2012, 2013; Thrane et al., 2014). The combination of CSF plus ECF represents the extracellular space of the brain. (4) Finally, the segregated blood compartment contains vessels of various diameters and circulating blood cells, which assist in delivering oxygen and nutrients to the brain tissue. All hydrophilic components (ions and nutrients) are transported inside the brain in a highly regulated manner via either the choroid plexus or the endothelial blood–brain barrier (BBB). The main message here is that in the brain, any change in the volume of an individual fluid compartment can happen only at the expense of its counterparts, unlike other tissues which can expand in a relatively uniform fashion (see Fig. 1C).

The total volume of the ECF is very small, which puts significant constraints on cell volume regulation. Due to the restricted transport of electrolytes across the BBB, brain tissue contains a rather small pool of extracellular osmolytes. Due to this fact, the transmembrane osmotic gradients can be easily disrupted even by limited changes in ionic fluxes. For the sake of argument, let us simplistically assume that the extracellular space occupies 20%, and the intracellular space 80%, of the total brain volume. A moderate, 10% increase in the volume of the cellular compartment (swelling) will occur at the expense of the extracellular space and reduce the latter from the initial 20% to 12% of (nearly by half). On its own, this reduction his would "concentrate" the extracellular ions and signaling molecules, including extracellular K^+. Now, consider that swollen brain cells engage in the process of RVD (see Section 1.1) and release 10% of their cytosolic K^+, whose levels inside the cells are in the range of 130–140 mM. Considering the above-mentioned volume ratios, we should expect a rise in the extracellular $[K^+]$ to well above 50 mM. Needless to say, this would drastically diminish the driving force for RVD and dramatically depolarize neuronal cells and glia. This overly-simplistic example can explain why cell volume regulation in the brain may not be as robust as in other tissues. It also shows that effective cell volume control may have a surprisingly negative impact on neural functions.

The morphological complexity of brain cells may allow for "local" fluctuations in the volume of their subcellular compartments. As an extreme example, we can consider projection neurons, which have exceptionally long and voluminous processes (both axonal and dendritic). The volume of the neuronal cell body represents <1% and in many cases as little as 0.1% of the total cell volume (Morfini, Stenoien, & Brady, 2006). Therefore, local ionic fluxes in the distal processes (such as during action potentials or pathological depolarization) are likely to change local volume independently of the cell body.

Release of typical cellular osmolytes (both inorganic and organic) has a strong physiological impact on brain cells. Since the brain is an excitable tissue, changes in the intracellular and the extracellular levels of K^+, Na^+, and Cl^- not only contribute to cell volume control, but also dramatically modify neuronal excitability (see example above). Furthermore, many of the "classical" compatible organic osmolytes are important signaling molecules in the brain. For instance, the most abundant cytosolic amino acid and organic osmolyte, glutamate, is the main excitatory neurotransmitter in the brain, playing a signaling role in 90% of the excitatory synapses (Meldrum, 2000). Another highly enriched cytosolic molecule, taurine, is an endogenous ligand for the two most important inhibitory receptors, GABA and glycine receptor-channels (Albrecht & Schousboe, 2005).

2. RVD IN THE CNS CELLS

Even in the face of the abovementioned limitations and caveats, brain cells are capable of regulating their volume much like the majority of cells in other tissues. The idea of active cell volume control within the brain was initially proposed based on early *in vivo* observations of the compensatory loss of inorganic ions and organic osmolytes from brain tissue in response to systemic hyponatremia (Melton & Nattie, 1983; Melton, Patlak, Pettigrew, & Cserr, 1987; Thurston & Hauhart, 1987; Verbalis & Gullans, 1991; Yannet, 1940). These latter studies were matched with numerous clinical and basic science findings that reductions in local or systemic osmolarity (hyponatremia) lead to the modification of neuronal excitability and the development of life-threatening neurological deficits, which tend to resolve over time as the brain adapts [see (Adrogue & Madias, 2000; Andrew, 1991; Sterns, 2015) and Section 6.1 for references and further discussion].

2.1 Neuronal RVD and Its Ionic Mechanisms

To the best of our knowledge, the first direct observation of neuronal RVD was made in cultures of dorsal root ganglion (DRG) cells by H. Horie et al.

(Horie, Ikuta, Takenaka, & Ito, 1989). Most of the subsequent early mechanistic studies were done in cultures of cerebellar granular cells by the group of H. Pasantes-Morales. They used a combination of Coulter counter and radiotracer assays to quantify RVD and osmolyte fluxes. Once subjected to hypoosmotic media, DRG and cerebellar neurons rapidly swell and then completely or partially regulate their volume to the same or a new steady-state level within 15 min (Horie et al., 1989; Pasantes-Morales, Chacon, Murray, & Moran, 1994; Pasantes-Morales, Maar, & Moran, 1993). The RVD process is driven by the loss of cytosolic K^+, because disruption of the transmembrane K^+ gradient with either high $[K^+]_o$ or high $[Rb^+]_o$ media completely blocks compensatory volume decrease (Pasantes-Morales et al., 1993). As in many other cells, K^+ permeability is the rate-limiting step in this process because the K^+ ionophore gramicidin dramatically increases the rate of RVD (Pasantes-Morales et al., 1994). Nevertheless, K^+ fluxes have to be supplemented by the efflux of Cl^- and other negatively charged organic and inorganic osmolytes to maintain electroneutrality (otherwise, hyperpolarization would strongly limit K^+ movement). Consistent with this idea, RVD is strongly inhibited by Cl^- channel blockers or by adding high millimolar levels of extracellular amino acids (Pasantes-Morales et al., 1993, 1994).

The initial observations of neuronal swelling and RVD have been reproduced in many neuronal cell types, including acutely isolated peripheral sympathetic neurons (Leaney, Marsh, & Brown, 1997), cerebellar granular cells (Patel, Lauritzen, Lazdunski, & Honore, 1998), cultured cortical neurons (Inoue, Mori, Morishima, & Okada, 2005), and the neuroblastoma N1E-115 cell line (Altamirano, Brodwick, & Alvarez-Leefmans, 1998; Falke & Misler, 1989). In these latter studies, cell volume dynamics in the substrate-attached cells was assessed using phase microscopy, or indirectly, based on the changes in intracellular calcein fluorescence. Several publications have explored the biophysical properties and possible mechanisms of activation of volume-sensitive Cl^- channels and $^{125}I^-$ efflux in neuroblastoma cell lines (Basavappa et al., 1995; Bond, Basavappa, Christensen, & Strange, 1999; Cheema, Pettigrew, & Fisher, 2007) and primary neuronal cells (Inoue et al., 2005; Leaney et al., 1997). The conclusions of these early reports were that (1) neurons respond to hypoosmotic gradients with predictable osmotic swelling, (2) in the majority of cases, neurons are capable of active cell volume control, (3) neuronal RVD is driven by volume-sensitive K^+ and Cl^- fluxes. However, the molecular identities of swelling-activated K^+ and Cl^- channels in neuronal cells remain poorly unexplored. They are inferred based on gene expression profiles and functional data obtained in other cell types. Due to the high background K^+ conductance in neurons, and the diverse assortment of K^+ channel proteins, multiple K^+ channels may assist in neuronal RVD. In contrast, since the resting Cl^- conductance is low, RVD is only possible with significant increases in Cl^- permeability, which are largely mediated by VRAC. Brain VRAC channels are assembled from the proteins belonging to the LRRC8 family, but, with the exception of one study in primary sensory neurons (Wang et al., 2017), this yet to be directly tested in diverse neuronal cells [see Section 8 and review (Mongin, 2016)].

Besides the functionally coupled K^+ and Cl^- channels, neuronal RVD can be additionally powered by one or more electroneutral K^+-Cl^- cotransporters. The "classical" KCC1, KCC3, and KCC4 are silent under basal conditions but are activated by increases in cell volume, while the neuron-specific KCC2 is basally active and additionally stimulated by cell swelling (Gamba, 2005; Lauf & Adragna, 2000). The contribution of KCCs to neuronal cell volume control has been suggested in the past but never definitively tested (Kahle et al., 2015; Wilson & Mongin, 2018). In one study, cultured hippocampal pyramidal neurons from KCC3-null mice showed a lack of RVD in hypoosmotic media (Boettger et al., 2003). Also, one of the phenotypic features of the KCC3-knockout animals is the enlarged diameter of peripheral axons, possibly due to defective cell volume regulation (Byun & Delpire, 2007). Conversely, expression of the constitutively-active KCC3 in the mouse is associated with shrunken nerve fibers, reinforcing the idea that the KCC3 cotransporter is involved in cell volume maintenance of the peripheral nerve processes (Flores, Schornak, & Delpire, 2018; Kahle et al., 2016).

The views on the osmotic behavior of neuronal cells have been strongly influenced by a number of studies coming from the groups of R.D. Andrew, G.G. Somjen, and S.A. Kirov, who found that pyramidal neurons in the hippocampus and cortex (acutely isolated or in brain slices), do not swell in response to moderate hypoosmotic challenge or swell with a significant delay (Aitken et al., 1998; Andrew, Labron, Boehnke, Carnduff, & Kirov, 2007; Caspi, Benninger, & Yaari, 2009; Somjen, Faas, Vreugdenhil, & Wadman, 1993). Interestingly, these publications also reported that the same neuronal cells readily increase their volume in response to spreading depolarizations or ischemia-like conditions. Another study performed a side-by-side comparison of acutely isolated hippocampal pyramidal cells and the hypothalamic magnocellular neurons and found that only magnocellular neurons swell to the predicted degree, while the pyramidal cell swelling was blunted (Zhang & Bourque, 2003). Likewise, imaging in brain slices from the transgenic animals expressing EGFP in neuronal cells, registered no swelling of dendritic processes in response to hypoosmotic gradients as large as −100 mOsm (Steffensen, Sword, Croom, Kirov, & MacAulay, 2015). Overall, these findings have been explained by the very low water permeability of neuronal membranes due to the lack of aquaporins (Andrew et al., 2007; Caspi et al., 2009), and/or by the paradoxical osmotic shutdown of the major ion permeability pathways in swollen pyramidal cells (Somjen et al., 1993). Although this narrative has become very dominant, is hard to say if low water permeability is a distinct property of pyramidal neurons, or the result of select experimental conditions. In the already mentioned study, which compared cultures of wild type and KCC3-null pyramidal neurons from the hippocampus, the wild type cells showed complete RVD within 20 min of exposure to hypoosmotic media (Boettger et al., 2003). Also, T. Fiacco and co-workers, found in brain slices that the thy1-EGFP-expressing pyramidal neurons swell and shrink to the same degree as the nearby situated "water-permeable" astrocytes (Murphy et al., 2017). One caveat of the latter work is that the majority of the relevant measurements have been done in relatively superficial neurons, which may have some degree of damage to their processes during slice preparation. The discrepancies between different slice studies as well as the experiments in cell cultures and *in situ* deserve further exploration. There have also been suggestions that cell volume control in brain slices relies on neuron—astrocyte interactions and is dependent on intracellular taurine content, as it can be restored by preincubation with taurine-rich media (Kreisman & Olson, 2003).

As briefly outlined in the introduction (Section 1.4), brain cells, specifically neurons, have such an elaborate morphology that the distant parts of the cell are likely to experience "local" hydration gradients and require "decentralized" mechanisms of cell volume control. Neuronal hyperexcitation is associated with significant fluxes of ions and water. Dendrites, which are endowed with a high density of postsynaptic ionotropic receptors, have especially high propensity to swell. This phenomenon is known as dendritic beading or the formation of synaptic varicosities; which is particularly pronounced under pathological conditions [reviewed in (Fiala, Spacek, & Harris, 2002; Greenwood & Connolly, 2007)]. Consequently, an important question is; are neuronal processes capable of "local" volume control? The few studies that have explored this topic suggest an affirmative answer.

For instance, swollen synaptosomes (isolated presynaptic nerve endings) completely recover their volume within 2 min after exposure to hypoosmotic media (Babila et al., 1990). Within the same time frame, swollen synaptosomes strongly upregulate their Ca^{2+} uptake via activation of unidentified cation channels (Mongin, Aksentsev, Orlov, & Konev, 1997). Hypoosmolarity-induced Ca^{2+} influx may explain the swelling-induced exocytosis of neurotransmitters, however, other mechanisms of release have also been suggested (Tuz & Pasantes-Morales, 2005; Tuz, Pena-Segura, Franco, & Pasantes-Morales, 2004; Waseem, Rakovich, Lavrukevich, Konev, & Fedorovich, 2005). In cortical neurons, acute exposure to glutamate or NMDA causes rapid (within 10 min), swelling of their somata and formation of dendritic varicosities (Inoue & Okada, 2007). For both compartments, this process is completely $[Na^+]_o$-dependent, but also inhibited by non-selective VRAC blockers. Once NMDA-receptor agonists are removed, swelling of the cell body and dendritic varicosities is resolved within 20–60 min, and the latter volume recovery is also sensitive to VRAC inhibitors (Inoue & Okada, 2007). These data suggest that in neuronal processes, VRAC contributes to both cellular swelling and compensatory RVD during and after treatment with excitotoxins. Collectively, studies in synaptosomes and neuronal cultures suggest that in neuronal cells, cell volume-sensitive ion channels are distributed throughout their distal cell processes, in both the dendritic and the axonal compartments, and mediate cell volume control. The Cl^- movement in local volume control appears to be largely mediated by VRAC.

2.2 Astrocytic RVD and Swelling-Activated Ion Channels

Although neurons represent the main cell type in the CNS, they are significantly outnumbered by several classes of non-neuronal cells — neuroglia. Among neuroglia, astrocytes are the most numerous and functionally complex, and the most studied in the context of cell volume regulation (Haydon & Carmignoto, 2006; Mongin & Kimelberg, 2009; Sofroniew & Vinters, 2010; Verkhratsky & Nedergaard, 2018). The keen interest in astrocytic cell volume homeostasis and its dysregulation was prompted by the very early knowledge that many diverse factors that lead to brain edema — such as toxins, hypoxia and metabolic inhibition, the Na^+/K^+-pump blocker ouabain, and brain traumas — all trigger profound swelling of astrocytes and corresponding reductions in the extracellular space [e.g., (Baethmann & Van Harreveld, 1973; Bakay & Lee, 1968; Bakay, Lee, Lee, & Peng, 1977; Bourke & Nelson, 1972; Cornog, Gonatas, & Feierman, 1967)]. The pathological significance of these and other relevant findings is discussed in Section 6, while the potential reasons for preferential swelling of astrocytes are considered in Section 5. The first observations of RVD in glial cells have been made in the C6 glioma cell line by the laboratory of A. Baethmann (Kempski, Chaussy, Gross, Zimmer, & Baethmann, 1983), and primary astrocyte cultures by H.K. Kimelberg et al. (Kimelberg & Frangakis, 1985; O'Connor, Kimelberg, Keese, & Giaever, 1993) and J.E. Olson et al. (Olson, Sankar, Holtzman, James, & Fleischhacker, 1986). The bulk of the subsequent mechanistic studies were carried out by the laboratories of H. Pasantes-Morales (Pasantes-Morales, Murray, Lilja, & Moran, 1994; Pasantes-Morales, Murray, et al., 1994; Sanchez-Olea, Pena, Moran, & Pasantes-Morales, 1993) and M.D. Norenberg (Bender, Neary, & Norenberg, 1993; Bender, Neary, Blicharska, Norenberg, & Norenberg, 1992).

Much like in neuronal cells, astrocytic RVD is mediated by the loss of K^+ and Cl^- via functionally coupled K^+ and Cl^- channels. These were initially explored using radiotracer assays and pharmacological inhibitors (Pasantes-Morales et al., 1994; Sanchez-Olea, Moran, Martinez, & Pasantes-Morales, 1993; Sanchez-Olea et al., 1993; Vitarella, DiRisio, Kimelberg, & Aschner, 1994). The biophysical properties and regulation of swelling-activated ion channels have been further explored using an electrophysiological approach. Surprisingly, the molecular nature and properties of swelling-activated glial K^+ channels has received little attention, perhaps because the membrane conductance in swollen cells is dominated by Cl^- channels. Yet, it is known that two types of channels dominate K^+ permeability in astroglia: the volume-insensitive, inward-rectifier channels, such as Kir4.1 (*KCNJ10*), and the swelling-activated two-pore domain TWIK-1 and TREK-1 channels [reviewed in (Kofuji & Newman, 2004; Olsen et al., 2015); but see the alternative opinion on Kir4.1 volume sensitivity in (Soe, MacAulay, & Klaerke, 2009)]. In contrast, numerous laboratories have studied the biophysical properties of astroglial VRAC as a pathway for Cl^- release (Abdullaev, Rudkouskaya, Schools, Kimelberg, & Mongin, 2006; Crepel, Panenka, Kelly, & MacVicar, 1998; Jackson & Strange, 1993; Jackson et al., 1994; Liu, Tashmukhamedov, Inoue, Okada, & Sabirov, 2006; Minieri, Pivonkova, Harantova, Anderova, & Ferroni, 2015; Olson & Li, 1997; Parkerson & Sontheimer, 2004). VRAC properties and functions in astrocytes and other CNS cells are summarized in two recent comprehensive reviews (Akita & Okada, 2014; Mongin, 2016). Here, we present for illustrative purposes our own electrophysiological recordings of typical VRAC currents in primary astrocytes (Fig. 2A–B). Besides VRAC, other types of swelling activated Cl^- channels, such as the voltage gated ClC-2 channels and maxi-Cl^- channels, also exist and are expressed in astrocytes. These may provide additional minor contributions to RVD (Liu et al., 2006; Parkerson & Sontheimer, 2004). The molecular nature of astrocytic VRAC has been uncovered only recently (Hyzinski-Garcia, Rudkouskaya, & Mongin, 2014; Schober, Wilson, & Mongin, 2017) and is further discussed in Section 8.

As already discussed, besides conducting inorganic anions, VRAC is permeable to a variety of organic osmolytes. The first publications on glial swelling-activated release of organic osmolytes putatively linked this process to VRAC based on sensitivity to diverse chloride channel blockers (Kimelberg, Goderie, Higman, Pang, & Waniewski, 1990; Martin, Madelian, Seligmann, & Shain, 1990; Pasantes Morales & Schousboe, 1988; Pasantes-Morales, Moran, & Schousboe, 1990; Strange & Morrison, 1992; Strange, Morrison, Shrode, & Putnam, 1993). The release of compatible osmolytes was initially considered a solely beneficial process, assisting in cell volume control. However, H.K. Kimelberg introduced a very counterintuitive at the time idea on the potentially harmful aspects of such release (Kimelberg et al., 1990). Because two cytosolic amino acids, glutamate and aspartate, are the excitatory neurotransmitters, their release can be toxic and drive tissue damage in a variety of brain pathologies. This aspect of VRAC activity has been the subject of several comprehensive reviews (Kimelberg & Mongin, 1998; Kimelberg, 2005; Mongin, 2007, 2016) and is further discussed in Sections 6 and 7. The link between organic osmolyte release and VRAC has been extensively tested in glial cells in several electrophysiological studies, which directly measured amino acid currents (Abdullaev et al., 2006; Jackson & Strange, 1993; Liu et al., 2006; Olson & Li, 1997; Roy, 1995). Here, we present our own data on the swelling-activated glutamate release in cultured astrocytes, and a microdialysis study of the VRAC-mediated glutamate release in the brain cortex *in vivo* (Fig. 2C and D). Interestingly, our recent molecular biology work strongly suggests that there are at least two different heteromeric VRAC complexes, which have preferential selectivity for uncharged and charged organic osmolytes [see (Schober et al., 2017) and additional discussion in Section 8].

Another important idea related to the functional consequences of cell volume control in the CNS is VRAC regulation by G protein-coupled receptors (GPCRs). In the brain, the degree of cell swelling under physiological conditions is estimated to be in the order of 2%–5% of "normal"

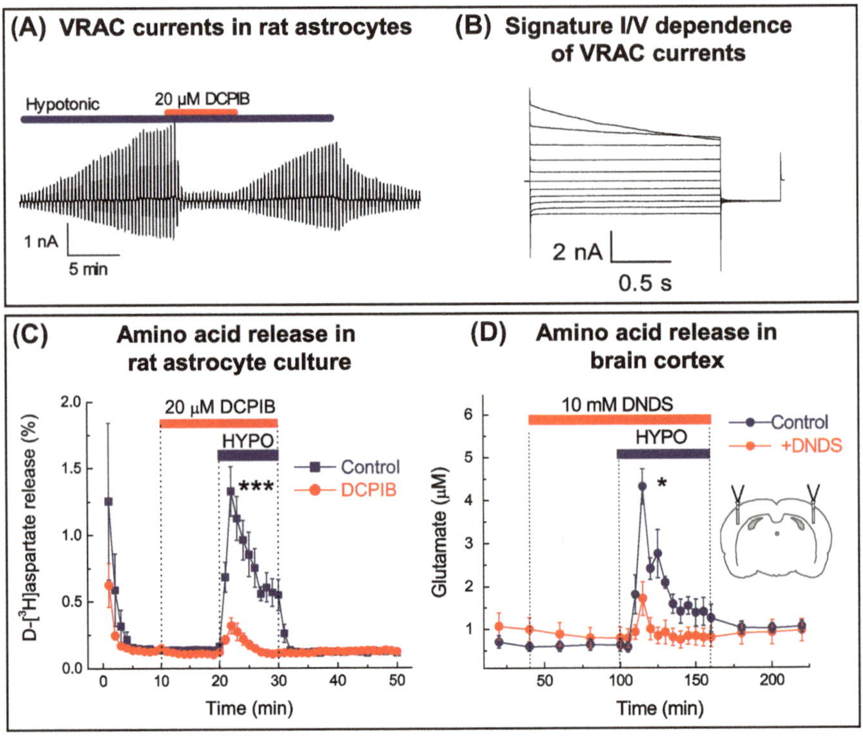

Figure 2 Swelling-activated Cl^- currents and amino acid release through VRAC *in vitro* and *in vivo*. (A) Representative whole-cell recordings of Cl^- currents in primary rat astrocytes exposed to hypoosmotic medium (−60 mOsm). Activity of Cl^- channels was measured by holding cells at 0 mV and alternately administering ± 40 mV voltage pulses. Currents were inhibited by treatment with the VRAC blocker DCPIB (20 μm). (B) Swelling-activated Cl^- currents in astrocytes in response to 20 mV step pulses from −100 to +100 mV, displaying the characteristic outward rectification and time-dependent inactivation at positive potentials. (C) Effect of DCPIB on swelling-activated glutamate release in primary astrocytes, traced with the non-metabolizable glutamate analog D-[^{3}H]aspartate. ***$p < 0.001$, effect of DCPIB. (D) Effect of the non-specific VRAC blocker DNDS on swelling-activated glutamate release in the rat cortex after stimulation with hypoosmotic artificial cerebrospinal fluid, measured by microdialysis approach and analyzed with HPLC. *$p < 0.05$, effect of DNDS. *(A–C) adapted from Abdullaev, I. F., Rudkouskaya, A., Schools, G. P., Kimelberg, H. K., & Mongin, A.A. (2006). Pharmacological comparison of swelling-activated excitatory amino acid release and Cl^- currents in rat cultured astrocytes.* Journal of Physiology, 572, *677–689, with permission. (D) Reproduced from Haskew-Layton, R. E., Rudkouskaya, A., Jin, Y., Feustel, P.J., Kimelberg, H.K., & Mongin, A.A. (2008). Two distinct modes of hypoosmotic medium-induced release of excitatory amino acids and taurine in the rat brain in vivo.* Plos One, 3, *e3543, under the Creative Commons Attribution (CC BY) license.*

cell volume. This is unlikely to be sufficient to open VRAC to any significant degree. However, we and others found that astroglial VRAC can be potently modulated by GPCR signaling. E.g., in astrocytes, activation of purinergic P2Y receptors causes partial, Ca^{2+}-dependent stimulation of VRAC, even when cells are not swollen (Akita, Fedorovich, & Okada, 2011; Darby, Kuzmiski, Panenka, Feighan, & MacVicar, 2003; Mongin and Kimelberg, 2002, Mongin & Kimelberg, 2005b; Takano et al., 2005). ATP and ADP act at $P2Y_1$, $P2Y_2$ and/or $P2Y_4$ receptors and prompt the release of intracellular Ca^{2+} via activation of classical PKC α and βI, and additionally Ca^{2+}/calmodulin-dependent PK II (Akita et al., 2011; Mongin and Kimelberg, 2002, Mongin & Kimelberg, 2005b; Rudkouskaya, Chernoguz, Haskew-Layton, & Mongin, 2008). Interestingly, pharmacological inhibitors and molecular biology tools that block the GPCR-driven VRAC opening have little to no effect on VRAC activation by hypoosmotic media. Based on our work, we proposed a model in which purinergic signaling activates or modulates channel activity via phosphorylation of a small fraction of VRAC channels that are already open or "primed" for opening. Although ATP effects are measurable in non-swollen cells, the main, larger pool of VRAC molecules remains inaccessible for purinergic stimulation. The concurrent cell swelling, even as small as 5%, recruits from the "silent," volume-sensitive pool of VRAC, and acts synergistically with the purinergic receptors (Mongin & Kimelberg, 2005a, 2005b).

Following our findings on the purinergic modulation of VRAC activity in astrocytes, several other groups uncovered very similar synergistic effects of cell swelling and diverse GPCR receptors, linked to either Ca^{2+} or cAMP signaling, on VRAC-like Cl^- currents and organic osmolyte release. For astrocytes and astrocytic cell lines, the reader can be referred to the following significant studies (Akita & Okada, 2011; Cheema, Ward, & Fisher, 2005; Liu, Akita, Shimizu, Sabirov, & Okada, 2009; Ramos-Mandujano, Vazquez-Juarez, Hernandez-Benitez, & Pasantes-Morales, 2007; Rosso, Peteri-Brunback, Poujeol, Hussy, & Mienville, 2004). Findings in non-CNS cells are covered by three comprehensive reviews (Fisher, Cheema, Foster, & Heacock, 2008; Fisher, Heacock, Keep, & Foster, 2010; Franco, Panayiotidis, & de La Paz, 2008). Recently, to definitively discriminate between VRAC and other potential mechanisms of ATP-stimulated amino acid release (such as vesicular release and others), we demonstrated that ATP-stimulated glutamate and taurine release is abolished by knocking down the essential VRAC subunit LRRC8A (Hyzinski-Garcia et al., 2014).

3. RVI IN THE CNS CELLS

Severe systemic hyperosmolarity, such as increases from ~290 to 410–420 mOsm, places mammals (studied mostly in rodent models) at the danger of death (Arieff, Guisado, & Lazarowitz, 1977; Gullans & Verbalis, 1993). Milder elevations in systemic osmolarity are typically associated with behavior abnormalities, such as restlessness and increased startle response (Arieff et al., 1977; Cserr et al., 1991; Gullans & Verbalis, 1993). Numerous early studies, most notably by H.F. Cserr and colleagues, found that in response to systemic hypertonic perturbations the brain as an organ shrinks but then regulates its volume via the accumulation of osmolytes and osmotically obligated water on the scale of a few hours to days [e.g., (Cserr, DePasquale, & Patlak, 1987a, 1987b; Cserr et al., 1991; DePasquale, Patlak, & Cserr, 1989; Gullans & Verbalis, 1993)]. The measurable tissue volume regulation occurs as rapidly as 90 min after hypertonic insult and is associated with the uptake of Na^+, Cl^-, and K^+, with the bulk of these ions accumulating in the intracellular rather than extracellular compartment (Cserr et al., 1987a, 1987b; Cserr et al., 1991). These results strongly suggest that brain cells undergo successful RVI *in vivo*. The secondary phase of adaptation to chronic hyperosmotic conditions involves the accumulation of idiogenic (compatible) organic osmolytes, which slowly replace the excess of inorganic ions within one or two weeks [see for example (Chan & Fishman, 1979; Lohr, McReynolds, Grimaldi, & Acara, 1988) and review (Gullans & Verbalis, 1993)].

In contrast to the whole brain, numerous studies in cultured and acutely isolated neurons (Aitken et al., 1998; Horie et al., 1989; Pasantes-Morales et al., 1993; Zhang & Bourque, 2003) and the majority of publications in cultured astrocytes [e.g., (Kimelberg & Frangakis, 1985; O'Connor et al., 1993)] found no evidence of RVI when cells have been exposed to hyperosmotic media for 10–90 min. Similarly, in brain slices prepared from either neocortex or hippocampus, pyramidal neurons and their dendritic trees shrunk under hyperosmotic conditions but did not demonstrate RVI, at least during incubation times under 1 h (Andrew & MacVicar, 1994; Andrew, Lobinowich, & Osehobo, 1997). To the best of our knowledge, there are only a few reports of RVI in brain cells, which originate from the studies in a C6 glioma cell line and immature rat neurons. Once exposed to the sucrose-adjusted hyperosmotic media, immature rat neurons restored their volume promptly (within 20 min) and effectively (60% RVI)

(Schomberg et al., 2003). In C6 cells, Kemplski et al. found complete but very slow (taking 3 h) RVI when osmolarity was increased with a non-ionic compound mannitol, but not after the addition of NaCl (Kempski et al., 1983). In contrast, the group of K. Strange detected in the same C6 cell line the rapid (within 20 min) and complete RVI in high NaCl medium, and this process was partially blocked by the NKCC inhibitor bumetanide (McManus & Strange, 1993; Strange & Morrison, 1992). In a less conclusive study, murine astrocytes demonstrated a trend for RVI upon expose to sucrose-containing hyperosmotic media, but this process was efficacious only when cells were co-stimulated with adrenergic agonists (Song, Xu, Hertz, & Peng, 2015). Overall, it appears that cultured neuronal and glial cells do not have transmembrane ionic gradients which would favor RVI (see Section 1.2). The apparent mismatch between the effective RVI in the whole brain *in vivo* and the lack of hyperosmotic cell volume regulation in cultured or acutely isolated primary cells deserves additional exploration.

Due to the apparent lack of RVI in model experiments, and because hypertonicity represents a relatively rare phenomenon in the CNS, there have been surprisingly few studies exploring osmotic regulation of the volume-sensitive ion transporters in the brain cells. We could recall only one paper on hypertonic activation of the Na^+/H^+ exchanger in rodent astrocytes (Shrode, Klein, O'Neill, & Putnam, 1995) and another already mentioned study on the hyperosmotic stimulation of NKCC1 in immature rat neurons (Schomberg et al., 2003). In contrast to "physiological" cell volume control, there have been numerous studies linking NHE transporters and NKCC1 to adverse outcomes in various neuropathologies, particularly stroke [reviewed in (Kintner, Wang, & Sun, 2007; Mongin, 2007)]. Again, further work in this area would be helpful for understanding how disruptions in RVI may contribute to neurological outcomes.

4. CELL VOLUME CHANGES DURING NORMAL BRAIN FUNCTIONING

Brain tissue exists in an environment of near constant osmolarity, which is accomplished via the tight control of systemic water-electrolyte homeostasis and is due to restrictions of electrolyte transport across the blood–brain barrier [reviewed in (Bourque & Oliet, 1997; Kimelberg, 2004; Strange, 1992)]. Nevertheless, brain cells experience constant fluctuations of their volume due to ionic and osmolyte fluxes resulting from electrical activities, reuptake and recycling of neurotransmitter molecules, and dynamic changes in tissue metabolism. The first study directly demonstrating nerve swelling upon electrical stimulation was performed in isolated nerve fibers of invertebrates (Iwasa, Tasaki, & Gibbons, 1980). This pioneer work was followed by numerous observations in higher organisms. One approach, which has been widely utilized in brain slices, is based on detecting variations in light scattering or the intrinsic optical signal (IOS) of the tissue, which indirectly reflect changes in cell volume (Aitken, Fayuk, Somjen, & Turner, 1999; Andrew & MacVicar, 1994; Lipton, 1973). Using IOS signals, several groups demonstrated that electrical stimulation of cortical and hippocampal brain slices causes cellular swelling (Andrew & MacVicar, 1994; Holthoff & Witte, 1996; MacVicar & Hochman, 1991).

An alternative technique allowing for the measurement of interstitial volume is dynamic monitoring of the extracellular concentrations of poorly membrane permeant cations, such as tetramethylammonium (TMA^+) or tetraethylammonium (TEA^+) [reviewed in (Nicholson & Sykova, 1998; Sykova & Nicholson, 2008; Sykova, 2004)]. The relevant chemical probes are introduced into the CNS electrophoretically, and then detected with ion-selective electrodes. The rate of diffusion of TMA^+/TEA^+ from the point of injection to the registering electrode gives information on the tortuosity of the extracellular space. Fluctuations of TMA^+/TEA^+ concentrations under steady-state conditions reflects changes in the volume of the interstitial liquid. Using this type of assay, several groups found that electrical stimulation of neurons causes significant reductions in the extracellular space (cell swelling) in brain preparations from lower vertebrates and mammals [see for example (Jing, Aitken, & Somjen, 1994; Larsen & MacAulay, 2017; Larsen et al., 2014; Svoboda & Sykova, 1991) and comprehensive review by (Sykova & Nicholson, 2008)].

The oldest method for monitoring changes in the volume of interstitial liquid is the electrical impedance assay, which was developed by A. Van Harreveld et al. (Van Harreveld, 1972; Van Harreveld, Dafny, & Khattab, 1971). This approach employs measuring changes in the extracellular resistance (impedance) upon the delivery of electric currents with their parameters selected to not allow charge transfer across cellular membranes. Using this technique or its modifications, several groups demonstrated dynamic activity-dependent changes in ion concentrations and volume of the extracellular space (Amzica & Neckelmann, 1999; Dietzel, Heinemann, & Lux, 1989; Dietzel, Heinemann, Hofmeier, & Lux, 1982). F. Amzica and D. Neckelmann were able to deduct which cell type undergoes swelling, by adding an additional approach and measuring the activity-dependent changes in capacitance (proportional to cell volume) in both neurons and astrocytes in the cortex of anesthetized cats. The capacitance changes of patched astrocytes closely followed neuronal activity and were most pronounced during seizure-like events, suggesting that swelling of astroglia modifies ECF volume. In contrast, apparent changes in neuronal cell volume were smaller and did not correlate with the degree of excitation (Amzica & Neckelmann, 1999).

In recent exciting advancements, the laboratory of U.V. Nagerl has brought to life the new imaging technology, super-resolution shadow imaging (SUSHI). SUSHI combines the 3D-stimulated emission depletion (STED) microscopy and the fluorescent labeling of the extracellular fluid with non-permeable fluorophores to allow for the high-resolution visualization of the structure of the extracellular space in brain slices and organotypic preparations (Tonnesen, Inavalli, & Nagerl, 2018). Besides the exceptional 3-D resolution, this extracellular labeling strategy dramatically reduces photobleaching and phototoxicity, which are the common problems with many traditional imaging techniques. Using this approach, the inventors detected expansion of the extracellular space under hyperosmotic conditions and, conversely, its shrinkage (i.e., cell swelling) in response to epileptiform activity and local glutamate release (Tonnesen et al., 2018).

Overall, each of the abovementioned methodologies reveal measurable changes in the volume of the intracellular and extracellular compartments in response to neuronal activation. These are mediated by the activity-dependent fluxes of ions and movement of osmotically obligated water. Changes in the volume and geometry of the extracellular compartment are sufficient to impact neuronal excitability in a normal brain, but are particularly profound in neuropathologies [see review by (Sykova & Nicholson, 2008) and Section 6]. Depending on the study, cellular swelling has been ascribed to either neurons or astrocytes, but most likely multiple cell types are affected. Thus, N.N. Haj-Yasein et al. analyzed ECF shrinkage (inversely related to cell swelling) in response to high-frequency stimulation in hippocampal slices prepared from either wild type animals or animals carrying deletion of the astrocytic water channel aquaporin-4 (AQP4). In the regions enriched with astrocytic processes, nerve stimulation triggered shrinkage of the extracellular space that was strongly influenced by the deletion of AQP4, suggesting that it occurred in astrocytes. In contrast, in the areas devoid of astrocytic processes, changes in the extracellular space were more profound but independent of AQP4 expression, pointing to their neuronal origin (Haj-Yasein et al., 2012).

Another important example of physiologically relevant cell volume changes can be found in hypothalamic neuroendocrine structures, which sense minor osmolarity changes in systemic circulation. The supraoptic (SON) and the paraventricular (PVN) nuclei of the hypothalamus contain magnocellular neurons, whose main function is regulation of the body's water-electrolyte homeostasis (Bourque & Oliet, 1997; Bourque, 2008; Hussy, Deleuze, Desarmenien, & Moos, 2000). These neurons project to the pituitary gland and secrete the antidiuretic hormones, vasopressin and oxytocin. Once released into systemic circulation, vasopressin in humans and vasopressin and oxytocin in rodents regulate water reabsorption in the kidney. Neuronal activity in the SON and PVN and release of antidiuretics are activated by hyperosmolarity and inhibited by hypoosmolarity via several distinct mechanisms, which are comprehensively discussed in several specialized reviews (Bourque, 2008; Hussy et al., 2000; Noda & Hiyama, 2015). Briefly: (1) Magnocellular neurons in the SON and PVN are regulated via the excitatory inputs from the upstream osmosensitive neurons located in the organum vasculosum lamina terminalis (OVLT) and the subfornical organ (SFO). (2) Both the OVLT and SFO cells and the SON and PVN neurons are intrinsically osmosensitive, via the mechanism deciphered in the laboratory of C.W. Bourque. This group discovered that hyperosmolarity directly stimulates osmosensitive neurons via opening the volume-sensitive splice variant of the TRPV1 channel, which lacks a portion of its amino terminus (ΔN-TRPV1) (Bourque, Oliet, & Richard, 1994; Ciura & Bourque, 2006; Ciura, Liedtke, & Bourque, 2011; Voisin, Chakfe, & Bourque, 1999). (3) An alternative, non-cell-autonomous mechanism responsible for sensing hypoosmolarity was identified by the group of N. Hussy, who found that even moderate reductions in medium osmolarity stimulate specialized populations of SON and PVN astrocytes, which contain extremely high cytosolic levels of taurine. Once swollen, astrocytes release taurine via a VRAC-like channel (Bres et al., 2000; Deleuze, Duvoid, & Hussy, 1998). Taurine inhibits the activity of SON and PVN magnocellular neurons by acting as the endogenous ligand for their inhibitory glycine receptors (Hussy, Deleuze, Pantaloni, Desarmenien, & Moos, 1997). A similar, astrocyte-specific mechanism was found in OVLT, where swelling triggers Ca^{2+} influx via TRPV4 and amplifies astrocytic taurine release via VRAC (Ciura et al., 2018). Overall, the direct activation of neurons by shrinkage and their indirect inhibition by taurine release from swollen astrocytes mediate the CNS osmolarity sensing.

5. A SPECIAL CASE FOR ASTROCYTIC SWELLING AND CELL VOLUME CONTROL

A large variety of brain pathologies – including hyponatremia, stroke, traumatic brain and spinal cord injuries, epilepsy, hepatic failure, and hypoglycemia – are associated with the prominent pathological swelling of astrocytes [see reviews (Kimelberg, 1995; Mongin & Kimelberg, 2005a; Norenberg, 1994; Stokum, Gerzanich, & Simard, 2016) and Section 6 for further discussion and references]. The phenomenon of pathological swelling in the CNS has been historically termed cytotoxic brain edema (Kimelberg, 1995; Klatzo, 1967; Stokum et al., 2016). This unusual name originates from the early studies on the effects of certain toxins, such as dinitrophenol, triethyltin, and hexachlorophene, on brain tissue. The latter toxic substances trigger the shift of electrolytes and water from the extracellular to the intracellular space, without the net accumulation of water in the brain or swelling of brain tissue. To avoid confusion, H.K. Kimelberg introduced the more accurate term of cellular edema (Kimelberg, 1995). Cellular/cytotoxic edema is usually explained by oncotic (Donnan) swelling of metabolically compromised cells. This is in contrast to vasogenic edema, which is caused by disruption of the integrity of the BBB at the blood–brain interface. Vasogenic edema causes the accumulation of plasma proteins and net water influx into the brain tissue; and is associated with massive and potentially life-threatening brain swelling. Cellular edema typically precedes vasogenic edema, but these two processed are connected. Despite much work in the field, the molecular mechanisms underlying cellular edema are incompletely understood. The major unanswered question is why cell swelling is preferentially seen in one cell type – astrocytes. In this section text and Fig. 3, we present several prevalent hypotheses.

5.1 Water Permeability and Aquaporin Channels

As shown in Fig. 3A, one popular explanation for the preferential swelling of astrocytes is the high water permeability of their membranes, which is facilitated by the presence of water channels, aquaporins (AQP). Astrocytes abundantly express the AQP4 protein, particularly in their processes contacting brain capillaries and larger vessels (Amiry-Moghaddam & Ottersen, 2003). Deletion of AQP4 in astrocytes reduces their water permeability by ~sevenfold at 12°C but only twofold at 37°C (Solenov, Watanabe, Manley, & Verkman, 2004). It is well known that water can move in and out of the cell through many alternative routes, passively or via assorted

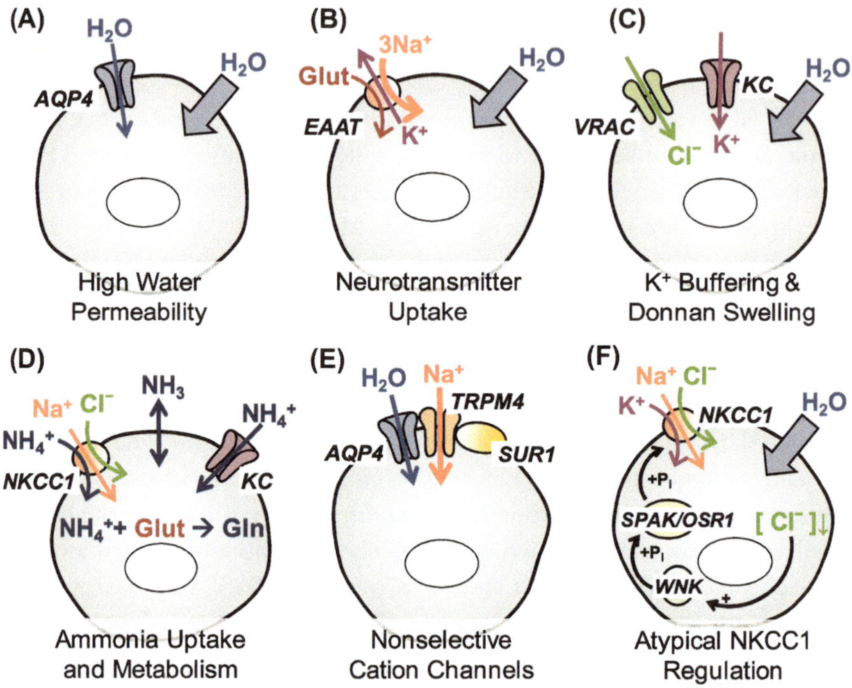

Figure 3 Astrocytic properties that may be responsible for preferential propensity of this cell type to cell swelling in neuropathologies (A) Astrocyte membranes are highly permeable to water due to expression of the water channel aquaporin-4 (AQP4); (B) Astrocytes take up neurotransmitters from the extracellular space in order to maintain normal neuronal activity. Excitatory amino acid transporters (EAATs) take one glutamate in together with 3 Na^+ and in exchange of 1 K^+, and promote the accumulation of osmotically obligated water; (C) Buffering of $[K^+]_o$ through potassium channels (KC) can be associated with the concomitant influx of Cl^- through VRAC and promote Donnan cell swelling; (D) In normal brain and neuropathologies, astrocytes accumulate extracellular ammonia (NH_3) via passive transmembrane diffusion and ammonium ions (NH_4^+) through potassium channels (KCs) and NKCC1. Inside the cell, NH_3/NH_4^+ is then assimilated to produce glutamine from glutamate (E) Upregulation and activation of the nonselective SUR1-TRPM4 channels and their assembly with AQP4 promotes Na^+ uptake and water accumulation; (F) Astrocytic NKCC1 cotransporter contributes to electrolyte and water accumulation in response to high $[K^+]_o$. Cell swelling may paradoxically activate NKCC1 via the low $[Cl^-]_i$-sensing WNK/SPAK/OSR1 cascade and in such a way further amplify the persistent astrocytic swelling.

channels and transporters conducting major osmolytes [see (MacAulay, Hamann, & Zeuthen, 2004) and Section 5.2]. Nonetheless, despite the relatively modest decrease in water permeability at physiological temperatures, AQP4-null animals show high resistance to systemic water intoxication, (hyponatremia) as compared to their wild-type counterparts (Manley et al., 2000). Specific enrichment of AQP4 in astrocytic endfeet is dependent on the adapter proteins, α-syntrophin and dystrophin (Amiry-Moghaddam et al., 2003). Deletion of α-syntrophin impedes water movement from the blood to the brain and reduces tissue edema in an animal model of stroke (Amiry-Moghaddam et al., 2003). Because brain and cell swelling in neuropathologies develops on a slow scale, it is not clear if the twofold changes in water permeability in AQP-null astrocytes or disruption of AQP4 localization in the astrocytic endfeet seen in α-syntrophin knockouts can explain the observed resistance to cellular and/or tissue edema. Perhaps, AQP4 plays more complex roles in regulating water movement and osmotic gradients, via modulating activities of ion channels and transporters. E.g., AQP4 can form functional complexes with the non-selective cation channel TRPV4 (Benfenati et al., 2011), the SUR1-TRPM4 channel complex (Stokum et al., 2018), and also regulates the activity of VRAC (Benfenati et al., 2007). On the other hand, cell swelling caused by spreading depolarizations is known to occur on a timescale of seconds and strongly contribute to neurological outcomes in stroke and traumatic brain injury (Dreier, Lemale, Kola, Friedman, & Schoknecht, 2018; Hartings et al., 2017).

5.2 Neurotransmitter Uptake

A well-supported mechanism for astrocytic swelling is related to the critical role for these cells in the re-uptake of the major amino acid neurotransmitters: glutamate, aspartate, and GABA (Danbolt, 2001; Zhou & Danbolt, 2013). Excitatory glutamate and aspartate molecules are taken inside the cell by excitatory amino transporters (EAATs) with 3 Na^+ ions in exchange for 1 K^+ out per each transport cycle, along with the uncoupled conduction of Cl^- (Fahlke, Kortzak, & Machtens, 2016; Machtens et al., 2015) (Fig. 3B). Uptake of the inhibitory amino acid GABA occurs through GAT transporters along with ~2–3 Na^+ and 1 Cl^- (Willford, Anderson, Spencer, & Eskandari, 2015). There is an extensive literature on glutamate-induced astrocyte swelling in cultures and *in situ*, indicating that such swelling involves the accumulation of Na^+ and Cl^-, and is Ca^{2+}-dependent (Bender, Schousboe, Reichelt, & Norenberg, 1998; Izumi, Kirby, Benz, Olney, & Zorumski, 1999; Koyama, Sugimoto, Shigenaga, Baba, & Iwata, 1991; Koyama et al., 2000; Schneider, Baethmann, & Kempski, 1992). Under pathologic conditions such as epilepsy or ischemic stroke, high extracellular glutamate levels can significantly contribute to astrocytic swelling via transporter or metabotropic receptor-mediated mechanisms (Bender et al., 1998; Hansson, Johansson, Westergren, & Ronnback, 1994). Importantly, in addition to increasing the osmotic load of the cell, the neurotransmitter transporters also readily move water (MacAulay et al., 2004; Zeuthen & MacAulay, 2002). E.g., the astrocytic glutamate transporter EAAT1/GLAST has the water permeability of roughly one-tenth of that of the water channel AQP1 (MacAulay, Gether, Klaerke, & Zeuthen, 2001).

5.3 K^+ Buffering and Donnan Swelling

Another well-recognized function of astroglial cells is the buffering of extracellular K^+, which is accomplished via the activity of K^+ channels, the Na^+,K^+-pump and the NKCC1 cotransporter, and further facilitated by redistributing the intracellular K^+ throughout the astrocytic syncytium (Kofuji & Newman, 2004; MacAulay & Zeuthen, 2012). Their high K^+ uptake capacity and negative membrane potential make astrocytes prone to K^+-induced swelling (Fig. 3C). Moderate elevations in $[K^+]_o$ during physiological excitation are thought to be the key reason for the dynamic changes in astrocyte volume in healthy brain tissue [Section 4, but see alternative view in (Larsen & MacAulay, 2017)]. Work in astrocyte cultures suggests that supraphysiologic increases in $[K^+]_o$ (up to 10–15 mM) trigger cell swelling via stimulation of the bumetanide-sensitive NKCC1 (Larsen et al., 2014). The same mechanism has been identified for the activity-dependent astrocyte swelling in the optic nerve (MacVicar, Feighan, Brown, & Ransom, 2002). However, in hippocampal slices, bumetanide does not prevent the activity-dependent shrinkage of the interstitial space (and by extension, astrocytic swelling), suggesting that other mechanisms may dominate this process in situ and in vivo (Larsen et al., 2014).

In pathologies such as cerebral hypoxia and ischemia, extracellular K^+ levels can rise as high as 80 mM (Hansen, 1985). This leads to robust astrocytic swelling via the Donnan mechanism: membrane depolarization creates an inward driving force for Cl^-, which in turn sustains the electromotive force for the additional influx of Na^+ and K^+ and continued cell swelling (Mongin, Cai, & Kimelberg, 1999; Rutledge & Kimelberg, 1996). In astrocyte cultures, high K^+-induced cell swelling depends on the extracellular $[Cl^-]$, and is completely blocked by Cl^- channel blockers and strongly reduced by $[Ca^{2+}]_i$ buffering, perhaps because $[Ca^{2+}]_i$ facilitates K^+ accumulation via Ca^{2+}-dependent K^+ channels or other Ca^{2+}-dependent pathways (Mongin et al., 1999; Rutledge, Aschner, & Kimelberg, 1998). There is experimental evidence that the NKCC1-dependent mechanism appears as equally important as the role of Donnan swelling in promoting astrocyte swelling in situ and in vivo (see Section 5.6).

5.4 Ammonia Uptake and Metabolism

Acute and chronic liver diseases frequently lead to increases in the systemic levels of ammonia, which at physiological pH exists in a balance between the charged and uncharged forms ($NH_4^+ \leftrightarrow NH_3$). NH_3 freely crosses the blood–brain barrier, and NH_4^+/NH_3 buildup in the brain tissue results in the development of hepatic encephalopathy (see Section 6.3). Astrocytic swelling is a major feature of this disease and its degree highly correlates with blood ammonia levels (Norenberg, 1977, 1998; Swain, Butterworth, & Blei, 1992). Uncharged NH_3 enters the cell via passive diffusion through the cell membrane, in a manner dependent on the intracellular pH (Antonenko, Pohl, & Denisov, 1997). In contrast, the charged NH_4^+ enters the cell through NKCC1 (taken up at the K^+-binding site) and via K^+ channels (Allert, Koller, & Siebler, 1998; Kinne, Kinne-Saffran, Schutz, & Scholermann, 1986; Marcaggi & Coles, 2001; Nagaraja & Brookes, 1998) (see Fig. 3D). Ammonia influx through K^+ channels causes astrocytic membrane depolarization and leads to Donnan swelling (Allert et al., 1998). Extensive *in vitro* and *in vivo* studies, the majority of which have been done in the laboratory of M.D. Norenberg, have shown that elevated levels of ammonia cause persistent astrocytic swelling (Norenberg, 1998; Norenberg et al., 1991). The mechanisms of such swelling are multifaceted and complex. However, it is known that they largely depend on ammonia assimilation by the astrocytic enzyme glutamine synthetase. For further information on ammonia transport, metabolism, and their relationship to cell swelling, see (Marcaggi & Coles, 2001; Norenberg, 1998) and Section 6.3.

5.5 SUR1/TRPM4 Non-selective Cation Channels

A series of studies performed by J.M. Simard and colleagues identified an additional mechanism for astrocytic swelling, which may be selectively relevant to pathological states. Biophysical studies in reactive astrocytes revealed a nonselective cation permeability ($NC_{Ca\text{-}ATP}$), which was activated in metabolically inhibited cells and caused their swelling and blebbing (Chen & Simard, 2001). The $NC_{Ca\text{-}ATP}$ channel contains the sulfonylurea receptor SUR1 that is typically coupled with Kir6.x proteins to form K_{ATP} channels (Chen, Dong, & Simard, 2003). However, in the case of $NC_{Ca\text{-}ATP}$ the non-selective cation permeability is mediated by the transient receptor potential melastatin 4 (TRPM4) channel (Woo, Kwon, Ivanov, Gerzanich, & Simard, 2013). In an animal stroke model, SUR1 expression steadily increased up to 6 h after the initiation of ischemia, and treatment with the sulfonylurea inhibitor glibenclamide reduced brain edema and tissue damage (Simard et al., 2006). Their most recent work demonstrated that SUR1-TRPM4 can form a tripartite complex with AQP4, and such association mediates the pathological, high-capacity influx of water that is ~7.6-fold higher than that stimulated by TRPM4 alone, and ~3.2-fold higher than that induced by the SUR1-TRPM4 assembly (Stokum et al., 2018) (Fig. 3E). TRPM4 knockout mice displayed reduced astrocytic cell swelling following cold-injury (Stokum et al., 2018), but additional studies are needed to determine if this heteromeric complex is critical for cell swelling in other pathologies.

5.6 NKCC1 Activity

As mentioned in Section 1.2, NKCC1 is one of the primary shrinkage-activated transporters, which contribute to RVI in a variety of cell types (Russell, 2000). In astrocyte cultures, pharmacological inhibition or genetic deletion of NKCC1 prevents the high $[K^+]_o$-induced astrocytic swelling and swelling-activated glutamate release (Su, Kintner, & Sun, 2002; Su, Kintner, Flagella, Shull, & Sun, 2002). This is paradoxical because NKCC1 is inhibited by cell swelling. This transporter is regulated by a protein kinase cascade, which incorporates the With No Lysine (K) (WNK) protein kinases WNK1-4 and the two homologous downstream protein kinases, Sterile20-related Proline-Alanine-rich Kinase (SPAK) and Oxidative Stress Responsive kinase 1 (OSR1) (Alessi et al., 2014; Gagnon & Delpire, 2012; Kahle, Ring, & Lifton, 2008). Typically, the activity of the WNK/SPAK/OSR is augmented by cell shrinkage and inhibited by cell swelling with matching effects on NKCC1 (Hoffmann et al., 2009; Russell, 2000). However, astrocytes represent an interesting exception: in these cells, swelling activates NKCC1 nearly as potently as cell shrinkage (Mongin et al., 1994, 1996). One potential mechanism is the activation of WNK kinases by low intracellular Cl^-. Astrocytes are high-Cl^- cells, in which swelling and subsequent RVD likely reduce $[Cl^-]_i$ and relieve the known inhibition of WNK kinases by this anion (Kahle et al., 2008a; Pacheco-Alvarez & Gamba, 2011; Piala et al., 2014; Terker et al., 2016). The end result of this unusual regulation would be an amplification of cell swelling due to additional osmolyte intake via NKCC1 (Fig. 3F).

5.7 Na^+-Bicarbonate Cotransporters and Other Emerging Mechanisms

Recent studies from the N. MacAulay laboratory (Larsen & MacAulay, 2017; Larsen et al., 2014), proposed alternative mechanisms of astrocytic swelling in hippocampal slices. They found that electrical stimulation of neuronal networks causes increases in intracellular $[K^+]$ and shrinkage of the interstitial space, reflecting cellular swelling. In brain slices, cell swelling was largely insensitive to the inhibitors of glutamate transporters (TBOAs), K^+ channels (Ba^{2+}), and NKKC1 (bumetanide). Instead, it was lessened by the non-selective inhibitor of the Na^+-bicarbonate cotransporter NBCe1, DIDS, and poorly selective monocarboxylate transporter blocker, α-cyano-4-hydroxycinnamate (Larsen & MacAulay, 2017), suggesting the contribution of these two latter transport mechanisms. Although this work remains to be validated using more selective molecular biology and genetics tools, the underlying ideas are supported by the literature. Thus, S.J. Mulligan and co-workers found that the K^+-induced astrocytic swelling in hippocampal slices is largely dependent on the presence of bicarbonate in the media, and is strongly reduced by blockers of NBCe1 and $GABA_A$ receptors (Florence, Baillie, & Mulligan, 2012). To what extent these additional mechanisms can contribute to pathological swelling of astroglia remains to be tested.

6. FAILURE OF CELL VOLUME CONTROL IN BRAIN PATHOLOGIES

In this section, we discuss several examples of neuropathologies which are associated with defective cell volume control (persistent cell swelling), as well as the potential relevance of cell volume changes to disease outcomes.

6.1 Hyponatremia

Hyponatremia is the most prevalent electrolyte disorder, which manifests as a drop in serum Na^+ levels below the threshold of ~135 mM (Adrogue & Madias, 2000). On average, this condition affects 1.72% of the U.S. population, but hospitalized patients, elderly people, and individuals with neurological disorders are disproportionally affected (Mohan, Gu, Parikh, & Radhakrishnan, 2013; Upadhyay, Jaber, & Madias, 2006). Most often, hyponatremia develops as a result of impaired capacity of the kidney to excrete water. It could be caused by underlying kidney pathologies, severe hormonal imbalance, or due to the syndrome of inappropriate antidiuretic hormone secretion (SIADH). SIADH is very common and in hospital settings; it can be triggered by anesthesia and several commonly prescribed medications or may develop in an idiogenic fashion. In neurological patients and endurance athletes, hyponatremia can also be caused by excessive water intake (Adrogue & Madias, 2000; Fraser & Arieff, 1997).

Although all tissues are affected, the brain represents the target organ of this disorder and is responsible for all major symptoms. Acute hyponatremia is typically associated with headaches, nausea, fatigue, confusion and hallucinations. In its most severe form, acute hyponatremia progresses to seizures and coma and puts patients at significant risk for brain damage and death. The latter outcomes are driven by expansion of the brain tissue and compression of blood vessels, particularly in the brainstem, leading to the life-threatening dysregulation of blood pressure, heart rate, thermal regulation and respiratory control (Adrogue & Madias, 2000; Fraser & Arieff, 1997; Podesta et al., 2015). Cell swelling in hyponatremia is the closest analog of the model osmotic cell swelling utilized in cell cultures and other model studies. As already mentioned, in the hyponatremic brain swelling is mainly seen in astrocytes, particularly in the astrocytic processes surrounding blood vessels [e.g., (Manley et al., 2000; Risher, Andrew, & Kirov, 2009; Wasterlain & Torack, 1968)]. There is conflicting information on neuronal swelling, with some brain slice and *in vivo* studies showing no swelling of neuronal somata or processes (Andrew et al., 2007; Steffensen et al., 2015), while others reporting either pronounced or blunted hypoosmotic swelling in the areas containing dendritic arbors and/or neuronal cell bodes (Andrew & MacVicar, 1994; Andrew et al., 1997; Murphy et al., 2017).

Irrespective of which cell types are swollen (or not), rapid adaptation of the brain to hyponatremia is mediated by the loss of inorganic ions, Na^+, K^+, and Cl^-, which move from the cytosol to the interstitial space to the circulatory system (Gullans & Verbalis, 1993; Melton et al., 1987). The loss of K^+ is the main driving mechanism of RVD but, when it occurs in the brain within the constraints of a small extracellular space, it leads to rapid dissipation of K^+ gradients, membrane depolarization, and neuronal excitation that can precipitate seizure development (Andrew, 1991; Andrew, Fagan, Ballyk, & Rosen, 1989; Somjen, 2004). Swollen cells also release a variety of small organic molecules — including glutamate, aspartate, *myo*-inositol, and the amino sulfonic acid taurine — and these can be measured in the extracellular space using a microdialysis approach (Haskew-Layton et al., 2008; Lehmann, 1989; Verbalis & Gullans, 1991;

Wade, Olson, Samson, Nelson, & Pazdernik, 1988). On a short time-scale, organic osmolytes are released predominantly via VRAC, but the long-term adaptation to hyponatremia likely involves additional processes (Estevez, O'Regan, Song, & Phillis, 1999; Haskew-Layton et al., 2008; Hyzinski-Garcia et al., 2011). It has been estimated that the loss of organic osmolytes is responsible for >50% of the regulatory loss of osmotically obligated water during compensatory changes in hyponatremia (Sterns et al., 1993). Yet, because aspartate and glutamate are the excitatory neurotransmitters, their release is not benign as it causes activation of neuronal glutamate receptors, with the NMDA subtype likely being the main target (Lauderdale et al., 2015; Mongin, 2016). Furthermore, activation of VRAC in astrocytes severely impairs extracellular glutamate scavenging because it interrupts its removal and metabolic conversion into the non-toxic glutamine (Haskew-Layton et al., 2008; Hyzinski-Garcia et al., 2011).

One interesting aspect of brain adaptation to hyponatremia is the preferential loss of one organic osmolyte, taurine. This phenomenon has been reported *in vitro* and *in vivo* and interpreted in the context of the privileged role of taurine in brain cell volume control and with the implied existence of taurine-selective volume regulatory pathways (Lehmann, 1989; Moran, Maar, & Pasantes-Morales, 1994; Pasantes Morales & Schousboe, 1988; Pasantes-Morales, Lezama, Ramos-Mandujano, & Tuz, 2006; Verbalis & Gullans, 1991; Wade et al., 1988). We have recently tested this hypothesis and found that the differences between the osmotic behavior of taurine and excitatory amino acids are determined by the rates of their reuptake and synthesis. Once glutamate uptake and de novo synthesis via transamination are pharmacologically blocked, astrocytes release cytosolic taurine and aspartate at very similar rates (Schober & Mongin, 2015). This suggests that the preservation of cytosolic glutamate and aspartate during RVD are homeostatic mechanisms related to their numerous roles in metabolism and brain signaling.

In contrast to the acute disease, chronic hyponatremia, which develops and/or persists for more than 48 h, is typically "asymptomatic". This happens because the brain tissue slowly adapts to new osmotic conditions via the already mentioned loss of inorganic and particularly organic osmoles. Despite an apparent lack of symptoms, chronic hyponatremia still requires correction because it is strongly associated with a host of mild neurological deficits and increases morbidity and mortality of co-existing clinical conditions (Adrogue & Madias, 2000; Podesta et al., 2015). However, its treatment often comes with a separate set of dangers. Rapid correction of systemic and Na^+ levels leads to osmotic demyelination syndrome, a demyelinating pathology that most commonly affects the central basis of pontis (pontine myelinosis) but also the nearby regions of thalamus and subcortical areas (Adrogue & Madias, 2000; Podesta et al., 2015; Sterns & Silver, 2006). To avoid demyelination, correction of chronic hyponatremia is performed by introducing additional electrolytes at a very slow rate. The etiology of demyelination is likely related to lowered levels of organic osmolytes, particularly *myo*-inositol, which play incompletely understood protective roles in the preservation of myelin integrity. In one experimental study, the intravenous supplementation of *myo*-inositol dramatically decreased mortality rates and preserved myelin structures in rats subjected to rapid correction of chronic hyponatremia (Silver, Schroeder, Sterns, & Rojiani, 2006; Sterns & Silver, 2006).

6.2 Epilepsy

Epilepsy is the fourth most common chronic neurologic disorder, which affects more than 2 million individuals in the U.S. and 65 million people worldwide (Hirtz et al., 2007; Thurman et al., 2011). The term epilepsy encompasses a constellation of diseases with somewhat different etiologies, all of which manifest as recurrent, unprovoked seizures which are usually initiated in the cortical areas or the hippocampus and then spread throughout the brain (McNamara, 1994; McNamara, Huang, & Leonard, 2006). While the main focus of epilepsy research is on changes in neuronal cells, emerging studies also implicate the complex interactions between neurons and various classes of glia, with the potentially strong involvement of changes in glial gene expression, glutamate metabolism, cellular morphology and cell volume (David et al., 2009; Eid, Williamson, Lee, Petroff, & de Lanerolle, 2008; McNamara et al., 2006; Murphy, Binder, & Fiacco, 2017). As mentioned in the prior section, a hypoosmotic challenge on its own can cause brain hyperexcitability and seizures, strongly suggesting that cell volume is a highly relevant factor.

The experimental evidence for cellular swelling in epilepsy is derived from three types of findings: (1) electron microscopy studies in fixed brain tissue after chemically-induced status epilepticus, (2) intravital magnetic resonance measurements of water mobility, and (3) intravital or *ex vivo* monitoring of volume dynamics in the extracellular space. In the study by (Fabene et al., 2006), the first two parameters were compared side-by-side. As reproduced in Fig. 4, rats with status epilepticus develop profound swelling of astrocytic cell bodies and processes, and additionally show

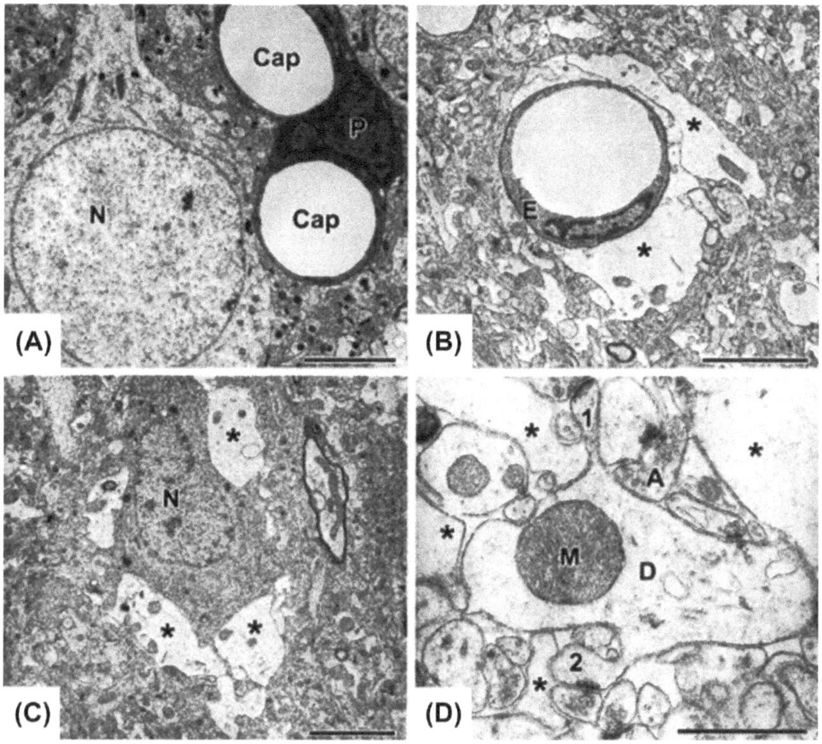

Figure 4 Cell swelling in status epilepticus revealed by electron microscopy (EM) (A) Structural analysis of the control rat neocortex reveals neuronal cell bodies [N], capillaries [Cap], and pericytes [P] (B and C) Two hours following injection of 4-amino-pyridine, EM shows swollen perivascular astrocytic endfeet [asterisks in (B)], which surround an endothelial cell [E]. Swollen astrocytic processes [asterisks in (C)] also border an apparently shrunken pyramidal neuronal body [N] (D) A higher magnification image in the CA3 layer of the hippocampus displays a swollen dendrite [D] with apparently normal mitochondrion [M], and adjacent electron-transparent swollen astrocytic processes [asterisks]. Additionally labeled are: an axon terminal [A] and dendritic spines [1,2]. The scale bars in the fields (A—C) = 5 μm, and in (D) = 1 μm. *Reproduced from Fabene, P. F., Weiczner, R., Marzola, P., Nicolato, E., Calderan, L., Andrioli, A. et al. (2006). Structural and functional MRI following 4-aminopyridine-induced seizures: A comparative imaging and anatomical study.* Neurobiology of Disease, 21, *80—89, with permission.*

swelling of dendritic spines. Interestingly, neuronal cell bodies appear to maintain their volume, or even moderately shrink with an apparent condensation of cytoplasm (Fabene et al., 2006). The same study correlated the phenomenon of cellular swelling with MRI-registered changes in water mobility. Increased signal of diffusion-weighted imaging (DWI) and the inversely related apparent diffusion coefficient (ADC) of water show a consistent and long lasting (for at least 24 h) decrease in water mobility in the epileptogenic brain areas (Fabene et al., 2006). Similar in nature and amplitude DWI and ADC changes are reliably associated with the movement of water from the extracellular to the intracellular space (Sotak, 2004). Other publications demonstrated that the development of seizure activity is preceded by shrinkage of the extracellular space, signifying increases in cellular volume [e.g., (Binder, Papadopoulos, Haggie, & Verkman, 2004; Slais et al., 2008)]. In brain slice studies, application of hyperosmotic media ameliorates the K^+-induced electrographic seizures; and, conversely, introduction of mild hypoosmolar media dramatically increases epileptogenic activity, again suggesting that cell volume and/or volume of extracellular space are highly relevant (Traynelis & Dingledine, 1989). Likely causes for hyperexcitability and cellular swelling are discussed below.

The core reason for neuronal dysfunction in epilepsy is the modification of cellular excitability, which is complex in its nature. Still, the initial triggers of disease are ionic in nature. Excessive neuronal activity leads to the accumulation of K^+ and neurotransmitter molecules in the extracellular space. High $[K^+]_o$ causes partial membrane depolarization, increases neuronal excitation, and removes the Mg^{2+} block of the glutamate NMDA receptors, which are all strongly associated with epileptogenic activities (McNamara, 1994). Additionally, K^+-induced depolarization alters the electrochemical driving force for Cl^- in neurons and dampens the activity of the inhibitory $GABA_A$ and glycine receptors (McNamara, 1994). The importance of changes in Cl^- gradients is strongly supported by studies of the functional roles for electroneutral KCC transporters. In neuronal cells, these transporters set low intracellular Cl^- levels, and in such a way allow for the inhibitory actions of GABA and glycine (Kahle et al., 2008b; Kaila, Price, Payne, Puskarjov, & Voipio, 2014). Deletion or downregulation of the activities of KCC2 and KCC3 result in neuronal hyperexcitability in humans and model organisms (Huberfeld et al., 2007; Kahle et al., 2014; Kaila et al., 2014; Woo et al., 2002). KCC2b-null mice develop spontaneous seizures and die shortly after birth (Woo et al., 2002), whereas KCC3-null phenotypes are more diverse. In humans, KCC3 mutations cause mental retardation, severe peripheral neuropathy, and loss of the corpus callosum (Howard et al., 2002). In mice, the loss of KCC3 has been alternatively associated with defects in peripheral myelination and neurological deficits, or alternatively, hyperexcitability, reduced seizure threshold, and deafness (Boettger et al., 2003; Howard et al., 2002). The KCC3 phenotypes have been interpreted in the context of defective ability of neurons to regulate their volume

(Kahle et al., 2015), but this remains to be proven. Many of the mechanisms listed in Section 5 can contribute to astrocytic swelling in epilepsy. These may include neurotransmitter uptake, the $[K^+]_o$-induced Donnan swelling, and anomalous NKCC1 activity [e.g., (Bender et al., 1998; Hubel & Ullah, 2016; Palma et al., 2006)].

An additional interesting topic is the link between the astrocytic water channel AQP4 and seizure susceptibility. Several studies have identified changes in the volume of the extracellular compartment and its dynamics during hyperexcitability, and found modified seizure thresholds in AQP4 knockout animals (Binder, Oshio, Ma, Verkman, & Manley, 2004; Binder et al., 2006; Lee et al., 2012). These data have been interpreted in the context of reduced water fluxes and/or defective cell volume homeostasis in AQP4-null mice (Binder, Nagelhus, & Ottersen, 2012; Murphy et al., 2017; Verkman, Binder, Bloch, Auguste, & Papadopoulos, 2006). However, even without AQP4, astrocytes retain fairly high water permeability (Solenov et al., 2004). Besides direct effect on water fluxes, the observed changes in AQP4-null animals can be explained by other factors, such as adaptive changes in the expression of other genes or the indirect effects of AQP4 on activities of ion transporters and channels. This protein has been shown to physically associate with or functionally regulate activities of the K^+ channel Kir4.1, volume-sensitive VRAC, and nonselective cation channels TRPV4 and TRPM4/SUR1 [(Benfenati et al., 2007, 2011; Nagelhus, Mathiisen, & Ottersen, 2004; Stokum et al., 2018), but see an alternative opinion for Kir4.1 (Zhang & Verkman, 2008)]. Therefore, effects of AQP4 deletion may develop independent of or be upstream of cell volume changes [see (Assentoft, Larsen, & MacAulay, 2015) for detailed discussion].

6.3 Hepatic Encephalopathy

Another syndrome that disproportionally affects the CNS and is strongly linked to astrocytic swelling is hyperammonemia — systemic increases in ammonia levels above the approximate threshold of $\sim$100 μM, which lead to neurologic deficits described as hyperammonemia or hepatic encephalopathy (Blei & Larsen, 1999; Brusilow, Koehler, Traystman, & Cooper, 2010; Haussinger & Schliess, 2008; Norenberg, Rao, & Jayakumar, 2005). In hyperammonemias, arterial ammonia levels frequently reach 1 mM, and can be as high as 5 mM, leading to severe brain edema and high mortality rates due to herniation of the brainstem (Blei & Larsen, 1999; Brusilow et al., 2010; Norenberg et al., 2005). Most frequently, hyperammonemia develops due to acute liver failure or chronic liver disease. Depending on the clinical population, 20% to 80% of patients with liver cirrhosis suffer from moderate to severe hepatic encephalopathy. However, other conditions — such as inherited deficiencies of the urea cycle, certain bacterial infections of the gut, and Reye's syndrome — may also lead to the same type of neuropathology (Brusilow et al., 2010; Felipo & Butterworth, 2002).

The main mechanism of the hepatic encephalopathy is thought to be osmolyte accumulation in the brain due to fixation of ammonia by the astrocytic enzyme glutamine synthetase (Brusilow et al., 2010; Norenberg et al., 2005). Glutamine synthetase produces glutamine from ammonia and glutamate in an ATP-dependent fashion, as a part of its normal physiological activity in the brain, the recycling of excitatory neurotransmitters (McKenna, 2007; Sofroniew & Vinters, 2010; Verkhratsky & Nedergaard, 2018). Because ammonia circulates systemically and freely diffuses into the brain, in hyperammonemia this process serves as an unlimited osmotic sink as long as the brain has energy reserves. The mechanisms of ammonia transport across cell membranes are summarized in Section 5 and Fig. 3D. The osmotic burden due to the activity of glutamine synthetase is further accentuated by the accumulation of brain alanine produced via the work of the alternative detoxification mechanism incorporating glutamate dehydrogenase and alanine aminotransferase (Dadsetan et al., 2013; Tofteng et al., 2006). In experimental hyperammonemias, treatments with the glutamine synthetase inhibitor methionine sulfoximine reduced the amount of brain edema and cell swelling, supporting the causal link of glutamine synthetase to disease development (Blei, Olafsson, Therrien, & Butterworth, 1994; Tanigami et al., 2005).

Because glutamine and alanine synthesis occur in astrocytes, it is not surprising that the main cellular pathology is astrocytic. Human pathological studies show the appearance of swollen astrocytes in acute disease or so-called Alzheimer type II astrocytes in chronic pathology, with the latter characterized by enlarged cell somata and nuclei with significant nucleolus. The frequency of observed pathological cells correlates with the severity of encephalopathy (Norenberg, 1998). The mechanisms of astrocytic swelling have been thoroughly explored in primary astrocyte cultures with main contributions coming from the laboratory of M.D. Norenberg. Surprisingly, these cell culture studies indicate that the osmotic effect of glutamine accumulation is not the leading reason for cell swelling. Instead, they propose a variety of alternative mechanisms, including membrane depolarization via ammonia uptake through K^+ channels followed by Donnan swelling (Allert et al., 1998), osmotic load due to NKCC1 activity (Jayakumar, Valdes, & Norenberg, 2011), and secondary swelling due to opening of mitochondrial permeability transition pore (Bai et al., 2001) followed by production of reactive oxygen and nitrogen species (Murthy, Rama Rao, Bai, & Norenberg, 2001). To reconcile the conflicting results on the role of astrocytic glutamine in cellular swelling, J. Albrecht and M.D. Norenberg proposed the "Trojan horse" hypothesis, in which the glutamine synthetase-derived glutamine causes astrocytic swelling via its secondary metabolic effects within the cells, particularly in mitochondria (Albrecht & Norenberg, 2006; Norenberg, Jayakumar, Rama Rao, & Panickar, 2007).

6.4 Stroke

Stroke is the fifth most prevalent source of mortality in the U.S. and the second most common cause of death worldwide, with 6.5 million deaths reported globally in 2013 (Feigin, Norrving, & Mensah, 2017). A stroke occurs due to the interruption of blood flow to an area of the brain. Because brain metabolism is almost exclusively dependent on oxidation of blood-derived glucose, reductions in cerebral blood flow below a threshold of ~50% of their normal levels cause a rapid decline in tissue metabolism, and anything below 20% of normal blood levels is followed by severe tissue damage (Brott & Bogousslavsky, 2000; Dirnagl, Iadecola, & Moskowitz, 1999; Lo, Dalkara, & Moskowitz, 2003). The majority of strokes (>85%) are ischemic and involve embolic or thrombotic occlusion of a larger blood vessel. A smaller (<15%), but deadlier fraction of strokes are hemorrhagic and result from bleeding through a rupture in the parenchyma or subarachnoid space. The mechanisms of tissue damage are not identical between the two types of stroke because of the nature of the initial insult, but they all share characteristics of ionic dysfunction and cell swelling. In this section we focus on ischemic stroke.

There is a significant body of work demonstrating astrocytic cell swelling in ischemic stroke, through (1) electron microscopy in animal stroke models, (2) measurements of brain interstitial volume in ischemia, (3) two-photon microscopy imaging of cell swelling through a cranial window, and (4) diffusion and perfusion weighted MRI imaging, which are widely utilized in clinical settings. The most commonly used animal model of stroke is focal ischemia induced by transient or permanent occlusion of the middle cerebral artery (MCAo) (Longa, Weinstein, Carlson, & Cummins, 1989).

EM analysis reveals astrocytic swelling in fixed MCAo tissue after only 30 min of ischemia and such swelling persists up to 24 h after reperfusion (Garcia et al., 1993). Under physiological conditions, the extracellular space makes up ~20% of the brain volume (Thrane et al., 2014). Early methods of measuring interstitial space in the brain used radiotracer diffusion studies or the ion-selective microelectrodes detecting the extracellular marker tetramethylammonium (TMA^+), and found that during severe ischemia the volume fraction drops to as low as 5% [reviewed in (Sykova & Nicholson, 2008)]. More recently, studies using 2-photon microscopy allowed for the direct observation of cell swelling in both animal slices and *in vivo*. The laboratory of S.A. Kirov found that the cross-sectional area of astrocytes in hippocampal slices increased by ~40% upon 3-min depolarization with high extracellular [K^+], and by ~30% when slices were exposed to oxygen/glucose deprivation (Risher et al., 2009). The same study used a cranial window and 2-photon microscopy to demonstrate that global ischemia, which was induced by cardiac arrest, prompted very rapid swelling of astrocytic soma by ~30% (Risher et al., 2009). Another publication, which used similar methodology for measuring astrocytic swelling in the ischemic penumbra of the photothrombotic stroke model, found the degree of astrocytic swelling to be in the range of 20%–35% (Risher, Croom, & Kirov, 2012).

In clinical settings, tissue edema is visualized using the MRI-derived DWI or ADC signals. The early hyperintensity in MRI signal or matching hypointensity of the ADC signal serve as a reliable representation of the ischemic area and are fairly predictive of the final infarction volume (Kucinski et al., 2002; Neumann-Haefelin et al., 1999; Thijs, Adami, Neumann-Haefelin, Moseley, & Albers, 2002). Acute ischemia-induced changes in DWI and ADC are shown in Fig. 5. The human MRI data have been extensively validated by crosschecking DWI imaging with histological outcomes in animal stroke models [e.g., (Neumann-Haefelin et al., 2000)]. The time-dependent development of DWI signal in human stroke correlates well with the evolution of infarction, suggesting that cellular edema contributes to the progression of tissue damage (Battey et al., 2014).

The reasons for ischemic swelling of astroglial cells are likely plural and may vary significantly depending on the proximity to the ischemic core. In the ischemic core, cessation of the blood flow and resulting lack of oxygen and glucose lead to the rapid (within 5 min) and complete depletion of energy reserves, inhibition of the Na^+,K^+-pump, and collapse of ionic

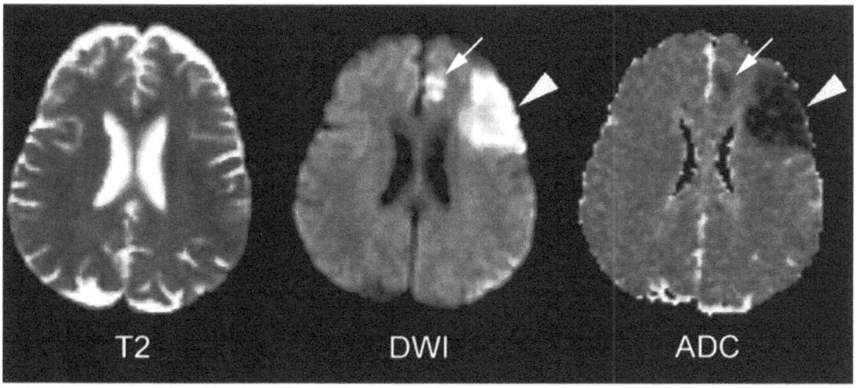

Figure 5 MRI imaging modes which are used in clinical settings to visualize changes in water mobility (cell swelling) following ischemic stroke. *Left panel*: At early times after the onset of focal ischemic stroke, fast spin-echo T2-weighted MRI imaging has yet to show brain damage. *Center panel*: In contrast to the T2 imaging, diffusion weighted imaging (DWI) shows a large hyperintense area in the territories of the prefrontal artery (arrowhead) and a smaller signal in the territory of the anterior cerebral artery (arrow). Hyperintensity reflects a decrease in water mobility upon its shift from the extracellular to the intracellular space. *Right panel*: Apparent diffusion coefficient (ADC) of water identifies the regions of low water mobility as hypointense signals. In 2–3 days, the DWI/ADC positive areas will likely develop full infarction that will be apparent in T2 imaging (not shown). *Reproduced from Lansberg M.G., Tong D.C., Norbash A.M., Yenari M.A., & Moseley M.E. (1999). Intra-arterial rtPA treatment of stroke assessed by diffusion- and perfusion-weighted MRI, Stroke 30, 678–680, with permission.*

gradients (Dirnagl et al., 1999). A dramatic rise in the intracellular [Na^+] and [Ca^{2+}] and extracellular [K^+] lead to Donnan cell swelling, which for poorly understood reasons is largely restricted to astrocytes. Astrocytes may retain their membrane potentials, and consequently the electrochemical driving force for ion accumulation, for a longer time than neurons. Furthermore, they are linked into syncytium by gap junctions, and at least in the beginning of an ischemic episode, work as one unified ion sink. Neurons also swell, but their swelling is largely limited to beading in the dendritic arbors, where a high density of ionotropic receptor-channels and water-permeable electroneutral transporters (NKKC1, KCC2, MCT, etc.) facilitate rapid osmotic loading (Steffensen et al., 2015). In contrast, neuronal cell bodies in the ischemic core typically retain their initial volume or even appear to shrink, perhaps due to the rapid redistribution of osmolytes into astrocytic compartments (Garcia et al., 1993).

In the ischemic penumbra, tissue ATP levels are partially preserved, or even elevated in the areas distal to occlusion. Nevertheless, the penumbral tissue is electrically silent as a result of presynaptic failure (Bolay & Dalkara, 1998; Bolay et al., 2002; Neumann-Haefelin & Witte, 2000). This is explained by the difficulties in sustaining metabolism in remote axonal processes and oxidative/nitrosative modification in presynaptic processes [reviewed in (Mongin, Dohare, & Jourd'heuil, 2012)]. Astrocytic swelling in the penumbra likely involves numerous mechanisms, including but not limited to osmolyte fluxes via the Na^+-dependent transporters (GLT-1, GLAST, and others). Moreover, astrocytic swelling is promoted by NKCC1 activity. Treatment with the NKCC1 inhibitor bumetanide given prior to the induction of animal ischemia reduces the infarction volume and tissue edema in a MCAo model of stroke (Yan, Dempsey, & Sun, 2001). In astrocyte cultures, bumetanide and genetic deletion of NKCC1 make astrocytes resilient to K^+-induced cell swelling and obliterates the K^+-swelling-induced release of excitatory amino acids (Su et al., 2002; Su et al., 2002). The major physiological signal for NKCC1 activation is phosphorylation by the WNK signaling cascade [see Section 5.6 and reviews (Kahle et al., 2008a; McCormick & Ellison, 2011)]. This cascade is typically inhibited by cell swelling, rendering NKCC1 silent. Yet, paradoxically, astrocytes and glioma C6 cells respond to hypoosmotic cell swelling with an increase in NKCC1 activity (Mongin et al., 1994, 1996). As already mentioned, one possible explanation is that cell swelling and RVD reduce the intracellular Cl^- levels and relieve Cl^- inhibition of protein kinases belonging to the WNK family (Kahle et al., 2008a; Pacheco-Alvarez & Gamba, 2011; Piala et al., 2014; Terker et al., 2016). Genetic deletion of the brain-enriched WNK3 reduces cerebral edema and tissue injury following MCAo stroke (Begum et al., 2015).

Another important cause for ischemic cell swelling is the pathological opening of ATP-sensitive nonselective cation channels. One type, which is assembled from the TRP channels belonging to the melastatin family (TRPM4) and the sulfonylurea receptor-1 (SUR1) appears to dominate brain edema and stroke outcomes (Chen & Simard, 2001; Simard et al., 2006, 2010). The SUR1 subunit expression is upregulated during ischemic hypoxia by HIF1-α activation, and in an animal model of stroke can be detected in neurons, astrocytes, and endothelial cells (Simard et al., 2006). Treatment with the sulfonylurea glibenclamide inhibits these channels, and has been shown to reduce cerebral edema and stroke volume by as much as 50% and increase viability in the malignant MCAo model of stroke (Simard et al., 2006, 2010). Recently, it has been shown that in astrocytic SUR1-TRPM4 forms a tripartite complex with AQP4, and in such a way significantly increases the rate of water influx into cells exposed to hypoosmotic media (Stokum et al., 2018) (see Section 5.5). Importantly, sulfonylureas, the SUR1 antagonists, protect not only animals but also humans and are now in clinical trials for treatment of ischemic stroke (Kunte et al., 2007; Sheth et al., 2016).

7. VRAC AS A VILLAIN IN NEUROPATHOLOGIES

Persistent cell swelling in neuropathologies disrupts normal neural functions via numerous mechanisms, including volume-dependent changes in ionic gradients and cell excitability. However, there are strong reasons to believe that among many relevant processes, opening of VRAC is the key to the pathological tissue damage. The uncontrolled release of glutamate and aspartate activates highly abundant receptors on neurons and all classes of glial cells, including the Na^+/K^+ permeable ionotropic AMPA and kainate receptors, the $Na^+/K^+/Ca^{2+}$-permeable NMDA receptors, and a variety of metabotropic mGluRs (Meldrum, 2000). Their combined actions cause sustained membrane depolarization and increases in the intracellular $[Ca^{2+}]$, which set in motion multiple neurodegenerative cascades causing immediate or delayed neuronal damage and demise. This process is termed excitotoxicity (Arundine & Tymianski, 2004; Choi, 1988, 1992). In principle, activation of VRAC can impact extracellular excitotoxin levels via at least three interconnected mechanisms: (1) VRAC serves as a conduit for excitatory amino acid release from swollen astrocytes and other cell types; (2) opening of VRAC shunts glutamate to the extracellular space and in such a way renders astrocytic glutamate uptake ineffective; (3) shunting of glutamate to the extracellular space interrupts glutamate processing via the glutamate–glutamine cycle, thus disrupting another major mechanism regulating extracellular glutamate levels (Haskew-Layton et al., 2008; Hyzinski-Garcia et al., 2011). Pathological contributions of VRAC have been most extensively studied in stroke.

In vivo studies in animal stroke models consistently demonstrated dramatic increases in extracellular glutamate levels both in the ischemic core and the penumbra (Benveniste, Drejer, Schousboe, & Diemer, 1984; Phillis & O'Regan, 1996; Phillis, Song, & O'Regan, 1997; Seki, Feustel, Keller, Jr., Tranmer, & Kimelberg, 1999; Feustel, Jin, & Kimelberg, 2004; Dohare et al., 2014). This pathological glutamate can be contributed to by several mechanisms, which are not limited to VRAC. Besides VRAC, the best candidates for release routes include reversal of glutamate transporters, opening of connexin hemichannels, and stimulation of P2X7-receptor associated pannexin hemichannels (Duan et al., 2003; Orellana, Avendano, & Montero, 2014; Rossi, Oshima, & Attwell, 2000; Szatkowski, Barbour, & Attwell, 1990; Ye, Wyeth, Baltan-Tekkok, & Ransom, 2003). It should be noted that not all of these mechanisms have been extensively tested *in vivo*. The relative contribution of VRAC was probed with diverse anion channel blockers, including the most selective among the available VRAC inhibitors, DCPIB. These compounds strongly reduced the intraischemic glutamate levels, clearly pointing to VRAC activity in the pathological brain (Phillis et al., 1997; Phillis, Song, & O'Regan, 1998; Seki, Feustel, Keller, Tranmer, & Kimelberg, 1999; Feustel et al., 2004; Zhang, Zhang, Feustel, & Kimelberg, 2008). However, because the specificity of VRAC blockers, including DCPIB, has been questioned (Benfenati et al., 2009; Bowens, Dohare, Kuo, & Mongin, 2013; Ye, Oberheim, Kettenmann, & Ransom, 2009), further work in transgenic animals and/or with more specific molecular biology tools, is needed to unequivocally link VRAC to glutamate release in stroke.

Its is important to note the fact that VRAC contributions to pathological glutamate levels are likely context-specific. The inhibitory effects of VRAC blockers and the inferred mechanisms of intraischemic glutamate release vary depending on the study. The major determinant appears to be the site of detection, e.g., position of microdialysis probe within ischemic brain. In the ischemic core, the main source of glutamate release is the reversal of the dihydrokainate-sensitive glutamate transporter GLT1 (Seki et al., 1999). In contrast, in the penumbra, glutamate release is dominated by the tamoxifen- and DCPIB-sensitive VRAC (Feustel et al., 2004; Zhang et al., 2008). There are several potential explanations for the paradoxical "silence" of VRAC the ischemic core, where cell swelling is the most pronounced. VRAC opening requires the non-hydrolytic binding of ATP, and therefore its activity and associated glutamate release are abolished in metabolically-inhibited cells (Jackson et al., 1994; Oike, Droogmans, & Nilius, 1994; Rutledge, Mongin, & Kimelberg, 1999). VRAC is also potently inhibited by arachidonic acid and other polyunsaturated fatty acids, which are abundantly produced in the ischemic tissue (Kubo & Okada, 1992; Nilius, Sehrer, & Droogmans, 1994). In contrast, on the periphery of the ischemic tissue, collateral blood flow partially sustains brain cells metabolism, allowing for sustained VRAC activity. It is in the ischemic penumbra that tamoxifen and DCPIB suppress glutamate release by up to 70% (Feustel et al., 2004; Zhang et al., 2008).

Even if we accept the link between VRAC and pathological glutamate release, the potential impact of this process on tissue viability is not certain due to the very complex nature of ischemic brain injury. Due to the role of VRAC in cell volume control, its activity in stroke tissue may beneficial, harmful, or net-neutral. The question about the therapeutic value of this channel has been addressed in animal stroke models by H.K. Kimelberg and co-workers. The systemic administration of the blood–brain barrier-permeable tamoxifen potently protected against brain injury in both the transient and the permanent cerebral ischemia, with maximal protection reaching >80% and an extended therapeutic window up to 3 h after occlusion of MCA (Kimelberg, Jin, Charniga, & Feustel, 2003; Kimelberg et al., 2000). Due to the limited selectivity of tamoxifen, its protective effects have been further verified with the more specific (but not blood–brain barrier-permeable) VRAC blocker DCPIB (Zhang et al., 2008). In Fig. 6A, we present representative data from the latter study illustrating the DCPIB-mediated protection in the ischemic brain. Because DCPIB is not BBB permeable, it required intracerebral injection to achieve neuroprotection. As shown in Fig. 6B, systemic administration of the same compound was ineffective, further suggesting that it acts in the brain parenchyma but not on blood vessels. The protective effects of VRAC blockers (tamoxifen) have been next validated in larger animal species, in a canine model of stroke (Boulos et al., 2011).

Altogether, animal data support the major pathological significance for VRAC in stroke and warrant giving it the title of the "villain" in the ischemic brain. The idea of the therapeutic utility of VRAC in stroke and other neuropathologies has been extensively covered in several relatively recent reviews (Kimelberg, 2005; Mongin, 2007, 2016).

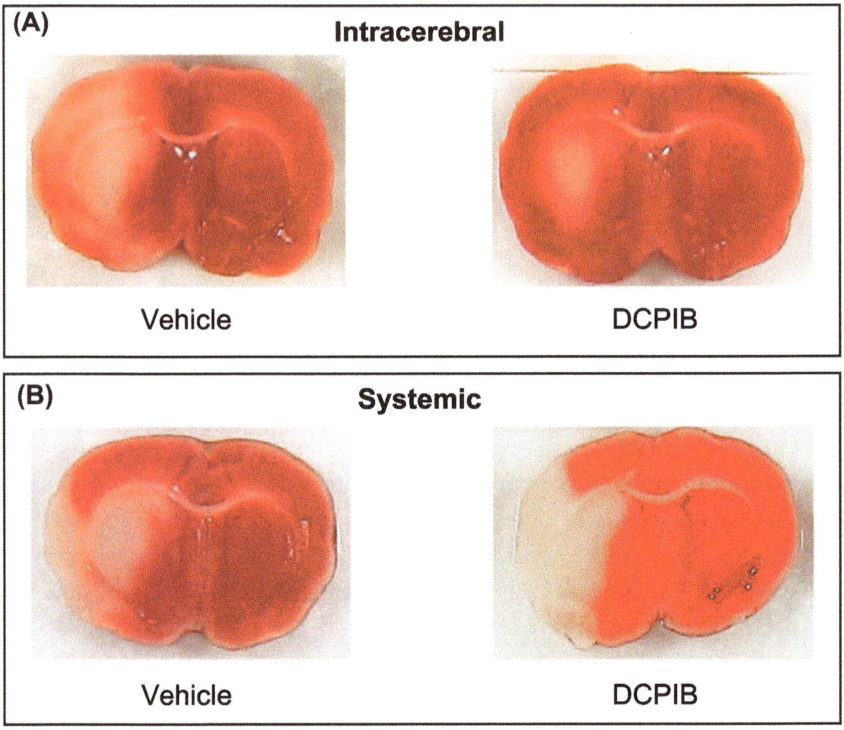

Figure 6 Intracerebral treatment with the VRAC blocker DCPIB potently protects against ischemic brain damage in a rat model of stroke induced by 2-hour occlusion of the middle cerebral artery (MCAo). (A) Representative images of brain sections prepared from MCAo animals after intracerebral injection of vehicle (DMSO) or 20 μg/kg DCPIB. (B) Brain sections from MCAo animal which were treated systemically with intravenous injection of vehicle or 10 mg/kg DCPIB. Animals were euthanized 3 days after ischemia and the extent of brain infarction was revealed by staining brain sections with 2,3,5-triphenyltetrazolium chloride (viable tissue is red). *Modified from Zhang, Y., Zhang, H., Feustel, P. J., & Kimelberg, H. K. (2008). DCPIB, a specific inhibitor of volume regulated anion channels (VRACs), reduces infarct size in MCAo and the release of glutamate in the ischemic cortical penumbra.* Experimental Neurology, 210, *514–520, with permission.*

8. THE EVOLVING UNDERSTANDING OF VRAC IDENTITY: PROGRESS AND PERSPECTIVES

Despite much progress in the field, many aspects of cell volume regulation in the brain require further mechanistic exploration. One avenue which carries the most promise is the recent and ongoing work on establishing the molecular nature of VRAC and the mechanisms contributing to its activation in swollen cells.

For nearly thirty years, the molecular identity of VRAC remained unknown. In 1988, two electrophysiological studies resulted in the first recordings of swelling-activated Cl^- currents in human lymphocytes and intestinal epithelial 407 cells (Cahalan & Lewis, 1988; Hazama & Okada, 1988). Since these two pioneer publications, the biophysical properties of VRAC have been a subject of intense interest and were characterized in numerous cell and tissue types [reviewed in (Mongin, 2016; Nilius et al., 1997; Okada, 1997; Pedersen et al., 2016; Strange et al., 1996)]. Nevertheless, despite numerous attempts to clone VRAC, its molecular identity was established only in 2014. The laboratories of A. Patapoutian and T.J. Jentsch performed independent genome-wide siRNA screens and identified VRAC as heteromeric protein complexes homologous to mammalian pannexins and assembled from the gene products belonging to the leucine-rich repeat containing family 8 (LRRC8) (Qiu et al., 2014; Voss et al., 2014). Both labs discovered that VRAC activity is critically dependent on the expression of the LRRC8A isoform, while the four other members of the same family, LRRC8B through E, play dispensable roles (Qiu et al., 2014; Voss et al., 2014). However, T.J. Jentsch and colleagues made an additional conceptual step and established that LRRC8A has to heteromerize with at least one alternative type of LRRC8 subunit in order to form a fully active VRAC (Voss et al., 2014). Moreover, they found that LRRB8C-E partners modify the biophysical properties of VRAC (Voss et al., 2014). These latter finding addressed the long-standing question of why VRACs have similar but not identical profiles in different cell types (Nilius & Droogmans, 2003; Okada, 1997; Okada, Sato, & Numata, 2009). Also importantly, deletion of LRRC8A and disruption of VRAC function blocked the RVD process in model cell lines (Qiu et al., 2014; Voss et al., 2014). The amazing progress in our knowledge of VRAC is covered in the two invited reviews in this book (Okada, 2018; Qiu, 2018). The most recent development in this area is the report on the 3D structure of VRAC resolved with X-ray crystallography and cryo-electron microscopy (Deneka, Sawicka, Lam, Paulino, & Dutzler, 2018).

In this chapter, we narrowly focus on the recent discovery of the heterogeneous VRAC complexes with distinct permeability profiles for organic molecules. As seen in the previous sections, our interest in this topic is motivated by the high relevance of organic osmolyte release to normal brain signaling and pathological states. The two original papers which linked VRAC to LRRC8 proteins, based their work on measurements of currents and fluxes of Cl^- and other halides. However, they also additionally found that the LRRC8A-containing channels are responsible for swelling-activated release of the zwitterionic taurine (Qiu et al., 2014; Voss et al., 2014). Our laboratory was the first to affirm the critical role of LRRC8A for VRAC activity in brain cells (astrocytes), but also made the side-by-side comparisons of the LRRC8A-dependent release of taurine and the negatively charged amino acids, glutamate and aspartate (Hyzinski-Garcia et al., 2014). The principal finding in our study was that the presence of LRRC8A is equally critical for the release of both taurine and glutamate/aspartate (Hyzinski-Garcia et al., 2014). On one hand, this was highly encouraging because the VRAC involvement in the release of organic osmolytes has been questioned in the past (Franco, 2003; Kirk, 1997; Shennan, 2008). On the other hand, the finding of a uniform dependence on LRRC8A expression contradicted the enduring idea about the existence of plural swelling-activated release pathways for inorganic anions and organic osmolytes. Over the years, many groups accrued strong, albeit indirect evidence that Cl^- and certain organic molecules, such as taurine, may be transported via distinct volume-sensitive pathways [e.g., (Lambert & Hoffmann, 1994; Mongin, Reddi, Charniga, & Kimelberg, 1999; Shennan, McNeillie, & Curran, 1994; Stutzin et al., 1999)]. The hypothesis that taurine has its own independent release route was particularly influential, but not rooted in any molecular knowledge. This has changed with the discovery of the LRRC8 proteins.

The first molecular evidence for multiple VRAC channels responsible for the release of inorganic and organic osmolytes came from the laboratory of T.J. Jentsch. They discovered that LRRC8A + LRRC8D-containing VRAC plays a unique role in transport of the chemotherapy drugs, cisplatin and carboplatin (Planells-Cases et al., 2015). The same study established that expression of the LRRC8D subunit is essential for the release of taurine, but dispensable for swelling-activated Cl^- currents (Planells-Cases et al., 2015). These findings led to the conclusion that LRRC8A + LRRC8D-containing heteromers are the long-sought after swelling-activated taurine channel. Several subsequent publications explored the requirements of LRRC8 composition for the release of taurine and other structurally diverse organic osmolytes, including glutamate, GABA glycine, lactate, and *myo*-inositol (Gaitan-Penas et al., 2016; Lutter, Ullrich, Lueck, Kempa, & Jentsch, 2017; Schober et al., 2017). The emerging picture is interesting but rather complex. In the most definitive study, Lutter et al. used an erase-and-replace approach in model cell lines to establish that (1) LRRC8A + LRRC8C-containing VRAC can fully sustain swelling-activated Cl^- currents but are poorly permeable to organic osmolytes, (2) LRRC8A + LRRC8D heteromers have very limited Cl^- conductance but transport all charged and uncharged organic osmolytes, and (3) LRRC8A + LRRC8E heteromers exert slightly reduced Cl^- conductance and are permeable to anionic amino acids, but not uncharged organic osmolytes (Lutter et al., 2017). As in the LRRC8 discovery studies, the combination of LRRC8A + LRRC8B did not support Cl^- currents or organic osmolyte release. More complex permutations of the LRRC8A, C, D, and E subunits containing three or more LRRC8 isoforms demonstrated some quantitative differences, suggesting that introduction of additional subunits (with the exception of B) modifies the permeability properties of VRAC. Yet, the take home message from this model work was that LRRC8D is indispensable for for release of uncharged and net-neural molecules, while LRRC8D and LRRC8E or their combination can sustain the release of excitatory amino acids.

Our work explored the requirements of the endogenous LRRC8 subunits for organic osmolyte release in brain astrocytes (Schober et al., 2017). The principal observations of this study are presented in Fig. 7. Although there were strong similarities between our study and the report of the T.J. Jentsch group (Lutter et al., 2017), we also found interesting qualitative differences, which may reflect the cell type-specific variations in VRAC composition and properties. Much like Lutter et al., we found that swelling activated release of non-polar and uncharged zwitterionic osmolytes, *myo*-inositol and taurine, was either strongly (taurine) or fully (*myo*-inositol) suppressed by siRNA knockdown of LRRC8D. The same transport processes were insensitive to the deletion of LRRC8C, LRRC8E, or their combination. In contrast, the release of negatively charged D-aspartate demonstrated a very different sensitivity profile. It was strongly (but not completely) reduced by the combined deletion of LRRC8C + LRRC8E. Unlike *myo*-inositol and taurine, D-aspartate fluxes were insensitive to the downregulation of LRRC8D. Finally and surprisingly, D-aspartate release was fully sensitive to the combined deletion of

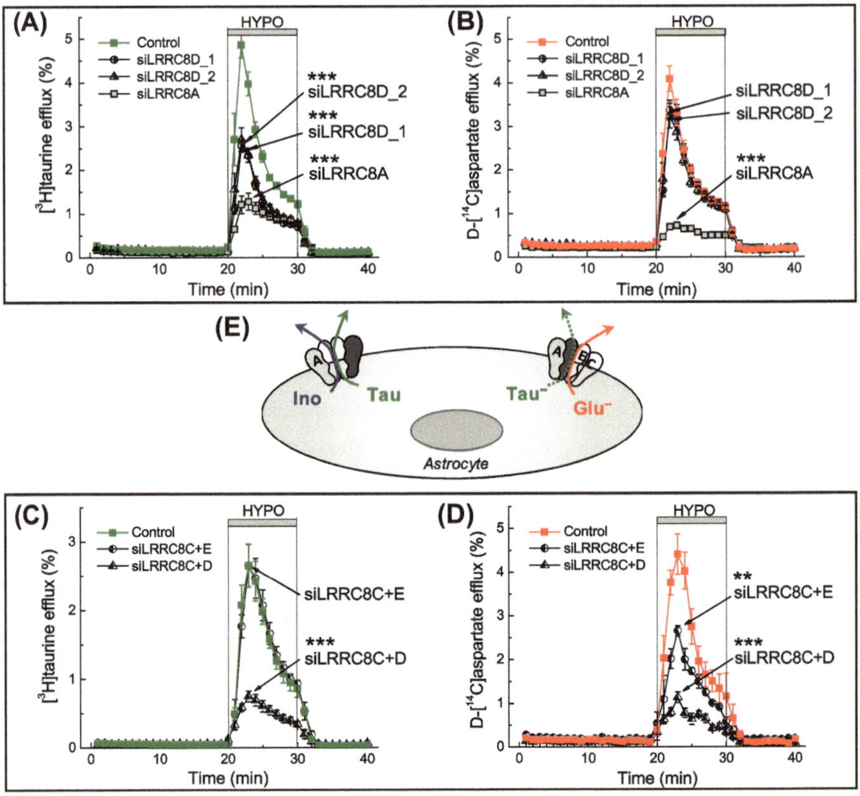

Figure 7 RNAi analysis reveals that astrocytes express at least two types of heteromeric LRRC8-containing VRAC channels with differential permeability to charged and uncharged organic osmolytes (A–D) Primary rat astrocytes were transfected with negative control siRNA (Control), or the gene-specific siRNA targeting the expression of one or more members of the proteins belonging to the family of leucine-rich repeat-containing 8 (LRRC8) proteins. After a 4-day incubation with the gene specific siRNA constructs, astrocytes were simultaneously loaded with [^{3}H]taurine and the glutamate analog D-[^{14}C]aspartate, and superfused with isosmotic or hypoosmotic (HYPO) media to activate VRAC and quantify the amino acid release. Taurine and glutamate fluxes have been measured simultaneously but, for clarity, are presented in separate panels (A–B) and (C–D). Knockdown of LRRC8A reduced release of all amino acids. Knockdown of LRRC8D reduced [^{3}H]taurine release by ~50% (A), but had no effect on the release of D-[^{14}C]aspartate (B). Conversely, the combined knockdown of LRRC8C + LRRC8E inhibited the efflux of D-[^{14}C]aspartate (D), but not [^{3}H]taurine (C). Finally, and surprisingly, the combined knockdown of LRRC8C + LRRC8D inhibited the release of all amino acids to the same extent as the knockdown of LRRC8A (E) The proposed composition of the two populations of LRRC8-containing VRAC heteromers which explains the data in (A–D). The LRRC8A+D combination is responsible for the movement of uncharged molecules such as taurine and *myo*-inositol (Ino) (data not shown). Another channel composed of LRRC8A+C+D+E is permeable to

LRRC8C + LRRC8D. The last bit of information is at odds with the findings of Lutter et al., and strongly suggests that LRRC8D is present in VRAC heteromers that are responsible for release of negatively charged amino acids. The simplest explanation for these rather complex findings is presented in Fig. 7E. We hypothesize that astrocytes express two types of VRAC channels: (1) the LRRC8A + LRRC8D-containing heteromers, which dominate the release of uncharged molecules, and (2) the LRRC8A + LRRC8C + LRRC8D + LRRC8E-containing heteromers, which mediate the movement of charged amino acids. It appears that in the latter "LRRC8A+C+D+E" channel the C, D and E subunits can partially substitute for each other, because their individual deletions had little effect on D-aspartate fluxes. The functional role for the LRRC8E subunit is particularly interesting and unexpected. In our hands, astrocytes have at least ten-fold lower expression of the LRRC8E mRNA as compared to other LRRC8 transcripts (Hyzinski-Garcia et al., 2014; Schober et al., 2017). Similarly, transcriptomics assays reveal extremely low LRRC8E expression levels in astrocyte and other brain cells (Zhang et al., 2014). Further work is needed to establish the LRRC8 stoichiometry for astrocytic VRAC, and whether the post-transcriptional stability of LRRC8E is different from other LRRC8 family members.

It is fitting that this admittedly non-exhaustive review is concluded with the discussion of structure, functions, and (patho)physiological roles of VRAC. In our opinion, the discovery of the LRRC8 proteins has been and continues to be a powerful catalyst for further progress in this field. Although VRAC studies can be dated back to the late 1980s, only now do we have the molecular biology and genetics tools to tackle some of the perennial questions about this channel and cell volume control in the brain. Perhaps, this newly acquired knowledge will open a door for finding the last and most significant piece in the puzzle of cell volume regulation: the molecular nature of enigmatic cell volume sensor(s).

negatively charged molecules, including glutamate. Because zwitterionic taurine is partially negatively charged, it shares both pathways. **$p < 0.01$, ***$p < 0.001$, amino acid release rates versus cells treated with the negative control siRNA. *Adapted from Schober, A. L., Wilson, C. S., & Mongin, A. A. (2017). Molecular composition and heterogeneity of the LRRC8-containing swelling-activated osmolyte channels in primary rat astrocytes.* Journal of Physiology, 595, *6939–6951, with permission.*

ACKNOWLEDGMENTS

The work in Authors' laboratory was funded in part by NIH/NINDS grant R01 NS061953.

REFERENCES

Abdullaev, I. F., Rudkouskaya, A., Schools, G. P., Kimelberg, H. K., & Mongin, A. A. (2006). Pharmacological comparison of swelling-activated excitatory amino acid release and Cl^- currents in rat cultured astrocytes. *Journal of Physiology, 572*, 677–689.

Adrogue, H. J., & Madias, N. E. (2000). Hyponatremia. *The New England Journal of Medicine, 342*, 1581–1589.

Aitken, P. G., Borgdorff, A. J., Juta, A. J., Kiehart, D. P., Somjen, G. G., & Wadman, W. J. (1998). Volume changes induced by osmotic stress in freshly isolated rat hippocampal neurons. *Pflügers Archiv, 436*, 991–998.

Aitken, P. G., Fayuk, D., Somjen, G. G., & Turner, D. A. (1999). Use of intrinsic optical signals to monitor physiological changes in brain tissue slices. *Methods, 18*, 91–103.

Akita, T., Fedorovich, S. V., & Okada, Y. (2011). Ca^{2+} nanodomain-mediated component of swelling-induced volume-sensitive outwardly rectifying anion current triggered by autocrine action of ATP in mouse astrocytes. *Cellular Physiology and Biochemistry: International Journal of Experimental Cellular Physiology, Biochemistry, and Pharmacology, 28*, 1181–1190.

Akita, T., & Okada, Y. (2011). Regulation of bradykinin-induced activation of volume-sensitive outwardly rectifying anion channels by Ca^{2+} nanodomains in mouse astrocytes. *Journal of Physiology, 589*, 3909–3927.

Akita, T., & Okada, Y. (2014). Characteristics and roles of the volume-sensitive outwardly rectifying (VSOR) anion channel in the central nervous system. *Neuroscience, 275*, 211–231.

Albrecht, J., & Norenberg, M. D. (2006). Glutamine: A Trojan horse in ammonia neurotoxicity. *Hepatology, 44*, 788–794.

Albrecht, J., & Schousboe, A. (2005). Taurine interaction with neurotransmitter receptors in the CNS: An update. *Neurochemical Research, 30*, 1615–1621.

Alessi, D. R., Zhang, J., Khanna, A., Hochdorfer, T., Shang, Y., & Kahle, K. T. (2014). The WNK-SPAK/OSR1 pathway: Master regulator of cation-chloride cotransporters. *Science Signaling [electronic Resource], 7*, re3.

Allert, N., Koller, H., & Siebler, M. (1998). Ammonia-induced depolarization of cultured rat cortical astrocytes. *Brain Research, 782*, 261–270.

Alper, S. L. (1991). The band 3-related anion exchanger (AE) gene family. *Annual Review of Physiology, 53*, 549–564.

Altamirano, J., Brodwick, M. S., & Alvarez-Leefmans, F. J. (1998). Regulatory volume decrease and intracellular Ca^{2+} in murine neuroblastoma cells studied with fluorescent probes. *The Journal of General Physiology, 112*, 145–160.

Amiry-Moghaddam, M., Otsuka, T., Hurn, P. D., Traystman, R. J., Haug, F. M., Froehner, S. C., et al. (2003). An alpha-syntrophin-dependent pool of AQP4 in astroglial end-feet confers bidirectional water flow between blood and brain. *Proceedings of the National Academy of Sciences of the United States of America, 100*, 2106–2111.

Amiry-Moghaddam, M., & Ottersen, O. P. (2003). The molecular basis of water transport in the brain. *Nature Reviews. Neuroscience, 4*, 991–1001.

Amzica, F., & Neckelmann, D. (1999). Membrane capacitance of cortical neurons and glia during sleep oscillations and spike-wave seizures. *Journal of Neurophysiology, 82*, 2731–2746.

Andrew, R. D. (1991). Seizure and acute osmotic change: Clinical and neurophysiological aspects. *Journal of Neurological Sciences, 101*, 7–18.

Andrew, R. D., Fagan, M., Ballyk, B. A., & Rosen, A. S. (1989). Seizure susceptibility and the osmotic state. *Brain Research, 498*, 175–180.

Andrew, R. D., Labron, M. W., Boehnke, S. E., Carnduff, L., & Kirov, S. A. (2007). Physiological evidence that pyramidal neurons lack functional water channels. *Cerebral Cortex, 17*, 787–802.
Andrew, R. D., Lobinowich, M. E., & Osehobo, E. P. (1997). Evidence against volume regulation by cortical brain cells during acute osmotic stress. *Experimental Neurology, 143*, 300–312.
Andrew, R. D., & MacVicar, B. A. (1994). Imaging cell volume changes and neuronal excitation in the hippocampal slice. *Neuroscience, 62*, 371–383.
Antonenko, Y. N., Pohl, P., & Denisov, G. A. (1997). Permeation of ammonia across bilayer lipid membranes studied by ammonium ion selective microelectrodes. *Biophysical Journal, 72*, 2187–2195.
Arieff, A. I., Guisado, R., & Lazarowitz, V. C. (1977). Pathophysiology of hyperosmolar states. In T. E. Andereoli, J. J. Grantham, & F. C. Rector (Eds.), *Disturbances in body fluid osmolality* (pp. 227–250). Bethesda, MD: American Physiological Society.
Arundine, M., & Tymianski, M. (2004). Molecular mechanisms of glutamate-dependent neurodegeneration in ischemia and traumatic brain injury. *Cellular and Molecular Life Sciences: CMLS, 61*, 657–668.
Assentoft, M., Larsen, B. R., & MacAulay, N. (2015). Regulation and function of AQP4 in the central nervous system. *Neurochemical Research, 40*, 2615–2627.
Babila, T., Atlan, H., Fromer, I., Schwalb, H., Uretzky, G., & Lichtstein, D. (1990). Volume regulation of nerve terminals. *Journal of Neurochemistry, 55*, 2058–2062.
Baethmann, A., & Van Harreveld, A. (1973). Water and electrolyte distribution in gray matter rendered edematous with a metabolic inhibitor. *Journal of Neuropathology and Experimental Neurology, 32*, 408–423.
Bai, G., Rama Rao, K. V., Murthy, C. R., Panickar, K. S., Jayakumar, A. R., & Norenberg, M. D. (2001). Ammonia induces the mitochondrial permeability transition in primary cultures of rat astrocytes. *Journal of Neuroscience Research, 66*, 981–991.
Bakay, L., & Lee, J. C. (1968). The effect of acute hypoxia and hypercapnia on the ultrastructure of the central nervous system. *Brain, 91*, 697–706.
Bakay, L., Lee, J. C., Lee, G. C., & Peng, J. R. (1977). Experimental cerebral concussion. Part 1: An electron microscopic study. *Journal of Neurosurgery, 47*, 525–531.
Banderali, U., & Roy, G. (1992). Anion channels for amino-acids in Mdck cells. *American Journal of Physiology, 263*, C1200–C1207.
Basavappa, S., Chartouni, V., Kirk, K., Prpic, V., Ellory, J. C., & Mangel, A. W. (1995). Swelling-induced chloride currents in neuroblastoma cells are calcium dependent. *The Journal of Neuroscience: the Official Journal of the Society for Neuroscience, 15*, 3662–3666.
Battey, T. W., Karki, M., Singhal, A. B., Wu, O., Sadaghiani, S., Campbell, B. C., et al. (2014). Brain edema predicts outcome after nonlacunar ischemic stroke. *Stroke, 45*, 3643–3648.
Begum, G., Yuan, H., Kahle, K. T., Li, L., Wang, S., Shi, Y., et al. (2015). Inhibition of WNK3 kinase signaling reduces brain damage and accelerates neurological recovery after stroke. *Stroke, 46*, 1956–1965.
Bender, A. S., Neary, J. T., Blicharska, J., Norenberg, L. O., & Norenberg, M. D. (1992). Role of calmodulin and protein kinase C in astrocytic cell volume regulation. *Journal of Neurochemistry, 58*, 1874–1882.
Bender, A. S., Neary, J. T., & Norenberg, M. D. (1993). Role of phosphoinositide hydrolysis in astrocyte volume regulation. *Journal of Neurochemistry, 61*, 1506–1514.
Bender, A. S., Schousboe, A., Reichelt, W., & Norenberg, M. D. (1998). Ionic mechanisms in glutamate-induced astrocyte swelling: Role of K^+ influx. *Journal of Neuroscience Research, 52*, 307–321.

Benfenati, V., Caprini, M., Dovizio, M., Mylonakou, M. N., Ferroni, S., Ottersen, O. P., et al. (2011). An aquaporin-4/transient receptor potential vanilloid 4 (AQP4/TRPV4) complex is essential for cell-volume control in astrocytes. *Proceedings of the National Academy of Sciences of the United States of America, 108*, 2563–2568.

Benfenati, V., Caprini, M., Nicchia, G. P., Rossi, A., Dovizio, M., Cervetto, C., et al. (2009). Carbenoxolone inhibits volume-regulated anion conductance in cultured rat cortical astroglia. *Channels (Austin), 3*, 323–336.

Benfenati, V., Nicchia, G. P., Svelto, M., Rapisarda, C., Frigeri, A., & Ferroni, S. (2007). Functional down-regulation of volume-regulated anion channels in AQP4 knockdown cultured rat cortical astrocytes. *Journal of Neurochemistry, 100*, 87–104.

Benveniste, H., Drejer, J., Schousboe, A., & Diemer, N. H. (1984). Elevation of the extracellular concentrations of glutamate and aspartate in rat hippocampus during transient cerebral ischemia monitored by intracerebral microdialysis. *Journal of Neurochemistry, 43*, 1369–1374.

Binder, D. K., Nagelhus, E. A., & Ottersen, O. P. (2012). Aquaporin-4 and epilepsy. *Glia, 60*, 1203–1214.

Binder, D. K., Oshio, K., Ma, T., Verkman, A. S., & Manley, G. T. (2004). Increased seizure threshold in mice lacking aquaporin-4 water channels. *Neuroreport, 15*, 259–262.

Binder, D. K., Papadopoulos, M. C., Haggie, P. M., & Verkman, A. S. (2004). In vivo measurement of brain extracellular space diffusion by cortical surface photobleaching. *The Journal of Neuroscience: the Official Journal of the Society for Neuroscience, 24*, 8049–8056.

Binder, D. K., Yao, X., Zador, Z., Sick, T. J., Verkman, A. S., & Manley, G. T. (2006). Increased seizure duration and slowed potassium kinetics in mice lacking aquaporin-4 water channels. *Glia, 53*, 631–636.

Blei, A. T., & Larsen, F. S. (1999). Pathophysiology of cerebral edema in fulminant hepatic failure. *Journal of Hepatology, 31*, 771–776.

Blei, A. T., Olafsson, S., Therrien, G., & Butterworth, R. F. (1994). Ammonia-induced brain edema and intracranial hypertension in rats after portacaval anastomosis. *Hepatology, 19*, 1437–1444.

Boettger, T., Rust, M. B., Maier, H., Seidenbecher, T., Schweizer, M., Keating, D. J., et al. (2003). Loss of K-Cl co-transporter KCC3 causes deafness, neurodegeneration and reduced seizure threshold. *The EMBO Journal, 22*, 5422–5434.

Bohmer, C., Wagner, C. A., Beck, S., Moschen, I., Melzig, J., Werner, A., et al. (2000). The shrinkage-activated Na^+ conductance of rat hepatocytes and its possible correlation to rENaC. *Cellular Physiology and Biochemistry: International Journal of Experimental Cellular Physiology, Biochemistry, and Pharmacology, 10*, 187–194.

Bolay, H., & Dalkara, T. (1998). Mechanisms of motor dysfunction after transient MCA occlusion: Persistent transmission failure in cortical synapses is a major determinant. *Stroke, 29*, 1988–1993.

Bolay, H., Gursoy-Ozdemir, Y., Sara, Y., Onur, R., Can, A., & Dalkara, T. (2002). Persistent defect in transmitter release and synapsin phosphorylation in cerebral cortex after transient moderate ischemic injury. *Stroke, 33*, 1369–1375.

Bondarava, M., Li, T., Endl, E., & Wehner, F. (2009). alpha-ENaC is a functional element of the hypertonicity-induced cation channel in HepG2 cells and it mediates proliferation. *Pflügers Archiv, 458*, 675–687.

Bond, T., Basavappa, S., Christensen, M., & Strange, K. (1999). ATP dependence of the ICl, swell channel varies with rate of cell swelling. Evidence for two modes of channel activation. *The Journal of General Physiology, 113*, 441–456.

Boulos, A. S., Deshaies, E. M., Dalfino, J. C., Feustel, P. J., Popp, A. J., & Drazin, D. (2011). Tamoxifen as an effective neuroprotectant in an endovascular canine model of stroke. *Journal of Neurosurgery, 114*, 1117–1126.

Bourke, R. S., & Nelson, K. M. (1972). Further studies on the K^+ -dependent swelling of primate cerebral cortex in vivo: The enzymatic basis of the K^+ -dependent transport of chloride. *Journal of Neurochemistry, 19*, 663–685.
Bourque, C. W. (2008). Central mechanisms of osmosensation and systemic osmoregulation. *Nature Reviews. Neuroscience, 9*, 519–531.
Bourque, C. W., & Oliet, S. H. (1997). Osmoreceptors in the central nervous system. *Annual Review of Physiology, 59*, 601–619.
Bourque, C. W., Oliet, S. H., & Richard, D. (1994). Osmoreceptors, osmoreception, and osmoregulation. *Front Neuroendocrinol, 15*, 231–274.
Bowens, N. H., Dohare, P., Kuo, Y. H., & Mongin, A. A. (2013). DCPIB, the proposed selective blocker of volume-regulated anion channels, inhibits several glutamate transport pathways in glial cells. *Molecular Pharmacology, 83*, 22–32.
Bres, V., Hurbin, A., Duvoid, A., Orcel, H., Moos, F. C., Rabie, A., et al. (2000). Pharmacological characterization of volume-sensitive, taurine permeable anion channels in rat supraoptic glial cells. *British Journal of Pharmacology, 130*, 1976–1982.
Brott, T., & Bogousslavsky, J. (2000). Treatment of acute ischemic stroke. *The New England Journal of Medicine, 343*, 710–722.
Brusilow, S. W., Koehler, R. C., Traystman, R. J., & Cooper, A. J. (2010). Astrocyte glutamine synthetase: Importance in hyperammonemic syndromes and potential target for therapy. *Neurotherapeutics, 7*, 452–470.
Burg, M. B., Ferraris, J. D., & Dmitrieva, N. I. (2007). Cellular response to hyperosmotic stresses. *Physiological Reviews, 87*, 1441–1474.
Byun, N., & Delpire, E. (2007). Axonal and periaxonal swelling precede peripheral neurodegeneration in KCC3 knockout mice. *Neurobiology of Disease, 28*, 39–51.
Cahalan, M. D., & Lewis, R. S. (1988). Role of potassium and chloride channels in volume regulation by T lymphocytes. *Soc. Gen. Physiol Ser., 43*, 281–301.
Caspi, A., Benninger, F., & Yaari, Y. (2009). KV7/M channels mediate osmotic modulation of intrinsic neuronal excitability. *The Journal of Neuroscience: the Official Journal of the Society for Neuroscience, 29*, 11098–11111.
Chamberlin, M. E., & Strange, K. (1989). Anisosmotic cell volume regulation: A comparative view. *American Journal of Physiology, 257*, C159–C173.
Chan, P. H., & Fishman, R. A. (1979). Elevation of rat brain amino acids, ammonia and idiogenic osmoles induced by hyperosmolality. *Brain Research, 161*, 293–301.
Cheema, T. A., Pettigrew, V. A., & Fisher, S. K. (2007). Receptor regulation of the volume-sensitive efflux of taurine and iodide from human SH-SY5Y neuroblastoma cells: Differential requirements for Ca^{2+} and protein kinase C. *The Journal of Pharmacology and Experimental Therapeutics, 320*, 1068–1077.
Cheema, T. A., Ward, C. E., & Fisher, S. K. (2005). Subnanomolar concentrations of thrombin enhance the volume-sensitive efflux of taurine from human 1321N1 astrocytoma cells. *The Journal of Pharmacology and Experimental Therapeutics, 315*, 755–763.
Chen, M., Dong, Y., & Simard, J. M. (2003). Functional coupling between sulfonylurea receptor type 1 and a nonselective cation channel in reactive astrocytes from adult rat brain. *The Journal of Neuroscience: the Official Journal of the Society for Neuroscience, 23*, 8568–8577.
Chen, M., & Simard, J. M. (2001). Cell swelling and a nonselective cation channel regulated by internal Ca^{2+} and ATP in native reactive astrocytes from adult rat brain. *The Journal of Neuroscience: the Official Journal of the Society for Neuroscience, 21*, 6512–6521.
Choi, D. W. (1988). Glutamate neurotoxicity and diseases of the nervous system. *Neuron, 1*, 623–634.
Choi, D. W. (1992). Excitotoxic cell death. *Journal of Neurobiology, 23*, 1261–1276.

Ciura, S., & Bourque, C. W. (2006). Transient receptor potential vanilloid 1 is required for intrinsic osmoreception in organum vasculosum lamina terminalis neurons and for normal thirst responses to systemic hyperosmolality. *The Journal of Neuroscience: the Official Journal of the Society for Neuroscience, 26*, 9069–9075.

Ciura, S., Liedtke, W., & Bourque, C. W. (2011). Hypertonicity sensing in organum vasculosum lamina terminalis neurons: A mechanical process involving TRPV1 but not TRPV4. *The Journal of Neuroscience: the Official Journal of the Society for Neuroscience, 31*, 14669–14676.

Ciura, S., Prager-Khoutorsky, M., Thirouin, Z. S., Wyrosdic, J. C., Olson, J. E., Liedtke, W., et al. (2018). Trpv4 mediates hypotonic inhibition of central osmosensory neurons via taurine gliotransmission. *Cell Reports [electronic Resource], 23*, 2245–2253.

Cornog, J. L., Gonatas, N. K., & Feierman, J. R. (1967). Effects of intracerebral injection of ouabain on the fine structure of rat cerebral cortex. *The American Journal of Pathology, 51*, 573–590.

Crepel, V., Panenka, W., Kelly, M. E., & MacVicar, B. A. (1998). Mitogen-activated protein and tyrosine kinases in the activation of astrocyte volume-activated chloride current. *The Journal of Neuroscience: the Official Journal of the Society for Neuroscience, 18*, 1196–1206.

Cserr, H. F., DePasquale, M., Nicholson, C., Patlak, C. S., Pettigrew, K. D., & Rice, M. E. (1991). Extracellular volume decreases while cell volume is maintained by ion uptake in rat brain during acute hypernatremia. *Journal of Physiology, 442*, 277–295.

Cserr, H. F., DePasquale, M., & Patlak, C. S. (1987a). Regulation of brain water and electrolytes during acute hyperosmolality in rats. *American Journal of Physiology, 253*, F522–F529.

Cserr, H. F., DePasquale, M., & Patlak, C. S. (1987b). Volume regulatory influx of electrolytes from plasma to brain during acute hyperosmolality. *American Journal of Physiology, 253*, F530–F537.

Dadsetan, S., Kukolj, E., Bak, L. K., Sorensen, M., Ott, P., Vilstrup, H., et al. (2013). Brain alanine formation as an ammonia-scavenging pathway during hyperammonemia: Effects of glutamine synthetase inhibition in rats and astrocyte-neuron co-cultures. *Journal of Cerebral Blood Flow & Metabolism, 33*, 1235–1241.

Danbolt, N. C. (2001). Glutamate uptake. *Progress in Neurobiology, 65*, 1–105.

Darby, M., Kuzmiski, J. B., Panenka, W., Feighan, D., & MacVicar, B. A. (2003). ATP released from astrocytes during swelling activates chloride channels. *Journal of Neurophysiology, 89*, 1870–1877.

David, Y., Cacheaux, L. P., Ivens, S., Lapilover, E., Heinemann, U., Kaufer, D., et al. (2009). Astrocytic dysfunction in epileptogenesis: Consequence of altered potassium and glutamate homeostasis? *The Journal of Neuroscience: the Official Journal of the Society for Neuroscience, 29*, 10588–10599.

Deleuze, C., Duvoid, A., & Hussy, N. (1998). Properties and glial origin of osmotic-dependent release of taurine from the rat supraoptic nucleus. *Journal of Physiology, 507*, 463–471.

Delpire, E. (2018). Cell volume maintenance: Water and salt homeostasis. *Current Topics in Membranes, 81*. In press.

Deneka, D., Sawicka, M., Lam, A. K. M., Paulino, C., & Dutzler, R. (2018). Structure of a volume-regulated anion channel of the LRRC8 family. *Nature*. https://doi.org/10.1038/s41586-018-0134-y. Epub ahead of print.

DePasquale, M., Patlak, C. S., & Cserr, H. F. (1989). Brain ion and volume regulation during acute hypernatremia in Brattleboro rats. *American Journal of Physiology, 256*, F1059–F1066.

Dietzel, I., Heinemann, U., Hofmeier, G., & Lux, H. D. (1982). Stimulus-induced changes in extracellular Na^+ and Cl^- concentration in relation to changes in the size of the extracellular space. *Experimental Brain Research, 46*, 73–84.

Dietzel, I., Heinemann, U., & Lux, H. D. (1989). Relations between slow extracellular potential changes, glial potassium buffering, and electrolyte and cellular volume changes during neuronal hyperactivity in cat brain. *Glia, 2*, 25–44.
Dirnagl, U., Iadecola, C., & Moskowitz, M. A. (1999). Pathobiology of ischaemic stroke: An integrated view. *Trends in Neurosciences, 22*, 391–397.
Dohare, P., Hyzinski-Garcia, M. C., Vipani, A., Bowens, N. H., Nalwalk, J. W., Feustel, P. J., et al. (2014). The neuroprotective properties of the superoxide dismutase mimetic tempol correlate with its ability to reduce pathological glutamate release in a rodent model of stroke. *Free Radical Biology and Medicine, 77*, 168–182.
Dreier, J. P., Lemale, C. L., Kola, V., Friedman, A., & Schoknecht, K. (2018). Spreading depolarization is not an epiphenomenon but the principal mechanism of the cytotoxic edema in various gray matter structures of the brain during stroke. *Neuropharmacology, 134*, 189–207.
Duan, S., Anderson, C. M., Keung, E. C., Chen, Y., Chen, Y., & Swanson, R. A. (2003). P2X7 receptor-mediated release of excitatory amino acids from astrocytes. *The Journal of Neuroscience: the Official Journal of the Society for Neuroscience, 23*, 1320–1328.
Eid, T., Williamson, A., Lee, T. S., Petroff, O. A., & de Lanerolle, N. C. (2008). Glutamate and astrocytes–key players in human mesial temporal lobe epilepsy? *Epilepsia, 49*(Suppl 2), 42–52.
Estevez, A. Y., O'Regan, M. H., Song, D., & Phillis, J. W. (1999). Effects of anion channel blockers on hyposmotically induced amino acid release from the in vivo rat cerebral cortex. *Neurochemical Research, 24*, 447–452.
Fabene, P. F., Weiczner, R., Marzola, P., Nicolato, E., Calderan, L., Andrioli, A., et al. (2006). Structural and functional MRI following 4-aminopyridine-induced seizures: A comparative imaging and anatomical study. *Neurobiology of Disease, 21*, 80–89.
Fahlke, C., Kortzak, D., & Machtens, J. P. (2016). Molecular physiology of EAAT anion channels. *Pflügers Archiv, 468*, 491–502.
Falke, L. C., & Misler, S. (1989). Activity of ion channels during volume regulation by clonal N1E115 neuroblastoma cells. *Proceedings of the National Academy of Sciences of the United States of America, 86*, 3919–3923.
Feigin, V. L., Norrving, B., & Mensah, G. A. (2017). Global burden of stroke. *Circulation Research, 120*, 439–448.
Felipo, V., & Butterworth, R. F. (2002). Neurobiology of ammonia. *Progress in Neurobiology, 67*, 259–279.
Fettiplace, R., & Haydon, D. A. (1980). Water permeability of lipid membranes. *Physiological Reviews, 60*, 510–550.
Feustel, P. J., Jin, Y., & Kimelberg, H. K. (2004). Volume-regulated anion channels are the predominant contributors to release of excitatory amino acids in the ischemic cortical penumbra. *Stroke, 35*, 1164–1168.
Fiala, J. C., Spacek, J., & Harris, K. M. (2002). Dendritic spine pathology: Cause or consequence of neurological disorders? *Brain Research Reviews, 39*, 29–54.
Finkelstein, A. (1987). *Water movement through lipid bilayers, pores, and plasma membranes: Theory and reality* (1 ed.). New York: Willey Interscience.
Fisher, S. K., Cheema, T. A., Foster, D. J., & Heacock, A. M. (2008). Volume-dependent osmolyte efflux from neural tissues: Regulation by G-protein-coupled receptors. *Journal of Neurochemistry, 106*, 1998–2014.
Fisher, S. K., Heacock, A. M., Keep, R. F., & Foster, D. J. (2010). Receptor regulation of osmolyte homeostasis in neural cells. *Journal of Physiology, 588*, 3355–3364.
Florence, C. M., Baillie, L. D., & Mulligan, S. J. (2012). Dynamic volume changes in astrocytes are an intrinsic phenomenon mediated by bicarbonate ion flux. *PLos One, 7*. e51124.

Flores, B., Schornak, C. C., & Delpire, E. (2018). A role for KCC3 in maintaining cell volume of peripheral nerve fibers. *Neurochemistry International*. https://doi.org/10.1016/j.neuint.2018.01.009. In press.

Franco, R. (2003). Osmosensitive taurine release: Does taurine share the same efflux pathway with chloride and other amino acid osmolytes? *Advances in Experimental Medicine and Biology, 526*, 189–196.

Franco, R., Panayiotidis, M. I., & de La Paz, L. D. (2008). Autocrine signaling involved in cell volume regulation: The role of released transmitters and plasma membrane receptors. *Journal of Cellular Physiology, 216*, 14–28.

Fraser, C. L., & Arieff, A. I. (1997). Epidemiology, pathophysiology, and management of hyponatremic encephalopathy. *The American Journal of Medicine, 102*, 67–77.

Gagnon, K. B., & Delpire, E. (2012). Molecular physiology of SPAK and OSR1: Two Ste20-related protein kinases regulating ion transport. *Physiological Reviews, 92*, 1577–1617.

Gaitan-Penas, H., Gradogna, A., Laparra-Cuervo, L., Solsona, C., Fernandez-Duenas, V., Barrallo-Gimeno, A., et al. (2016). Investigation of LRRC8-mediated volume-regulated anion currents in Xenopus oocytes. *Biophysical Journal, 111*, 1429–1443.

Gamba, G. (2005). Molecular physiology and pathophysiology of electroneutral cation-chloride cotransporters. *Physiological Reviews, 85*, 423–493.

Garcia, J. H., Yoshida, Y., Chen, H., Li, Y., Zhang, Z. G., Lian, J., et al. (1993). Progression from ischemic injury to infarct following middle cerebral artery occlusion in the rat. *The American Journal of Pathology, 142*, 623–635.

Greenwood, S. M., & Connolly, C. N. (2007). Dendritic and mitochondrial changes during glutamate excitotoxicity. *Neuropharmacology, 53*, 891–898.

Groulx, N., Boudreault, F., Orlov, S. N., & Grygorczyk, R. (2006). Membrane reserves and hypotonic cell swelling. *The Journal of Membrane Biology, 214*, 43–56.

Gullans, S. R., & Verbalis, J. G. (1993). Control of brain volume during hyperosmolar and hypoosmolar conditions. *Annual Review of Medicine, 44*, 289–301.

Haj-Yasein, N. N., Jensen, V., Ostby, I., Omholt, S. W., Voipio, J., Kaila, K., et al. (2012). Aquaporin-4 regulates extracellular space volume dynamics during high-frequency synaptic stimulation: A gene deletion study in mouse hippocampus. *Glia, 60*, 867–874.

Hansen, A. J. (1985). Effect of anoxia on ion distribution in the brain. *Physiological Reviews, 65*, 101–148.

Hansson, E., Johansson, B. B., Westergren, I., & Ronnback, L. (1994). Glutamate-induced swelling of single astroglial cells in primary culture. *Neuroscience, 63*, 1057–1066.

Hartings, J. A., Shuttleworth, C. W., Kirov, S. A., Ayata, C., Hinzman, J. M., Foreman, B., et al. (2017). The continuum of spreading depolarizations in acute cortical lesion development: Examining Leao's legacy. *Journal of Cerebral Blood Flow & Metabolism, 37*, 1571–1594.

Haskew-Layton, R. E., Rudkouskaya, A., Jin, Y., Feustel, P. J., Kimelberg, H. K., & Mongin, A. A. (2008). Two distinct modes of hypoosmotic medium-induced release of excitatory amino acids and taurine in the rat brain in vivo. *PLos One, 3*. e3543.

Haussinger, D., & Schliess, F. (2008). Pathogenetic mechanisms of hepatic encephalopathy. *Gut, 57*, 1156–1165.

Haydon, P. G., & Carmignoto, G. (2006). Astrocyte control of synaptic transmission and neurovascular coupling. *Physiological Reviews, 86*, 1009–1031.

Hazama, A., & Okada, Y. (1988). Ca^{2+} sensitivity of volume-regulatory K^+ and Cl^- channels in cultured human epithelial cells. *Journal of Physiology, 402*, 687–702.

Hirtz, D., Thurman, D. J., Gwinn-Hardy, K., Mohamed, M., Chaudhuri, A. R., & Zalutsky, R. (2007). How common are the "common" neurologic disorders? *Neurology, 68*, 326–337.

Hoffmann, E. K., Lambert, I. H., & Pedersen, S. F. (2009). Physiology of cell volume regulation in vertebrates. *Physiological Reviews, 89*, 193–277.

Holthoff, K., & Witte, O. W. (1996). Intrinsic optical signals in rat neocortical slices measured with near-infrared dark-field microscopy reveal changes in extracellular space. *The Journal of Neuroscience: the Official Journal of the Society for Neuroscience, 16*, 2740–2749.

Horie, H., Ikuta, S., Takenaka, T., & Ito, S. (1989). Adaptation of cultured mammalian neurons to a hypotonic environment with age-related response. *Brain Research, 477*, 233–240.

Howard, H. C., Mount, D. B., Rochefort, D., Byun, N., Dupre, N., Lu, J., et al. (2002). The K-Cl cotransporter KCC3 is mutant in a severe peripheral neuropathy associated with agenesis of the corpus callosum. *Nature Genetics, 32*, 384–392.

Hubel, N., & Ullah, G. (2016). Anions govern cell volume: A case study of relative astrocytic and neuronal swelling in spreading depolarization. *PLos One, 11*. e0147060.

Huberfeld, G., Wittner, L., Clemenceau, S., Baulac, M., Kaila, K., Miles, R., et al. (2007). Perturbed chloride homeostasis and GABAergic signaling in human temporal lobe epilepsy. *The Journal of Neuroscience: the Official Journal of the Society for Neuroscience, 27*, 9866–9873.

Hussy, N., Deleuze, C., Desarmenien, M. G., & Moos, F. C. (2000). Osmotic regulation of neuronal activity: A new role for taurine and glial cells in a hypothalamic neuroendocrine structure. *Progress in Neurobiology, 62*, 113–134.

Hussy, N., Deleuze, C., Pantaloni, A., Desarmenien, M. G., & Moos, F. (1997). Agonist action of taurine on glycine receptors in rat supraoptic magnocellular neurones: Possible role in osmoregulation. *Journal of Physiology, 502*, 609–621.

Hyzinski-Garcia, M. C., Rudkouskaya, A., & Mongin, A. A. (2014). LRRC8A protein is indispensable for swelling-activated and ATP-induced release of excitatory amino acids in rat astrocytes. *Journal of Physiology, 592*, 4855–4862.

Hyzinski-Garcia, M. C., Vincent, M. Y., Haskew-Layton, R. E., Dohare, P., Keller, R. W., Jr., & Mongin, A. A. (2011). Hypoosmotic swelling modifies glutamate-glutamine cycle in the cerebral cortex and in astrocyte cultures. *Journal of Neurochemistry, 118*, 140–152.

Iliff, J. J., Lee, H., Yu, M., Feng, T., Logan, J., Nedergaard, M., et al. (2013). Brain-wide pathway for waste clearance captured by contrast-enhanced MRI. *Journal of Clinical Investigation, 123*, 1299–1309.

Iliff, J. J., Wang, M., Liao, Y., Plogg, B. A., Peng, W., Gundersen, G. A., et al. (2012). A paravascular pathway facilitates CSF flow through the brain parenchyma and the clearance of interstitial solutes, including amyloid beta. *Science Translational Medicine, 4*, 147ra111.

Inoue, H., Mori, S., Morishima, S., & Okada, Y. (2005). Volume-sensitive chloride channels in mouse cortical neurons: Characterization and role in volume regulation. *The European Journal of Neuroscience, 21*, 1648–1658.

Inoue, H., & Okada, Y. (2007). Roles of volume-sensitive chloride channel in excitotoxic neuronal injury. *The Journal of Neuroscience: the Official Journal of the Society for Neuroscience, 27*, 1445–1455.

Iwasa, K., Tasaki, I., & Gibbons, R. C. (1980). Swelling of nerve fibers associated with action potentials. *Science, 210*, 338–339.

Izumi, Y., Kirby, C. O., Benz, A. M., Olney, J. W., & Zorumski, C. F. (1999). Muller cell swelling, glutamate uptake, and excitotoxic neurodegeneration in the isolated rat retina. *Glia, 25*, 379–389.

Jackson, P. S., Morrison, R., & Strange, K. (1994). The volume-sensitive organic osmolyte-anion channel VSOAC is regulated by nonhydrolytic ATP binding. *American Journal of Physiology, 267*, C1203–C1209.

Jackson, P. S., & Strange, K. (1993). Volume-sensitive anion channels mediate swelling-activated inositol and taurine efflux. *American Journal of Physiology, 265*, C1489–C1500.

Jayakumar, A. R., Valdes, V., & Norenberg, M. D. (2011). The Na-K-Cl cotransporter in the brain edema of acute liver failure. *Journal of Hepatology, 54*, 272–278.

Jentsch, T. J. (2016). VRACs and other ion channels and transporters in the regulation of cell volume and beyond. *Nature Reviews. Molecular Cell Biology, 17*, 293–307.

Jing, J., Aitken, P. G., & Somjen, G. G. (1994). Interstitial volume changes during spreading depression (SD) and SD-like hypoxic depolarization in hippocampal tissue slices. *Journal of Neurophysiology, 71*, 2548–2551.

Kahle, K. T., Flores, B., Bharucha-Goebel, D., Zhang, J., Donkervoort, S., Hegde, M., et al. (2016). Peripheral motor neuropathy is associated with defective kinase regulation of the KCC3 cotransporter. *Science Signaling [electronic Resource], 9*, ra77.

Kahle, K. T., Khanna, A. R., Alper, S. L., Adragna, N. C., Lauf, P. K., Sun, D., et al. (2015). K-Cl cotransporters, cell volume homeostasis, and neurological disease. *Trends in Molecular Medicine, 21*, 513–523.

Kahle, K. T., Merner, N. D., Friedel, P., Silayeva, L., Liang, B., Khanna, A., et al. (2014). Genetically encoded impairment of neuronal KCC2 cotransporter function in human idiopathic generalized epilepsy. *EMBO Reports, 15*, 766–774.

Kahle, K. T., Ring, A. M., & Lifton, R. P. (2008a). Molecular physiology of the WNK kinases. *Annual Review of Physiology, 70*, 329–355.

Kahle, K. T., Staley, K. J., Nahed, B. V., Gamba, G., Hebert, S. C., Lifton, R. P., et al. (2008b). Roles of the cation-chloride cotransporters in neurological disease. *Nature Clinical Practice Neurology, 4*, 490–503.

Kaila, K., Price, T. J., Payne, J. A., Puskarjov, M., & Voipio, J. (2014). Cation-chloride cotransporters in neuronal development, plasticity and disease. *Nature Reviews. Neuroscience, 15*, 637–654.

Kempski, O., Chaussy, L., Gross, U., Zimmer, M., & Baethmann, A. (1983). Volume regulation and metabolism of suspended C6 glioma cells: An in vitro model to study cytotoxic brain edema. *Brain Research, 279*, 217–228.

Kimelberg, H. K. (1995). Current concepts of brain edema. Review of laboratory investigations. *Journal of Neurosurgery, 83*, 1051–1059.

Kimelberg, H. K. (2004). Water homeostasis in the brain: Basic concepts. *Neuroscience, 129*, 851–860.

Kimelberg, H. K. (2005). Astrocytic swelling in cerebral ischemia as a possible cause of injury and target for therapy. *Glia, 50*, 389–397.

Kimelberg, H. K., Feustel, P. J., Jin, Y., Paquette, J., Boulos, A., Keller, R. W., Jr., et al. (2000). Acute treatment with tamoxifen reduces ischemic damage following middle cerebral artery occlusion. *Neuroreport, 11*, 2675–2679.

Kimelberg, H. K., & Frangakis, M. V. (1985). Furosemide- and bumetanide-sensitive ion transport and volume control in primary astrocyte cultures from rat brain. *Brain Research, 361*, 125–134.

Kimelberg, H. K., Goderie, S. K., Higman, S., Pang, S., & Waniewski, R. A. (1990). Swelling-induced release of glutamate, aspartate, and taurine from astrocyte cultures. *The Journal of Neuroscience: the Official Journal of the Society for Neuroscience, 10*, 1583–1591.

Kimelberg, H. K., Jin, Y., Charniga, C., & Feustel, P. J. (2003). Neuroprotective activity of tamoxifen in permanent focal ischemia. *Journal of Neurosurgery, 99*, 138–142.

Kimelberg, H. K., & Mongin, A. A. (1998). Swelling-activated release of excitatory amino acids in the brain: Relevance for pathophysiology. *Contributions To Nephrology, 123*, 240–257.

Kinne, R., Kinne-Saffran, E., Schutz, H., & Scholermann, B. (1986). Ammonium transport in medullary thick ascending limb of rabbit kidney: Involvement of the Na^+,K^+,Cl^--cotransporter. *The Journal of Membrane Biology, 94*, 279–284.

Kintner, D. B., Wang, Y., & Sun, D. (2007). Role of membrane ion transport proteins in cerebral ischemic damage. *Frontiers in Bioscience, 12*, 762–770.
Kirk, K. (1997). Swelling-activated organic osmolyte channels. *The Journal of Membrane Biology, 158*, 1–16.
Klatzo, I. (1967). Presidential Address: Neuropathological aspects of brain edema. *Journal of Neuropathology and Experimental Neurology, 26*, 1–13.
Kofuji, P., & Newman, E. A. (2004). Potassium buffering in the central nervous system. *Neuroscience, 129*, 1045–1056.
Koos, B., Christmann, J., Plettenberg, S., Kading, D., Becker, J., Keteku, M., et al. (2018). Hypertonicity-induced cation channels in HepG2 cells: Architecture and role in proliferation vs. apoptosis. *Journal of Physiology, 596*, 1227–1241.
Koyama, Y., Ishibashi, T., Okamoto, T., Matsuda, T., Hashimoto, H., & Baba, A. (2000). Transient treatments with L-glutamate and threo-beta-hydroxyaspartate induce swelling of rat cultured astrocytes. *Neurochemistry International, 36*, 167–173.
Koyama, Y., Sugimoto, T., Shigenaga, Y., Baba, A., & Iwata, H. (1991). A morphological study on glutamate-induced swelling of cultured astrocytes: Involvement of calcium and chloride ion mechanisms. *Neuroscience Letters, 124*, 235–238.
Kreisman, N. R., & Olson, J. E. (2003). Taurine enhances volume regulation in hippocampal slices swollen osmotically. *Neuroscience, 120*, 635–642.
Kubo, M., & Okada, Y. (1992). Volume-regulatory Cl^- channel currents in cultured human epithelial cells. *Journal of Physiology, 456*, 351–371.
Kucinski, T., Vaterlein, O., Glauche, V., Fiehler, J., Klotz, E., Eckert, B., et al. (2002). Correlation of apparent diffusion coefficient and computed tomography density in acute ischemic stroke. *Stroke, 33*, 1786–1791.
Kunte, H., Schmidt, S., Eliasziw, M., del Zoppo, G. J., Simard, J. M., Masuhr, F., et al. (2007). Sulfonylureas improve outcome in patients with type 2 diabetes and acute ischemic stroke. *Stroke*.
Lambert, I. H., & Hoffmann, E. K. (1994). Cell swelling activates separate taurine and chloride channels in Ehrlich mouse ascites tumor cells. *The Journal of Membrane Biology, 142*, 289–298.
Lang, F., Busch, G. L., Ritter, M., Volkl, H., Waldegger, S., Gulbins, E., et al. (1998). Functional significance of cell volume regulatory mechanisms. *Physiological Reviews, 78*, 247–306.
Lansberg, M. G., Tong, D. C., Norbash, A. M., Yenari, M. A., & Moseley, M. E. (1999). Intra-arterial rtPA treatment of stroke assessed by diffusion-and perfusion-weighted MRI. *Stroke, 30*, 678–680.
Larsen, B. R., Assentoft, M., Cotrina, M. L., Hua, S. Z., Nedergaard, M., Kaila, K., et al. (2014). Contributions of the Na^+/K^+-ATPase, NKCC1, and Kir4.1 to hippocampal K^+ clearance and volume responses. *Glia, 62*, 608–622.
Larsen, B. R., & MacAulay, N. (2017). Activity-dependent astrocyte swelling is mediated by pH-regulating mechanisms. *Glia, 65*, 1668–1681.
Lauderdale, K., Murphy, T., Tung, T., Davila, D., Binder, D. K., & Fiacco, T. A. (2015). Osmotic edema rapidly increases neuronal excitability through activation of NMDA receptor-dependent slow inward currents in juvenile and adult hippocampus. *ASN Neuro, 7*.
Lauf, P. K., & Adragna, N. C. (2000). K-Cl cotransport: Properties and molecular mechanism. *Cellular Physiology and Biochemistry: International Journal of Experimental Cellular Physiology, Biochemistry, and Pharmacology, 10*, 341–354.
Leaf, A. (1956). On the mechanism of fluis exchange of tissues in vitro. *The Biochemical Journal, 62*, 241–248.
Leaney, J. L., Marsh, S. J., & Brown, D. A. (1997). A swelling-activated chloride current in rat sympathetic neurones. *Journal of Physiology, 501*, 555–564.

Lee, D. J., Amini, M., Hamamura, M. J., Hsu, M. S., Seldin, M. M., Nalcioglu, O., et al. (2012). Aquaporin-4-dependent edema clearance following status epilepticus. *Epilepsy Research, 98*, 264–268.

Lehmann, A. (1989). Effects of microdialysis-perfusion with anisoosmotic media on extracellular amino acids in the rat hippocampus and skeletal muscle. *Journal of Neurochemistry, 53*, 525–535.

Lipton, P. (1973). Effects of membrane depolarization on light scattering by cerebral cortical slices. *Journal of Physiology, 231*, 365–383.

Liu, H. T., Akita, T., Shimizu, T., Sabirov, R. Z., & Okada, Y. (2009). Bradykinin-induced astrocyte-neuron signalling: Glutamate release is mediated by ROS-activated volume-sensitive outwardly rectifying anion channels. *Journal of Physiology, 587*, 2197–2209.

Liu, H. T., Tashmukhamedov, B. A., Inoue, H., Okada, Y., & Sabirov, R. Z. (2006). Roles of two types of anion channels in glutamate release from mouse astrocytes under ischemic or osmotic stress. *Glia, 54*, 343–357.

Lo, E. H., Dalkara, T., & Moskowitz, M. A. (2003). Mechanisms, challenges and opportunities in stroke. *Nature Reviews. Neuroscience, 4*, 399–415.

Lohr, J. W., McReynolds, J., Grimaldi, T., & Acara, M. (1988). Effect of acute and chronic hypernatremia on myoinositol and sorbitol concentration in rat brain and kidney. *Life Sciences, 43*, 271–276.

Longa, E. Z., Weinstein, P. R., Carlson, S., & Cummins, R. (1989). Reversible middle cerebral artery occlusion without craniectomy in rats. *Stroke, 20*, 84–91.

Lutter, D., Ullrich, F., Lueck, J. C., Kempa, S., & Jentsch, T. J. (2017). Selective transport of neurotransmitters and modulators by distinct volume-regulated LRRC8 anion channels. *Journal of Cell Science, 130*, 1122–1133.

MacAulay, N., Gether, U., Klaerke, D. A., & Zeuthen, T. (2001). Water transport by the human Na^{+}-coupled glutamate cotransporter expressed in Xenopus oocytes. *Journal of Physiology, 530*, 367–378.

MacAulay, N., Hamann, S., & Zeuthen, T. (2004). Water transport in the brain: Role of cotransporters. *Neuroscience, 129*, 1031–1044.

MacAulay, N., & Zeuthen, T. (2012). Glial K^{+} clearance and cell swelling: Key roles for cotransporters and pumps. *Neurochemical Research, 37*, 2299–2309.

Machtens, J. P., Kortzak, D., Lansche, C., Leinenweber, A., Kilian, P., Begemann, B., et al. (2015). Mechanisms of anion conduction by coupled glutamate transporters. *Cell, 160*, 542–553.

Macknight, A. D. C., & Leaf, A. (1977). Regulation of cellular volume. *Physiological Reviews, 57*, 510–573.

MacVicar, B. A., Feighan, D., Brown, A., & Ransom, B. (2002). Intrinsic optical signals in the rat optic nerve: Role for K^{+} uptake via NKCC1 and swelling of astrocytes. *Glia, 37*, 114–123.

MacVicar, B. A., & Hochman, D. (1991). Imaging of synaptically evoked intrinsic optical signals in hippocampal slices. *The Journal of Neuroscience: the Official Journal of the Society for Neuroscience, 11*, 1458–1469.

Manley, G. T., Fujimura, M., Ma, T., Noshita, N., Filiz, F., Bollen, A. W., et al. (2000). Aquaporin-4 deletion in mice reduces brain edema after acute water intoxication and ischemic stroke. *Nature Medicine, 6*, 159–163.

Manolopoulos, V. G., Voets, T., Declercq, P. E., Droogmans, G., & Nilius, B. (1997). Swelling-activated efflux of taurine and other organic osmolytes in endothelial cells. *American Journal of Physiology, 273*, C214–C222.

Marcaggi, P., & Coles, J. A. (2001). Ammonium in nervous tissue: Transport across cell membranes, fluxes from neurons to glial cells, and role in signalling. *Progress in Neurobiology, 64*, 157–183.

Martin, D. L., Madelian, V., Seligmann, B., & Shain, W. (1990). The role of osmotic pressure and membrane potential in K^+-stimulated taurine release from cultured astrocytes and LRM55 cells. *The Journal of Neuroscience: the Official Journal of the Society for Neuroscience, 10*, 571–577.
McCormick, J. A., & Ellison, D. H. (2011). The WNKs: Atypical protein kinases with pleiotropic actions. *Physiological Reviews, 91*, 177–219.
McKenna, M. C. (2007). The glutamate-glutamine cycle is not stoichiometric: Fates of glutamate in brain. *Journal of Neuroscience Research, 85*, 3347–3358.
McManus, M. L., & Strange, K. (1993). Acute volume regulation of brain cells in response to hypertonic challenge. *Anesthesiology, 78*, 1132–1137.
McNamara, J. O. (1994). Cellular and molecular basis of epilepsy. *The Journal of Neuroscience: the Official Journal of the Society for Neuroscience, 14*, 3413–3425.
McNamara, J. O., Huang, Y. Z., & Leonard, A. S. (2006). Molecular signaling mechanisms underlying epileptogenesis. *Science's STKE, 2006*, re12.
Meldrum, B. S. (2000). Glutamate as a neurotransmitter in the brain: Review of physiology and pathology. *The Journal of Nutrition, 130*, 1007S–1015S.
Melton, J. E., & Nattie, E. E. (1983). Brain and CSF water and ions during dilutional and isosmotic hyponatremia in the rat. *American Journal of Physiology, 244*, R724–R732.
Melton, J. E., Patlak, C. S., Pettigrew, K. D., & Cserr, H. F. (1987). Volume regulatory loss of Na, Cl, and K from rat brain during acute hyponatremia. *American Journal of Physiology, 252*, F661–F669.
Minieri, L., Pivonkova, H., Harantova, L., Anderova, M., & Ferroni, S. (2015). Intracellular Na(+) inhibits volume-regulated anion channel in rat cortical astrocytes. *Journal of Neurochemistry, 132*, 286–300.
Mohan, S., Gu, S., Parikh, A., & Radhakrishnan, J. (2013). Prevalence of hyponatremia and association with mortality: Results from NHANES. *The American Journal of Medicine, 126*, 1127–1137.
Mongin, A. A. (2007). Disruption of ionic and cell volume homeostasis in cerebral ischemia: The perfect storm. *Pathophysiology, 14*, 183–193.
Mongin, A. A. (2016). Volume-regulated anion channel–a frenemy within the brain. *Pflügers Archiv, 468*, 421–441.
Mongin, A. A., Aksentsev, S. L., Orlov, S. N., & Konev, S. V. (1997). Hypoosmotic shock activates Ca^{2+} channels in isolated nerve terminals. *Neurochemistry International, 31*, 835–843.
Mongin, A. A., Aksentsev, S. L., Orlov, S. N., Kvacheva, Z. B., Mezen, N. I., Fedulov, A. S., et al. (1996). Swelling-induced activation of Na^+,K^+,$2Cl^-$ cotransport in C6 glioma cells: Kinetic properties and intracellular signalling mechanisms. *Biochimica et Biophysica Acta, 1285*, 229–236.
Mongin, A. A., Aksentsev, S. L., Orlov, S. N., Slepko, N. G., Kozlova, M. V., Maximov, G. V., et al. (1994). Swelling-induced K^+ influx in cultured primary astrocytes. *Brain Research, 655*, 110–114.
Mongin, A. A., Cai, Z. H., & Kimelberg, H. K. (1999). Volume-dependent taurine release from cultured astrocytes requires permissive $[Ca^{2+}]_i$ and calmodulin. *American Journal of Physiology. Cell Physiology, 277*, C823–C832.
Mongin, A. A., Dohare, P., & Jourd'heuil, D. (2012). Selective vulnerability of synaptic signaling and metabolism to nitrosative stress. *Antioxidants and Redox Signaling, 17*, 992–1012.
Mongin, A. A., & Kimelberg, H. K. (2002). ATP potently modulates anion channel-mediated excitatory amino acid release from cultured astrocytes. *American Journal of Physiology. Cell Physiology, 283*, C569–C578.
Mongin, A. A., & Kimelberg, H. K. (2005a). Astrocytic swelling in neuropathology. In H. Kettenmann, & B. R. Ransom (Eds.), *Neuroglia* (2nd ed., pp. 550–562). Oxford/New York: Oxford University Press.

Mongin, A. A., & Kimelberg, H. K. (2005b). ATP regulates anion channel-mediated organic osmolyte release from cultured rat astrocytes via multiple Ca^{2+}-sensitive mechanisms. *American Journal of Physiology. Cell Physiology, 288*, C204–C213.

Mongin, A. A., & Kimelberg, H. K. (2009). Regulation of cell volume in neural cells. In L. R. Squire (Ed.), *Encyclopedia of neuroscience* (pp. 81–87). Amsterdam/Heidelberg/London/New York/Paris/Sidney/Tokyo: Elsevier.

Mongin, A. A., & Orlov, S. N. (2001). Mechanisms of cell volume regulation and possible nature of the cell volume sensor. *Pathophysiology, 8*, 77–88.

Mongin, A. A., Reddi, J. M., Charniga, C., & Kimelberg, H. K. (1999). [^{3}H]Taurine and D-[^{3}H]aspartate release from astrocyte cultures are differently regulated by tyrosine kinases. *American Journal of Physiology. Cell Physiology, 276*, C1226–C1230.

Moran, J., Maar, T., & Pasantes-Morales, H. (1994). Cell volume regulation in taurine deficient cultured astrocytes. *Advances in Experimental Medicine and Biology, 359*, 361–367.

Morfini, G. A., Stenoien, D. L., & Brady, S. T. (2006). Axonal transport. In G. J. Siegel, R. W. Albers, S. T. Brady, & D. L. Price (Eds.), *Basic neurochemistry: Molecular, cellular and medical aspects* (7th ed., pp. 485–501). London: Elsevier Academic Press.

Morris, C. E., & Homann, U. (2001). Cell surface area regulation and membrane tension. *The Journal of Membrane Biology, 179*, 79–102.

Murphy, T. R., Binder, D. K., & Fiacco, T. A. (2017). Turning down the volume: Astrocyte volume change in the generation and termination of epileptic seizures. *Neurobiology of Disease, 104*, 24–32.

Murphy, T. R., Davila, D., Cuvelier, N., Young, L. R., Lauderdale, K., Binder, D. K., et al. (2017). Hippocampal and cortical pyramidal neurons swell in parallel with astrocytes during acute hypoosmolar stress. *Frontiers in Cellular Neuroscience, 11*, 275.

Murthy, C. R., Rama Rao, K. V., Bai, G., & Norenberg, M. D. (2001). Ammonia-induced production of free radicals in primary cultures of rat astrocytes. *Journal of Neuroscience Research, 66*, 282–288.

Nagaraja, T. N., & Brookes, N. (1998). Intracellular acidification induced by passive and active transport of ammonium ions in astrocytes. *American Journal of Physiology, 274*, C883–C891.

Nagelhus, E. A., Mathiisen, T. M., & Ottersen, O. P. (2004). Aquaporin-4 in the central nervous system: Cellular and subcellular distribution and coexpression with KIR4.1. *Neuroscience, 129*, 905–913.

Neumann-Haefelin, T., Kastrup, A., de, C. A., Yenari, M. A., Ringer, T., Sun, G. H., et al. (2000). Serial MRI after transient focal cerebral ischemia in rats: Dynamics of tissue injury, blood-brain barrier damage, and edema formation. *Stroke, 31*, 1965–1972.

Neumann-Haefelin, T., & Witte, O. W. (2000). Periinfarct and remote excitability changes after transient middle cerebral artery occlusion. *Journal of Cerebral Blood Flow & Metabolism, 20*, 45–52.

Neumann-Haefelin, T., Wittsack, H. J., Wenserski, F., Siebler, M., Seitz, R. J., Modder, U., et al. (1999). Diffusion- and perfusion-weighted MRI. The DWI/PWI mismatch region in acute stroke. *Stroke, 30*, 1591–1597.

Nicholson, C., & Sykova, E. (1998). Extracellular space structure revealed by diffusion analysis. *Trends in Neurosciences, 21*, 207–215.

Nilius, B., & Droogmans, G. (2003). Amazing chloride channels: An overview. *Acta Physiologica Scandinavica, 177*, 119–147.

Nilius, B., Eggermont, J., Voets, T., Buyse, G., Manolopoulos, V., & Droogmans, G. (1997). Properties of volume-regulated anion channels in mammalian cells. *Progress in Biophysics and Molecular Biology, 68*, 69–119.

Nilius, B., Sehrer, J., & Droogmans, G. (1994). Permeation properties and modulation of volume-activated Cl^{-}-currents in human endothelial cells. *British Journal of Pharmacology, 112*, 1049–1056.

Noda, M., & Hiyama, T. Y. (2015). Sodium sensing in the brain. *Pflügers Archiv, 467*, 465–474.

Norenberg, M. D. (1977). A light and electron microscopic study of experimental portal-systemic (ammonia) encephalopathy. Progression and reversal of the disorder. *Laboratory Investigation; a Journal of Technical Methods and Pathology, 36*, 618–627.

Norenberg, M. D. (1994). Astrocyte responses to CNS injury. *Journal of Neuropathology and Experimental Neurology, 53*, 213–220.

Norenberg, M. D. (1998). Astroglial dysfunction in hepatic encephalopathy. *Metabolic Brain Disease, 13*, 319–335.

Norenberg, M. D., Baker, L., Norenberg, L. O., Blicharska, J., Bruce-Gregorios, J. H., & Neary, J. T. (1991). Ammonia-induced astrocyte swelling in primary culture. *Neurochemical Research, 16*, 833–836.

Norenberg, M. D., Jayakumar, A. R., Rama Rao, K. V., & Panickar, K. S. (2007). New concepts in the mechanism of ammonia-induced astrocyte swelling. *Metabolic Brain Disease, 22*, 219–234.

Norenberg, M. D., Rao, K. V., & Jayakumar, A. R. (2005). Mechanisms of ammonia-induced astrocyte swelling. *Metabolic Brain Disease, 20*, 303–318.

O'Connor, E. R., Kimelberg, H. K., Keese, C. R., & Giaever, I. (1993). Electrical resistance method for measuring volume changes in monolayer cultures applied to primary astrocyte cultures. *American Journal of Physiology, 264*, C471–C478.

O'Neill, W. C. (1999). Physiological significance of volume-regulatory transporters. *American Journal of Physiology, 276*, C995–C1011.

Oike, M., Droogmans, G., & Nilius, B. (1994). The volume-activated chloride current in human endothelial cells depends on intracellular ATP. *Pflügers Archiv, 427*, 184–186.

Okada, Y. (1997). Volume expansion-sensing outward-rectifier Cl^- channel: Fresh start to the molecular identity and volume sensor. *American Journal of Physiology. Cell Physiology, 273*, C755–C789.

Okada, Y. (2018). Molecular identities of volume-regulatory anion channels. *Current Topics in Membranes, 81*. In press.

Okada, Y., Sato, K., & Numata, T. (2009). Pathophysiology and puzzles of the volume-sensitive outwardly rectifying anion channel. *Journal of Physiology, 587*, 2141–2149.

Olsen, M. L., Khakh, B. S., Skatchkov, S. N., Zhou, M., Lee, C. J., & Rouach, N. (2015). New insights on astrocyte ion channels: Critical for homeostasis and neuron-Glia signaling. *The Journal of Neuroscience: the Official Journal of the Society for Neuroscience, 35*, 13827–13835.

Olson, J. E., & Li, G. Z. (1997). Increased potassium, chloride, and taurine conductances in astrocytes during hypoosmotic swelling. *Glia, 20*, 254–261.

Olson, J. E., Sankar, R., Holtzman, D., James, A., & Fleischhacker, D. (1986). Energy-dependent volume regulation in primary cultured cerebral astrocytes. *Journal of Cellular Physiology, 128*, 209–215.

Orellana, J. A., Avendano, B. C., & Montero, T. D. (2014). Role of connexins and pannexins in ischemic stroke. *Current Medicinal Chemistry, 21*, 2165–2182.

Pacheco-Alvarez, D., & Gamba, G. (2011). WNK3 is a putative chloride-sensing kinase. *Cellular Physiology and Biochemistry: International Journal of Experimental Cellular Physiology, Biochemistry, and Pharmacology, 28*, 1123–1134.

Palma, E., Amici, M., Sobrero, F., Spinelli, G., Di, A. S., Ragozzino, D., et al. (2006). Anomalous levels of Cl^- transporters in the hippocampal subiculum from temporal lobe epilepsy patients make GABA excitatory. *Proceedings of the National Academy of Sciences of the United States of America, 103*, 8465–8468.

Parkerson, K. A., & Sontheimer, H. (2004). Biophysical and pharmacological characterization of hypotonically activated chloride currents in cortical astrocytes. *Glia, 46*, 419–436.

Pasantes Morales, H., & Schousboe, A. (1988). Volume regulation in astrocytes: A role for taurine as an osmoeffector. *Journal of Neuroscience Research, 20*, 503–509.

Pasantes-Morales, H., Chacon, E., Murray, R. A., & Moran, J. (1994). Properties of osmolyte fluxes activated during regulatory volume decrease in cultured cerebellar granule neurons. *Journal of Neuroscience Research, 37*, 720–727.

Pasantes-Morales, H., Lezama, R. A., Ramos-Mandujano, G., & Tuz, K. L. (2006). Mechanisms of cell volume regulation in hypo-osmolality. *The American Journal of Medicine, 119*, S4–S11.

Pasantes-Morales, H., Maar, T. E., & Moran, J. (1993). Cell volume regulation in cultured cerebellar granule neurons. *Journal of Neuroscience Research, 34*, 219–224.

Pasantes-Morales, H., Moran, J., & Schousboe, A. (1990). Volume-sensitive release of taurine from cultured astrocytes: Properties and mechanism. *Glia, 3*, 427–432.

Pasantes-Morales, H., Murray, R. A., Lilja, L., & Moran, J. (1994). Regulatory volume decrease in cultured astrocytes. I. Potassium- and chloride-activated permeability. *American Journal of Physiology, 266*, C165–C171.

Pasantes-Morales, H., Murray, R. A., Sanchez-Olea, R., & Moran, J. (1994). Regulatory volume decrease in cultured astrocytes. II. Permeability pathway to amino acids and polyols. *American Journal of Physiology, 266*, C172–C178.

Patel, A. J., Lauritzen, I., Lazdunski, M., & Honore, E. (1998). Disruption of mitochondrial respiration inhibits volume-regulated anion channels and provokes neuronal cell swelling. *The Journal of Neuroscience: the Official Journal of the Society for Neuroscience, 18*, 3117–3123.

Pedersen, S. F., Okada, Y., & Nilius, B. (2016). Biophysics and physiology of the volume-regulated anion channel (VRAC)/volume-sensitive outwardly rectifying anion channel (VSOR). *Pflügers Archiv, 468*, 371–383.

Phillis, J. W., & O'Regan, M. H. (1996). Mechanisms of glutamate and aspartate release in the ischemic rat cerebral cortex. *Brain Research, 730*, 150–164.

Phillis, J. W., Song, D., & O'Regan, M. H. (1997). Inhibition by anion channel blockers of ischemia-evoked release of excitotoxic and other amino acids from rat cerebral cortex. *Brain Research, 758*, 9–16.

Phillis, J. W., Song, D., & O'Regan, M. H. (1998). Tamoxifen, a chloride channel blocker, reduces glutamate and aspartate release from the ischemic cerebral cortex. *Brain Research, 780*, 352–355.

Piala, A. T., Moon, T. M., Akella, R., He, H., Cobb, M. H., & Goldsmith, E. J. (2014). Chloride sensing by WNK1 involves inhibition of autophosphorylation. *Science Signaling [electronic Resource], 7*, ra41.

Planells-Cases, R., Lutter, D., Guyader, C., Gerhards, N. M., Ullrich, F., Elger, D. A., et al. (2015). Subunit composition of VRAC channels determines substrate specificity and cellular resistance to Pt-based anti-cancer drugs. *The EMBO Journal, 34*, 2993–3008.

Podesta, M. A., Faravelli, I., Cucchiari, D., Reggiani, F., Oldani, S., Fedeli, C., et al. (2015). Neurological counterparts of hyponatremia: Pathological mechanisms and clinical manifestations. *Current Neurology and Neuroscience Reports, 15*, 18.

Qiu, Z. (2018). Molecular biology and physiology of volume-regulated anion channel (VRAC). *Current Topics in Membranes, 81*.

Qiu, Z., Dubin, A. E., Mathur, J., Tu, B., Reddy, K., Miraglia, L. J., et al. (2014). SWELL1, a plasma membrane protein, is an essential component of volume-regulated anion channel. *Cell, 157*, 447–458.

Ramos-Mandujano, G., Vazquez-Juarez, E., Hernandez-Benitez, R., & Pasantes-Morales, H. (2007). Thrombin potently enhances swelling-sensitive glutamate efflux from cultured astrocytes. *Glia, 55*, 917–925.

Risher, W. C., Andrew, R. D., & Kirov, S. A. (2009). Real-time passive volume responses of astrocytes to acute osmotic and ischemic stress in cortical slices and in vivo revealed by two-photon microscopy. *Glia, 57*, 207–221.

Risher, W. C., Croom, D., & Kirov, S. A. (2012). Persistent astroglial swelling accompanies rapid reversible dendritic injury during stroke-induced spreading depolarizations. *Glia*.

Rossi, D. J., Oshima, T., & Attwell, D. (2000). Glutamate release in severe brain ischaemia is mainly by reversed uptake. *Nature, 403*, 316–321.

Rosso, L., Peteri-Brunback, B., Poujeol, P., Hussy, N., & Mienville, J. M. (2004). Vasopressin-induced taurine efflux from rat pituicytes: A potential negative feedback for hormone secretion. *Journal of Physiology, 554*, 731–742.

Roy, G. (1995). Amino-acid current through anion channels in cultured human glial-cells. *The Journal of Membrane Biology, 147*, 35–44.

Rudkouskaya, A., Chernoguz, A., Haskew-Layton, R. E., & Mongin, A. A. (2008). Two conventional protein kinase C isoforms, alpha and betaI, are involved in the ATP-induced activation of volume-regulated anion channel and glutamate release in cultured astrocytes. *Journal of Neurochemistry, 105*, 2260–2270.

Russell, J. M. (2000). Sodium-potassium-chloride cotransport. *Physiological Reviews, 80*, 211–276.

Rutledge, E. M., Aschner, M., & Kimelberg, H. K. (1998). Pharmacological characterization of swelling-induced D-[^{3}H]aspartate release from primary astrocyte cultures. *American Journal of Physiology, 274*, C1511–C1520.

Rutledge, E. M., & Kimelberg, H. K. (1996). Release of [^{3}H]-D-aspartate from primary astrocyte cultures in response to raised external potassium. *The Journal of Neuroscience: the Official Journal of the Society for Neuroscience, 16*, 7803–7811.

Rutledge, E. M., Mongin, A. A., & Kimelberg, H. K. (1999). Intracellular ATP depletion inhibits swelling-induced D-[^{3}H]aspartate release from primary astrocyte cultures. *Brain Research, 842*, 39–45.

Sanchez-Olea, R., Moran, J., Martinez, A., & Pasantes-Morales, H. (1993). Volume-activated Rb^+ transport in astrocytes in culture. *American Journal of Physiology, 264*, C836–C842.

Sanchez-Olea, R., Pena, C., Moran, J., & Pasantes-Morales, H. (1993). Inhibition of volume regulation and efflux of osmoregulatory amino acids by blockers of Cl^- transport in cultured astrocytes. *Neuroscience Letters, 156*, 141–144.

Schneider, G. H., Baethmann, A., & Kempski, O. (1992). Mechanisms of glial swelling induced by glutamate. *Canadian Journal of Physiology and Pharmacology, 70*(Suppl), S334–S343.

Schober, A. L., & Mongin, A. A. (2015). Intracellular levels of glutamate in swollen astrocytes are preserved via neurotransmitter reuptake and de novo synthesis: Implications for hyponatremia. *Journal of Neurochemistry, 135*, 176–185.

Schober, A. L., Wilson, C. S., & Mongin, A. A. (2017). Molecular composition and heterogeneity of the LRRC8-containing swelling-activated osmolyte channels in primary rat astrocytes. *Journal of Physiology, 595*, 6939–6951.

Schomberg, S. L., Bauer, J., Kintner, D. B., Su, G., Flemmer, A., Forbush, B., et al. (2003). Cross talk between the GABA(A) receptor and the Na-K-Cl cotransporter is mediated by intracellular Cl^-. *Journal of Neurophysiology, 89*, 159–167.

Seki, Y., Feustel, P. J., Keller, R. W., Jr., Tranmer, B. I., & Kimelberg, H. K. (1999). Inhibition of ischemia-induced glutamate release in rat striatum by dihydrokinate and an anion channel blocker. *Stroke, 30*, 433–440.

Shennan, D. B. (2008). Swelling-induced taurine transport: Relationship with chloride channels, anion-exchangers and other swelling-activated transport pathways. *Cellular Physiology and Biochemistry: International Journal of Experimental Cellular Physiology, Biochemistry, and Pharmacology, 21*, 15–28.

Shennan, D. B., McNeillie, S. A., & Curran, D. E. (1994). The effect of a hyposmotic shock on amino acid efflux from lactating rat mammary tissue: Stimulation of taurine and glycine efflux via a pathway distinct from anion exchange and volume-activated anion channels. *Experimental Physiology, 79*, 797–808.

Sheth, K. N., Elm, J. J., Molyneaux, B. J., Hinson, H., Beslow, L. A., Sze, G. K., et al. (2016). Safety and efficacy of intravenous glyburide on brain swelling after large hemispheric infarction (GAMES-RP): A randomised, double-blind, placebo-controlled phase 2 trial. *The Lancet Neurology, 15*, 1160–1169.

Shrode, L. D., Klein, J. D., O'Neill, W. C., & Putnam, R. W. (1995). Shrinkage-induced activation of Na^{+}/H^{+} exchange in primary rat astrocytes: Role of myosin light-chain kinase. *American Journal of Physiology, 269*, C257–C266.

Silver, S. M., Schroeder, B. M., Sterns, R. H., & Rojiani, A. M. (2006). Myoinositol administration improves survival and reduces myelinolysis after rapid correction of chronic hyponatremia in rats. *Journal of Neuropathology and Experimental Neurology, 65*, 37–44.

Simard, J. M., Chen, M., Tarasov, K. V., Bhatta, S., Ivanova, S., Melnitchenko, L., et al. (2006). Newly expressed SUR1-regulated NC(Ca-ATP) channel mediates cerebral edema after ischemic stroke. *Nature Medicine, 12*, 433–440.

Simard, J. M., Tsymbalyuk, N., Tsymbalyuk, O., Ivanova, S., Yurovsky, V., & Gerzanich, V. (2010). Glibenclamide is superior to decompressive craniectomy in a rat model of malignant stroke. *Stroke, 41*, 531–537.

Slais, K., Vorisek, I., Zoremba, N., Homola, A., Dmytrenko, L., & Sykova, E. (2008). Brain metabolism and diffusion in the rat cerebral cortex during pilocarpine-induced status epilepticus. *Experimental Neurology, 209*, 145–154.

Soe, R., MacAulay, N., & Klaerke, D. A. (2009). Modulation of Kir4.1 and Kir4.1-Kir5.1 channels by small changes in cell volume. *Neuroscience Letters, 457*, 80–84.

Sofroniew, M. V., & Vinters, H. V. (2010). Astrocytes: Biology and pathology. *Acta Neuropathologica, 119*, 7–35.

Solenov, E., Watanabe, H., Manley, G. T., & Verkman, A. S. (2004). Sevenfold-reduced osmotic water permeability in primary astrocyte cultures from AQP-4-deficient mice, measured by a fluorescence quenching method. *American Journal of Physiology. Cell Physiology, 286*, C426–C432.

Somjen, G. G. (2004). *Ions in the brain: Normal functions, seizures, and stroke*. New York: Oxford University Press.

Somjen, G. G., Faas, G. C., Vreugdenhil, M., & Wadman, W. J. (1993). Channel shutdown: A response of hippocampal neurons to adverse environments. *Brain Research, 632*, 180–194.

Song, D., Xu, J., Hertz, L., & Peng, L. (2015). Regulatory volume increase in astrocytes exposed to hypertonic medium requires beta1 -adrenergic Na^{+}/K^{+}-ATPase stimulation and glycogenolysis. *Journal of Neuroscience Research, 93*, 130–139.

Sotak, C. H. (2004). Nuclear magnetic resonance (NMR) measurement of the apparent diffusion coefficient (ADC) of tissue water and its relationship to cell volume changes in pathological states. *Neurochemistry International, 45*, 569–582.

Stauber, T. (2015). The volume-regulated anion channel is formed by LRRC8 heteromers - molecular identification and roles in membrane transport and physiology. *Biological Chemistry, 396*, 975–990.

Steffensen, A. B., Sword, J., Croom, D., Kirov, S. A., & MacAulay, N. (2015). Chloride cotransporters as a molecular mechanism underlying spreading depolarization-induced dendritic beading. *The Journal of Neuroscience: the Official Journal of the Society for Neuroscience, 35*, 12172–12187.

Sterns, R. H. (2015). Disorders of plasma sodium–causes, consequences, and correction. *The New England Journal of Medicine, 372*, 55–65.

Sterns, R. H., Baer, J., Ebersol, S., Thomas, D., Lohr, J. W., & Kamm, D. E. (1993). Organic osmolytes in acute hyponatremia. *American Journal of Physiology, 264*, F833–F836.

Sterns, R. H., & Silver, S. M. (2006). Brain volume regulation in response to hypo-osmolality and its correction. *The American Journal of Medicine, 119*, S12–S16.
Stokum, J. A., Gerzanich, V., & Simard, J. M. (2016). Molecular pathophysiology of cerebral edema. *Journal of Cerebral Blood Flow & Metabolism, 36*, 513–538.
Stokum, J. A., Kwon, M. S., Woo, S. K., Tsymbalyuk, O., Vennekens, R., Gerzanich, V., et al. (2018). SUR1-TRPM4 and AQP4 form a heteromultimeric complex that amplifies ion/water osmotic coupling and drives astrocyte swelling. *Glia, 66*, 108–125.
Strange, K. (1992). Regulation of solute and water balance and cell volume in the central nervous system. *Journal of the American Society of Nephrology: JASN, 3*, 12–27.
Strange, K., Emma, F., & Jackson, P. S. (1996). Cellular and molecular physiology of volume-sensitive anion channels. *American Journal of Physiology, 270*, C711–C730.
Strange, K., & Morrison, R. (1992). Volume regulation during recovery from chronic hypertonicity in brain glial cells. *American Journal of Physiology, 263*, C412–C419.
Strange, K., Morrison, R., Shrode, L., & Putnam, R. (1993). Mechanism and regulation of swelling-activated inositol efflux in brain glial cells. *American Journal of Physiology, 265*, C244–C256.
Stutzin, A., Torres, R., Oporto, M., Pacheco, P., Eguiguren, A. L., Cid, L. P., et al. (1999). Separate taurine and chloride efflux pathways activated during regulatory volume decrease. *American Journal of Physiology. Cell Physiology, 277*, C392–C402.
Su, G., Kintner, D. B., Flagella, M., Shull, G. E., & Sun, D. (2002). Astrocytes from Na^+-K^+-Cl^- cotransporter-null mice exhibit absence of swelling and decrease in EAA release. *American Journal of Physiology. Cell Physiology, 282*, C1147–C1160.
Su, G., Kintner, D. B., & Sun, D. (2002). Contribution of Na^+-K^+-Cl^- cotransporter to high-[K(+)](o)- induced swelling and EAA release in astrocytes. *American Journal of Physiology. Cell Physiology, 282*, C1136–C1146.
Svoboda, J., & Sykova, E. (1991). Extracellular space volume changes in the rat spinal cord produced by nerve stimulation and peripheral injury. *Brain Research, 560*, 216–224.
Swain, M., Butterworth, R. F., & Blei, A. T. (1992). Ammonia and related amino acids in the pathogenesis of brain edema in acute ischemic liver failure in rats. *Hepatology, 15*, 449–453.
Sykova, E. (2004). Extrasynaptic volume transmission and diffusion parameters of the extracellular space. *Neuroscience, 129*, 861–876.
Sykova, E., & Nicholson, C. (2008). Diffusion in brain extracellular space. *Physiological Reviews, 88*, 1277–1340.
Szatkowski, M., Barbour, B., & Attwell, D. (1990). Non-vesicular release of glutamate from glial cells by reversed electrogenic glutamate uptake. *Nature, 348*, 443–446.
Takano, T., Kang, J., Jaiswal, J. K., Simon, S. M., Lin, J. H., Yu, Y., et al. (2005). Receptor-mediated glutamate release from volume sensitive channels in astrocytes. *Proceedings of the National Academy of Sciences of the United States of America, 102*, 16466–16471.
Tanigami, H., Rebel, A., Martin, L. J., Chen, T. Y., Brusilow, S. W., Traystman, R. J., et al. (2005). Effect of glutamine synthetase inhibition on astrocyte swelling and altered astroglial protein expression during hyperammonemia in rats. *Neuroscience, 131*, 437–449.
Terker, A. S., Zhang, C., Erspamer, K. J., Gamba, G., Yang, C. L., & Ellison, D. H. (2016). Unique chloride-sensing properties of WNK4 permit the distal nephron to modulate potassium homeostasis. *Kidney International, 89*, 127–134.
Thijs, V. N., Adami, A., Neumann-Haefelin, T., Moseley, M. E., & Albers, G. W. (2002). Clinical and radiological correlates of reduced cerebral blood flow measured using magnetic resonance imaging. *Archives of Neurology, 59*, 233–238.
Thrane, A. S., Rangroo Thrane, V., & Nedergaard, M. (2014). Drowning stars: Reassessing the role of astrocytes in brain edema. *Trends in Neurosciences, 37*, 620–628.
Thurman, D. J., Beghi, E., Begley, C. E., Berg, A. T., Buchhalter, J. R., Ding, D., et al. (2011). Standards for epidemiologic studies and surveillance of epilepsy. *Epilepsia, 52*(Suppl 7), 2–26.

Thurston, J. H., & Hauhart, R. E. (1987). Brain amino acids decrease in chronic hyponatremia and rapid correction causes brain dehydration: Possible clinical significance. *Life Sciences, 40*, 2539–2542.

Tofteng, F., Hauerberg, J., Hansen, B. A., Pedersen, C. B., Jorgensen, L., & Larsen, F. S. (2006). Persistent arterial hyperammonemia increases the concentration of glutamine and alanine in the brain and correlates with intracranial pressure in patients with fulminant hepatic failure. *Journal of Cerebral Blood Flow & Metabolism, 26*, 21–27.

Tonnesen, J., Inavalli, V. V. G. K., & Nagerl, U. V. (2018). Super-resolution imaging of the extracellular space in living brain tissue. *Cell, 172*, 1108–1121.

Traynelis, S. F., & Dingledine, R. (1989). Role of extracellular space in hyperosmotic suppression of potassium-induced electrographic seizures. *Journal of Neurophysiology, 61*, 927–938.

Tuz, K., & Pasantes-Morales, H. (2005). Hyposmolarity evokes norepinephrine efflux from synaptosomes by a depolarization- and Ca^{2+} -dependent exocytotic mechanism. *The European Journal of Neuroscience, 22*, 1636–1642.

Tuz, K., Pena-Segura, C., Franco, R., & Pasantes-Morales, H. (2004). Depolarization, exocytosis and amino acid release evoked by hyposmolarity from cortical synaptosomes. *The European Journal of Neuroscience, 19*, 916–924.

Upadhyay, A., Jaber, B. L., & Madias, N. E. (2006). Incidence and prevalence of hyponatremia. *The American Journal of Medicine, 119*, S30–S35.

Van Harreveld, A. (1972). The extracellular space in the vertebrate central nervous system. In G. H. Bourne (Ed.), *The structure and function of nervous system* (pp. 449–512). Academic Press.

Van Harreveld, A., Dafny, N., & Khattab, F. I. (1971). Effects of calcium on the electrical resistance and the extracellular space of cerebral cortex. *Experimental Neurology, 31*, 358–367.

Verbalis, J. G., & Gullans, S. R. (1991). Hyponatremia causes large sustained reductions in brain content of multiple organic osmolytes in rats. *Brain Research, 567*, 274–282.

Verkhratsky, A., & Nedergaard, M. (2018). Physiology of astroglia. *Physiological Reviews, 98*, 239–389.

Verkman, A. S., Binder, D. K., Bloch, O., Auguste, K., & Papadopoulos, M. C. (2006). Three distinct roles of aquaporin-4 in brain function revealed by knockout mice. *Biochimica et Biophysica Acta, 1758*, 1085–1093.

Vitarella, D., DiRisio, D. J., Kimelberg, H. K., & Aschner, M. (1994). Potassium and taurine release are highly correlated with regulatory volume decrease in neonatal primary rat astrocyte cultures. *Journal of Neurochemistry, 63*, 1143–1149.

Voisin, D. L., Chakfe, Y., & Bourque, C. W. (1999). Coincident detection of CSF Na^{+} and osmotic pressure in osmoregulatory neurons of the supraoptic nucleus. *Neuron, 24*, 453–460.

Voss, F. K., Ullrich, F., Munch, J., Lazarow, K., Lutter, D., Mah, N., et al. (2014). Identification of LRRC8 heteromers as an essential component of the volume-regulated anion channel VRAC. *Science, 344*, 634–638.

Wade, J. V., Olson, J. P., Samson, F. E., Nelson, S. R., & Pazdernik, T. L. (1988). A possible role for taurine in osmoregulation within the brain. *Journal of Neurochemistry, 51*, 740–745.

Wang, R., Lu, Y., Gunasekar, S., Zhang, Y., Benson, C. J., Chapleau, M. W., et al. (2017). The volume-regulated anion channel (LRRC8) in nodose neurons is sensitive to acidic pH. *JCI Insight, 2*. e90632.

Waseem, T. V., Rakovich, A. A., Lavrukevich, T. V., Konev, S. V., & Fedorovich, S. V. (2005). Calcium regulates the mode of exocytosis induced by hypotonic shock in isolated neuronal presynaptic endings. *Neurochemistry International, 46*, 235–242.

Wasterlain, C. G., & Torack, R. M. (1968). Cerebral edema in water intoxication. II. An ultrastructural study. *Archives of Neurology, 19*, 79–87.

Wehner, F., Bondarava, M., ter, V. F., Endl, E., Nurnberger, H. R., & Li, T. (2006). Hypertonicity-induced cation channels. *Acta Physiologica, 187*, 21–25.
Wehner, F., Olsen, H., Tinel, H., Kinne-Saffran, E., & Kinne, R. K. H. (2003). Cell volume regulation: Osmolytes, osmolyte transport, and signal transduction. *Reviews of Physiology, Biochemistry and Pharmacology, 148*, 1–80.
Wehner, F., Shimizu, T., Sabirov, R., & Okada, Y. (2003). Hypertonic activation of a non-selective cation conductance in HeLa cells and its contribution to cell volume regulation. *FEBS Letters, 551*, 20–24.
Willford, S. L., Anderson, C. M., Spencer, S. R., & Eskandari, S. (2015). Evidence for a revised ion/substrate coupling stoichiometry of GABA transporters. *The Journal of Membrane Biology, 248*, 795–810.
Wilson, T. H. (1954). Ionic permeability and osmotic swelling of cells. *Science, 120*, 104–105.
Wilson, C. S., & Mongin, A. A. (2018). The signaling role for chloride in the bidirectional communication between neurons and astrocytes. *Neuroscience Letters*. https://doi.org/10.1016/j.neulet.2018.01.012. Epub ahead of print.
Woo, S. K., Kwon, M. S., Ivanov, A., Gerzanich, V., & Simard, J. M. (2013). The sulfonylurea receptor 1 (Sur1)-transient receptor potential melastatin 4 (Trpm4) channel. *Journal of Biological Chemistry, 288*, 3655–3667.
Woo, N. S., Lu, J., England, R., McClellan, R., Dufour, S., Mount, D. B., et al. (2002). Hyperexcitability and epilepsy associated with disruption of the mouse neuronal-specific K-Cl cotransporter gene. *Hippocampus, 12*, 258–268.
Yan, Y., Dempsey, R. J., & Sun, D. (2001). Na^+-K^+-Cl^- cotransporter in rat focal cerebral ischemia. *Journal of Cerebral Blood Flow & Metabolism, 21*, 711–721.
Yannet, H. (1940). Changes in the brain resulting from depletion of extracellular electrolytes. *American Journal of Physiology, 128*, 683–689.
Ye, Z. C., Oberheim, N., Kettenmann, H., & Ransom, B. R. (2009). Pharmacological "cross-inhibition" of connexin hemichannels and swelling activated anion channels. *Glia, 57*, 258–269.
Ye, Z. C., Wyeth, M. S., Baltan-Tekkok, S., & Ransom, B. R. (2003). Functional hemichannels in astrocytes: A novel mechanism of glutamate release. *The Journal of Neuroscience: the Official Journal of the Society for Neuroscience, 23*, 3588–3596.
Zeuthen, T., & MacAulay, N. (2002). Cotransporters as molecular water pumps. *International Review of Cytology, 215*, 259–284.
Zhang, Z., & Bourque, C. W. (2003). Osmometry in osmosensory neurons. *Nature Neuroscience, 6*, 1021–1022.
Zhang, Y., Chen, K., Sloan, S. A., Bennett, M. L., Scholze, A. R., O'Keeffe, S., et al. (2014). An RNA-sequencing transcriptome and splicing database of glia, neurons, and vascular cells of the cerebral cortex. *The Journal of Neuroscience: the Official Journal of the Society for Neuroscience, 34*, 11929–11947.
Zhang, H., & Verkman, A. S. (2008). Aquaporin-4 independent Kir4.1 K^+ channel function in brain glial cells. *Molecular and Cellular Neurosciences, 37*, 1–10.
Zhang, Y., Zhang, H., Feustel, P. J., & Kimelberg, H. K. (2008). DCPIB, a specific inhibitor of volume regulated anion channels (VRACs), reduces infarct size in MCAo and the release of glutamate in the ischemic cortical penumbra. *Experimental Neurology, 210*, 514–520.
Zhou, Y., & Danbolt, N. C. (2013). GABA and Glutamate transporters in brain. *Frontiers in Endocrinology, 4*, 165.

CHAPTER TWELVE

Cytotoxic Swelling of Sick Excitable Cells — Impaired Ion Homeostasis and Membrane Tension Homeostasis in Muscle and Neuron

Catherine E. Morris[1]

Senior Scientist Emeritus, Neuroscience, Ottawa Hospital Research Institute, Ottawa, ON, Canada
[1]Corresponding author: E-mail: cmorris@uottawa.ca

Contents

Current Topics in Membranes, Volume 81
ISSN 1063-5823
https://doi.org/10.1016/bs.ctm.2018.06.001

Abstract

When they become simultaneously leaky to both Na^+ and Cl^-, excitable cells are vulnerable to potentially lethal cytotoxic swelling. Swelling ensues in spite of an isosmotic milieu because the entering ions add osmolytes to the cytoplasm's high concentration of impermeant anionic osmolytes. An influx of osmotically-obliged water is unavoidable. A cell that cannot stanch at least one the leaks will succumb to death by Donnan effect. "Sick excitable cells" are those injured through ischemia, trauma, inflammation, hyperactivity, genetically-impaired membrane skeletons and other insults, all of which foster bleb-damage to regions of the plasma membrane. Nav channels resident in damaged membrane exhibit left-shifted kinetics; the corresponding Nav window conductance constitutes a Na^+-leak. In cortical neurons, sustained depolarization to ~ -20 mV elicits a sustained lethal gCl. Underlying V_{rest} in skeletal muscle is a constitutively active gCl; not surprisingly therefore, dystrophic muscle fibers, which are prone to bleb damage and which exhibit Nav-leak and Na^+-overload, are prone to cytotoxic swelling. To restore viability in cytotoxically swelling neurons and muscle, the imperative of fully functional ion homeostasis is well-recognized. However, as emphasized here, in a healthy excitable cell, fully functional membrane tension homeostasis is also imperative. ATPase-pumps keep plasma membrane batteries charged, and ATPase-motor proteins maintain membrane tone. In sick excitable cells, neither condition prevails.

Abbreviations

AFM	atomic force microscopy
CLS	coupled left-shift
CNS	central nervous system
DHGP16	short for Dijkstra, Hofmeijer, van Gils, and van Putten (2016).
FES	free energy starvation
gCl	chloride conductance
HD14, HSD14, HU16	short for model systems in 3 Hübel first author papers (see References)
IH	ion homeostasis
KCC2	a KCl cotransporter
MD	muscular dystrophy
MPS	membrane-associated periodic skeleton
MTH	membrane tension homeostasis
Nav, Nav1.X	voltage-gated sodium channel (isoforms are signified as 1.X; X is a digit)
Na^+/K^+-ATPase (or Na/K pump)	the "sodium pump"

NMDA	agonist after which NMDA glutamate channels are named
P4-ATPase	a membrane flippase (same family as Na^+/K^+-ATPase)
SLC26A11	a neuronal V-regulated chloride conductance
SD	spreading depression or spreading depolarization
TTX	tetrodotoxin
$\mathbf{V_m}$	membrane voltage
$\mathbf{V_{rest}}$	resting membrane voltage
VLD	vacuole-like dilation

1. MEMBRANE INJURY. NA^+-LOADING. SWELLING DEATH. OR NOT?

Excitotoxic swelling is potentially lethal for both neurons and muscle fibers. A deeper understanding of the mechanisms of cytotoxic swelling in these excitable cells is emerging, particularly from studies of excitotoxic brain injury and dystrophic muscle. In both brain and skeletal muscle, voltage-gated sodium (Nav) channels underlie electrical excitability. For some years "acquired sodium channelopathies" was a term for excitable cell pathologies involving malfunction of genetically normal Nav channels (Waxman 2001, 2005; Novak et al., 2009). In 2009 (Wang et al., 2009) we noted a commonality: bleb-type damage of the plasma membrane was a pre-existing condition for all the acquired channelopathies. Bleb-damage to the plasma membrane involves regions where the bilayer, having become detached from the membrane skeleton to which it is normally adherent, fails to re-attach and remodel; in effect, the active membrane maintenance processes (e.g., Barthélémy, Defour, Lévy, & Krahn, 2018; Norman, Brugués, Sengupta, Sens, & Aranda-Espinoza, 2010) are impaired for one of many reasons. Not surprisingly, bleb-damage to the plasma membrane is a feature of ischemically, excitotoxically and traumatically injured neurons and of muscle fibers with genetically impaired membrane skeletons (as in dystrophic muscle fibers) (Morris, Boucher, & Joós, 2012a). Nav channels reside in bleb-damaged membrane, and though fully functional, their voltage-dependent activation process occurs at abnormally hyperpolarized voltages, and fast inactivation follows suit because it is so tightly coupled kinetically, resulting in coupled a hyperpolarizing shift or coupled left-shift or CLS, yielding a small window conductance closer to the excitable cell's normal V_{rest} (Boucher, Joós, & Morris, 2012). The term "*sick excitable* cell" (Morris et al., 2012a, 2016) signifies **a**) a pre-existing bleb-conducive sick state (= "*sick* cell") and **b**) functioning but left-shifted Nav channels (= "*excitable* cell").

Spectrin arranges plasma membrane Nav channels and a reliable signal of neuronal death is calpain-mediated spectrin fragmentation (e.g., Siman et al., 2015). The involvement of excitable cell spectrin in membrane mechanics was little examined (though see Herring, Juranka, Mcnally, Lesiuk, and Morris (2000) and/or Figure 4 in Morris (2001) or Figure 5k in Morris and Homann (2001)) until recent super resolution fluorescence nanoscopy of axons revealed spectrin, F-actin and the actin-capping protein adducin arranged in a striking periodic manner (Xu et al., 2013); this is termed the **m**embrane-associated **p**eriodic **s**keleton (MPS). The axonal MPS is a regular series of actin rings connected by spectrin filaments held under tension. The dynamic responses of such a latticework to mechanical stresses in the different compartments of a neuron immediately raised interest (see Rasband, 2013). The MPS is ubiquitous in neurons, but idiosyncratic arrangements in different compartment are apparent (Han, Zhou, Xia, & Zhuang, 2017). Importantly, responses of the axonal MPS to degenerative stimuli (Unsain et al., 2018) and MPS actin interactions with membrane myosin (Berger et al., 2018) are being studied. How the MPS fares during cytotoxic swelling of neurons has yet to be addressed.

Osmotic fragility tests monitor resistance of the plasma membrane to rupture and or blebbing (Ganote & Vander Heide 1988; Lo et al., 2015; Shimizu-Motohashi et al., 2015; Wan, Harris, & Morris, 1995). In muscular dystrophy (MD), sarcolemmal fragility is a pervasive characteristic, but as Hutter (1992) first spelled out, knowing that loss of dystrophin was implicated in Duchenne MD was not enough to explain fragility. MD results from mutations in dystrophin, or in several of its binding associates or in other membrane-associated proteins, but the list of mutant proteins (García-Pelagio, Bloch, Ortega, & González-Serratos, 2011; Le Rumeur, Winder, & Hubert, 2010; Nicolas, Delalande, Hubert, & Le Rumeur, 2014) did not in itself yield a global mechanistic explanation for poor sarcolemmal mechanical resilience across the various forms of MD. However, cell biological tools garnered from MD mutants plus biophysical and imaging approaches point to caveolar dynamics as a key factor in sarcolemma resilience, as is also the case in endothelial cells (see Echarri & Del Pozo, 2015; Lo, Hall, & Parton, 2016). The elegant simplicity of healthy fiber caveolar dynamics is of course underpinned by a daunting array of molecular players (see proteomic and lipidomic analyses of MD sarcolemma, Murphy et al., 2018; Srivastava, Yadav, Mukherjee, Pal, & Sinha, 2017).

As this article tours through some issues germane to cytotoxic swelling in excitable cells, I emphasize repeatedly that the plasma membrane is not only the locus for electrical excitability and ion homeostasis (IH). Energy-dependent IH is mediated across a plasma membrane that needs to be regarded a dynamic quasi 3-dimensional energy-consuming organelle (e.g., Liu et al., 2016; Lopez-Marques, Theorin, Palmgren, & Pomorski, 2014; Norman et al., 2010; Sheetz, Sable, & Döbereiner, 2006; Taylor, Lampe, & Merrifield, 2012; Wu, Su, Tong, Wu, & Liu, 2018). To capture these attributes of the plasma membrane I introduce the term "**membrane tension homeostasis**". "Membrane" in this phrase refers to the plasma membrane as organelle: bilayer, integral membrane proteins and glycoproteins, scaffolding proteins and the lattice of membrane skeleton proteins that associate with bilayer either directly or via linker proteins, and membrane-associated motor proteins. More specifically, in sketching how faltering IH yields cytotoxic swelling in dystrophic muscle and in cortical neurons during spreading depression (SD), I will emphasize that faltering membrane tension homeostasis (MTH) is also a key feature of cytotoxic swelling in excitable cells.

In a generic "sick excitable cell", therefore, death by cytotoxic swelling can be said to go as follows: impaired MTH → Nav-based leak → impaired IH → cytotoxic swelling (See Figure 1). With ATP supplies low and swelling underway, both IH and MTH will falter and vicious feedbacks along the path will worsen outcomes. Effective interventions would address IH but also accommodate the need for the sick excitable cells to fortify, remove and/or repair structurally compromised Nav-channel bearing membrane.

2. MEMBRANE TENSION HOMEOSTASIS AND MEMBRANE FRAGILITY

2.1 The Concept of Membrane Tension Homeostasis

Homeostasis is a far-from-equilibrium state dependent on a free energy supply. IH is achieved across the osmotic barrier of a bilayer that, if expanded $> \sim 3\%$, will rupture. MTH acts to prevent bilayer tension from becoming lytic. Although various mechanical parameters (tensile strength, stiffness, rupture tension, etc) provide an index of membrane robustness, MTH itself involves multiple stochastic events, some passive, some energy-dependent, with homeostatic feedback operating through passive tension-reducing steps and active tension-increasing steps (at least one of each).

This concept, MTH, echoes that of transmembrane IH, with its dynamic balance of passive and active ion fluxes, and with ion concentration-dependent steps to provide feedback (in IH models this is achieved by assigning

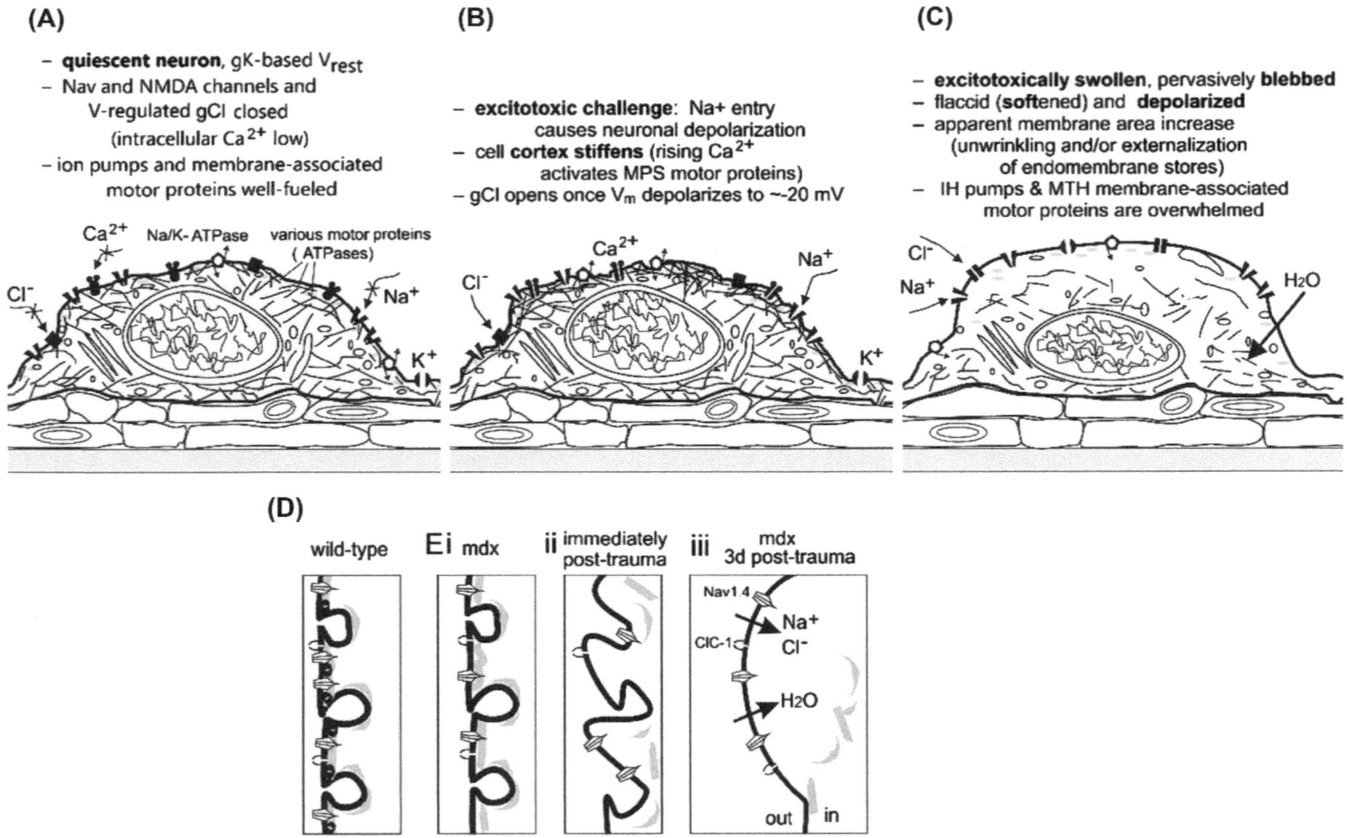

Figure 1 ***Death by cytotoxic swelling.*** Energy-consuming proteins are like nugget of negative-entropy because they embody evolutionary know-how that cells use to temporarily snub the second law of thermodynamics. A cell dies and the second law wins when free energy that was stored courtesy of ATP-consuming proteins gets outmatched by thermal energy. A universally recognized hallmark of the living state

is the ATPase-dependent storage of free energy across the plasma membrane as ion gradients. But there are also ATPases (motor proteins) that generate mechanical energy stores within the plasma membrane: healthy membranes could, therefore, be described as "spring-loaded". As a cell dies from cytotoxic swelling, it jiggles to thermal energy death, its ion gradients dissipating and its mechanical-energy stores frittering away. To survive the onset of cytotoxic swelling, a cell needs to both recharge its batteries and re-establish good "membrane tone". Schematics here depict some aspects of cytotoxic swelling as discussed in the text. **A**, **B**, **C** is a neuron responding to an excitotoxic challenge (Sections 7 and 8). Once neuronal bleb damage develops (between B and C), CLS-Nav leak (Section 6.2) is expected and would sustain potentially lethal cytotoxic swelling (even with NMDA channel blockers). **D**, **Ei-iii** shows a small patch of sarcolemma from an intact wild type versus a mechanically traumatized dystrophin-minus (mdx) skeletal muscle fiber (see Section 6). Several caveolae are shown (Section 3.1). Because mdx fibers are prone to bleb damage but wild type fibers are not, bleb damage occurs only in mdx, I suggest, eliciting CLS-Nav leak (**Eiii**) and so, with open Na^+ and Cl^- channels present, cytotoxic swelling and depolarization ensues. The net water influx (**C**, **Eiii**) under isosmotic conditions is taking the cell towards a Donnan equilibrium (Section 5). Note that neurons have negligible resting gCl (water influx only develops once the depolarization-activated gCl opens) whereas the skeletal muscle V_{rest} is based on a steady-state gCl.

Michaelis–Menten constants for the Na_{in} and K_{out} binding sites of the Na^+/K^+-ATPase).

MTH would encompass the tension-dependent balance of caveolar disassembly/assembly or the tension-dependent balance of exocytosis/endocytosis or bleb formation/recovery as well as combinations of these and tension-dependent bilayer modifying reactions. And just as fleshed-out descriptions of IH have to bring in Na^+/Ca^{2+} or K^+/Cl^-, Na^+/H^+ exchangers etc, full-fledged MTH will have multiple facets. Bilayer structure itself is dynamic and its molecular components affect the ability of the bilayer to withstand rupture. Some lipid molecules enter or exit or relocate in the plasma membrane bilayer, modifying its mechanics (e.g., Vásquez, Krieg, Lockhead, & Goodman, 2014). Which steps are passively tension-sensitive? Actively tension-sensitive? For example, are leaflet asymmetry controlling reactions of plasma membrane-resident lipid flippases tension-sensitive (Karasawa et al., 2013; Lopez-Marques et al., 2014)?

2.2 Intersecting Homeostatic Processes as Seen in Vicious Feedback Loops

Homeostasis is a powerful concept, but the specific processes it describes lack discrete functional boundaries because diverse homeostatic process are strongly interactive. Intersecting actions among IH and MTH become alarmingly evident as vicious feedback loops that plague sick excitable cells. For example, excitotoxic swelling of neurons is accompanied by rising $[Ca^{2+}]_{intracellular}$ that fosters bleb-damage, that saturates the Na^+/Ca^{2+} exchangers reliant on the Na^+/K^+ pumps that are themselves overburdened because the bleb-damaged Nav-bearing membrane fosters a continual CLS-Nav based Na^+ leak.

Another intersecting homeostatic process that is jeopardized in cytotoxic swelling is the regulation of intracellular pH. For example, the Na^+/H^+ exchanger-1 is a mechanosensitive transporter whose activity is activated and inhibited by cell shrinking and swelling respectively (Fuster, Moe, & Hilgemann, 2004). The exchanger's responsiveness to osmotic pressure and to membrane lipid composition arises from its sensitivity to local bilayer leaflet asymmetry (Pang et al., 2012). Glutamate excitotoxicity causes intracellular acidosis, activating the Na^+/H^+ exchanger-1 (Lee & Jung, 2017). But exchanger operation occurs at the expense of Na^+ influx, thereby increasing the burden on Na^+/K^+ pumps.

When considering therapeutic strategies it is obviously advisable to remain cognizant of the fuzzy-edged complexity of homeostatic processes.

To prevent death by cytotoxic swelling after, say, an excitotoxic or traumatic insult, an excitable cell must re-establish appropriate interactions among (at least) MTH, IH and pH homeostasis.

3. SAFEGUARDING THE BILAYER — TENSION BUFFERING

3.1 Tension Buffering in Sarcolemma — Caveolae

Caveolae, dynamic cup-shaped 50 nm membrane invaginations abundant in healthy sarcolemma, depend on membrane bound caveolin. Caveolae interact extensively with actin but are not part of the membrane-spanning dystrophin glycoprotein complex (Crosbie, Yamada, Venzke, Lisanti, & Campbell, 1998; Schilling, Head, & Patel, 2018).

Early electron microscopy of frog sarcolemma under mechanical stress (Dulhunty & Franzini-Armstrong (1975)) showed that caveolae flatten with extreme stress and that membrane rupture occurs once the reserve of caveolar invaginations depletes. Recent electron microscopy of mice and zebrafish muscle (Lo et al., 2015) confirms this but shows that the events also occur with normal muscle contractions; flattening under stress maintains the integrity of the sarcolemma, and caveolae clustered at the neck of T-tubules would likely deal with forces generated there as fiber diameter changed.

Sinha et al. (2011) demonstrated for non-muscle cells, the rapid disappearance of caveolae with either uniaxial stretch or hypoosmotic swelling. Tether-pulling force data showed that caveolar flattening and disassembly is the primary actin- and ATP-independent cell process responsible for buffering surges of membrane tension during mechanical stress whereas the return to "resting" caveolar density is actin- and ATP-dependent.

Caveolae in endothelial cells of lung and of cardiac muscle (Cheng et al., 2015) disassemble in response to higher shear forces during acutely increased cardiac output. Microvascular endothelial cells from caveolin-1 knockout versus control mice subjected to increasing cardiac output in vivo; electron microscopic and 2-photon imaging of the plasma membrane revealed greater susceptibility to acute membrane rupture in the knockout. Caveolar tension buffering operating during routine physiological activity in this manner shows it to be membrane homeostatic process, not solely an emergency measure.

A 1975 freeze fracture study of molluscan (*Aplysia*) sarcolemma (Prescott and Brightman, 1976) used flash-freezing to capture the natural state of the sarcolemma. It showed caveolae to be non-uniform-sized, with some in a flattened state. The smooth muscles studied ensheathed interganglionic connectives; when the connectives were stretched to 2-3X in situ length, caveolae in the sarcolemma lost their invaginated shape. This flattening was reversible upon release of stretch. Caveolae, the authors suggested, might serve as "miniature stretch-receptors" for the sarcolemma. Recently, a molluscan caveolin considered to be an ortholog of caveolin-1 was identified (Udayantha et al., 2018).

In accord with this collection of findings are "mechanical stress test" data from myotubes isolated from MD patients carrying a loss of function mutation, caveolin-3-(P28L); there, the lack of a functional caveolar reservoir was associated with elevated membrane fragility (Sinha et al., 2011). Myocytes of Limbe-Girdle MD mice express caveolin-3 at various quantifiable densities; in a membrane indentation assay (relaxation times of an unidentified stretch-activated whole-cell cation current), greater caveolin density was linked to greater membrane stiffness (Huang, Bae, Sachs, & Suchyna, 2013). Since upregulation of cavolin-3 is noted in, for example Duchenne MD (e.g., Kunert-Keil et al., 2011), Huang et al. (2013) posited that in those situations augmented stiffness may partially compensate for other mechanical deficiencies.

3.2 Tension Buffering in Neurons – Open Question

3.2.1 Neuron-Specific Requirements

Central neurons can be caveolin-rich (MacDonald, Fame, Gillis-Buck, & Macklis, 2018) and neuron-specific caveolin facilitates recovery from traumatic brain injury (Egawa et al., 2017), but though neuronal caveolin binds to cholesterol-rich planar domains, neurons have no morphological caveolae (Schilling et al., 2018). Neuronal tension buffering will therefore differ from that in muscle.

The morphologically and functionally distinct compartments—soma, dendrites, axon initial segment, axon (myelinated or not), motile growth cones, pre- and post-synapse – are mechanically different. The soma needs to last a lifetime. Dendritic and axonal arborizations need to be morphologically plastic. Unsurprisingly, distinctive MPS arrangements are found in soma and dendrite (Han et al., 2017).

Van Essen hypothesized (1999) that neurite tension is critical for morphogenesis and compact wiring in the CNS. Neurons in situ would thus need to sustain a small hydrostatic pressure. This general hypothesis remains influential (e.g. Kroenke & Bayly et al., 2018) but quantitative evidence of neuronal pressures are lacking. Kelly and Macklem (1991) measured cell pressure for *Xenopus* oocytes and subsequently those authors and I unsuccessfully attempted measurements in pond snail (*Lymnaea*) giant neurons. As these neurons resist rupture in distilled water for ~ 1 h, a contractile membrane skeleton protects the bilayer (Wan et al., 1995). Direct (servo-null method) pressure measurements from migrating fibroblasts (Petrie & Koo, 2014) suggest that measurements should be possible for, say large pyramidal cortical neurons undergoing hypoosmotic and cytotoxic swelling. Anti-myosin agents of greater specificity than N-ethylmaleimide (Wan et al., 1995) could help establish if a dynamic MPS (Berger et al., 2018) regulates neuronal pressure and volume.

Cellular anti-rupture strategies used, for example, by lung epithelial cells (Groulx, Boudreault, Orlov, & Grygorczyk, 2006) that involve maintenance of a substantial microfolded actin-bound surface bilayer (see also Figard & Sokac 2014), would be inappropriate for axons since charging/discharging of the excess capacitance would slow action potentials and be metabolically expensive.

3.2.2 Neuronal Pre-stress

Goodman and colleagues (Krieg, Dunn, & Goodman, 2014), using AFM force spectroscopy, identified a quantifiable pre-stress in neurons and hypothesized that this an emergent property of the neuronal cortical cytoskeleton (MPS) would help protect neurons from damage. In the special case they studied (worms with genetically impaired spectrin and impaired touch responses) pre-stress serves as molecular stress distributor for the sensory neuron.

If pre-stress is maintained in ubiquitous, bilayer-adherent (Sheetz et al., 2006), compartmentally distinctive cortical cytoskeletons (i.e., the MPS), this could satisfy the need identified by Van Essen (1997) for neuronal processes (under steady-state tension) to sustain a hydrostatic pressure for morphological purposes. A pre-stressed membrane-adherent MPS would counterbalance a hydrostatic pressure, dynamically stabilizing cell morphology. The remarkable osmo-resistance of healthy neurons could reflect a dynamic (ATP-consuming) MPS.

If this appealing picture of a pre-stressed MPS holds, then one of the putative tension buffering system of neurons—tension-sensitive regulation of neuronal surface area—would need to operate within that framework regulation (see Morris & Homman, 2002). Tension sensitive surface area regulation remains best understood in plants, which like neurons, lack caveolae but need tension buffering. In plant protoplasts, exocytosis and clathrin-mediated endocytosis are both modulated by osmotic stress. Balancing the two processes preserves the integrity of membranes during routinely encountered stresses, such as the stresses integral to osmotically driven guard cell opening and closing (Meckel, Hurst, Thiel, & Homann, 2005; Zwiewka, Nodzyński, Robert, Vanneste, & Friml, 2015).

Tension-sensitive exocytosis would allow for recruitment of electrically sequestered excess capacitance when a neurite became too taut. Tension-sensitive endocytosis (faster at low tension) would facilitate retrieval of excess membrane in a retracting neurite. Although tension effects on transmitter-related exocytosis and endocytosis (Wen et al., 2016; Wu et al., 2017) are being examined, issues of tension-sensitivity of membrane trafficking at non-synaptic neurite regions where exocytosis of enlargeosomes occurs (Colombo, Racchetti, & Meldolesi, 2014) have yet to be addressed.

3.2.3 Neuronal Tension Buffering at a Snail's Pace

In *Lymnaea neurons,* tension-sensitive exo/endocytosis would explain membrane capacitance increases during hypoosmotic swelling and decreases upon reshrinking (Wan et al., 1995; Dai, Sheetz, Wan, & Morris, 1998). Tether force measurements from those neurons (Dai et al., 1998) suggest more intense adhesion between bilayer and MPS in the molluscan neurons than in mammalian neurons. In osmotically swelling *Lymnaea* neurons, sudden spontaneous tether force drops are suggestive of exocytosis of the molluscan neuron equivalent (seen in high voltage electron microscopy of *Aplysia* neurons) of a mammalian enlargeosome (Colombo et al., 2014). When comparable experiments are undertaken for mammalian neurons, it would be desirable to acquire capacitance (and, possibly, cell volume) data simultaneously with membrane mechanical data.

Excitotoxic injury to mammalian neurons elevates $[Ca^{2+}]_{intracellular}$ and produces bleb-type damage. *Lymnaea* neurons subjected to bleb-damage from elevated $[Ca^{2+}]_{intracellular}$ show a dramatic drop in their ability to buffer bilayer tension. The assay that revealed this drop tested the susceptibility of the endogenous "background K channels" (presumably 2-pore domain channels) to stretch-activation (Small & Morris, 1994; Wan, Juranka, & Morris, 1999). The procedure: make a gently-formed gigaohm-sealed cell-attached patch, apply a 'first hit" of negative pipette pressure (fixed intensity and duration). In healthy neurons, the first hit may elicit no channel activity but with further hits (and thus successively worse mechanically-induced bleb damage) the stretch-activated channels activate more and more readily and eventually turn on immediately and fully with each hit. In neurons pre-treated to elevate $[Ca^{2+}]_{intracellular}$ to 0.5 μm the first hit elicited immediate and maximal stretch-channel activation. The interpretation: the bilayer-embedded stretch-sensitive channels experienced increased bilayer tension only when loss of adherent MPS left the bilayer unprotected.

Neurons (molluscan, hippocampal (Dai et al., 1998)) rapidly lose cellular water when reshrinking after a hypoosmotic shock. Water exiting the cell into the constricted volume between cell and adherent substrate generates a hydrostatic back-pressure (Morris, Wang, & Markin, 2003) that causes invaginations (=vacuole-like dilations or VLDs) (Figs. 6–10, Morris & Homann, 2001) to form triggering dramatic rearrangements of membrane associated F-actin (Mills & Morris, 1998). Some invaginations acquire a membrane skeleton (F-actin, spectrin) within seconds (see Figure 5j,k of Morris & Homann (2001)). Probing MPS dynamics in a healthy state via the rapid and reproducible phenomenon of VLD formation/recovery in hippocampal neurons (Dai et al., 1998) could provide an ideal baseline for studying the MPS in those same neurons during excitoxic swelling.

4. MAMMALIAN NEURON OSMORESISTIVITY REVISITED

For much of the last decade, healthy cortical neurons in situ have been regarded as being so osmoresistant that they can steadfastly retain their volume in the face of hyposmotic challenges (e.g. Andrew, Labron, Boehnke, Carnduff, & Kirgeov, 2007; Steffensen, Sword, Croom, Kirov, & MacAulay, 2015). However a recent no-potential-artifact-left-unturned brain slice imaging study entitled "*Hippocampal and cortical pyramidal neurons swell in parallel with astrocytes during acute hypoosmolar stress*" (Murphy et al., 2017) showed neurons rapidly (~1 min) swelling in hypoosmotic media then reshrinking on return to normosmolar medium. Swelling and shrinking is independent of aquaporin.

Whether the membrane capacitance cortical and hippocampal neurons changes (as in molluscan neurons) when they swell and shrink will need to be tested. If reversible area changes are detected, many putative molecular components would need to be queried regarding a tension buffering system in mammalian neurons. For instance, two decades ago diCamilli et al. (Butler et al., 1997) reported amphiphysin II, a BAR domain protein, at high levels in the "cortical cytomatrix" of axon initial segments and bordering CNS nodes of Ranvier (and at skeletal muscle T-tubules). These are Nav-channel rich sites vulnerable to ischemic and traumatic injury and thus, prone to excitotoxic swelling. It has become feasible to ask how the neuronal MPS latticework interacts with tension-sensitive BAR domain proteins (Nishimura, Morone, & Suetsugu, 2018) that are known to regulate endocytosis (Schöneberg et al., 2017; Drager et al., 2017).

5. DONNAN THE ELEPHANT (IN THE SICK EXCITABLE ROOM)

When cellular ion fluxes and cell volume are under consideration, there is typically an elephant in the room, its presence politely acknowledged then politely ignored. That elephant is the cytoplasm's substantial quantity of impermeant charged — mostly anionic - molecules. Ignoring these anions is ill-advised when seeking to understand differences between healthy and cytotoxically swelling excitable cells — i.e., cells that use cation entry to generate action potentials. The impermeant anions of the cytoplasm (plus small quantities of potentially permeant anions) have counterions, mostly K^+ plus a little Na^+. Outside, Na^+ (plus a little K^+) and Cl^- are abundant and balanced. Thus, there are two isosmotic electroneutral compartments. The membrane is permeable to water and has ion selective channels for Cl^- and K^+ and Na^+ (gCl, gK and gNa) — for the moment, imagine them all closed. The permeant ions obey electro-diffusion, so with just one open conductance, V_m will be E_{Cl} or E_K or E_{Na} (i.e., the Nernst equilibrium potential). If gK and gNa are both open, both gradients dissipate and V_m becomes 0 mV (this is not a Donnan situation since ALL anions, including Cl^- are impermeant). If, however, gCl and gK are both opened, V_m depolarizes as Cl^- and K^+ move (inward and outward, respectively) down their concentration gradients. At some point the impermeant anions electrostatically discourage any further net influx of Cl^- or net efflux of K^+ and Donnan equilibrium is attained. The impermeant cytoplasmic anions have governed the redistribution of permeant ions such that $E_{Cl} = E_K$ (any small quantity of impermeant extracellular anions will also contribute to the final state). If we recommence and open just a smallish gK, then $V_m = E_K$ (rather hyperpolarized). Next, gCl and gNa are both opened; this, in simplified form, starts a scenario of the neuronal version of cytotoxic swelling. Electrodiffusive net inward fluxes of Na^+ and Cl^- down their concentrations gradients begin. The impermeant cytoplasmic anions are osmolytes. So too are the incoming Na^+ and Cl^- and thus the cell becomes hyperosmotic. The response is a net influx of "osmotically obliged" water molecules down their activity gradient and, if the cell has a rigid membrane, a hydrostatic counter-pressure develops. Plant cell walls allow turgor pressure to build, but animal cells, lacking a rigid barrier, swell. The cell depolarizes as gradients of the permeant ions diminish. If the cell started ovoid, it gets increasingly spherical and would rupture only once the bilayer stretched by a few percent. V_m can have only one value, so at thermodynamic equilibrium, with all three conductances open, $E_{Cl} = E_{Na} = E_K$ will be close to 0 mV and death by Donnan effect is imminent (Nelson 2003) solves for V_m in a 3-conductances example). This, in simplified version, is the scenario for cytotoxic swelling. In skeletal muscle, with their continuously open resting gCl, there is special urgency to avoid inward Na^+ leaks; this is addressed in the next section.

Cell physiology textbook accounts of Donnan equilibrium generally make Cl^- and just one of the cations permeant, with initially-equal compartment volumes. This facilitates calculation of V_m and cell volume at thermodynamic equilibrium. Thereafter, Donnan the E(lephant) becomes the answer to an exam question, but apart from that gets dismissed as being somewhat cumbersome in excitability models for healthy neurons or even in accounts of healthy-cell volume regulation. But cytotoxic swelling is not a healthy cell state.

The death-by-Donnan-effect scenario that yields cytotoxic swelling just described is conceptually quite simple. It is evident that Donnan-driven concentration changes and swelling could be very slow if chloride conductance values were small as they tend to be in neurons. Swelling and gradient dissipation could thus be ignored for short time scales. But over tens of minutes to hours and days, as in excitable cell disease states, ignoring Donnan effects is unjustified. Inclusion of Donnan effects in models of sick excitable cell behavior can be finessed (HD14) or made explicit (HU16); either way, the math gets trickier.

Healthy excitable cells operate far from Donnan equilibrium and typically have multiple conductances operating at various times, so they clearly must be taking measures that effectively counter the ever-present Donnan effects. For instance, a healthy cell with, 3 conductances as sketched above could:

A. Actively fortify its bilayer in various ways to better withstand hydrostatic pressures that developed due to influx of osmolytes.

B. Open its Nav channels as infrequently as feasible (consistent with making its voltage signals), thereby staying near that "original" $E_{Cl} \sim = E_K$ much of the time and

C. Actively pump Na^+ and Cl^- back out and K^+ back in to minimize transmembrane osmotic and hydrostatic pressures and to maintain ion gradients.

Suggestion A alludes to membrane tension homeostasis (MTH), B alludes to Nav channel kinetics, and C alludes to ion homeostasis (IH). All three are themes in the next sections on cytotoxic swelling in muscle and neurons.

6. CYTOTOXIC SWELLING OF SKELETAL MUSCLE

6.1 Swelling of Muscle Fibers in Duchenne MD

Muscle fiber swelling in hypoosmotic fluids is a complex process because various regulatory responses develop (Brunmair et al., 2000; Usher-Smith, Huang, & Fraser, 2009). Muscle fiber swelling in normosmotic fluids – the pathological phenomenon of cytotoxic swelling–can seem even less clear, as when human muscle fibers swell several days after exercise-induced sarcolemmal injury (Yu, Liu, Carlsson, Thornell, & Stål, 2013).

Swollen muscle tissue, a standard feature in Duchenne MD, was long interpreted as an interstitial edema linked to tissue inflammation (e.g., Kunert-Keil, 2011). In recent years, it was revealed however, that the swollen state of dystrophic muscle tissues occurs because the individual muscle fibres themselves are inflated. Clarification came through ^{23}Na MRI showing myoplasm to have elevated Na^+ levels (Weber, Nagel, Jurkat-Rott, & Lehmann-Horn, 2011). This evidence of cytoplasmic Na^+-loading indicated that dystrophic muscles are enlarged because their individual fibers are cytotoxically swollen, with aberrantly high $[Na^+]_{intracellular}$ being the rate-limiting osmotic (cytotoxic) agent.

Na^+-driven cytotoxic swelling would not occur without concomitant anion-loading. In skeletal muscle (unlike in neurons) the resting conductance is based on an abundantly expressed Cl^- channel, ClC-1 (Pusch, Steinmeyer, & Jentsch, 1994). In human muscle, ClC-1 accounts for ~80% of the resting membrane conductance and V_{rest} is nearly at E_{Cl} (Pedersen, Riisager, de Paoli, Chen, & Nielsen, 2016). Presumably, during pathological Na^+-loading, the steady-state ClC-1 conductance is able to accommodates excess (i.e., osmotically-coupled) Cl^- influx. A secondary active Cl^- transport mechanism has been identified in skeletal muscle fibers (Lindinger, Leung, Trajcevski, & Hawke, 2011) but is considered to have a rather limited capacity, in contrast to central neurons where it is suggested that secondary active Cl^- extrusion would mitigate the severity of cytotoxic swelling (Gagnon et al., 2013) as discussed in Rungta et al. (2015).

What precipitates cyto-toxic Na^+-loading in dystrophic muscle? Unidentified mechanosensitive cation channels susceptible to inhibition by 5–10 μM GsMTx4 has been one suggestion (see Huang et al., 2013). As I argued previously however (Morris, 2011a, 2011b) GsMTx4 inhibition of voltage-gated channels, especially in damaged membrane, is not ruled out. This is non-trivial because by chemical structure and provenance (ICK peptide from tarantula venom) GsMTx4 is extremely close to many voltage sensor toxins from the same venom (e.g., Chen & Chung 2013; Suchyna & Sachs 2007). Moreover, as Schmidt and MacKinnon (2008) showed, the affinity/efficacy of a voltage sensor toxin (they examined VsTx1 interactions with a Kv channel) can dramatically increase after the plasma membrane in which the channel is embedded has been subjecting to bleb-type damage (graphically summarized in Morris 2011a, Figure 8).

Thus, while unidentified cation channels might contribute to Na^+-loading in dystrophic skeletal muscle, here I will instead focus on the known involvement of an identified and abundantly expressed channel for which tetrodotoxin is a selective high affinity inhibitor and for which venom-based peptides might act as lead compounds for clinically useful inhibitors.

The channel in question is the skeletal muscle voltage-gated Na^+ channel isoform, Nav1.4. Isolated dystrophic mouse (*mdx*^{5cv}) fibers exhibit elevated rate of death (tested over 3 days) against which 3 nM tetrodotoxin is protective (Hirn et al., 2008). In control fibers, Na^+ levels are unaffected by levels of tetrodotoxin that in *mdx*^{5cv} muscle reduce the influx enough to bring Na^+ to the control fiber level. Nav1.4 expression is lower in *mdx*^{5cv} fibers than wild-type. Fibers were not spontaneously contractile and so Na^+-loading due to action potentials is ruled out; *mdx*^{5cv} fibers evidently had a steady-state Nav-leak.

The isolation procedures for the survival experiments would, I suggest, have been more traumatic for the *mdx*^{5cv} sarcolemma than for wild-type; isolation involves passing collagenase-digested tissue fragments repeatedly through pipettes of decreasing size until single intact fibers are obtained. Biomechanical measurements of the sarcolemma (García-Pelagio et al., 2011) show that a lack of dystrophin renders the *mdx* sarcolemma more compliant and susceptible to mechanically-induced bleb-damage than wild-type; specifically, the intercostamer region of dystrophin-less sarcolemma detaches more readily from superficial myofibrils and ruptures at lower pipette aspiration pressures. Hirn et al. (2008) suggest that *mdx*^{5cv} myocyte Nav channels leakiness is due to slower inactivation that is also right-shifted in *mdx*^{5cv}. However, the understandable diffidence evinced about their own patch clamp data in their discussion is also reflected in their choice to not to include kinetic information in the abstract. Earlier, a comprehensive study of mdx versus wildtype fiber Nav1.4 channel kinetics by Bezanilla and colleagues (Mathes, Bezanilla, & Weiss, 1991), who examined gating current, macroscopic and unitary current, revealed no difference. Suffice it to say, then, that Hirn et al. (2008) found that for 3 days after an unavoidably traumatic isolation procedure, electrically quiescent dystrophin-lacking muscle fibers (with lower plasma membrane Nav1.4 density than control fibers) were imperiled by a Nav1.4 channel-based leak not present in control fibers.

Given the assertion that GsMTx4 protects mdx fibers (Suchyna & Sachs, 2007) it seems urgent to perform survival experiments like those of Hirn et al. (2008) testing both GsMtx4 and tetrodotoxin. If micromolar levels of GsMtx4 proved as myoprotective as nanomolar levels of tetrodotoxin in such tests, an exciting possible explanation would be that toxin can bind Nav1.4 voltage-sensor domains in bleb-damaged membranes of mdx fibers while remaining ineffective as a Nav1.4 channel inhibitor in healthy plasma membrane. Tetrodotoxin is untenable clinically, but GsMTx4 administered systemically to mice is reportedly non-toxic (Gottlieb, Suchyna, & Sachs, 2007).

6.2 Nav Channel Gating in Bleb-Damaged Membrane

Effects of bleb damage on the gating of Nav 1.4 and Nav1.6 channels (node of Ranvier isoform) have been reviewed extensively (Joos, Barlow, & Morris, 2017; Morris et al., 2012a, 2012b, Morris & Joos, 2016). Briefly, findings are based on channels expressed in *Xenopus* oocytes at levels that yield macroscopic currents in cell attached patches. After gentle gigaohm seal formation (e.g. see Figure 1A in Morris, Prikryl, & Joós, 2015) 'intact'-membrane currents are obtained, then sustained pipette aspiration is used to induce bleb-type membrane damage. Aspiration is released, currents are again recorded, and so on. Depending on the intensity of the mechanically-induced bleb damage, Nav1.4 channel operation is progressively and irreversibly left-shifted (i.e., leftward or hyperpolarized on an I/V plot). Because of the kinetic characteristics of Nav channels, tails of the sigmoid activation and inactivation (availability) plots overlap on the voltage axis, yielding a voltage region where a steady-state "window conductance" operates. With damage, that window conductance is left-shifted, bringing it closer to the excitable cell's resting potential. Under conditions of mild damage in an excitable cell (the cell has a near normal V_{rest} but small regions of membrane are damaged) the left-shifted window conductance constitutes a steady-state, tetrodotoxin-sensitive Na^{+}-leak.

The shorthand for this phenomenon is coupled left shift (CLS) (Boucher et al., 2012) (or Nav-CLS or CLS-Nav-leak). CLS happens because activation of Nav channels gets easier (requires less depolarization to trigger voltage sensor movement) in bleb-damaged membrane. Fast inactivation is not inherently voltage-sensitive, but it is so tightly coupled to activation, kinetically speaking, that the two process undergo the same-fold hyperpolarizing shift in bleb-damaged membranes. Computations (Boucher et al., 2012) show that, provided IH is maintaining an excitable cell's ion gradients within tolerable limits, CLS of increasing intensity will produce a wide array of dysfunctional behaviors. Depending on the severity of damage, mild (1–2 mV of CLS) through to very severe (e.g. 20–40 V of CLS) CLS-damaged would produce dysfunctional excitability presenting as hyper-excitability, then as spontaneous subthreshold oscillations and ectopic activity (bursts, then tonic firing, including small amplitude voltage oscillations) and finally, as depolarizing block (see Joos et al., 2017; Morris & Joos 2016). The burden placed on IH by CLS leak is abundantly clear; the pump goes from over-taxed through to overwhelmed (Boucher et al., 2012; Yu, Morris, Joós, & Longtin, 2012).

Excitotoxic cell swelling in brains changes the volumes of both cellular and extracellular compartments. Computations show that even if CLS injury intensity was stable, when IH is unable to maintain the normal Na^+ and K^+ gradients, the ongoing volume changes would be associated with changing patterns of dysfunctional ectopic CLS-Nav trigged firing (Boucher et al., 2012).

6.3 Nav Leak and the Na/K Pump Response in Dystrophic Muscle

Returning specifically to dystrophic muscle fibers, the prevalence of exercise-induced membrane damage (e.g. see Thakur, Swiderski, Ryall, & Lynch, 2018) suggests that Nav-CLS leak can be expected, regardless of disease genotype specifics. Inflammatory states common in dystrophic muscle could further exacerbate sarcolemma damage. In the frankly inflammatory conditions of septic infections, critical illness myopathy, or other situations that expose muscle to endotoxin, Nav gating becomes left-shifted (Haeseler et al., 2008); likewise for neuropathies that develop under similar conditions (Morris et al., 2012a). In muscle fibers, the resting ClC-1 conductance would support I_{Cl} to accompany any Na^+ loading, and osmotically obliged water would cause the fiber to swell in a normosmotic milieu.

Healthy skeletal muscle fibers are not at risk from their Nav-window conductance because its voltage range falls well to "the right" (i.e., more depolarized) of muscle fiber V_{rest} so there is virtually no steady-state Nav-leak. In a generally healthy fiber, a small membrane injury (and thus mild Nav-CLS and relatively minor Na^+-loading) is presumably handled by well-fueled Na^+/K^+ pumps and that can restore polarization and allow the muscle fibers to methods to repair the membrane injury (e.g. see Barthélémy et al., 2018).

Lehmann-Horn and colleagues reported an unexpected clinical outcome that seems directly relevant, in dystrophic fibers, to the scenario described at the end of Section 6.2, wherein pump capacity becomes insufficient to handle IH during cell swelling. To provide general benefit from the anti-edema action of a diuretic, a Duchenne MD patient was administered the aldosterone antagonist, eplerenone. As hoped, skeletal muscle tissue swelling decreased but, unexpectedly, muscle fiber repolarization was also promoted (Lehmann-Horn et al., 2012). This "side-effect" outcome was gratifying but initially baffling. Years later it was shown that eplerenone promotes tyrosine dephosphorylation at an identified site on the Na^+/K^+-ATPase $\alpha1$ subunit and that this dephosphorylation upregulates extant but previously inactive pumps in the sarcolemma (Breitenbach, Lehmann-Horn, & Jurkat-Rott, 2016). In later sections, we will see that neurons that are depolarized and cytotoxically swollen experience 'free energy starvation" (i.e., an inability repolarize and reshrink even with pumps working at full capacity) that seems not unlike this scenario in depolarized dystrophic muscle (Section 9.4 below; HD14, HSD14).

7. EXCITOTOXIC INSULTS STIFFEN NEURONS. THEN THEY SWELL, BLEB, SOFTEN, DIE

Loss of the adherent F-actin-spectrin membrane skeleton from the bilayer formally defines bleb-formation, and, as seen by AFM spectroscopy, this produces a softening of the cell cortex (Steltenkamp, Rommel, Wegener, & Janshoff, 2006). Cortical neurons die slowly when exposed to excitotoxic agents NMDA or veratridine (activators of cation-selective glutamate channels and Nav channels, respectively). After ~50 min in 20 μM NMDA cortical neurons are inflated by cytotoxic swelling to about 130% and in this moribund state, they are soft, as gauged by AFM force spectroscopy (Zou et al., 2013) and thus bleb-damaged.

Cortical neurons do not, however, soften immediately; monitored (AFM) starting ~2 min after exposure to NMDA or veratridine, they are stiffer than normal. Stiffening was near maximal at that ~2 min point so stiffening had evidently commenced earlier. In veratridine application experiments, Rungta et al. (2015) monitored cortical cell volume and Na^+ continuously; Na^+ increased detectably for ~1.5 min and cytotoxic swelling began at ~1.5 min (half maximal swelling was at ~5 min). A rapidly occurring secondary effect of veratridine-activation of neuronal Nav channels is a rise in $[Ca^{2+}]_{intracellular}$ (Mohammed, Doran, Grundy, & Nassar, 2017; e.g., V-gated Ca^{2+} channels get activated). Thus, the AFM force spectroscopy findings, while preliminary, are consistent with the following scenario: the initial cortical neuron response to an excitotoxic insult (NMDA or veratridine) is a Ca^{2+}-influx dependent stiffening the cortex, likely via Ca^{2+}-induced activation of MPS-actomyosin, but then, as swelling gets underway so too does bleb-damage because Ca^{2+}-mediated proteolysis of the MPS spectrin and F-actin is also stimulated. Accordingly, the dying cortical neurons, become inflated and soft (they are non-spherical, and so do not become turgid).

The softness of cytotoxically dying neurons is non-trivial. It signifies that when healthy, they had been maintaining a passively-accessible bilayer reserve (e.g., invaginations, bilayer adjacent exocytotic vesicles, MPS-supported undulations (see Figure 1 of Morris and Homann, 2001). When a prolonged rise in Ca^{2+} damages the MPS, the bilayer detaches and the excess bilayer makes the neuron flaccid, even as it swells. Only if inflation continued till the neuron attained a truly spherical state would it momentarily stiffen then rupture (Rupture of grossly swollen spherical neurons is not unfamiliar to those who have voltage-clamped neurons with sharp 3 M KCl electrodes).

AFM spectroscopy measurements of healthy neurons by Zhang et al. (2017) show membrane stiffness of axon to be greater than cell bodies or dendrites. Their coarse-grained molecular dynamics simulation based on the reported axonal MPS accurately simulates the AFM-measured axolemma stiffness. They point out that in the MPS, spectrin filaments are held under tension (they are "pre-stressed"), and so severing spectrin filaments should permanently disrupt the MPS (because spectrin filaments could not spontaneously re-form their initial under-tension configuration). This would pertain during, say hyper-activated calpain mediated bleb damage (as invoked above). In healthy cells however, membrane remodeling activity of myosin II (Berger et al., 2018) would presumably be able re-tension a detached (though not a proteolytically-cleaved) spectrin filament.

Brief exposure to hypotonic medium also stiffened cortical neurons (Zou et al., 2013). As mentioned in Section 4, some labs report that neurons in slice preparations do not respond to hypoosmotic challenges with discernible swelling (e.g., Andrew et al., 2007; Steffensen et al., 2015); it would be useful to know if those apparently osmo-resistant neurons stiffened in response. Likewise for molluscan neurons exhibiting actomyosin "membrane tone" that constraints their rate of swelling in distilled water (Wan et al., 1995).

8. CYTOTOXIC SWELLING OF NEURONS

8.1 Excitotoxic Swelling of Cortical Neurons — The "Missing" Cl^--Leak Identified

In ischemic and traumatic brain injury, epilepsy and migraine, cytotoxic swelling occurs in association with cortical spreading depolarization (or depression) (SD) which is a slow (several mm/min) depolarizing wave that passes through neurons and glia, swelling the neurons, distorting dendritic spines, and silencing electrical excitability (Dreier, Lemale, Kola, Friedman, & Schoknecht, 2017; Zhang, Shuttleworth, Moskal, & Stanton, 2015). In these conditions, the cytotoxic swelling of neurons is considered the major cause of cell death (Rungta et al., 2015).

The major initiating factor for SD and cytotoxic swelling is aberrant "excitotoxic" entry of Na^+ through the channels of chemical and electrical excitability, i.e., NMDA glutamate-activated cation channels and Nav channels (Zhou et al., 2013; Wu & Fujikawa, 2002; Risher et al., 2011; Rungta et al., 2015). Resting gCl in most CNS neurons is vanishingly small, making it unclear how Na^+ entry triggers swelling. During the isotonic swelling process, a transport pathway is needed to support an electroneutralizing anion influx (as per Section 5). In 2015 an anion pathway that supports excitotoxic swelling of neurons was rigorously identified by the MacVicar lab (Rungta et al., 2015). A battery of pharmacological agents and ion-dyes was used in this comprehensive live-cell imaging brain slice study along with whole-cell voltage clamp. Astrocytes, imaged with a rhodamine dye, showed no obvious volume changes and served as useful landmarks. The Na^+-indicator dye used, CoroNa-Green, is used preferentially taken up by the cortical and hippocampal neurons; whole-cell clamp at known $[Na^+]_{intracellular}$ was used to calibrate the dye signal in the environment or the neuronal cytoplasm.

The pivotal finding is that once excitotoxically injured neurons become depolarized to ~-20 mV, a non-inactivating V-regulated anion conductance permeant to Cl^- becomes active. Thereafter, as cytoplasmic $[Na^+]$ and $[Cl^-]$ both rise, osmotically obliged water follows and the neurons swell. Cells die after ~ 90 min. Lethal swelling occurs even if aberrant Ca^{2+} entry is prevented. The gCl responsible is requires SLC26A11, a protein from a family of transporters first identified from kidney. Knock-down via siRNA inhibited cytotoxic swelling in slices and voltage-regulated I_{Cl} (of the appropriate inhibitory pharmacology) in neurons under whole-cell clamp.

SLC26A11 protein is expressed in neurons throughout the brain, suggesting that similar mechanisms of cytotoxic swelling and neuronal death can be expected wherever its expression is high enough. Functionally, however, the SLC26A11 protein is a shape-shifter; identified first as a sulfate transporter, in various tissues it participates in diverse ion exchange pairings (see Rungta et al., 2015 for references). Its physiological roles in the neurons studied by the MacVicar group are unknown, but in cerebellar neurons it does act as a gCl, moderating Cl^- homeostasis and neuronal activity in conjunction with strong inhibitory (GABAergic) input (Rahmati, Vinueza Veloz, Xu, & Barone, 2016). A biophysical-biochemical analysis of native neuronal SLC26A11 protein to establish the general structure of the gCl and its binding partners is still to come. Likewise for the V-regulated activation, the ion selectivity mechanism, and the molecular basis of inhibitory pharmacology in the native channel.

8.2 Beading of Dendrites During SD Differs From Cytotoxic Swelling of the Soma

Since SD engenders beading of the dendrites along with cytotoxic swelling of the soma, it might be expected that these two phenomena have similar mechanisms. Not so, according to Steffensen et al. (2015), who report that although dendritic beading is strictly dependent on Cl^- influx, there is no accompanying osmotic water influx. Combined blockade of several Cl^--coupled transporters reduces dendritic beading. Rearrangements of neither actin nor microtubule are required for beading. The authors hypothesize that SD-induced dendritic beading is a consequence of altered driving forces for Cl^--dependent cotransporters, and that in a manner independent of osmotic forces, water entry occurs as part of cotransport mechanism. In this study, dendrites resisted swelling during severe osmotic challenge (see also Andrew et al., 2007).

Monnerie et al. (2010) reported dendritic beading as a "bystander" effect in an experimental model system that traumatically stretch-injures the axons of rat cortical neurons without mechanical injury to dendrites. Minutes after axon stretch, dendrites develop beads that subside over time. Tetrodotoxin (to block axonal Nav channels) plus MK-801 (to block dendritic NMDA glutamate channels) prevents beading. Thus mechanical injury confined to axons impacts dendrite morphology via an excitotoxic mechanism involving dysregulated leaky Nav channels. It would clearly be useful to determine if blockade of the above-mentioned Cl^-cotransporters diminishes trauma-bystander beading of dendrites, and to examine the axonal MPS following bystander injury.

9. MODELING EXCITOTOXIC SWELLING IN NEURONS

9.1 Accurate Depiction of Mild Injury Is the Goal

"Mild brain injury" implies the possibility of restoration of a normal physiological state given appropriate conditions. To identify "appropriate conditions" in connection with SD, models that predict the major features of SD, including the cytotoxic swelling sequelae are needed. Particular challenging is the fact that electrical excitability is a millisecond-range phenomenon, but mildly injured cells either worsen and die or recover over periods of hours or even days. A model by Dijkstra et al. (2016) (**DHGP16** here) for neurons undergoing cytotoxic swelling meets the important criterion of representing phenomena from milliseconds to several days and is especially interesting because it incorporates the critical new mechanistic aspects of cytotoxic swelling uncovered by Rungta et al. (2015) (see Section 8.1 above). The DHGP16 model (discussed in Section 10 below) builds directly on important concepts recently formalized by others in the broader context of SD and so some of that work is discussed first in Sections 9.2—9.5.

9.2 Modeling Sick Excitable Cells That Leak Both Cl^- and Na^+

A model that addresses mild injury in sick excitable cells needs, at a minimum, the ion channels underlying excitability and the pumps that keep cellular batteries charged. Likely pathological electrophysiological consequences of CLS Nav-leak were explored in a model like that (Boucher et al., 2012; Lachance, Longtin, Morris, Yu, & Joós, 2014; Yu et al., 2012). The channel/pump combination yields fast (electrical excitability) and slow (IH-dependent) voltage dynamics (Joos et al., 2017). But with cytotoxic swelling, pathological IH is the issue. In two papers exploring IH in spreading depression (SD), Hübel and Dahlem (2014) and Hübel, Schöll and Dahlem (2014) (hereafter, HD14, HSD14) developed a modeling framework appropriate for a constellation of sick excitable cell conditions that can trigger SD (e.g. epilepsy, ischemia, trauma). Low energy conditions (ischemia) deplete ATP; this is modeled by diminishing pump (ATPase) activity. Because cytotoxic swelling is central to the SD phenomenon, HD14 and HSD14 took the critically important step of incorporating Donnan effects. Doing so ensures that IH has to meet the challenges posed by the cytoplasm's impermeant anions (in Section 5 a Donnan scenario was outlined, emphasizing the unpleasant consequences of concomitant Na^+/Cl^- influx). Once Donnan effects are included in a model, a new patho-physiologically realistic slow IH phenomenon emerges with all the hallmarks of SD. Graphical analysis of these slower properties defines a pathophysiologically important system-attribute that they term "ionic excitability".

9.3 Bifurcation Plots Are to Homeostatic Modeling What Gels Are to Biochemistry

Behaviors of the modeled homeostatic systems are examined using so-called bifurcation analyses of parameter space. For modelers of nonlinear systems, bifurcation and null-cline plots are gold, but for most of us they are as much fun as suites of peptide or nucleotide gels, which also hold treasure chests of information. The scientific act-of-faith in which one scientist says to another, "I accept your account about what information is contained in your gels and bifurcation plots, so can we now cut to the chase?" is understandable. But to get a grasp of the critical issue of ionic excitability, I made myself plunge in a bit. The bifurcation analysis plots most central to cytotoxic swelling during SD are ones that plot pump current density versus the steady-state V_m. Given that the pathological nature of SD involves sudden depolarizations and then some minutes later (if cells have not actually died), abrupt repolarizations, a key aim of the bifurcation analysis is to seek out regions of IH stability (steady-states) and regions of IH instability and between them, thresholds that would be crossed during the depolarizing and repolarizing transitions (HD14, HSD14). An overview of this SD modeling up to 2015 is available (Zandt, ten Haken, van Putten, & Dahlem, 2015).

9.4 What Is "Ionic Excitability"?

"Excitability" in neurons typically refers to Hodkgin-Huxley type ***electrical excitability*** where the concepts of an unstable threshold voltage region, a stable V_{rest} and a stable "plateau" potential are standard electrophysiological parlance. The simplest example, which formally is a case of bistability rather than full excitability, would occur with E_K and E_{Na} separated by (say) ~120 mV where there is a background gK plus a non-inactivating gNa; V_m perturbations will trigger threshold-crossing behavior putting V_m either near E_K or near E_{Na}. Now, add Kv channels. Perturb V_m a little and V_m goes from near E_K to near E_{Na} and then returns to near E_K; this is a fully excitable system. What HSD14 and HD14 show is that analogous types of bistability and excitability occur in the realm of IH, i.e, ***ionic excitability***. This, it should be noted is an entirely undesirable kind of excitability, associated with low energy conditions. This type of ionic excitability, unchecked, will produce more sick excitable cells.

The HSD14/HD14 model framework is for neurons and glia plus of external fluids compartments. The computational neurons have Hodgkin-Huxley ion channels, a generic gCl, Donnan effects, and Na^+/K^+ pumps. Compartmental volumes are taken into account. Electrical excitability plays out in all the computations (e.g. see Figure 7 in HD14) but the bifurcation analyses focus on IH-related parameters. Distinctive IH stable states are separated from each other by unstable threshold regions. In HSD14, Figure 2, the cell stays near −70 mV over a wide range of pump current densities, but if the current density gets too low (transient ischemia, for example), V_m transits to ~-20 mV and now, the pump rate that had previously maintained the normal physiological state (green square in that figure) is unable to repolarize the cell. This stable IH state is called a free energy starvation (FES) state. During cytotoxic swelling, entry into a stably depolarized FES state is likely to be lethal.

In electrical excitability, V_m perturbations trigger threshold crossing. For ionic excitability, it is ion flux perturbations that trigger threshold crossings. Such perturbations can be beneficial (e.g., extra pump current due to upregulation of Na^+/K^+ pumps) or they can be deleterious (e.g., CLS-Nav leak). In light of this, it is good to know that, when healthy, some neurons can evidently detect an excess Nav leak and boost pump density (thalamic neurons that experience excessive persistent Nav leak respond by upregulating greater pump capacity (Senatorov, Stys, & Hu, 2000); this kind of response would clearly be out of the question for a neuron that was in FES.

In the "ionic excitability" characteristic of SD with its attendant cytotoxic swelling, the irreversible depolarization and swelling and would represent a bistable case: specifically the system irreversibly crosses the threshold from normal into the FES state (HSD14).

During the recurrent waves of SD, however, two-way threshold crossing occurs repetitively (full excitability with, in that case, "limit cycle" characteristics). To avoid bistable lethality, the global system needs to provide the neurons with recurrent beneficial ion flux perturbations that diminish the load on the pumps (the HD14, HSD14 models can do so because they include glia and circulatory ion sinks). Examples of IH systems that exhibit bistable and fully excitable characteristics are summarized in HSD14 using null-cline plots for $[K^+]_{intracellular}$ versus $[Na^+]_{intracellular}$ (their Figure 4). There, in the fully excitable case, the IH trajectory that is followed after a perturbation (redline loop, with the dotted section being a perturbation or "stimulus") brings the system back to its original high $[K^+]_{intracellular}$, low $[Na^+]_{intracellular}$ state. In the bistable model, the trajectory starts along a similar path then remains stuck in a Na^+-loaded condition – i.e. in the FES state.

The FES state is reminiscent of a Donnan equilibrium in that the membrane is depolarized, the ion gradients are reduced, the membrane is electrically inexcitable. In the systems here, the pathological state goes from being "metastable" to stable if IH has to rely exclusively on the neuron's existing population of Na^+/K^+ pumps to redress ion balances (without, say, having access to glial compartment uptake of K^+). The modeling approach in HD14 and HSD14 is taken to be broadly consistent with experimental findings because it links the typical course of SD pathological events—a breakdown of ion gradients, prolonged depolarization and abrupt repolarizations — to a metastable version of FES.

The efficacy for Duchenne MD patients of a pharmaceutical intervention that upregulates skeletal muscle Na^+/K^+-ATPase thereby redressing swelling and promoting repolarization (Breitenbach et al., 2016; discussed in Section 6.3) might be similar to these neuronal scenarios.

9.5 Modeling Volume Changes of SD-Associated Cytotoxic Swelling

Osmosis and cell volume changes that are inferred in HD14 and HSD14 are made explicit in a model by Hübel and Ullah (2016) (HU16 for short) for neural and glial volume dynamics. The explanatory power of this model emerges largely from the physics of osmosis and electroneutrality, not specific biophysical details. The fact that it nevertheless generates the phenomenon of FES is seen as a strong validation of the hypothesis that SD is a process of threshold transitions between a FES state and the normal state. In HU16, anion fluxes during cellular volume changes are shown to be inevitable, but an active role for Cl-exchange in volume control was shown to be useful but not necessary for the threshold transitions.

Of special note, HU16 considers the possible consequence of a stretch-sensitive increase in gCl for a swollen cell. Such a volume-activated gCl would not, they find, provide a way for a neuron to recover once depolarization—induced swelling had pushed it into FES. A mechanism to repolarize the membrane potential first is needed; only then could fluxes through the Cl^- channels decrease the volume of the swollen cell. In HU16, open Cl^- channels are necessary but not sufficient for regulatory volume decrease after SD.

In addressing how a tension-sensitive gCl could affect IH during SD, HU16 in effect have recognized that the bilayer is more than a capacitor and osmotic barrier, but the cell cartoons there do nevertheless serve as a reminder that MTH itself does not feature in the HD14, HSD14 or HU16 models (nor in DHGP16). Does this matter with respect to the emerging theories for SD?

A healthy cell bilayer is not floppy everywhere (as seemingly cartooned in HU16) but is organized as a 3D organelle that, at one extreme is strongly attached to the MPS and thus stiff and at the other extreme, in small patches (blebs), can be floppy. Insofar as the energy-dependent dynamic membrane acts to constrain swelling, it constrains volume increase and thus affects $[ion]_{intracellular}$ without recourse to transmembrane movement of ions or water, i.e., independent of IH.

ATP-dependent MTH processes almost certainly interact with slow ATP-dependent IH in establishing rates of cell-volume expansion. During the early recurrent waves of SD, increased membrane-tone arising from excitotoxin-induced stiffening of cortical neurons (Zou et al., 2013) might, for example, slow a threshold-crossing into FES. It would be good to ascertain for healthy neurons that are regulating their volume in response to a hypoosmotic challenge, what are the characteristic changes in stiffness (e.g. see Kowalsky et al. (2011) and Ayee, LeMaster, Teng, Lee, and Levitan (2018) for non-neural cell stiffness during regulatory volume decrease). During successive waves of SD, a regulatory volume decrease is required each time; likewise, it would be good to ascertain how stiffness changes in that pathological situation.

Non-neural mammalian cells and hippocampal neurons undergoing regulatory volume decrease transiently produce VLDs (vacuole-like dilations; see Figures 6a and 7, Morris & Homann, 2001). In the constricted and tortuous extracellular space of the CNS, formation of such invaginations would help preserve established cell–cell contacts even as a neuron was shrinking (Figure 10 in Morris & Homann, 2001, Figure 11 in Morris, Wang & Markin, 2003) and would also explain why tracer diffusion experiments (Chen & Nicholson, 2000) indicate that non-uniform cell shape changes occur in shrinking brain cells, creating residual extracellular spaces that trap diffusing molecules.

10. MECHANISTIC MODEL FOR EXCITOTOXIC NEURONAL SWELLING DURING SD

In a model broadly similar to HD14, HSD14 and particularly to HU16, but biophysically precise regarding ion flux proteins, Dijkstra et al. (2016) (referred to here as DHGP16) address the excitotoxic swelling of cortical neurons. Impermeant cytoplasmic anions, osmosis and volume changes are explicit in DHGP16, with model elements parameterized to reflect newly-available experimental findings of Rungta et al. (2015). In particular, unlike the generic gCl of HU16, the gCl of DHGP16 is voltage-regulated consistent with SLC26A11-based conductance responsible for excitotoxic swelling in the Rungta et al. (2015) experiments. A secondary active extrusion mechanism equivalent to KCC2 (an electroneutral K/Cl cotransporter expressed in cortical and hippocampal neurons (see Rungta et al., 2015) though mentioned in the abstract and operational in the model, is not manipulated computationally. The experimental situation mimicked inDHGP16 is: brain slice, neurons in an essentially infinite extracellular volume, pumps functioning (at different intensities including "off" in some cases), no capillary flow, SD triggered by aberrant Nav channel leak (veratridine induced).

It is readily evident in Figure 3 of DHGP16 (pumps functioning in this case), how remarkably well key findings from Rungta et al. (2015) are mimicked by the model. The Nav channel agonist, veratridine (mimicked by removing inactivation in the gNa(V)), abruptly augments $[Na^+]_{intracellular}$ and then moment later neuronal swelling begins, reaching ~ half maximal at ~ 4 min. Swelling proceeds with external osmolarity held constant. Cell swelling in DHGP16, as in experiments, is prevented by a Cl^--free external solution and is diminished by Cl^- channel blockers. The cytotoxic swelling time course plot, unfortunately, does not include traces showing $I_{Na}(t)$and $I_{Cl}(t)$ and the flux of water into the neuron. For various levels of low energy supply (mimicking ischemia of various severities) DHGP16 accurately predicts over extended periods, excitability (both electrical excitability and ionic excitability) within the neuron, and the swelling dynamics of the neuron (time scales from millisecond duration action potentials to cell size at 48 h are evident in their Figure 5). Importantly, the model substantiates the key experimental observation that Cl^- influx generates cytotoxic edema, whereas Na^+ entry alone does not.

Bifurcation analyses of DHGP16 reveals a region of instability in Na^+/K^+ATPase functionality, a tipping point below which cell volume rapidly increases (their Figure 6A,B, pump strength versus V_m or cell size). Here a Donnan-like equilibrium state similar to the FES state of HU16 is entered irreversibly. Once the system falls to this degraded steady-state, even pumping at full strength is insufficient for a return to the physiological baseline.

A therapeutic suggestion emerging from DHGP16 is that if the homeostatic system in extremis were de-stressed by temporary blockade of Nav channels, cell volume and membrane potential could normalize. The idea is to reduce the load on the Na^+/K^+ pumps until they can pull IH out of the FES zone, crossing the threshold into the normal physiological normal zone.

A prolonged temporary Nav blockade is also a measure previously deemed necessary to staunch CLS-Nav window current in sick excitable cells (Morris et al., 2012a). Remodeling of bleb-damaged membrane is a slow energy-dependent process, and is likely to be abnormally sluggish in neurons globally insulted by SD. Blockade long enough to give Na^+/K^+ pumps time to bring IH back to the normal zone would be insufficient if there was still bleb damaged membrane with producing CLS-Nav leak. CLS-Nav leak imperils both IH and MTH. A systemically administered clinical Nav blocker designed for this task would need to specifically target the Nav channels resident in bleb-damaged membrane (Morris & Joos, 2016).

11. CONCLUSION: IH AND MTH WORKING TOGETHER

Devising the experiments and models that are leading to an understanding cytotoxic swelling of neurons and muscles would be almost impossible without the powerful intellectual tool of IH as a formal concept. Our ability to learn about the condition of these sick excitable cells would benefit if MTH too, was recognized as a formal concept, not just seen piecemeal as is presently the case.

MTH I argue, is no less a "thing" for animal cells than IH. The most fundamental requirement of IH is to regulate transmembrane [ions], with channels and transporters dissipating ion gradients and ion pumps restoring them. This is all for nought if the plasma membrane bilayer ruptures. The most fundamental requirement of MTH is to prevent lytic bilayer tensions. Lysis can result when the bilayer becomes too taut or at the other extreme when a large floppy region of unsupported bilayer spontaneously sheds a vesicle or buds off. Too-high tension can be reduced by passive addition of reserve membrane (e.g., exocytosis, caveolar flattening, collapse of bilayer undulations by release of adhesive contacts to the MPS) and too-low tension can be elevated by active retrieval processes (e.g., endocytosis, reassembly of caveola, imposition of bilayer undulations by contractile action on the bilayer adherent MPS). All these processes will be tension-dependent, with steps that increase bilayer area being facilitated by high tension and steps that decrease bilayer area being facilitated by low tension.

Embedded in any homeostatic process there must be sources of feedback information that establish a set point defended by the homeostatic process. For MTH, the tension-sensitive steps provide feedback. A set point is not some quantity determined from outside the homeostatic process (the commonly used thermostat example notwithstanding). Biotic homeostatic set points are "merely" emergent properties of cellular systems with particular components; set points are continually worked out by evolutionary trial-and-error, constrained by the laws of physics, and specific to any given cell type. A cell's V_{rest}, for example, can be regarded as a simple IH set point. As depicted by Dahlem, Hübel, Ullah and colleagues, SD arises when system components for IH change (e.g., ischemia) with the result that a second extremely deleterious set point becomes available. In recurrent waves of SD, both set points are visited at intervals.

In the realm of MTH would new set points turn up in sick excitable cells? Cells routinely make and repair nano blebs as part of membrane remodeling and motility (Norman et al., 2010). When system conditions change (e.g. low ATP), is there a threshold crossing by which this normal dynamic physiological state transitions into a sick excitable cell version of blebbing? If so, what conditions must be in place to ensure that a transition back to normal membrane-remodeling type blebbing was possible? There is an immense need to study the blebbing condition of mildly injured sick excitable cells. Although there has been great attention to irreversible apoptotic blebbing, and very little is known about the mild bleb injury state. Studying this issue within an MTH framework would help move these questions forward with some rigor.

Sens and colleagues have put healthy cell blebbing on a sound theoretical and mathematical footing, though under the rubric of membrane *area* homeostasis (Norman et al., 2010). Importantly, though, their last words in the paper say the bleb/cytoskeleton interactions "play a critical role in the global mechanical ***homeostasis*** of the cell ***through regulation of membrane tension***" (italics mine). Membrane tension, indeed, is the critical entity. As discussed previously (Morris, 2002), cells cannot detect their membrane area and they cannot detect their volume but they can and must continually detect and adjust membrane tension. Cell biology/ physiology would benefit from formalizing the concept of MTH.

Homeostatic theory is already available for several other aspects of MTH, particularly for caveolae or similar invaginations (Sens & Turner, 2004; Norman et al., 2010; Sinha et al., 2011; Sens & Plastino 2015) and the organization role of membrane tension in cell structure and dynamics is increasingly recognized (Gauthier, Masters, & Sheetz, 2012; Diz-Muñoz, Fletcher, & Weiner, 2013; Zhang et al., 2017; Lou 2018; Nishimura et al., 2018). What needs to happen is a drawing together of these understandings under the encompassing tent of membrane tension homeostasis.

Modeling IH is attractive because well-defined physical principals (electrodiffusion equilibria, flux equations, osmosis; e.g., Hübel, Hosseini-Zare, Žiburkus, and Ullah (2017)) mesh with parameters that can be measured experimentally: ion concentrations, membrane voltages and membrane currents, osmolarity. By comparison, MTH can be distressingly untidy because its central parameter, membrane tension, while not inherently mysterious, is not generally accessible to direct measurement. New benchmark work from Levitan's group (Ayee & Levitan, 2018; Ayee et al., 2018) perfectly illustrates this point: simultaneous AFM and tether force measurements during regulatory volume decrease (with/without intact F-actin) showed that even as global stiffness of the plasma-membrane-as-organelle increased, tension in the bilayer did not. Is this evidence of MTH and IH successfully collaborating to safeguard bilayer tension? Do skeletal muscles (which have caveolar tension buffering) respond similarly? Do neurons, which do not use caveolae but which stiffen upon hypotonic challenge respond similarly? Clearly there is work to do. Exploiting AFM, tether force and other (e.g. Lab, Bhargava, Wright, & Gorelik, 2013) mechanical measurements in conjunction with new methods for probing forces in specific membrane skeleton proteins in situ (e.g. Yang et al., 2015) offers considerable promise in the near future.

The inescapable complexity of plasma membrane mechanics is all the more reason for developing rigorous theory for MTH. Membrane mechanical measurements obtained in conjunction with data from advanced live-cell imaging will need such a framework. From the points of view of cell vitality and viability, MTH is no less important than IH. Moreover, because IH requires a healthy membrane, the dynamic energy-dependent processes of IH and MTH always work together.

REFERENCES

Andrew, R. D., Labron, M. W., Boehnke, S. E., Carnduff, L., & Kirgeov, S. A. (2007). Physiological evidence that pyramidal neurons lack functional water channels. *Cerebral Cortex, 17*(4), 787–802.

Ayee, M. A. A., LeMaster, E., Teng, T., Lee, J., & Levitan, I. (2018). Hypotonic challenge of endothelial cells increases membrane stiffness with no effect on tether force. *Biophysical Journal, 114*(4), 929–938.

Ayee, M. A. A., & Levitan, I. (2018). Membrane stiffening in osmotic swelling: Elastic modulus versus membrane tension. *Current Topics in Pharmacology.*

Barthélémy, F., Defour, A., Lévy, N., Krahn, M., & Bartoli, M. (2018). Muscle Cells fix breaches by ochestrating a membrane repair ballet. *Journal of Neuromuscular Diseases, 5*(1), 21–28.

Berger, S. L., Leo-Macias, A., Yuen, S., Khatri, L., Pfennig, S., Zhang, Y., et al. (2018). Localized Myosin II activity regulates assembly and plasticity of the axon initial segment. *Neuron, 97*(3), 555e6–570e6.

Boucher, P. A., Joós, B., & Morris, C. E. (2012). Coupled left-shift of Nav channels: Modeling the Na -loading and dysfunctional excitability of damaged axons. *Journal of Computational Neuroscience, 33*(2), 301–319.

Breitenbach, S., Lehmann-Horn, F., & Jurkat-Rott, K. (2016). Eplerenone repolarizes muscle membrane through Na,K-ATPase activation by Tyr10 dephosphorylation. *Acta Myologica: Myopathies and Cardiomyopathies: Official Journal of the Mediterranean Society of Myology, 35*(2), 86–89.

Brunmair, B., Neschen, S., Gras, F., Roden, M., Nowotny, P., Waldhäusl, W., et al. (2000). Metabolic response to anisoosmolarity of rat skeletal muscle in vitro. *Hormone and Metabolic Research, 32*(7), 251–255.

Butler, M. H., David, C., Ochoa, G. C., Freyberg, Z., Daniell, L., Grabs, D., et al. (1997). Amphiphysin II (SH3P9; BIN1), a member of the amphiphysin/Rvs family, is concentrated in the cortical cytomatrix of axon initial segments and nodes of ranvier in brain and around T tubules in skeletal muscle. *Journal of Cell Biology, 137*(6), 1355–1367.

Chen, R., & Chung, S. H. (2013). Effect of gating modifier toxins on membrane thickness: Implications for toxin effect on gramicidin and mechanosensitive channels. *Toxins, 5*(2), 456–471.

Cheng, J. P., Mendoza-Topaz, C., Howard, G., Chadwick, J., Shvets, E., Cowburn, A. S., et al. (2015). Caveolae protect endothelial cells from membrane rupture during increased cardiac output. *Journal of Cell Biology, 211*(1), 53–61.
Chen, K. C., & Nicholson, C. (2000). Changes in brain cell shape create residual extracellular space volume and explain tortuosity behavior during osmotic challenge. *Proceedings of the National Academy of Sciences of the United States of America, 97*, 8306–8311.
Colombo, F., Racchetti, G., & Meldolesi, J. (2014). Neurite outgrowth induced by NGF or L1CAM via activation of the TrkA receptor is sustained also by the exocytosis of enlargeosomes. *Proceedings of the National Academy of Sciences of the United States of America, 111*(47), 16943–16948.
Crosbie, R. H., Yamada, H., Venzke, D. P., Lisanti, M. P., & Campbell, K. P. (1998). Caveolin-3 is not an integral component of the dystrophinglycoprotein complex. *FEBS Letters, 427*(2), 279–282.
Dai, J., Sheetz, M. P., Wan, X., & Morris, C. E. (1998). Membrane tension in swelling and shrinking molluscan neurons. *The Journal of Neuroscience: the Official Journal of the Society for Neuroscience, 18*(17), 6681–6692.
Dijkstra, K., Hofmeijer, J., van Gils, S. A., & van Putten, M. J. (2016). A biophysical model for cytotoxic cell swelling. *The Journal of Neuroscience: the Official Journal of the Society for Neuroscience, 36*(47), 11881–11890.
Diz-Muñoz, A., Fletcher, D. A., & Weiner, O. D. (2013). Use the force: Membrane tension as an organizer of cell shape and motility. *Trends Cell Biology, 23*(2), 47–53.
Dräger, N. M., Nachman, E., Winterhoff, M., Brühmann, S., Shah, P., Katsinelos, T., Boulant, S., Teleman, A. A., Faix, J., & Jahn, T. R. (2017). Bin1 directly remodels actin dynamics through its BAR domain. *EMBO Reports, 18*(11), 2051–2066.
Dreier, J. P., Lemale, C. L., Kola, V., Friedman, A., & Schoknecht, K. (2017). Spreading depolarization is not an epiphenomenon but the principal mechanism of the cytotoxic edema in various gray matter structures of the brain during stroke. *Neuropharmacology*. pii: S0028–3908(17)30444-6.
Dulhunty, A. F., & Franzini-Armstrong, C. (1975). The relative contributions of the folds and caveolae to the surface membrane of frog skeletal muscle fibres at different sarcomere lengths. *The Journal of Physiology, 250*(3), 513–539.
Echarri, A., & Del Pozo, M. A. (2015). Caveolae - mechanosensitive membrane invaginations linked to actin filaments. *Journal of Cell Science, 128*(15), 2747–2758.
Egawa, J., Schilling, J. M., Cui, W., Posadas, E., Sawada, A., Alas, B., et al. (2017). Neuron-specific caveolin-1 overexpression improves motor function and preserves memory in mice subjected to brain trauma. *The FASEB Journal, 31*(8), 3403–3411.
Figard, L., & Sokac, A. M. (2014). A membrane reservoir at the cell surface: Unfolding the plasma membrane to fuel cell shape change. *Bioarchitecture, 4*(2), 39–46.
Fuster, D., Moe, O. W., & Hilgemann, D. W. (2004). Lipid- and mechanosensitivities of sodium/hydrogen exchangers analyzed by electrical methods. *Proceedings of the National Academy of Sciences of the United States of America, 101*(28), 10482–10487.
Gagnon, M., Bergeron, M. J., Lavertu, G., Castonguay, A., Tripathy, S., Bonin, R. P., et al. (2013). Chloride extrusion enhancers as novel therapeutics for neurological diseases. *Nature Medicine, 19*(11), 1524–1528.
Ganote, C. E., & Vander Heide, R. S. (1988). Irreversible injury of isolated adult rat myocytes. Osmotic fragility during metabolic inhibition. *The American Journal of Pathology, 132*(2), 212–222.
García-Pelagio, K. P., Bloch, R. J., Ortega, A., & González-Serratos, H. (2011). Biomechanics of the sarcolemma and costameres in single skeletal muscle fibers from normal and dystrophin-null mice. *Journal of Muscle Research and Cell Motility, 31*(5–6), 323–336.
Gauthier, N. C., Masters, T. A., & Sheetz, M. P. (2012). Mechanical feedback between membrane tension and dynamics. *Trends Cell Biology, 22*(10), 527–535.

Gottlieb, P. A., Suchyna, T. M., & Sachs, F. (2007). Properties and mechanism of the mechanosensitive ion channel inhibitor GsMTx4, a therapeutic peptide derived from tarantula venom. *Current Topics in Membranes, 59*, 81–109.

Groulx, N., Boudreault, F., Orlov, S. N., & Grygorczyk, R. (2006). Membrane reserves and hypotonic cell swelling. *The Journal of Membrane Biology, 214*(1), 43–56.

Haeseler, G., Foadi, N., Wiegand, E., Ahrens, J., Krampfl, K., Dengler, R., et al. (2008). Endotoxin reduces availability of voltage-gated human skeletal muscle sodium channels at depolarized membrane potentials. *Critical Care Medicine, 36*(4), 1239–1247.

Han, B., Zhou, R., Xia, C., & Zhuang, X. (2017). Structural organization of the actin-spectrin-based membrane skeleton in dendrites and soma of neurons. *Proceedings of the National Academy of Sciences of the United States of America, 114*(32), E6678–E6685.

Herring, T. L., Juranka, P., Mcnally, J., Lesiuk, H., & Morris, C. E. (2000). The spectrin skeleton of newly-invaginated plasma membrane. *Journal of Muscle Research and Cell Motility, 21*(1), 67–77.

Hirn, C., Shapovalov, G., Petermann, O., Roulet, E., & Ruegg, U. T. (2008). Nav1.4 deregulation in dystrophic skeletal muscle leads to Na+ overload and enhanced cell death. *Journal of General Physiology, 132*(2), 199–208.

Huang, H., Bae, C., Sachs, F., & Suchyna, T. M. (2013). Caveolae regulation of mechanosensitive channel function in myotubes. *PLoS One, 8*(8), e72894.

Hübel, N., & Dahlem, M. A. (2014). Dynamics from seconds to hours in Hodgkin-Huxley model with time-dependent ion concentrations and buffer reservoirs. *PLoS Computational Biology, 10*(12), e1003941.

Hübel, N., Hosseini-Zare, M. S., Žiburkus, J., & Ullah, G. (2017). The role of glutamate in neuronal ion homeostasis: A case study of spreading depolarization. *PLoS Computational Biology, 13*(10), e1005804.

Hübel, N., Schöll, E., & Dahlem, M. A. (2014). Bistable dynamics underlying excitability of ion homeostasis in neuron models. *PLoS Computational Biology, 10*(5), e1003551.

Hübel, N., & Ullah, G. (2016). Anions govern cell colume: A case study of relative astrocytic and neuronal swelling in spreading depolarization. *PLoS One, 11*(3), e0147060.

Hutter, O. F. (1992). The membrane hypothesis of Duchenne muscular dystrophy: Quest for functional evidence. *Journal of Inherited Metabolic Disease, 15*(4), 565–577.

Joos, B., Barlow, B. M., & Morris, C. E. (2017). Calculating the consequences of left-shifted Nav channel activity in sick excitable cells. *Handbook of Experimental Pharmacology*.

Karasawa, A., Swier, L. J., Stuart, M. C., Brouwers, J., Helms, B., & Poolman, B. (2013). Physicochemical factors controlling the activity and energy coupling of an ionic strength-gated ATP-binding cassette (ABC) transporter. *Journal of Biological Chemistry, 288*(41), 29862–29871.

Kelly, S. M., & Macklem, P. T. (1991). Direct measurement of intracellular pressure. *American Journal of Physiology, 260*(3 Pt 1), C652–C657.

Kowalsky, G. B., Beam, D., Oh, M. J., Sachs, F., Hua, S. Z., & Levitan, I. (2011). Cholesterol depletion facilitates recovery from hypotonic cell swelling in CHO cells. *Cellular Physiology and Biochemistry: International Journal of Experimental Cellular Physiology, Biochemistry, and Pharmacology, 28*(6), 1247–1254.

Krieg, M., Dunn, A. R., & Goodman, M. B. (2014). Mechanical control of the sense of touch by β spectrin. *Nature Cell Biology, 16*(3), 224–233.

Kroenke, C. D., & Bayly, P. V. (2018). How forces fold the cerebral cortex. *The Journal of Neuroscience: the Official Journal of the Society for Neuroscience, 38*(4), 767–775.

Kunert-Keil, C., Gredes, T., Lucke, S., Morgenstern, S., Mielczarek, A., Sporniak-Tutak, K., et al. (2011). Caveolin-1, caveolin-3 and VEGF expression in the masticatory muscles of mdx mice. *Folia Histochemica et Cytobiologica, 49*(2), 291–298.

Lab, M. J., Bhargava, A., Wright, P. T., & Gorelik, J. (2013). The scanning ion conductance microscope for cellular physiology. *American Journal of Physiology. Heart and Circulatory Physiology, 304*(1), H1–H11.

Lachance, M., Longtin, A., Morris, C. E., Yu, N., & Joós, B. (2014). Stimulation-induced ectopicity and propagation windows in model damaged axons. *Journal of Computational Neuroscience, 37*(3), 523–531.

Le Rumeur, E., Winder, S. J., & Hubert, J. F. (2010). Dystrophin: More than just the sum of its parts. *Biochimica et Biophysica Acta, 1804*(9), 1713–1722.

Lee, B. K., & Jung, Y. S. (2017). Sustained intracellular acidosis triggers the Na /H exchager-1 *(sic)* activation in glutamate excitotoxicity. *Biomolecules and Therapeutics (Seoul), 25*(6), 593–598.

Lehmann-Horn, F., Weber, M. A., Nagel, A. M., Meinck, H. M., Breitenbach, S., Scharrer, J., & Jurkat-Rott, K. (2012). Rationale for treating oedema in Duchenne muscular dystrophy with eplerenone. *Acta Myologica: Myopathies and Cardiomyopathies: Official Journal of the Mediterranean Society of Myology, 31*(1), 31–39.

Lindinger, M. I., Leung, M., Trajcevski, K. E., & Hawke, T. J. (2011). Volume regulation in mammalian skeletal muscle: The role of sodium-potassium-chloride cotransporters during exposure to hypertonic solutions. *The Journal of Physiology, 589*(Pt 11), 2887–2899.

Liu, X., Shu, S., Billington, N., Williamson, C. D., Yu, S., Brzeska, H., et al. (2016). Mammalian nonmuscle myosin II binds to anionic phospholipids with concomitant dissociation of the regulatory light chain. *Journal of Biological Chemistry, 291*(48), 24828–24837.

Lo, H. P., Hall, T. E., & Parton, R. G. (2016). Mechanoprotection by skeletal muscle caveolae. *Bioarchitecture, 6*(1), 22–27.

Lo, H. P., Nixon, S. J., Hall, T. E., Cowling, B. S., Ferguson, C., Morgan, G. P., et al. (2015). The caveolin-cavin system plays a conserved and critical role in mechanoprotection of skeletal muscle. *Journal of Cell Biology, 210*(5), 833–849.

Lopez-Marques, R. L., Theorin, L., Palmgren, M. G., & Pomorski, T. G. (2014). P4-ATPases: Lipid flippases in cell membranes. *Pflügers Archiv, 466*, 1227–1240.

Lou, X. (2018). Sensing exocytosis and triggering endocytosis at synapses: Synaptic vesicle exocytosis-endocytosis coupling. *Frontiers in Cellular Neuroscience, 12*, 66.

MacDonald, J. L., Fame, R. M., Gillis-Buck, E. M., & Macklis, J. D. (2018). Caveolin1 identifies a specific subpopulation of cerebral cortex callosal projection neurons(CPN) including dual projecting cortical callosal/frontal projection neurons (CPN/FPN). *eNeuro, 5*(1).

Mathes, C., Bezanilla, F., & Weiss, R. E. (1991). Sodium current and membrane potential in EDL muscle fibers from normal and dystrophic (mdx) mice. *American Journal of Physiology, 261*, C718–C725.

Meckel, T., Hurst, A. C., Thiel, G., & Homann, U. (2005). Guard cells undergo constitutive and pressure-driven membrane turnover. *Protoplasma, 226*(1–2), 23–29.

Mills, L. R., & Morris, C. E. (1998). Neuronal plasma membrane dynamics evoked by osmomechanical perturbations. *The Journal of Membrane Biology, 166*(3), 223–235.

Mohammed, Z. A., Doran, C., Grundy, D., & Nassar, M. A. (2017). Veratridine produces distinct calcium response profiles in mouse Dorsal Root Ganglia neurons. *Scientific Reports,* 7, 45221.

Monnerie, H., Tang-Schomer, M. D., Iwata, A., Smith, D. H., Kim, H. A., & Le Roux, P. D. (2010). Dendritic alterations after dynamic axonal stretch injury in vitro. *Experimental Neurology, 224*(2), 415–423.

Morris, C. E., & Homann, U. (2001). Cell surface area regulation and membrane tension. *Journal of Membrane Biology, 179*(2), 79–102.

Morris, C. E. (2001). Mechanoprotection of the plasma membrane in neurons and other non-erythroid cells by the spectrin-based membrane skeleton. *Cellular & Molecular Biology Letters, 6*(3), 703–720.

Morris, C. E. (2002). How did cells get their size? *The Anatomical Record, 268*(3), 239–251.

Morris, C. E. (2011). Why are so many channels mechanosensitive? In N. Sperelakis (Ed.), *Cell physiology source book* (4th ed., pp. 493–505). Elsevier: Amsterdam.

Morris, C. E. (2011). Pacemaker, potassium, calcium, sodium: Stretch modulation of the voltage-gated channels. In P. Kohl, F. Sachs, & M. Franz (Eds.), *Cardiac mechano-electric coupling and arrhythmias: From pipette to patient* (2nd ed., pp. 43–49). Elsevier Saunders.

Morris, C. E., Boucher, P. A., & Joós, B. (2012). Left-shifted Nav channels in injured bilayer: Primary targets for neuroprotective Nav antagonists? *Front Pharmacology, 23*(3), 19.

Morris, C. E., & Homann, U. (2001). Cell surface area regulation and membrane tension. *The Journal of Membrane Biology, 179*(2), 79–102.

Morris, C. E., & Joos, B. (2016). Nav channels in damaged membranes. *Current Topics in Membranes, 78*, 561–597.

Morris, C. E., Juranka, P. F., & Joós, B. (2012). Perturbed voltage-gated channel activity in perturbed bilayers: Implications for ectopic arrhythmias arising from damaged membrane. *Progress in Biophysics and Molecular Biology, 110*(2–3), 245–256.

Morris, C. E., Prikryl, E. A., & Joós, B. (2015). Mechanosensitive gating of Kv channels. *PLoS One, 10*(2), e0118335.

Morris, C. E., Wang, J. A., & Markin, V. S. (2003). The invagination of excess surface area by shrinking neurons. *Biophysical Journal, 85*(1), 223–235.

Murphy, T. R., Davila, D., Cuvelier, N., Young, L. R., Lauderdale, K., Binder, D. K., et al. (2017). Hippocampal and cortical pyramidal neurons swell in parallel with astrocytes during acute hypoosmolar stress. *Frontiers in Cellular Neuroscience, 11*, 275.

Murphy, S., Zweyer, M., Henry, M., Meleady, P., Mundegar, R. R., Swandulla, D., et al. (2018). Proteomic analysis of the sarcolemma-enriched fraction from dystrophic mdx-4cv skeletal muscle. *Journal of Proteomics*. pii: S1874-3919(18)30046-0.

Nelson, P. (2003). *Biological physics: Energy, information, life*. New York: Freeman.

Nicolas, A., Delalande, O., Hubert, J. F., & Le Rumeur, E. (2014). The spectrin family of proteins: A unique coiled-coil fold for various molecular surface properties. *Journal of Structural Biology, 186*(3), 392–401.

Nishimura, T., Morone, N., & Suetsugu, S. (2018). Membrane re-modelling by BAR domain superfamily proteins via molecular and non-molecular factors. *Biochemical Society Transactions, 46*(2), 379–389.

Norman, L. L., Brugués, J., Sengupta, K., Sens, P., & Aranda-Espinoza, H. (2010). Cell blebbing and membrane area homeostasis in spreading and retracting cells. *Biophysical Journal, 99*(6), 1726–1733.

Novak, K. R., Nardelli, P., Cope, T. C., Filatov, G., Glass, J. D., Khan, J., et al. (2009). Inactivation of sodium channels underlies reversible neuropathy during critical illness in rats. *Journal of Clinical Investigation, 119*(5), 1150–1158.

Pang, V., Counillon, L., Lagadic-Gossmann, D., Poet, M., Lacroix, J., Sergent, O., et al. (2012). On the role of the difference in surface tensions involved in the allosteric regulation of NHE-1 induced by low to mild osmotic pressure, membrane tension and lipid asymmetry. *Cell Biochemistry and Biophysics, 63*(1), 47–57.

Pedersen, T. H., Riisager, A., de Paoli, F. V., Chen, T. Y., & Nielsen, O. B. (2016). Role of physiological ClC-1 Cl- ion channel regulation for the excitability and function of working skeletal muscle. *The Journal of General Physiology, 147*(4), 291–308.

Petrie, R. J., & Koo, H. (2014). Direct measurement of intracellular pressure. *Current Protocols in Cell Biology, 63*(12.9), 1–9.

Prescott, L., & Brightman, M. W. (1976). The sarcolemma of Aplysia smooth muscle in freeze-fracture preparations. *Tissue Cell, 8*(2), 241–258.

Pusch, M., Steinmeyer, K., & Jentsch, T. J. (1994). Low single channel conductance of the major skeletal muscle chloride channel, ClC-1. *Biophysical Journal, 66*(1), 149–152.

Rahmati, N., Vinueza Veloz, M. F., Xu, J., Barone, S., Rodolfo Ben Hamida, N., Schonewille, M., et al. (2016). SLC26A11 (KBAT) in Purkinje cells is critical for inhibitory transmission and contributes to locomotor coordination. *eNeuro, 3*(3).

Rasband, M. N. (2013). Cytoskeleton: Axons earn their stripes. *Current Biology: CB, 23*(5), R197–R198.

Risher, W. C., Lee, M. R., Fomitcheva, IV, Hess, D. C., & Kirov, S. A. (2011). Dibucaine mitigates spreading depolarization in human neocortical slices and prevents acute dendritic injury in the ischemic rodent neocortex. *PLoS One, 6*(7), e22351.

Rungta, R. L., Choi, H. B., Tyson, J. R., Malik, A., Dissing-Olesen, L., Lin, P. J. C., et al. (2015). The cellular mechanisms of neuronal swelling underlying cytotoxic edema. *Cell, 161*(3), 610–621.

Schilling, J. M., Head, B. P., & Patel, H. H. (2018). Caveolins as regulators of stress adaptation. *Molecular Pharmacology, 93*(4), 277–285.

Schmidt, D., & MacKinnon, R. (2008). Voltage-dependent K+ channel gating and voltage sensor toxin sensitivity depend on the mechanical state of the lipid membrane. *Proceedings of the National Academy of Sciences of the United States of America, 105*(49), 19276–19281.

Schöneberg, J., Lehmann, M., Ullrich, A., Posor, Y., Lo, W. T., Lichtner, G., Schmoranzer, J., Haucke, V., & Noé, F. (2017). Lipid-mediated PX-BAR domain recruitment couples local membrane constriction to endocytic vesicle fission. *Nat Commun, 8*, 15873.

Senatorov, V. V., Stys, P. K., & Hu, B. (2000). Regulation of Na+,K+-ATPase by persistent sodium accumulation in adult rat thalamic neurones. *The Journal of Physiology, 525*(Pt 2), 343–353.

Sens, P., & Plastino, J. (2015). Membrane tension and cytoskeleton organization in cell motility. *Journal of Physics: Condensed Matter, 27*(27), 273103.

Sens, P., & Turner, M. S. (2004). Theoretical model for the formation of caveolae and similar membrane invaginations. *Biophysical Journal, 86*(4), 2049–2057.

Sheetz, M. P., Sable, J. E., & Döbereiner, H. G. (2006). Continuous membrane-cytoskeleton adhesion requires continuous accommodation to lipid and cytoskeleton dynamics. *Annual Review of Biophysics and Biomolecular Structure, 35*, 417–434.

Shimizu-Motohashi, Y., Asakura, Y., Motohashi, N., Belur, N. R., Baumrucker, M. G., & Asakura, A. (2015). Pregnancy-induced amelioration of muscular dystrophy phenotype in mdx mice via muscle membrane stabilization effect of glucocorticoid. *PLoS One, 10*(3), e0120325.

Siman, R., Shahim, P., Tegner, Y., Blennow, K., Zetterberg, H., & Smith, D. H. (2015). Serum SNTF increases in concussed Professional ice Hockey players and relates to the severity of Postconcussion symptoms. *Journal of Neurotrauma, 32*(17), 1294–1300.

Sinha, B., Köster, D., Ruez, R., Gonnord, P., Bastiani, M., Abankwa, D., et al. (2011). Cells respond to mechanical stress by rapid disassembly of caveolae. *Cell, 144*(3), 402–413.

Small, D. L., & Morris, C. E. (1994). Delayed activation of single mechanosensitive channels in *Lymnaea* neurons. *American Journal of Physiology, 267*(2 Pt 1), C598–C606.

Srivastava, N. K., Yadav, R., Mukherjee, S., Pal, L., & Sinha, N. (2017). Abnormal lipid metabolism in skeletal muscle tissue of patients with muscular dystrophy: In vitro, high-resolution NMR spectroscopy based observation in early phase of the disease. *Magnetic Resonance in Imaging, 38*, 163–173.

Steffensen, A. B., Sword, J., Croom, D., Kirov, S. A., & MacAulay, N. (2015). Chloride cotransporters as a molecular mechanism underlying spreading depolarization-induced dendritic beading. *The Journal of Neuroscience: the Official Journal of the Society for Neuroscience, 35*(35), 12172–12187.

Steltenkamp, S., Rommel, C., Wegener, J., & Janshoff, A. (2006). Membrane stiffness of animal cells challenged by osmotic stress. *Small, 2*, 1016–1020.

Suchyna, T. M., & Sachs, F. (2007). Mechanosensitive channel properties and membrane mechanics in mouse dystrophic myotubes. *The Journal of Physiology, 581*(Pt 1), 369–387.

Taylor, M. J., Lampe, M., & Merrifield, C. J. (2012). A feedback loop between dynamin and actin recruitment during clathrin-mediated endocytosis. *PLoS Biology, 10*(4), e1001302.

Thakur, S. S., Swiderski, K., Ryall, J. G., & Lynch, G. S. (2018). Therapeutic potential of heat shock protein induction for muscular dystrophy and other muscle wasting conditions. *Philosophical Transactions of the Royal Society of London B Biological Sciences, 373*(1738).

Udayantha, H. M. V., Bathige, S. D. N. K., Priyathilaka, T. T., Lee, S., Kim, M. J., & Lee, J. (2018). Identification and characterization of molluscan caveolin-1 ortholog from Haliotis discus discus: Possible involvement in embryogenesis and host defense mechanism against pathogenic stress. *Gene Expression Patterns, 27*, 85–92.

Unsain, N., Bordenave, M. D., Martinez, G. F., Sami, J., von Bilderling, C., Barabas, F. M., et al. (2018). Remodeling of the actin/apectrin membrane-associated periodic skeleton, growth cone collapse and F-Actin decrease during axonal degeneration. *Scientific Reports, 8*(1), 3007.

Usher-Smith, J. A., Huang, C. L., & Fraser, J. A. (2009). Control of cell volume in skeletal muscle. *Biological Reviews of the Cambridge Philosophical Society, 84*(1), 143–159.

Van Essen, D. C. (1997). A tension-based theory of morphogenesis and compact wiring in the central nervous system. *Nature, 385*(6614), 313–318.

Vásquez, V., Krieg, M., Lockhead, D., & Goodman, M. B. (2014). Phospholipids that contain polyunsaturated fatty acids enhance neuronal cell mechanics and touch sensation. *Cell Reports [electronic Resource], 6*(1), 70–80.

Wang, J. A., Lin, W., Morris, T., Banderali, U., Juranka, P. F., & Morris, C. E. (2009). Membrane trauma and Na^+ leak from Nav1.6 channels. *American Journal of Physiology, 297*(4), C823–C834.

Wan, X., Harris, J. A., & Morris, C. E. (1995). Responses of neurons to extreme osmomechanical stress. *The Journal of Membrane Biology, 145*(1), 21–31.

Wan, X., Juranka, P., & Morris, C. E. (1999). Activation of mechanosensitive currents in traumatized membrane. *American Journal of Physiology, 276*(2 Pt 1), C318–C327.

Waxman, S. G. (2001). Acquired channelopathies in nerve injury and MS. *Neurology, 56*(12), 1621–1627.

Waxman, S. G. (2005). Cerebellar dysfunction in multiple sclerosis: Evidence for an acquired channelopathy. *Progress in Brain Research, 148*, 353–365.

Weber, M. A., Nagel, A. M., Jurkat-Rott, K., & Lehmann-Horn, F. (2011). Sodium (23Na) MRI detects elevated muscular sodium concentration in Duchenne muscular dystrophy. *Neurology, 77*(23), 2017–2024.

Wen, P. J., Grenklo, S., Arpino, G., Tan, X., Liao, H. S., Heureaux, J., et al. (2016). Actin dynamics provides membrane tension to merge fusing vesicles into the plasma membrane. *Nature Communications, 7*, 12604.

Wu, X. S., Elias, S., Liu, H., Heureaux, J., Wen, P. J., Liu, A. P., et al. (2017). Membrane tension inhibits rapid and slow endocytosis in secretory cells. *Biophysical Journal, 113*(11), 2406–2414.

Wu, A., & Fujikawa, D. G. (2002). Effects of AMPA-receptor and voltage-sensitive sodium channel blockade on high potassium-induced glutamate release and neuronal death in vivo. *Brain Research, 946*(1), 119–129.

Wu, Z., Su, M., Tong, C., Wu, M., & Liu, J. (2018). Membrane shape-mediated wave propagation of cortical protein dynamics. *Nature Communications, 9*(1), 136.

Xu, K., Zhong, G., & Zhuang, X. (2013). Actin, spectrin, and associated proteins form a periodic cytoskeletal structure in axons. *Science, 339*(6118), 452–456.

Yang, C., Zhang, X., Guo, Y., Meng, F., Sachs, F., & Guo, J. (2015). Mechanical dynamics in live cells and fluorescence-based force/tension sensors. *Biochimica et Biophysica Acta, 1853*(8), 1889–1904.

Yu, J. G., Liu, J. X., Carlsson, L., Thornell, L. E., & Stål, P. S. (2013). Re-evaluation of sarcolemma injury and muscle swelling in human skeletal muscles after eccentric exercise. *PLoS One, 8*(4), e62056.

Yu, N., Morris, C. E., Joós, B., & Longtin, A. (2012). Spontaneous excitation patterns computed for axons with injury-like impairments of sodium channels and Na/K pumps. *PLoS Computational Biology, 8*(9), e1002664.

Zandt, B. J., ten Haken, B., van Putten, M. J., & Dahlem, M. A. (2015). How does spreading depression spread? Physiology and modeling. *Reviews in the Neurosciences, 26*(2), 183–198.

Zhang, Y., Abiraman, K., Li, H., Pierce, D. M., Tzingounis, A. V., & Lykotrafitis, G. (2017). Modeling of the axon membrane skeleton structure and implications for its mechanical properties. *PLoS Computational Biology, 13*(2), e1005407.

Zhang, X. L., Shuttleworth, C. W., Moskal, J. R., & Stanton, P. K. (2015). Suppression of spreading depolarization and stabilization of dendritic spines by GLYX-13, an NMDA receptor glycine-site functional partial agonist. *Experimental Neurology, 273*, 312–321.

Zhou, N., Rungta, R. L., Malik, A., Han, H., Wu, D. C., & MacVicar, B. A. (2013). Regenerative glutamate release by presynaptic NMDA receptors contributes to spreading depression. *Journal of Cerebral Blood Flow & Metabolism, 33*(10), 1582–1594.

Zou, S., Chisholm, R., Tauskela, J. S., Mealing, G. A., Johnston, L. J., & Morris, C. E. (2013). Force spectroscopy measurements show that cortical neurons exposed to excitotoxic agonists stiffen before showing evidence of bleb damage. *PLoS One, 8*(8), e73499.

Zwiewka, M., Nodzyński, T., Robert, S., Vanneste, S., & Friml, J. (2015). Osmotic stress modulates the balance between exocytosis and clathrin-mediated endocytosis in *Arabidopsis thaliana*. *Molecular Plant, 8*(8), 1175–1187.

FSC
www.fsc.org
MIX
Papier | Fördert
gute Waldnutzung
FSC® C083411

Kolibri 360 GmbH
Ferdinand-Jühlke-Straße 7
99095 Erfurt, Deutschland
produktsicherheit@kolibri360.de